SAP PRESS e-books

Print or e-book, Kindle or iPad, workplace or airplane: Choose where and how to read your SAP PRESS books! You can now get all our titles as e-books, too:

- By download and online access
- For all popular devices
- And, of course, DRM-free

Convinced? Then go to www.sap-press.com and get your e-book today.

Migrating to SAP BW/4HANA®

SAP PRESS is a joint initiative of SAP and Rheinwerk Publishing. The know-how offered by SAP specialists combined with the expertise of Rheinwerk Publishing offers the reader expert books in the field. SAP PRESS features first-hand information and expert advice, and provides useful skills for professional decision-making.

SAP PRESS offers a variety of books on technical and business-related topics for the SAP user. For further information, please visit our website: *www.sap-press.com*.

Thorsten Lüdtke, Marina Lüdtke
SAP BW/4HANA 2.0: The Comprehensive Guide
2021, 663 pages, hardcover and e-book
www.sap-press.com/4544

Renjith Kumar Palaniswamy
Operational Data Provisioning with SAP BW/4HANA
2023, 608 pages, hardcover and e-book
www.sap-press.com/5676

Prathyusha Garimella, Shashidhar Garimella
SAP Landscape Transformation Replication Server: The Practical Guide
2024, 317 pages, hardcover and e-book
www.sap-press.com/5821

Colle, Dentzer, Hrastnik
Core Data Services for ABAP (3rd Edition)
2024, 754 pages, hardcover and e-book
www.sap-press.com/5642

Anil Bavaraju
Data Modeling for SAP HANA 2.0
2019, 432 pages, hardcover and e-book
www.sap-press.com/4722

Renjith Kumar Palaniswamy

Migrating to SAP BW/4HANA®

Editor Rachel Gibson
Acquisitions Editor Hareem Shafi
Copyeditor Julie McNamee
Cover Design Graham Geary
Photo Credit iStockphoto: 1498750593/© BlackJack3D
Layout Design Vera Brauner
Production Hannah Lane
Typesetting III-Satz, Germany
Printed and bound in the United States of America, on paper from sustainable sources

ISBN 978-1-4932-2576-7
© 2024 by Rheinwerk Publishing, Inc., Boston (MA)
1st edition 2024

Library of Congress Cataloging-in-Publication Control Number: 2024019988

All rights reserved. Neither this publication nor any part of it may be copied or reproduced in any form or by any means or translated into another language, without the prior consent of Rheinwerk Publishing, 2 Heritage Drive, Suite 305, Quincy, MA 02171.

Rheinwerk Publishing makes no warranties or representations with respect to the content hereof and specifically disclaims any implied warranties of merchantability or fitness for any particular purpose. Rheinwerk Publishing assumes no responsibility for any errors that may appear in this publication.

"Rheinwerk Publishing" and the Rheinwerk Publishing logo are registered trademarks of Rheinwerk Verlag GmbH, Bonn, Germany. SAP PRESS is an imprint of Rheinwerk Verlag GmbH and Rheinwerk Publishing, Inc.

All of the screenshots and graphics reproduced in this book are subject to copyright © SAP SE, Dietmar-Hopp-Allee 16, 69190 Walldorf, Germany.

SAP, ABAP, ASAP, Concur Hipmunk, Duet, Duet Enterprise, ExpenseIt, SAP ActiveAttention, SAP Adaptive Server Enterprise, SAP Advantage Database Server, SAP ArchiveLink, SAP Ariba, SAP Business ByDesign, SAP Business Explorer (SAP BEx), SAP BusinessObjects, SAP BusinessObjects Explorer, SAP BusinessObjects Web Intelligence, SAP Business One, SAP Business Workflow, SAP BW/4HANA, SAP C/4HANA, SAP Concur, SAP Crystal Reports, SAP EarlyWatch, SAP Fieldglass, SAP Fiori, SAP Global Trade Services (SAP GTS), SAP GoingLive, SAP HANA, SAP Jam, SAP Leonardo, SAP Lumira, SAP MaxDB, SAP NetWeaver, SAP PartnerEdge, SAPPHIRE NOW, SAP PowerBuilder, SAP PowerDesigner, SAP R/2, SAP R/3, SAP Replication Server, SAP Roambi, SAP S/4HANA, SAP S/4HANA Cloud, SAP SQL Anywhere, SAP Strategic Enterprise Management (SAP SEM), SAP SuccessFactors, SAP Vora, TripIt, and Qualtrics are registered or unregistered trademarks of SAP SE, Walldorf, Germany.

All other products mentioned in this book are registered or unregistered trademarks of their respective companies.

Contents at a Glance

1	Introduction to SAP BW/4HANA	21
2	Overview of SAP BW/4HANA Migration	49
3	System Preparation and Prerequisites	69
4	Remote Conversion	135
5	Shell Conversion	251
6	In-Place Conversion	279
7	SAP BW Bridge for SAP Datasphere	345

Contents

Introduction 15

1 Introduction to SAP BW/4HANA 21

1.1 SAP BW/4HANA Overview 21
- 1.1.1 A Brief History of SAP Business Warehouse 21
- 1.1.2 SAP Business Warehouse 7.x Objects 23
- 1.1.3 Evolution of SAP BW/4HANA 29
- 1.1.4 SAP BW/4HANA Design Principles 30

1.2 What's New in SAP BW/4HANA? 35
- 1.2.1 Eclipse-Based Modeling Tool 35
- 1.2.2 SAP BW/4HANA Data Modeling Objects 37

1.3 Summary 48

2 Overview of SAP BW/4HANA Migration 49

2.1 Business Scenarios 49
- 2.1.1 Organizations without an SAP Business Warehouse Landscape 49
- 2.1.2 Organization with an SAP Business Warehouse System Landscape 50

2.2 SAP BW/4HANA Conversion Types 53
- 2.2.1 Greenfield SAP BW/4HANA Implementation 54
- 2.2.2 In-Place Conversion 55
- 2.2.3 Remote Conversion 58
- 2.2.4 Shell Conversion 61
- 2.2.5 Landscape Transformation 62

2.3 Phases in SAP BW/4HANA Conversion 63
- 2.3.1 Conversion Prepare Phase 63
- 2.3.2 Conversion Realization Phase 65

2.4 Project Management 67

2.5 Summary 68

3   System Preparation and Prerequisites   69

3.1	System Requirements for SAP BW/4HANA	69
3.2	Impacts on the System Landscape	71
3.3	Preparing the System	74
	3.3.1 Minimum Supported Release for Original System	74
	3.3.2 Minimum Support Release for Target System	75
3.4	Supported Source Systems	76
	3.4.1 SAP Business Suite Applications	76
	3.4.2 SAP Business Warehouse	76
	3.4.3 SAP Source System and Operational Data Provisioning Readiness	77
	3.4.4 Other Source Systems	78
3.5	Maintenance Planner	78
3.6	Data Volume Management	83
3.7	SAP Business Warehouse Sizing	84
3.8	Installing the Note Analyzer Report	92
3.9	SAP Readiness Check for SAP BW/4HANA	94
	3.9.1 Readiness Check Basics	94
	3.9.2 Basics of the ST-A/PI Add-On	95
	3.9.3 Prerequisites	96
	3.9.4 Executing SAP Readiness Check for SAP BW/4HANA	98
	3.9.5 Functions Available in the SAP Readiness Check for SAP BW/4HANA	100
	3.9.6 Viewing the SAP Readiness Check for SAP BW/4HANA	102
3.10	Prechecks for SAP BW/4HANA Migration	105
	3.10.1 Installing Precheck Tools	106
	3.10.2 Executing Prechecks	107
	3.10.3 Fixing Issues in the Precheck Report	114
	3.10.4 Sizing Report	116
	3.10.5 Code Scan	117
	3.10.6 Cleanup Reports	119
3.11	SAP BW/4HANA Simplification List	119
	3.11.1 SAP Business Warehouse 3.X Data Flows	120
	3.11.2 Core Data Warehouse Modeling Objects	121
	3.11.3 Data Staging	126
	3.11.4 Source Systems	131
3.12	Summary	133

4 Remote Conversion 135

4.1 Overview of Remote Conversion 135
4.2 SAP BW/4HANA Target System Provisioning 136
- 4.2.1 Installing the SAP HANA Database 136
- 4.2.2 Installing SAP BW/4HANA 141
- 4.2.3 Post-Installation Steps 151
- 4.2.4 Creating Source Systems 155
- 4.2.5 Settings in the SAP BW/4HANA System 159
- 4.2.6 Installing Essential Add-Ons 163

4.3 SAP BW/4HANA Sender System Provisioning 166
- 4.3.1 Data Migration Integration Server Checks 166
- 4.3.2 Sender System Landscape Preparation with Note Analyzer 166

4.4 Source System Provisioning 168
4.5 Preparing for Remote Conversion 169
- 4.5.1 Project Planning and Understanding 169
- 4.5.2 User Roles and Authorizations 170

4.6 Basics of the SAP BW/4HANA Conversion Cockpit 175
- 4.6.1 Overview 176
- 4.6.2 Scope Check Tool 179
- 4.6.3 Task List Based on Transaction STC01 and Transaction STC02 183
- 4.6.4 Essential Checks before Remote Conversion 185
- 4.6.5 Evaluating Scope Objects 187

4.7 SAP Business Warehouse Data Model 187
4.8 Package Creation 190
4.9 Execution 192
- 4.9.1 Package Settings 192
- 4.9.2 Collect Scope and Metadata Transfer 194
- 4.9.3 Define Object Mapping and Store Object List 198
- 4.9.4 Prepare Propagate Request 201
- 4.9.5 Map InfoObject for InfoProvider 202
- 4.9.6 Select Main Partition Criteria for Semantically Partitioned 203
- 4.9.7 Determine Usage of Involved Objects 203
- 4.9.8 Checklist for Usage of Involved Objects 204
- 4.9.9 Confirm Metadata in Target System with Data Transfer Prep 205
- 4.9.10 Setup Phase 211
- 4.9.11 Select Conversion-Relevant InfoObjects 217
- 4.9.12 Define Mapping Values 218
- 4.9.13 Make Settings for Conversion-Relevant InfoObjects. 219

	4.9.14	Generation Phase	220
	4.9.15	Preprocessing Programs During System Lock	225
	4.9.16	Copy Delta Queues from SAPI to ODP Technology	226
	4.9.17	Confirm That Delta Was Loaded to Cloned DataSources	229
	4.9.18	Check Whether Delta Queues Can Be Copied from SAPI to ODP	231
	4.9.19	Confirm the INIT SIMU Procedure	231
	4.9.20	Make Settings to Prepare to Lock All Data Changes	231
	4.9.21	Migration Phase	235
	4.9.22	Make Runtime Settings for Migration	235
	4.9.23	Reset Cluster Data Deletion Date	236
	4.9.24	Start Migration	237
	4.9.25	Monitor: Transformation Status (Optional)	237
	4.9.26	Confirm Data Selection and Unlock Data Loading	239
	4.9.27	Verify Task Completion	239
	4.9.28	Post Processing Activities During System Lock	240
	4.9.29	Execute RSRV Checks in the Target System	241
	4.9.30	View Time-Related Migration Statistics (Optional)	241
	4.9.31	System Settings After Migration (Active Phase)	242
4.10	Post-Conversion		247
4.11	Useful Reports and Tables		248
4.12	Summary		250

5   Shell Conversion   251

5.1	Overview of Shell Conversion		252
	5.1.1	Business Cases	252
	5.1.2	Shell Conversion Flow	252
5.2	System Landscape for Shell Conversion		253
5.3	SAP Business Warehouse System Compatibility		254
5.4	SAP Business Warehouse System Provisioning		255
	5.4.1	Target SAP BW/4HANA System Provisioning	255
	5.4.2	Sender System Provisioning	255
	5.4.3	Sender and Target System Landscape Preparation	256
5.5	Prechecks and Simulation		257
5.6	Realization Phase		259
	5.6.1	Scope Object Identification	259
	5.6.2	Scope Selection Execution	263
	5.6.3	Define Object Mapping and Store Object List	266

		5.6.4	Map InfoObjects for InfoProvider	270
		5.6.5	Select Main Partition Criteria for Semantically Partitioned Objects	271
		5.6.6	Determine Usage of Involved Objects	271
		5.6.7	Checklist for Usage of Involved Objects	272
5.7	Post-Conversion Activities			274
5.8	Summary			277

6 In-Place Conversion   279

6.1	Overview of In-Place Conversion			279
6.2	System Landscape for In-Place Conversion			281
6.3	SAP Business Warehouse System Provisioning			282
6.4	Sender System Provisioning			286
6.5	Realization Phase			287
6.6	Introduction to the SAP BW/4HANA Starter Add-On			288
		6.6.1	Process during In-Place Conversion	288
		6.6.2	SAP BW/4HANA Starter Add-On	289
6.7	Operating Modes			295
		6.7.1	Switching Modes	296
		6.7.2	Mode Switch and Prerequisites	297
		6.7.3	Handling Unsupported Objects	298
		6.7.4	Landscape Considerations	300
6.8	Transferring Standard Authorization			301
		6.8.1	Conversion Rules for Authorization Objects	301
		6.8.2	Scope Transfer and Authorization Conversion	302
		6.8.3	Initial and Delta Run	303
		6.8.4	Simulation Mode	303
		6.8.5	Authorization Transfer Process (Initial Run)	304
6.9	Execution Overview			309
		6.9.1	General Features	309
		6.9.2	Cleanup Execution	310
		6.9.3	Data Transfer Logic	311
		6.9.4	SAP Business Warehouse Data Model	312
6.10	Starting In-Place Conversion			315
6.11	Scope Transfer			317
6.12	Executing Scope Selection			318

	6.12.1	Collect Scope for Transfer	318
	6.12.2	Define Object Mapping and Store Object List	320
	6.12.3	Prepare Propagate Requests	322
	6.12.4	Map InfoObjects for InfoProviders	324
	6.12.5	Select Main Partition Criteria for Semantically Partitioned Objects	324
	6.12.6	Determine Usage of Involved Objects	324
	6.12.7	Checklist for Usage of Involved Objects	324
	6.12.8	Copy Delta Queues from SAPI to ODP Technology	326
	6.12.9	Confirm Delta Was Loaded for the Cloned DataSources	328
	6.12.10	Synchronize Delta Queues between SAPI and ODP	328
	6.12.11	Extract from PSAs of DataSources and Error Stacks	329
	6.12.12	Master Data Activation	330
	6.12.13	Propagate Requests	330
	6.12.14	Lock All Data Changes	330
	6.12.15	Prepare Request Mapping for Transfer	331
	6.12.16	Save Data	331
	6.12.17	Transfer Metadata	331
	6.12.18	Transferring the Request Info to the New Status Management	332
	6.12.19	Retrieve Data	332
	6.12.20	Unlock Loading and Data Target Changes	332
	6.12.21	Post Transfer Checks	333
6.13	**Transfer Standard Authorization (Delta)**		336
6.14	**Objects Deletion**		337
	6.14.1	Technical Content	338
	6.14.2	Delete Other SAP Business Warehouse Objects	338
	6.14.3	Delete SAP Business Warehouse Queries	339
6.15	**Final System Conversion**		339
6.16	**Summary**		343

7 SAP BW Bridge for SAP Datasphere 345

7.1	**Introduction to SAP Datasphere**	345
7.2	**Introduction to SAP BW Bridge**	347
7.3	**Shell Conversion with SAP BW Bridge**	350
	7.3.1 Prepare Phase Activities	351
	7.3.2 System Preparation	356
	7.3.3 Realization Phase	360
	7.3.4 Connect the Sender System to SAP BW Bridge	362
	7.3.5 Create an SAP BW Bridge Project	366

	7.3.6	ABAP Cloud Project in ABAP Development Tools	369
	7.3.7	Object-Specific Simplification	373
	7.3.8	Prework	374
	7.3.9	Execution	375
7.4	**Remote Conversion with SAP BW Bridge**		**382**
	7.4.1	Preparation	382
	7.4.2	Execution	386
	7.4.3	Post-Migration Options in SAP Datasphere	404
7.5	**Summary**		**407**

The Author	409
Index	411

Introduction

You may have heard that in enterprise technology, change is the only constant. As a result, unlocking the full potential of data has become an exhilarating and challenging journey for organizations worldwide. The combination of technological advancements, business imperatives, and data-driven innovation has businesses finding themselves in uncharted territory. Within this context, the migration to SAP BW/4HANA is a transformative leap that not only keeps organizations abreast of the latest in technological evolution but also positions them at the forefront of data management and analytics.

As organizations grapple with the relentless pace of change in the digital era, the migration to SAP BW/4HANA takes center stage as a strategic imperative. SAP's sophisticated platform promises not only enhanced agility and efficiency in data processing but also a quantum leap in analytics capabilities. This book serves as a comprehensive guide and trusted companion for organizations embarking on this exciting and complex migration to the SAP BW/4HANA platform.

More than just a manual for technical implementation, this book is a repository of insights, strategies, and practical wisdom distilled from the collective experiences of seasoned experts and industry leaders. By demystifying the complexities of the migration process, the book offers a road map that goes beyond technical details to address the strategic and operational considerations that are integral to a successful migration.

Here, you'll discover both the technical details of SAP BW/4HANA and the strategies that underscore the decision to migrate from legacy data warehouse systems. The aim is to empower decision-makers, IT professionals, and project managers with the knowledge needed to make informed choices at every juncture of the migration journey. By combining technical expertise with strategic foresight, this book equips you to navigate migration challenges, ensuring that the transition to SAP BW/4HANA is not merely a technological upgrade but also a move toward sustained business excellence.

Before getting into SAP BW/4HANA, let's first discuss the history of data warehousing.

History of Data Warehousing

Understanding the changes and the need for migrating to a new data warehouse tool requires taking a deep dive into the history of data warehousing. This historical context sheds light on the transformative evolution of data warehousing and serves as a foundation for appreciating why migrating to newer technologies is so important. You'll gain insights into the dynamic forces that have influenced data warehousing, which will help you understand the evolving technological landscape and the persistent need for organizations to adapt. Migration aligns organizations with current data management and analytics requirements and positions them to leverage the potential of emerging technologies.

The trajectory of data warehousing from its formative years in the 1970s to the current day shows a journey with pivotal milestones and advancements. Let's see that in a very high-level overview:

- **1970s: Emergence of business intelligence (BI)**
  The groundwork for data warehousing was laid in the 1970s with the inception of business intelligence. This era witnessed the development of decision support systems (DSSs) aimed at providing tools to gather, organize, and present business data.

- **1980s: Decision support systems (DSS)**
  As the 1980s unfolded, the term "data warehousing" began to gain prominence. DSSs evolved to incorporate data warehousing components, prompting organizations to explore centralized data management for improved decision-making.

- **1990s: Rise of data warehousing**
  The 1990s witnessed the widespread recognition of data warehousing, with various technologies and methodologies coming to the forefront. Pioneers such as Teradata introduced parallel processing databases tailored for data warehousing. Bill Inmon (known as the "father of data warehousing") wrote *Building the Data Warehouse* (1991, QED Technical), which outlined the key principles for constructing data warehouses. Inmon's ideas, such as the concept of the data warehouse as a *single version of the truth*, aligned with the goals of SAP Business Warehouse (SAP BW) when it was introduced during that time to provide a unified and consistent view of organizational data for reporting and analysis. Organizations implementing SAP BW often follow data warehousing best practices influenced by Inmon's principles to ensure the success of their data management initiatives. Later, the commercial data warehouse appliances also emerged in the late 1990s, providing integrated solutions for faster and more efficient data processing.

- **2000s: Growth and maturity**
  The 2000s ushered in continued evolution, introducing features such as online analytical processing (OLAP), data mining, and enhanced query performance. Data warehousing solutions became more scalable and capable of handling substantial data volumes. The *data warehouse appliance* concept gained popularity, offering preconfigured bundles for expedited deployment.

- **2010s: Big data and cloud integration**
  The onset of big data in the 2010s presented new challenges and opportunities. Organizations had to integrate and analyze not just structured but also semi-structured and unstructured data. Cloud computing gained prominence, leading companies to explore cloud-based data warehousing solutions that offered flexibility, scalability, and cost-effectiveness.

- **2010s–Present: Modern data warehousing**
  Modern data warehousing solutions in the latter part of the decade leveraged advanced technologies such as in-memory processing, columnar databases, and machine learning for superior analytics and performance. The boundaries between traditional data warehousing and big data technologies blurred, giving rise to hybrid data warehouse architectures capable of handling diverse data types. Data warehouses evolved into integral components of comprehensive data management ecosystems, seamlessly integrating with data lakes, data marts, and other systems to provide a holistic view of organizational data.

In conclusion, the trajectory of data warehousing reflects a continual pursuit by organizations to leverage data for informed decision-making, adapt to technological advancements, and overcome challenges posed by the expanding volume and diversity of data.

Evolution of In-Memory-Based Data Warehouse Solutions

The evolution of in-memory data warehousing marks a pivotal advance in data management that commenced with the introduction of SAP HANA—SAP's groundbreaking in-memory database platform. This platform was engineered to revolutionize data processing by storing and processing information directly in RAM, as opposed to conventional disk storage. This shift aimed to overcome the limitations of disk-based databases, providing the capability for real-time analytics and significantly faster data access.

As the need for instantaneous insights and analytics gained prominence, SAP extended the capabilities of SAP HANA to support comprehensive data warehousing solutions. This led to the emergence of SAP BW on SAP HANA, an integration of SAP HANA's in-memory computing prowess into the SAP BW environment. This integration empowered organizations to harness the advantages of in-memory computing for advanced analytics, sophisticated data modeling, and agile reporting within the structured framework of a data warehouse.

Building on this foundation, SAP further refined its data warehousing strategy, giving rise to SAP BW/4HANA—a next-generation data warehouse solution. Developed natively on the SAP HANA platform, SAP BW/4HANA represents a paradigm shift in data warehousing architecture. It's designed to offer a simplified, open, and modern environment for data warehousing. This solution enables organizations to efficiently manage and analyze vast volumes of data while incorporating advanced features such as predictive analytics, machine learning, and seamless integration with diverse data sources. SAP BW/4HANA stands as a comprehensive response to the evolving landscape of data management and analytics, providing businesses with the tools they need to navigate the challenges of the digital era.

The Evolution of Cloud-Based Data Warehouse Systems

The evolution of SAP BW/4HANA on the cloud reflects a strategic response to the growing demand for flexible, scalable, and agile data solutions in the digital era. Initially, SAP BW/4HANA was primarily an on-premises solution, leveraging the in-memory computing capabilities of the SAP HANA database. Recognizing the shift toward cloud-centric architectures and the benefits they offer in terms of scalability, accessibility, and cost-effectiveness, SAP embarked on a journey to bring SAP BW/4HANA to the cloud.

The first significant step in this evolution was the introduction of SAP BW/4HANA Cloud, providing users with the option to deploy and manage their data warehouses in the cloud environment. This cloud-native version allows organizations to leverage the power of SAP BW/4HANA without the need for extensive on-premises infrastructure. Users can benefit from cloud features such as rapid provisioning, elastic scalability, and pay-as-you-go pricing models, aligning data warehousing resources with actual usage.

Furthermore, SAP extended its cloud offerings by integrating SAP BW/4HANA with platforms such as SAP HANA Cloud and SAP Data Warehouse Cloud, which is now called SAP Datasphere. This integration allows users to harness the advantages of a unified cloud ecosystem, seamlessly connecting data stored in different repositories and enabling collaborative analytics. The evolution of SAP BW/4HANA on the cloud aligns with SAP's commitment to providing a comprehensive suite of cloud-based solutions that cater to the diverse needs of modern businesses.

As organizations increasingly prioritize cloud-based solutions for enhanced flexibility and reduced infrastructure management overhead, SAP's evolution of SAP BW/4HANA on the cloud underscores its dedication to staying at the forefront of data warehousing innovation. This evolution not only meets the demands of the contemporary digital landscape but also positions SAP BW/4HANA as a versatile and future-ready solution for businesses embracing cloud-centric strategies.

Next, in the first chapter of this book, we'll start with the basics of SAP BW 7.x and gradually focus on SAP BW/4HANA, exploring its advanced features and innovations as a prelude to the subsequent exploration of migration concepts. As a result, you'll gain a comprehensive understanding of SAP BW/4HANA's capabilities in modern data management and analytics.

Acknowledgments

Many thanks to the SAP PRESS team, especially the acquisitions editor, Hareem Shafi, who helped with the chapter organization and book structure creation. She was very cooperative in addressing queries related to the book length and organization. Her constructive feedback during the initial proposal was key to completing this book effectively. Many thanks to the development editor, Rachel Gibson. She has been very helpful throughout all phases of the writing process, starting from the first draft and suggesting appropriate edits wherever required. Thanks to Julie McNamee, the copyeditor, who has been very helpful in the final stage of edits. Thanks to the cover designer, Graham Geary, for designing the beautiful cover for this book. Many thanks to all the SAP PRESS team members who helped bring this book from the draft phase to the publishing phase; the team has been extremely outstanding. Thanks to the SAP Americas Center of Expertise (CoE) leadership team, namely Thomas Walter (global head premium hub, SAP CoE), Balaji Rao Gaddam (VP and head, SAP CoE North America), and Suhan Hegde (director, CoE analytics, SAP Americas). I would like to thank my close friend Kalaiarasu Thirunavukkarasu from SAP America. Thanks also to Thomas Rinneberg and Baetz Udo from the SAP BW development team SAP SE, as well as to Ashish Bedi from the system landscape optimization development team at SAP Labs India. Thanks to all my colleagues and friends in the SAP CoE North America analytics team for all their support. Thanks to all my SAP BW/4HANA development team colleagues and SAP product support team colleagues. Finally, many thanks to my parents, Palaniswamy and Sarojini; my wife, Mathumathi; my children, Diya and Ram; my brother, Manivannan Palaniswamy; and all my family members and friends for their extended support while writing this book.

We hope you enjoy the book. Happy learning!

Chapter 1
Introduction to SAP BW/4HANA

This chapter serves as an introductory overview of SAP BW/4HANA, offering fundamental insights into its key concepts and functionalities. By the end of this chapter, you'll have a comprehensive understanding of SAP BW/4HANA at a high level, including its core components, architecture, and design principles. Furthermore, you'll acquire valuable knowledge about SAP BW/4HANA–based objects and the data modeling process, laying a solid foundation for learning in subsequent chapters that will focus on the conversion of objects to SAP BW/4HANA.

To effectively handle a substantial migration project, it's essential to grasp the core concepts of SAP BW/4HANA, including a comprehensive understanding of the diverse object types used in the previous SAP Business Warehouse (SAP BW) 7.x version. Familiarity with the modifications made to these objects in the new SAP BW/4HANA environment is crucial. Additionally, gaining insights into the novel features introduced in SAP BW/4HANA is imperative. This chapter is designed to provide a comprehensive exploration of these aspects, offering a thorough overview for seamless project execution.

This chapter provides you with an overview of SAP BW 7.x and SAP BW/4HANA. We'll start by exploring the history of SAP BW/4HANA, SAP BW Objects, and SAP BW/4HANA design principles in Section 1.1. Then, in Section 1.2, we'll discuss the new features that are available in SAP BW/4HANA.

1.1 SAP BW/4HANA Overview

In this section, we'll cover the details of SAP BW/4HANA migration, starting with the history of SAP BW.

1.1.1 A Brief History of SAP Business Warehouse

SAP BW has undergone a transformative journey, beginning with its inaugural version, SAP BW 1.0, which was introduced in 1998. This initial release marked SAP's entry into the data warehousing arena, providing organizations with a platform to consolidate and analyze their business data. SAP BW 1.0 laid the groundwork for subsequent developments, offering capabilities for data extraction, transformation, and loading (ETL).

In 2004, SAP released SAP BW 3.5, a significant milestone in the evolution of the platform. This version brought about substantial improvements in ETL processes and data management, addressing performance concerns and enhancing integration with other SAP solutions. The subsequent major release, SAP BW 7.0 (2006), introduced advanced features such as refined modeling capabilities and tighter integration with SAP NetWeaver, SAP's integrated technology stack.

As technology continued to advance, SAP responded with SAP BW 7.3 in 2011, focusing on usability enhancements, improved data modeling, and expanded integration options. Building upon this foundation, SAP BW 7.4, released in 2013, further optimized performance and extended support for big data scenarios. A pivotal moment came in 2015 with the introduction of SAP BW 7.5, a version that embraced SAP HANA as its foundational database. This marked a significant shift toward in-memory processing, enabling faster analytics and data processing.

SAP BW 7.5 not only leveraged the capabilities of SAP HANA but also brought additional advancements in data integration, modeling, and overall user experience. With a continued commitment to adaptability and innovation, SAP BW has solidified its position as a comprehensive data warehousing solution within the SAP ecosystem, evolving to meet the changing needs of businesses and leveraging the latest technologies for optimal performance and efficiency. The versions highlighted in this journey represent key milestones in the continual refinement and enhancement of SAP BW, as shown in Figure 1.1.

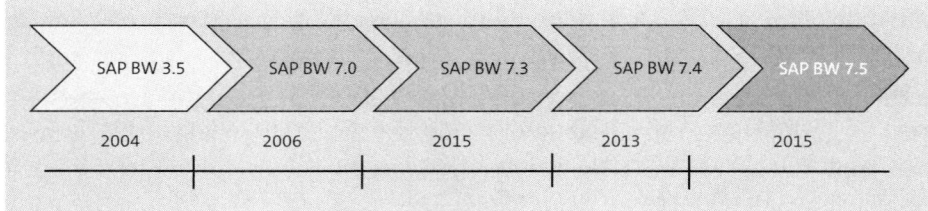

Figure 1.1 SAP BW 7.5 Evolution

SAP BW underwent a notable evolution from SAP BW version 7.3 to SAP BW/4HANA 1.0. Commencing with SAP BW 7.3 in 2011, the platform aimed to improve usability and broaden integration options. It's worth mentioning that the 7.3 version supported the SAP HANA database, earning the designation 7.3 powered by SAP HANA. Subsequent releases, such as SAP BW 7.4 in 2013, continued to support the SAP HANA database, with a focus on refining performance optimization and accommodating big data requirements. In 2015, SAP BW 7.5 marked a significant transition by adopting SAP HANA as its foundational database, harnessing its in-memory processing capabilities for accelerated analytics. The introduction of the SAP BW/4HANA starter add-on paved the way for users to prepare for migration to the SAP BW/4HANA platform, officially released in 2016. SAP BW/4HANA signifies a fundamental shift, embracing a simplified and modernized approach to data warehousing tailored explicitly for the SAP HANA platform, underscoring SAP's commitment to staying abreast of technological trends and meeting evolving business requirements. Figure 1.2 shows the evolution of SAP BW versions on SAP HANA.

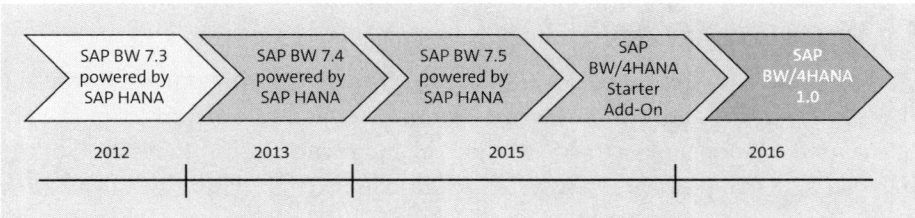

Figure 1.2 SAP BW on SAP HANA Versions

1.1.2 SAP Business Warehouse 7.x Objects

Most organizations planning to migrate to SAP BW/4HANA typically use SAP BW versions ranging from 7.0 to 7.5. Significant changes have occurred in SAP BW objects from version 3.x to 7.x, and, similarly, there are substantial differences in SAP BW objects when transitioning to SAP BW/4HANA. This section will delve into the most-used SAP BW objects in the 7.x system. Understanding these objects is crucial for a successful SAP BW/4HANA migration, as errors related to these objects are prevalent in the 7.x system. A clear understanding of these objects facilitates swift issue resolution and helps reduce the overall project timeline. Now, let's explore these objects and their utility from an extraction modeling and reporting perspective.

Extraction Objects in SAP BW 7.x

In the SAP environment, the process of extracting data involves the use of either SAP's standard extractor, specifically designed for this purpose, or the creation of a custom generic extractor tailored to the organization's unique requirements. To facilitate this data extraction from enterprise resource planning (ERP) systems, it's imperative to establish a source system in SAP BW 7.x.

The interaction between the ERP and SAP BW systems for data extraction was facilitated through the Service API (SAPI). This API played a pivotal role in streamlining the transfer of data from the ERP systems to the BW environment, ensuring efficient and accurate extraction processes. For connectivity with non-SAP systems, various options were available. Connecting to other database systems was achieved through options such as DB Connect (Database Connect) and UD Connect (Universal Data Connect), allowing seamless integration with diverse data sources outside the SAP ecosystem. In addition, a flat file–based source system proved instrumental in extracting data from flat files, providing flexibility in handling information stored in nondatabase file formats. This comprehensive approach to data extraction reflects the adaptability and

1 Introduction to SAP BW/4HANA

versatility of the SAP environment, offering tailored solutions for connecting and retrieving data from both SAP and non-SAP systems.

The source system type and its connectivity are important factors in the data staging and extraction into SAP BW systems.

Modeling Objects in SAP BW 7.x

Data modeling in SAP BW 7.0 involves the design and structuring of data within the SAP BW system to facilitate efficient reporting, analytics, and decision-making. The primary goal of data modeling is to create a logical and organized representation of the business data, making it easier to extract meaningful insights, which means defining how data elements, such as products, customers, and sales, are structured and related to each other in a way that reflects the business processes. By establishing a logical model, organizations can ensure that the data within the SAP BW system aligns with business concepts and workflows. This logical representation serves as the foundation for building other data modeling objects. Here are some of the objects that were used in SAP BW 7.x:

- **DataSources**

  DataSource is a critical component that defines the extraction of data from source systems into the SAP BW environment. It serves as a connection point between the source system and SAP BW, specifying the structure of the data to be extracted, including fields, relationships, and extraction logic. DataSources are used in the ETL process to bring data from source systems, such as SAP ERP, into the SAP BW system for reporting and analysis. A DataSource is essentially a metadata object that describes the data structure in the source system and provides the necessary information for extracting data.

- **InfoObjects**

  In SAP BW, InfoObjects are fundamental components used to structure and define data. They fall into two categories: characteristics (descriptive data, e.g., product, customer, or vendor) and key figures (quantitative data, e.g., sales revenue and quantity). Characteristics form dimensions for analysis, while key figures provide measurable values. InfoObjects are crucial in creating other InfoProviders such as InfoCubes, and DataStore objects (DSOs, see definition later in this list) for enabling efficient reporting and analytics. They play a central role in organizing and categorizing business data for meaningful insights.

- **InfoCubes**

  A standard InfoCube is a persistence InfoProvider where the data is physical stored using multidimensional modeling structures. InfoCubes in SAP BW serve as powerful tools for multidimensional modeling, enabling detailed reporting and analysis. The basic or standard cube, which is a key element of this multidimensional data storage approach, organizes data by integrating descriptive elements (characteristics), such as product or region, and measurable metrics (key figures) such as sales revenue. This multidimensional model creates a structure where data is stored at the intersections of these characteristics and key figures, allowing for thorough analysis from various perspectives. Essentially, it provides a dynamic space for exploring relationships and patterns within the data, enhancing the depth and clarity of analyses. Users can gain valuable insights more effectively in the SAP BW environment through this structured multidimensional modeling approach.

A *virtual cube* provides a way to report on data without physically storing it in the cube. Virtual cubes are useful for real-time reporting without the need to load data into persistent storage. It acts as a dynamic layer on top of existing data sources, offering a way to analyze information in real time without the necessity of loading it into a persistent storage structure. This virtual cube is particularly valuable for agile and responsive reporting, providing users with immediate access to insights without the constraints of traditional data storage. Essentially, it offers a flexible space for exploring and visualizing data, enhancing the speed and efficiency of real-time reporting in the SAP BW environment.

- **DataStore objects (DSOs)**
  In SAP BW 7.x, DSO refers to a pivotal component designed for the purpose of storing and managing detailed transactional data with a high level of granularity. Acting as a cornerstone in the data processing landscape, DSOs are fundamental to the ETL process within the SAP BW system, which is crucial for collecting, refining, and transferring data from source systems to the SAP BW environment. The versatility of DSOs lies in their ability to store data in different layers, including an active data table, a change log, and a data target, allowing for comprehensive management of transactional information.

  DSOs specialize in the storage of detailed and atomic-level transactional data, capturing every transactional event at its finest level of granularity. This level of detail is particularly valuable for operational reporting and analysis, where users often require access to individual transactions for insights into specific business events. Following are a few types of DSOs:
  - Standard DSO: Used for persistent storage of detailed transactional data, providing historical data retention capabilities and flexibility in data modeling.
  - Write-optimized DSO: Geared toward high-speed data loads, which is ideal for scenarios where historical data retention isn't a priority, and swift data acquisition is essential.
  - Direct update DSO: Enables real-time data acquisition and supports immediate availability for reporting without the need for a separate loading process.

- **Nonpersistent InfoProviders**
  Nonpersistent InfoProviders don't hold any physical data. These SAP BW objects can be used to access the underlying persistent InfoProviders to fetch the data for analytical reporting purposes:

- **MultiProvider**: Combines data from multiple InfoProviders (InfoCubes, DSOs) into a single virtual InfoProvider. It allows users to create reports on data from different sources.
- **CompositeProvider**: Combines multiple InfoProviders, including InfoObjects, InfoCubes, and DSOs, to create a unified view for reporting.

- **InfoSets**
  InfoSets provide a way to define a logical view of data across different InfoProviders without physically combining them. One option is the *logical query view*, which combines data from multiple InfoProviders for reporting purposes.

- **Transformation rules**
  Transformation rules define the mapping and conversion of data from a source structure to a target structure during the data loading process.

- **Start routines and end routines**
  These ABAP routines execute at the start and end of the transformation process, allowing for additional custom logic.

- **Open hub destination**
  An open hub destination is used for exporting data. It enables the extraction of data from SAP BW to external systems or flat files and supports various formats and destinations for data distribution.

- **Analysis authorizations**
  These authorizations are used to define access permissions and to specify which users or user roles have access to specific data in the SAP BW system. Analysis authorizations control data security during reporting.

- **Process chains**
  Process chains are sequences of linked processes. They automate end-to-end data loading, transformation, and reporting tasks. Process chains allow for the orchestration of various processes in a logical sequence.

Reporting Objects in SAP BW 7.x

The reporting objects in SAP BW are used to create the analytical reports based on the InfoProviders defined in the SAP BW system. You can create the report and view the analytics reports using these reporting objects. Some of the SAP BW reporting objects are listed here:

- **SAP BEx Query Designer**
  SAP BEx Query Designer is a graphical tool embedded in SAP BW, designed for creating queries tailored to InfoProviders such as InfoCubes. With an intuitive interface, it allows users to define key figures, characteristics, and filters. Offering advanced capabilities, the tool facilitates the creation of intricate queries, including calculated key figures and structures. Its graphical representation of data aids in the visualization of complex information, contributing to a comprehensive analysis of business data within the SAP BW system.

- **SAP BEx Analyzer**

  SAP BEx Analyzer seamlessly integrates with Microsoft Excel, providing a powerful interface for users to analyze SAP BW data directly within Excel. Users benefit from its intuitive functionalities, enabling easy execution and navigation of queries. The tool supports drag-and-drop operations, streamlining the manipulation of report components and enhancing the overall user experience. Through this Excel-based solution, organizations can leverage familiar spreadsheet capabilities while harnessing the robust analytical capabilities of SAP BW.

- **SAP BEx Web Analyzer**

  SAP BEx Web Analyzer is a web-based reporting tool within SAP BW, offering a dynamic and user-friendly interface for ad hoc reporting and analysis. With interactive features such as drilldowns and filters, users can explore and analyze data without the need for Excel. This browser-based solution provides flexibility and accessibility, catering to users who prefer web interfaces over traditional spreadsheet applications. SAP BEx Web Analyzer empowers users to generate insights and make informed decisions through a convenient and responsive web platform.

- **SAP Analysis for Microsoft Office**

  SAP Analysis for Microsoft Office is an Excel add-in that enhances the analytical capabilities of Microsoft Excel for SAP BW data. It seamlessly integrates with Excel, empowering users with advanced features such as hierarchies, calculated key figures, and dynamic formulas. With SAP Analysis for Microsoft Office, users can create dynamic and interactive reports directly within the familiar Excel environment, ensuring a smooth transition for users accustomed to Excel-based analytics. This add-in is instrumental in elevating the data analysis experience and facilitating deeper insights into SAP BW datasets.

- **SAP Lumira**

  SAP Lumira is a self-service data visualization tool that enables users to create visually compelling dashboards and reports. With features for data preparation and exploration, SAP Lumira facilitates the transformation of raw data into meaningful insights. Integration with SAP BW and other data sources enhances its versatility, allowing users to visualize and interpret data from various perspectives. Its wide range of visualization options, including charts, graphs, and maps, makes SAP Lumira a valuable tool for organizations seeking to convey complex information in an accessible and visually appealing manner.

- **SAP Lumira, designer edition**

  In 2016, there was an announcement from SAP that created a road map to combine SAP Lumira and SAP BusinessObjects Design Studio into a new product: SAP Lumira, designer edition. SAP Lumira, designer edition is a robust dashboard and application design tool that empowers users to create interactive and customized dashboards. Featuring a drag-and-drop interface, it facilitates the easy creation of visually appealing dashboards without extensive coding. The tool supports scripting for advanced customization, enabling users to tailor applications to specific business requirements. Real-time connectivity to SAP BW ensures that dashboards reflect the most current data, providing decision-makers with accurate and timely insights. SAP Lumira, designer edition stands out as a versatile solution for organizations seeking to develop sophisticated and engaging business intelligence applications.

- **SAP Crystal Reports**
SAP Crystal Reports is a powerful reporting tool renowned for creating pixel-perfect reports with extensive formatting options. Integrated seamlessly with SAP BW and diverse data sources, it supports dynamic linking, ensuring real-time updates of data in reports. With its robust capabilities, SAP Crystal Reports is a preferred choice for organizations requiring highly formatted and structured reports. Its ability to handle complex layouts and deliver precise, print-ready documents makes it a staple in business reporting, catering to users who prioritize precision and presentation in their reports.

- **SAP BusinessObjects Web Intelligence**
SAP BusinessObjects Web Intelligence is a web-based reporting tool designed for ad hoc reporting and exploration. With a user-friendly web interface, it allows users to create, view, and analyze reports with ease. Connecting seamlessly to SAP BW and other data sources, SAP BusinessObjects Web Intelligence facilitates the creation of dynamic and interactive reports that can adapt to changing business requirements. Its browser-based accessibility ensures widespread availability, enabling users to access and collaborate on reports without the need for specialized software installations.

Layered Scalable Architecture in SAP BW 7.x

SAP BW 7.x data modeling is based on the layered scalable architecture (LSA), where there will be multiple data persistences, as shown in Figure 1.3.

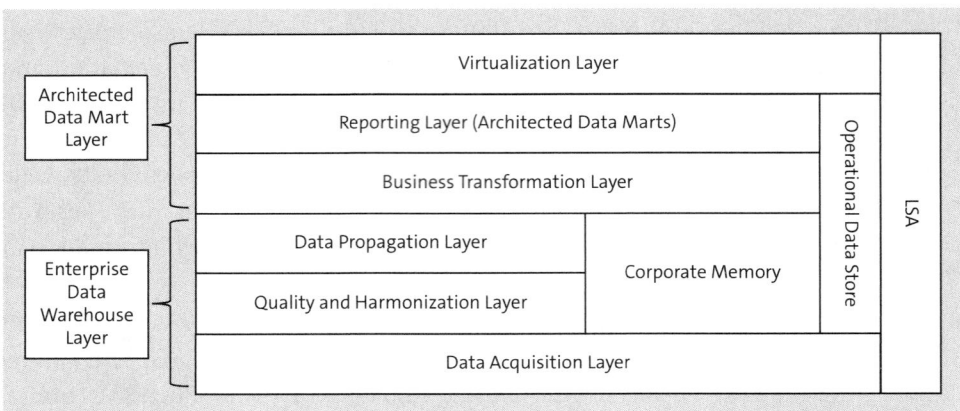

Figure 1.3 LSA with Multiple Data Persistences

The general data flow in SAP BW 7.x will follow the LSA process. A sample flow is shown in Figure 1.4.

1.1 SAP BW/4HANA Overview

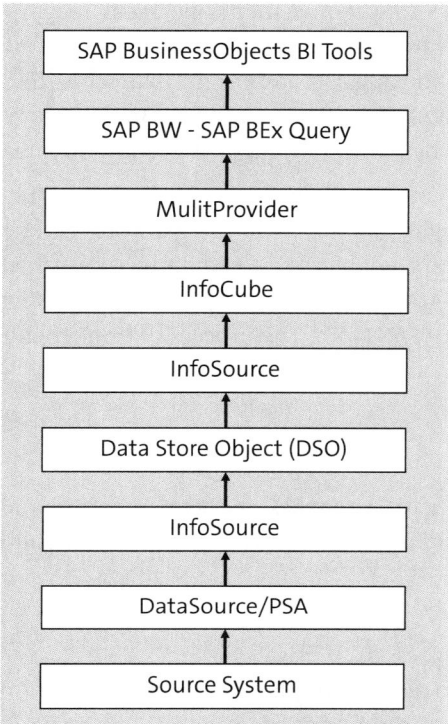

Figure 1.4 SAP BW 7.x Data Flow

1.1.3 Evolution of SAP BW/4HANA

Figure 1.5 shows the evolution of SAP BW/4HANA from previous versions. SAP BW/4HANA signifies a fundamental shift, embracing a simplified and modernized approach to data warehousing tailored explicitly for the SAP HANA platform. You need to see SAP BW/4HANA as a new product when compared to SAP BW on SAP HANA as the code has been optimized for the SAP HANA database. There are many simplifications in the extraction, modeling, and reporting components, which were optimized to use the SAP HANA database more effectively.

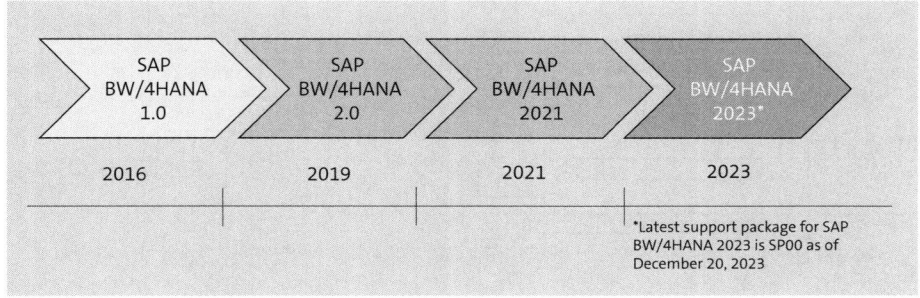

Figure 1.5 Evolution of SAP BW/4HANA

1 Introduction to SAP BW/4HANA

Let's consider an example: Organization A used SAP S/4HANA for its operational reporting and has no analytical reporting. With an eye firmly fixed on the future based on the real-time analytical reports for quick decision-making, Organization A has set forth a visionary plan: the implementation of SAP BW/4HANA, a next-generation data warehouse solution that promises to revolutionize their analytical reporting requirements.

By embracing SAP BW/4HANA, Organization A aims to elevate its analytical reporting capabilities to new heights. This transition will empower the company to extract actionable insights from its data more efficiently, enabling faster and more informed decision-making processes when compared to the online transaction processing (OLTP) reporting, which it's using based on SAP S/4HANA. The process of implementing SAP BW/4HANA from scratch is also known as a greenfield implementation.

1.1.4 SAP BW/4HANA Design Principles

When SAP BW/4HANA was introduced, simplicity, openness, modern interface, and high performance were the fundamental design principles. We'll discuss the design principles further in the following sections.

Simplicity

SAP BW/4HANA aims to be simple to use and to reduce complexity; the core focus is on reduced end-to-end development time and keep running at a lower cost. This resulted in a reduction of many objects from the previous version, which we'll cover in the next section. In the classical SAP BW versions, there were close to 10 objects, but with the simplified design principles, these objects were reduced to 4, as shown in Figure 1.6.

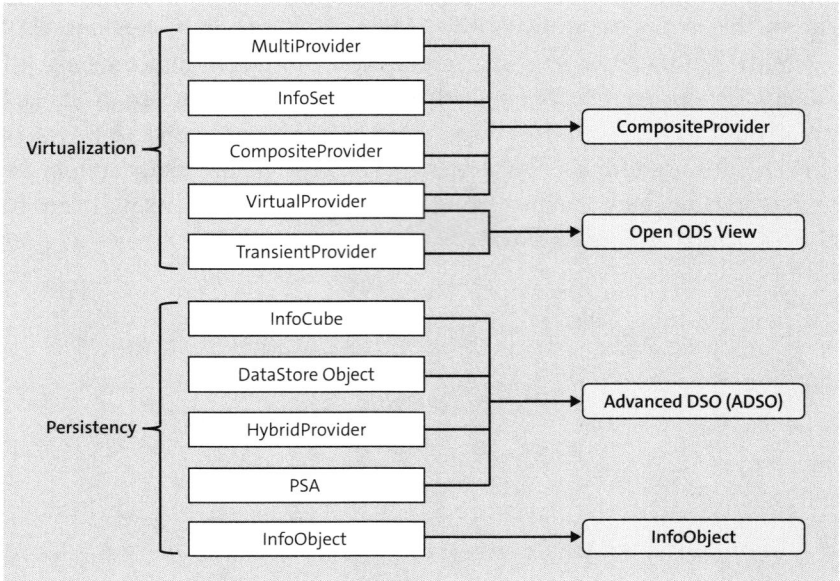

Figure 1.6 Simplicity in SAP BW/4HANA

Apart from this simplified design, SAP BW/4HANA has removed the other legacy SAP BW functionalities; for example, data modeling in SAP GUI is no longer supported (you need to use SAP BW modeling tools for the data modeling), and SAP BEx Query Designer isn't supported.

Openness

It's now easier to bring data from different sources into an SAP BW/4HANA system and consume data from the SAP BW system in order to manage all kinds of data and integration of data with all systems. The integration between SAP BW/4HANA and SAP HANA allows for the direct consumption of SAP BW/4HANA data via the generated SQL views. This enables exposing SAP BW/4HANA models as native SAP HANA views; these automatically generated views can be consumed in the SAP HANA platform to build new SAP HANA models, or they can be consumed by visualization tools. Figure 1.7 shows how SAP BW/4HANA objects are exposed to the SAP HANA layer.

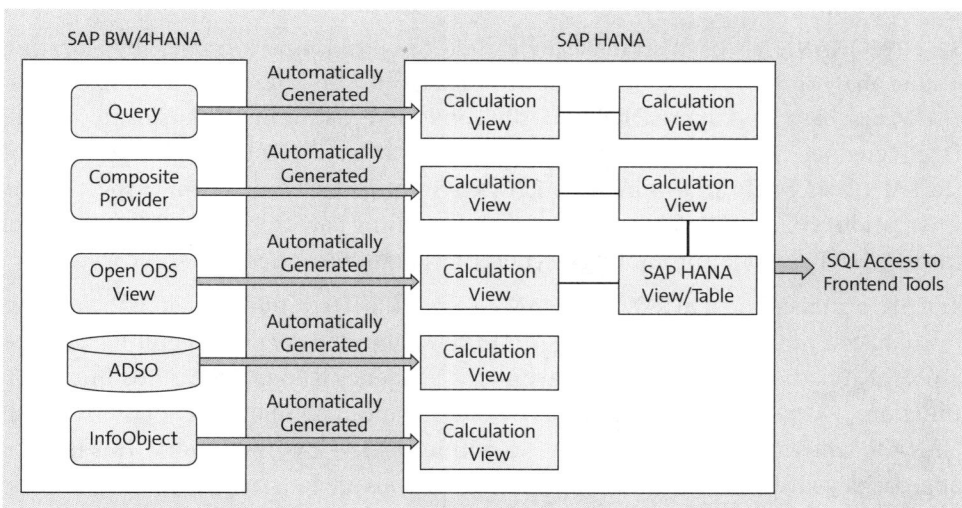

Figure 1.7 Openness in SAP BW/4HANA

Modern Interface

A modern interface ensures that users have simple access to data through easy-to-use tools with the new UI Eclipse. The modern interfaces are available for business users, developers, and administrators.

In the dynamic landscape of SAP BW/4HANA, a paradigm shift in interfaces has unfolded to meet the evolving needs of users. Business users experience a departure from the traditional SAP BEx interface, as SAP BW/4HANA emphasizes tighter integration with the SAP BusinessObjects BI toolset and extends collaboration to SAP Analytics Cloud. While SAP BEx queries persist as a crucial element, the revamped interfaces offer a more intuitive environment for data analysis and reporting, aligning with the modern expectations of business users.

Developers, on the other hand, engage with the system through advanced modeling tools seamlessly integrated into SAP HANA Studio and Eclipse. This developer-centric interface not only accelerates the development lifecycle but also provides a cohesive environment for constructing intricate data models and transformations. Simultaneously, administrators benefit from SAPUI5 tools, including the SAPUI5-based process chain monitor, which replaces the traditional ABAP process chain monitor. These SAPUI5-based tools offer a responsive and visually intuitive interface, enhancing the monitoring and management of data workflows for administrators within the SAP BW/4HANA framework. Together, these interface enhancements underscore SAP BW/4HANA's commitment to a more user-friendly, efficient, and contemporary analytics environment.

High Performance

SAP BW/4HANA has been strategically crafted to optimize performance by elevating online analytical processing (OLAP) functionality and computational tasks to the SAP HANA database. This approach extends beyond merely shifting workloads to the SAP HANA engine; it emphasizes comprehensive integration to harness the full potential of the SAP HANA platform within the SAP BW environment. One significant aspect of this integration is exemplified by seamlessly incorporating data from an SAP HANA model into an SAP BW model, promoting a synergistic and interconnected data ecosystem.

The performance boost in SAP BW/4HANA is a multifaceted strategy that involves both pushing computational tasks to the SAP HANA engine and deeply integrating with the SAP HANA platform's capabilities. Beyond the efficiency gained from offloading OLAP functions, SAP BW/4HANA capitalizes on the advanced computational power of SAP HANA for intricate data processing tasks within the SAP BW framework. This holistic approach ensures that the integration isn't merely about data transfer; it's a symbiotic relationship that enhances the overall efficiency and effectiveness of data processing.

Moreover, SAP BW/4HANA goes a step further by providing users with the capability to leverage a diverse array of advanced analytical libraries and algorithms. This encompasses sophisticated tools for data mining and predictive analytics, empowering organizations to explore complex data patterns and derive meaningful insights. By tapping into these libraries, users can construct robust predictive models, facilitating informed decision-making in various data-driven scenarios.

In essence, the performance optimization strategy in SAP BW/4HANA is twofold—pushing computational tasks to SAP HANA for efficiency and deeply integrating with the platform to unlock its full range of capabilities. This ensures that SAP BW/4HANA isn't just a platform for processing data; it's a dynamic environment that seamlessly combines OLAP functionality, computational power, and advanced analytics to drive

1.1 SAP BW/4HANA Overview

performance and innovation in the realm of business intelligence and data management.

SAP BW/4HANA is designed to use the full potential of SAP HANA to leverage large amount of data in real time. There have been designs to push down the OLAP functionality and complex computations to the SAP HANA database. The SAP BW/4HANA objects are designed to enable this pushdown to the database layer. SAP HANA–specific libraries can be used while running in the SAP HANA platform. There are many ways to do a code pushdown such as executing the data transfer process (DTP) in SAP HANA and creating SAP HANA transformations with the ABAP-managed database procedure (AMDP) to push down the complex ABAP logic, as shown in Figure 1.8.

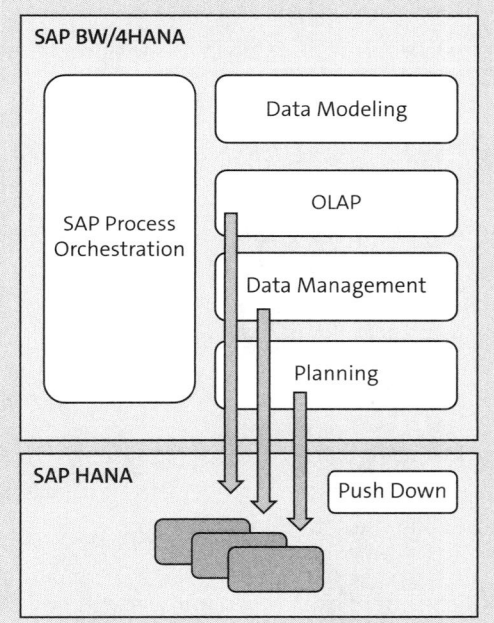

Figure 1.8 SAP HANA Pushdown

Apart from using the new objects, there have been many changes, such as moving from LSA to LSA++. In LSA++, the operational layer enables reporting on Advanced DSOs (ADSOs, described in the upcoming list). The virtual layer is based on the SAP HANA CompositeProvider to support the virtual data marts. Queries can be directly accessed in the SAP HANA database. These are the available objects:

- **Advanced DSO (ADSO)**
  These objects are used for standard data persistency.
- **Open ODS view**
  This defines the reusable SAP BW data warehouse semantics on field-based structures.

- **CompositeProvider**
  This defines virtual data marts on persistence objects (ADSO/InfoObjects) and/or open ODS views, SAP HANA calculation views, or a combination of both.

Layered Scalable Architecture++ Layers

Following is an overview of the different layers in LSA++, as shown in Figure 1.9:

- **Operational DataStore layer**
  This is an entry layer that offers services such as queries and reports. The services offered include the following:
  - Paired down and extended data acquisition layer services
  - Immediate query and reporting option
  - Early and simple integration of data in data warehouse using LSA++ for simplified data warehousing

- **EDW Propagation**
  This data propagation layer provides semantic and value standardization for data from different sources in a highly harmonized form. In the propagation layer, the "Extract once, deploy more" standard principle applies. The data propagation later serves as a persistent architected data mart and is the basis for the virtual data mart layer. The data is saved and consolidated in standard ADSOs and in InfoObjects. Note that all the major data warehousing services are implemented in the LSA++ propagation layer.

- **Corporate memory**
  This contains the complete history of the loaded data from the source systems. This is filled separately in the architected data marts. Corporate memory can be used as a source for the reconstruction without accessing the source system again.

- **Architect data mart**
  This layer serves as a query access layer if the query can't be built already in propagation or the Open ODS layer. It's also used when additional business logic is involved. Most of the business requirements can be built with the propagation layer and virtual data mart layer. A persistent architected data mart layer is built only if the propagation and virtual data mart layers aren't possible, and these architected data marts are defined based on ADSOs and InfoObjects. In the ADSO **Maintenance** screen, there is a separate template for the modeling the ADSO for the data mart layer: **Reporting on union of active table and inbound queue**. This option will make ADSOs behave like InfoCubes.

- **Virtual data mart layer**
  Combines data from the persistence provider (ADSO and InfoObject) with VirtualProviders (union/joins). Here, the layers don't represent strict limits, which means you can combine the objects from the Open ODS layer with objects from the propagation layer or architected data mart layer.

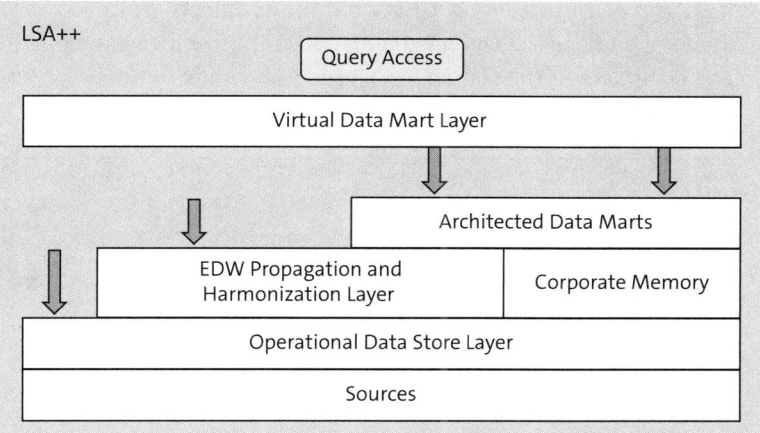

Figure 1.9 The Layers of LSA++

1.2 What's New in SAP BW/4HANA?

SAP BW/4HANA offers many new features. We'll walk through the essential features in the following sections.

1.2.1 Eclipse-Based Modeling Tool

In SAP BW/4HANA, the Eclipse-based modeling tool is a central component of the development landscape, serving as the primary interface for data modeling tasks. Eclipse, an open-source integrated development environment (IDE), provides developers with a unified platform to design, build, and maintain data models seamlessly. This tool facilitates graphical data modeling, enabling users to visually represent complex structures such as InfoObjects and transformations. It not only streamlines the creation of intricate data models but also offers advanced features for designing components such as CompositeProviders, ADSOs, and InfoObjects.

The Eclipse interface is designed to be user-friendly, enabling developers to be more efficient in their data modeling tasks within the SAP BW/4HANA environment. It supports version management, and transport functions ensure robust control over different versions of data models for seamless transitions between various environments. The Eclipse-based modeling tool in SAP BW/4HANA is a fundamental element in creating and managing sophisticated data structures in the SAP ecosystem. A sample eclipse tool is shown in Figure 1.10.

You can see that the **Data Sources** folder has the details of all the connected source systems. In SAP BW/4HANA, all the connected source systems are of type **ODP**, which stands for *operational data provisioning*. You can see the list of many InfoAreas under the **Favorites** section where you can group the SAP BW modeling objects based on the

1   Introduction to SAP BW/4HANA

application areas. For example, InfoArea **RKPODPBW** has other objects grouped based on the extraction. When you choose the **Open Perspective** icon, which can be seen among the icons in the right side of the screen, you'll see the details to open the **BW Modeling** tools, as shown in Figure 1.11.

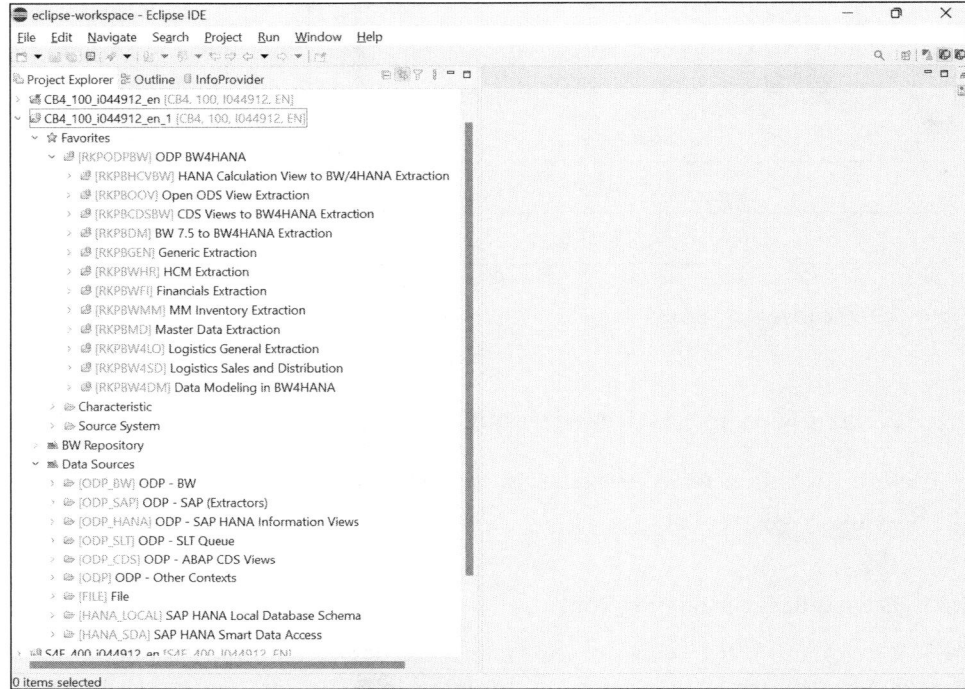

Figure 1.10 Eclipse-Based SAP BW Modeling Tools

Figure 1.11 Perspectives in Eclipse

You also can use the web cockpit, as shown in Figure 1.12.

1.2 What's New in SAP BW/4HANA?

Figure 1.12 SAP BW/4HANA Web Cockpit

1.2.2 SAP BW/4HANA Data Modeling Objects

In this section, we'll explore the objects that are used for SAP BW data modeling. We'll start with the InfoObjects-based data modeling, and then we'll discuss how to model ADSOs, which have persistent data. You'll learn about the modeling features of SAP HANA CompositeProviders, which are virtual entities used for union and join operations for analytical reporting. Then, you'll learn about the Core Data Services (CDS) view–based extraction, ADSO remodeling, real-time extraction basics, and SAP BW/4HANA statistics options.

InfoObject Modeling

You can only create InfoObjects in the Eclipse-based SAP BW modeling tools. Figure 1.13 shows the InfoObject maintenance screen. The default screen won't have any **Properties** checkboxes selected, and you'll see the three tabs **General**, **BI Clients**, and **Extended** by default.

Within the InfoObject maintenance screen, you'll find options to configure the **Properties** of the InfoObject. Upon selecting a **Properties** checkbox, various configuration choices become available for InfoObject maintenance across tabs, including **Master Data**, **Attributes**, and **Hierarchies**. Each tab provides insights into tables and crucial details pertaining to the metadata of the InfoObject. Notably, in SAP BW/4HANA, there's an option to expose InfoObjects as SAP HANA views, facilitating consumption within the SAP HANA database; this feature requires enabling the **External SAP HANA View for Master Data** checkbox, as shown in Figure 1.14.

37

1   Introduction to SAP BW/4HANA

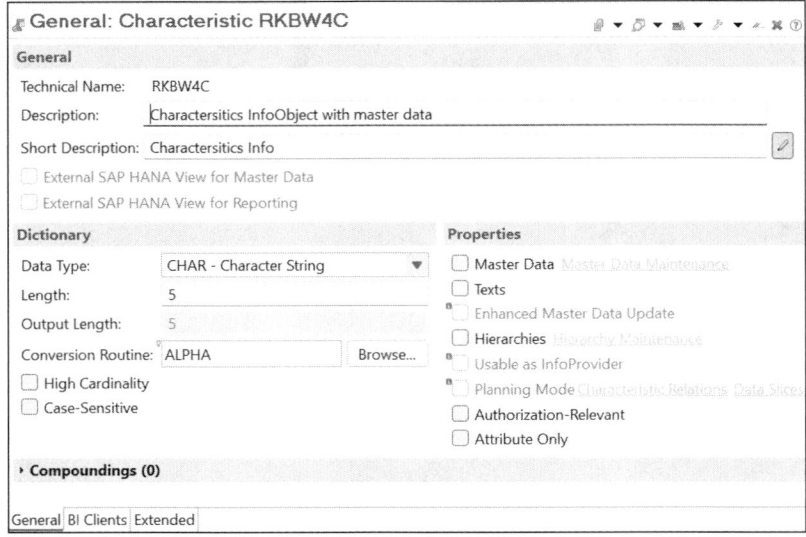

Figure 1.13 InfoObject Maintenance: Characteristics

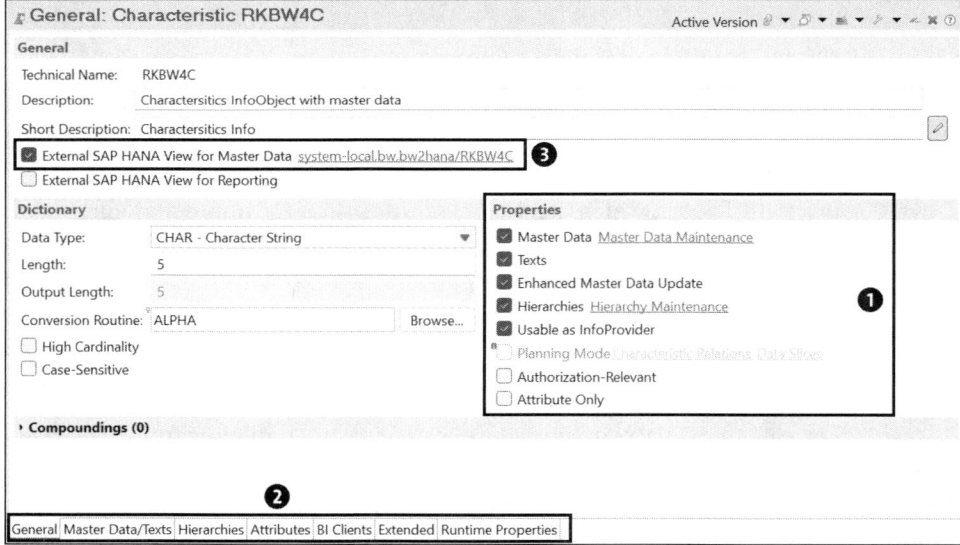

Figure 1.14 Characteristics InfoObject Properties

In Figure 1.14, you can see the following details:

❶ Using **Properties** settings allows you to select your preferences when generating InfoObjects.

❷ Based on the **Properties** selected, the tabs are shown here dynamically.

❸ When you choose the **External SAP HANA View for Master Data** option, the associated path (**system-local.bw.bw2hana/RKBW4C**) is shown here. When you log on to the database perspective of SAP HANA Studio and check the system-local path, you can see the equivalent view for the InfoObject.

Let's walk through some of the other options:

- **High Cardinality**
  Enabling the **High Cardinality** property permits the creation of more than 2 billion attributes for a characteristic. Given the constraints associated with a characteristic having high cardinality, it's advisable to activate this property only when a substantial number of characteristic attributes is anticipated. A characteristic marked with high cardinality lacks persistent Surrogate ID values and doesn't possess an surrogate ID table. Furthermore, it can't be employed in hierarchies or as a navigational attribute.

- **Case-Sensitive**
  When the **Case-Sensitive** option is activated, the system distinguishes between uppercase and lowercase letters in entered values. Conversely, if this option isn't enabled, the system automatically converts all letters to uppercase. Importantly, this conversion doesn't take place during the loading process or transformation, so values containing lowercase letters can't be uploaded to an InfoObject that doesn't permit the use of lowercase letters.

- **Master Data/Texts**
  You can specify whether the characteristic should have master data and/or texts. You can then see the **Master Data/Texts** tab pages and can edit the properties displayed there.

- **Hierarchies**
  You can specify whether the characteristic should have hierarchies. You can then see the **Hierarchies** tab page and edit the properties displayed there.

- **Usable as InfoProvider**
  If you select **Usable as InfoProvider**, then queries can be executed on the characteristic and the flag can be used in a CompositeProvider. Then, the additional **Runtime Properties** tab will appear, where you can configure various settings at query runtime.

- **Authorization-Relevant**
  If you flag the characteristic as authorization-relevant, the authorization check will run whenever the query is worked on. Set the **Authorization-Relevant** flag for a characteristic if you want to create authorizations that restrict the selection conditions for this characteristic to a single characteristic value.

- **Attribute Only**
  The characteristic can be used only as a display attribute for another characteristic, not as a navigation attribute.

Now let's see the details of creating a key figure and its properties in SAP BW/4HANA, as shown in Figure 1.15. Where you have the **InfoObject Type** set as **KYF - Key Figure**, there will be multiple data types for the key figure such as integer, date, time, currency, and unit. You need to choose the related data type to create the key figure. In our example, we're creating a key figure of type **CURR** (currency).

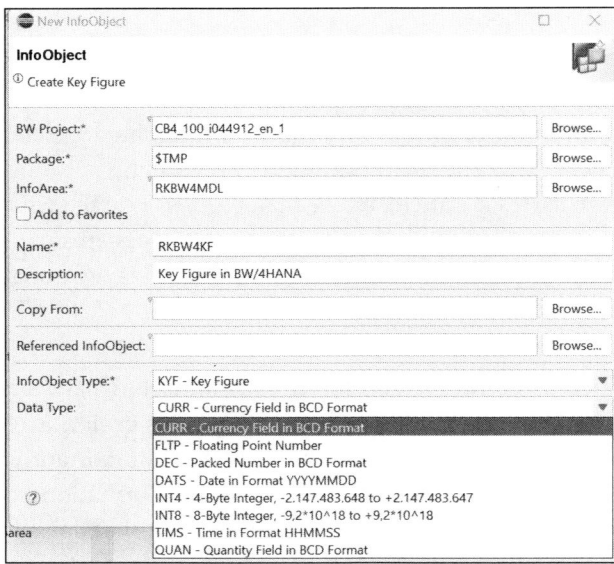

Figure 1.15 Key Figure InfoObject Options

Once you choose the **Data Type** of **Key Figure** and press Enter, you'll land in the key figure maintenance window with the options shown in Figure 1.16. You can see multiple **Aggregation** options such as **Summation**, and as with SAP BW 7.x, there is an **Exception Aggregation** dropdown. In the **Properties** area are several checkboxes, including **Non-Cumulative** for handling special scenarios such as inventory data.

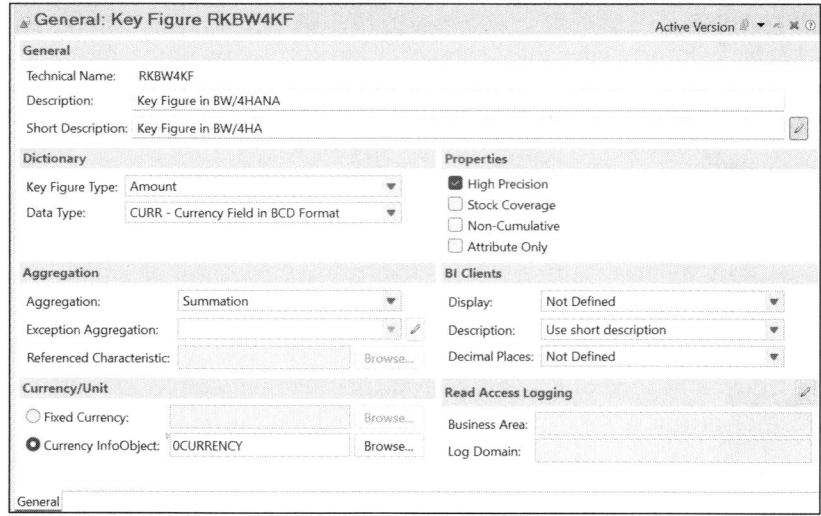

Figure 1.16 Key Figure Maintenance

Modeling Advanced DataStore Objects

In SAP BW/4HANA, the exclusive persistence object is the ADSO. During the migration from SAP BW to version 7.x, all existing persistent staging areas (PSAs), DSOs, and InfoCubes undergo conversion into various types of ADSOs. Figure 1.17 shows the ADSO maintenance screen, which can be seen when you create a new ADSO in the SAP BW modeling tools. The **Modeling Properties** section provides the flexibility to specify the type of ADSO you want to create. The inclusion of **Special Properties** enables the configuration of ADSOs tailored for specific scenarios, such as those pertaining to inventory and planning. Additionally, the **Data Tiering Properties** section allows you to define and manage the temperature schema, enabling the storage of data across **Hot**, **Warm**, and **Cold** storage tiers.

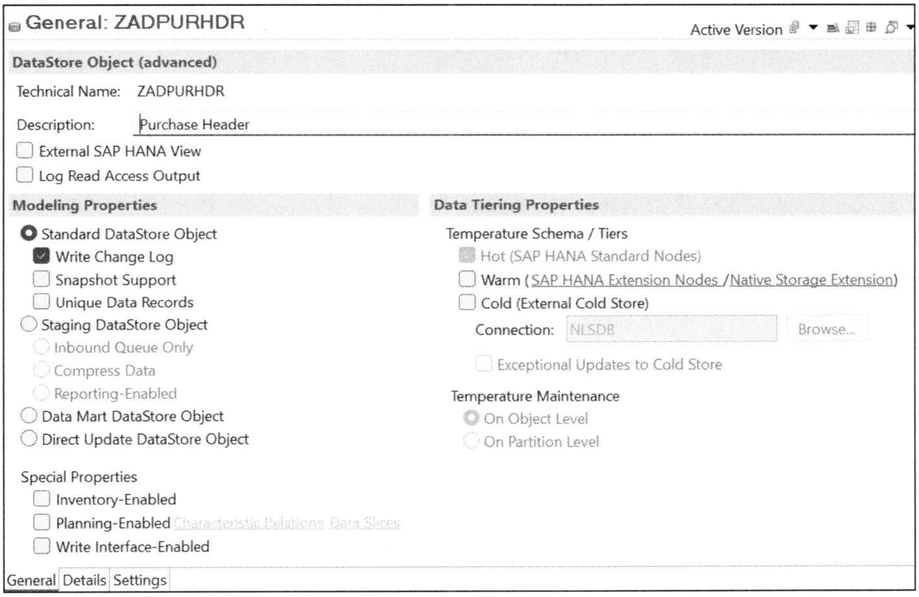

Figure 1.17 ADSO Maintenance

In the **Special Properties** area, you can see the **Write Interface-Enabled** option, which allows data to be written directly into the inbound queue of the DSO in SAP BW/4HANA. The inbound queue is a part of the data loading process in SAP BW/4HANA and is responsible for temporarily storing data before it's written to the actual data tables of the DSO. Enabling **Write Interface-Enabled** means that you're configuring the DSO to accept data from external sources or interfaces directly into its inbound queue, bypassing certain processing steps. This setting is particularly useful in scenarios where you want to streamline and accelerate the data loading process, especially when dealing with high-frequency or real-time data updates. It helps in integration of SAP Data Services, Cloud Integration in SAP Integration Suite, and SAP Data Intelligence.

1   Introduction to SAP BW/4HANA

The activation of a standard ADSO initiates the creation of tables, as depicted in Figure 1.18. In SAP BW/4HANA, three additional tables are generated for the views, aligning with the other three same structures in SAP BW 7.x: **Inbound Table**, **Active Data Table**, and **Change Log**. It's crucial to note that in SAP BW/4HANA, these tables follow a new naming convention: **Inbound Table** ends in "1", **Change Log** ends in "2", and **Active Data Table** ends in "3". For instance, when conducting a reconciliation of InfoCube data post-convention, it's imperative to reconcile with the active table of the newly created ADSO, identified by the "3" suffix.

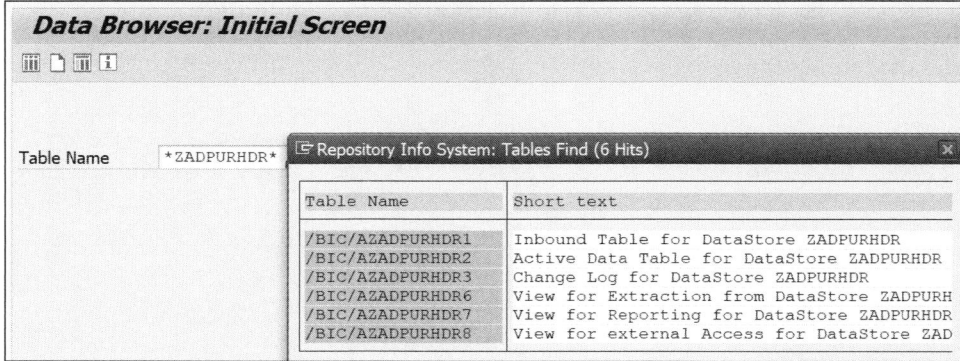

Figure 1.18 Generated Standard ADSO Tables in SAP BW/4HANA

When you create the ADSO for handling inventory scenarios, you'll use the options shown in Figure 1.19. Note that the **Special Properties** section has the **Inventory-Enabled** checkbox selected.

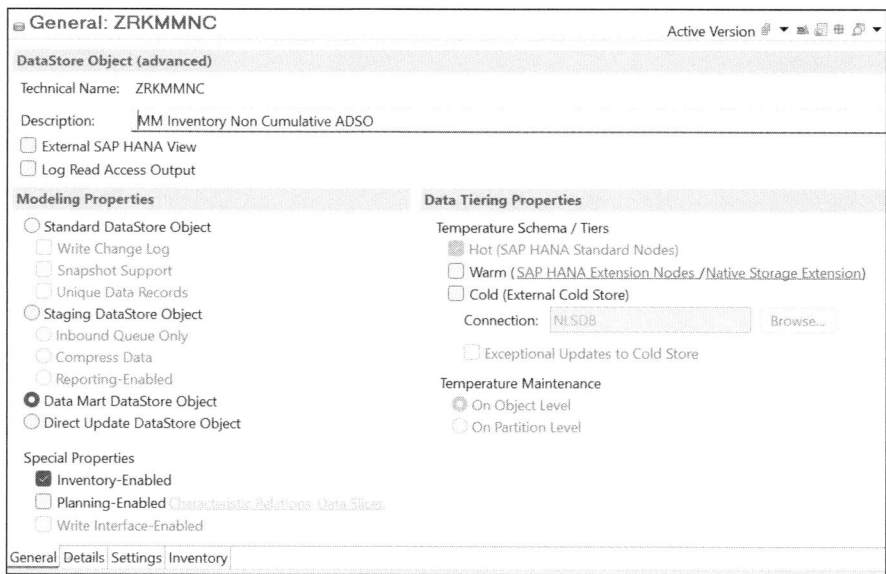

Figure 1.19 Inventory Model ADSO

42

Upon activating the ADSO, the tables that will be generated are shown in Figure 1.20. **Validity Table** ends in "4", and **Reference Point Table** ends in "5".

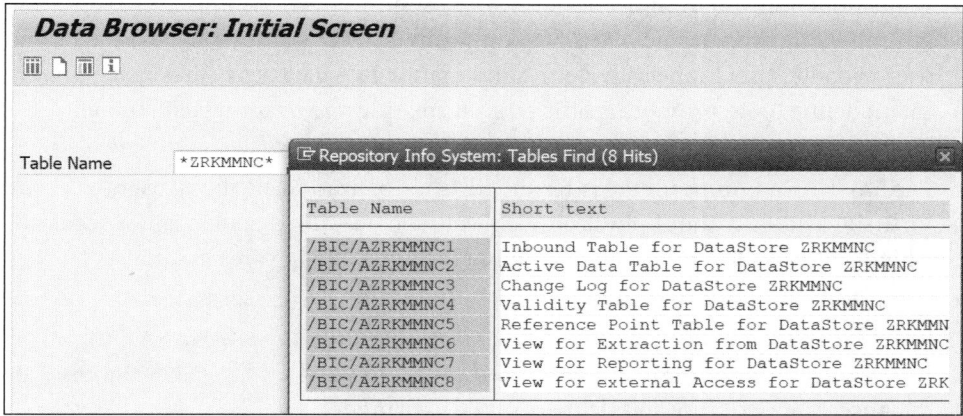

Figure 1.20 Inventory ADSO Tables in SAP BW/4HANA

Modeling SAP HANA CompositeProvider

In SAP BW/4HANA, a CompositeProvider is a modeling object that allows users to combine data from different source objects, providing a unified view for reporting and analysis. The SAP HANA CompositeProvider is an enhanced version of the CompositeProvider specifically designed to leverage the capabilities of the SAP HANA database.

Key features and characteristics of the SAP HANA CompositeProvider in SAP BW/4HANA include the following:

- **Unified data view**
  The primary purpose of the SAP HANA CompositeProvider is to bring together data from various source objects, such as InfoProviders, ADSOs, and InfoObjects. This consolidation allows for the creation of a unified and harmonized view of the data.

- **In-memory processing**
  Leveraging the power of the SAP HANA in-memory database, the SAP HANA CompositeProvider takes advantage of the high-performance processing capabilities for faster data retrieval and improved query performance.

- **Advanced join and union operations**
  The SAP HANA CompositeProvider supports advanced join and union operations directly within the SAP HANA database. This enables users to perform complex data manipulations and aggregations, reducing the need for extensive data transformations in the modeling layer.

- **SAP HANA smart data access**
  SAP HANA smart data access is a feature that allows the SAP HANA CompositeProvider to access and integrate data from external sources in real time. This extends

1   Introduction to SAP BW/4HANA

the scope of data integration beyond the SAP landscape, supporting a more comprehensive view of the business data.

- **Agile modeling and design**
  The modeling process for SAP HANA CompositeProviders is designed to be agile and user-friendly. Users can easily define and manipulate data structures using graphical modeling tools, promoting efficiency in creating and adapting data models.

- **Support for virtualization**
  The SAP HANA CompositeProvider supports virtualization, enabling users to create VirtualProviders without physically replicating data. This is beneficial for scenarios where real-time data access and flexibility in data modeling are essential.

- **Enhanced calculation capabilities**
  Users can perform advanced calculations directly within the SAP HANA CompositeProvider using features such as calculated key figures and restricted key figures. This enhances the flexibility and richness of the data model.

- **Integration with SAP BW/4HANA Tools**
  The SAP HANA CompositeProvider seamlessly integrates with other SAP BW/4HANA tools and features, providing a cohesive environment for data modeling, reporting, and analytics.

By combining the capabilities of the SAP HANA database with the flexibility of CompositeProviders, the SAP HANA CompositeProvider in SAP BW/4HANA offers a robust solution for creating sophisticated data models that cater to the evolving needs of modern BI and analytics.

Figure 1.21 shows a sample SAP HANA CompositeProvider (HCPR) maintenance screen.

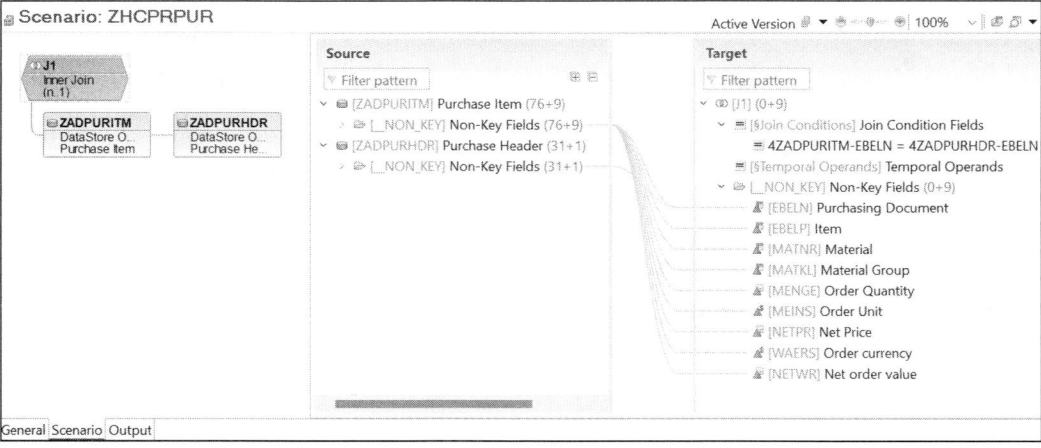

Figure 1.21 HCPR Modeling

1.2 What's New in SAP BW/4HANA?

SAP HANA CompositeProvider Based on the SAP HANA Calculation View

You can create the HCPR based on a calculation view, for example, in Figure 1.22, where the HCPR is calling the calculation view. If you choose the **Scenario** tab, you can see the mapping between the **Source** and **Target** fields.

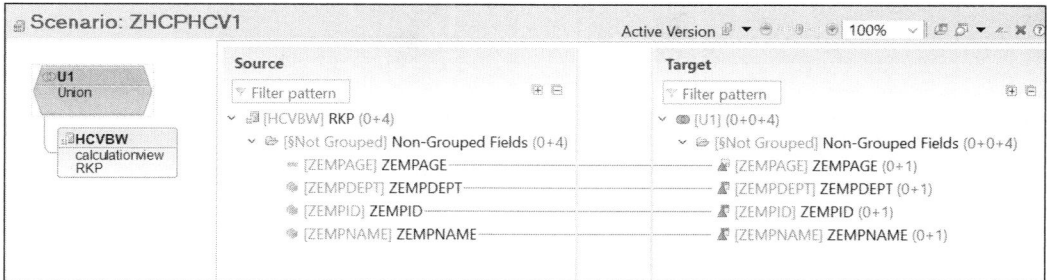

Figure 1.22 HCPR Based on a Calculation View

Data Extraction to SAP BW/4HANA Based on the Core Data Services View

You can use the CDS views with analytics annotations to send data to SAP BW/4HANA. Just replicate the CDS views as the DataSource, and create the DTP and transformation based on the CDS views, as shown in Figure 1.23.

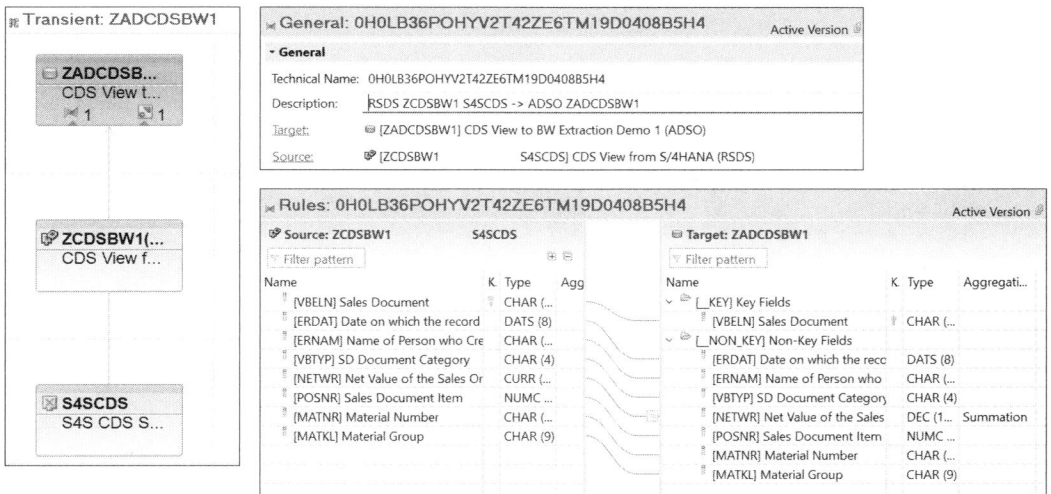

Figure 1.23 CDS View for SAP BW/4HANA Extraction

The DTP for CDS-based extraction will have the settings shown in Figure 1.24. Ensure that under **Adapter Settings**, **ODP Context** is set to **ABAP Core Data Services**.

1   Introduction to SAP BW/4HANA

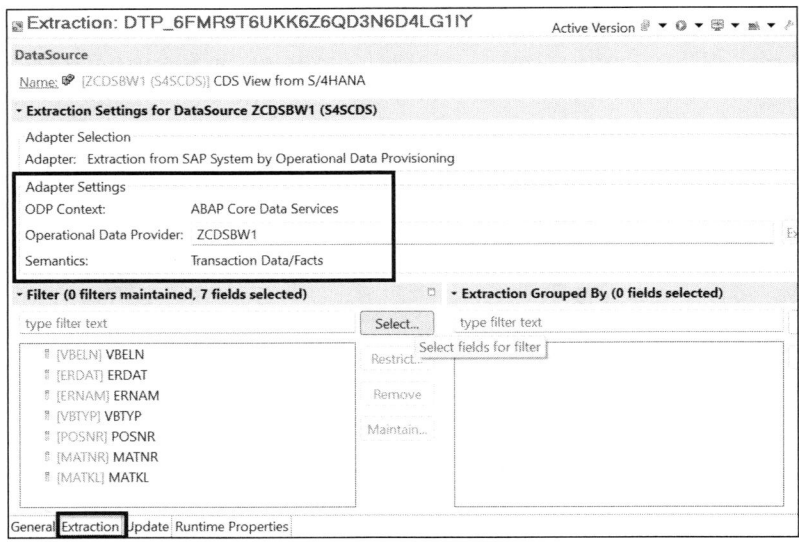

Figure 1.24 CDS-Based DTP for Extraction

Remodeling in Advanced DataStore Objects

You have the option to remodel in ADSOs, as shown in Figure 1.25. To remodel an ADSO in SAP BW/4HANA, begin by accessing the SAP BW modeling environment and selecting the targeted ADSO.

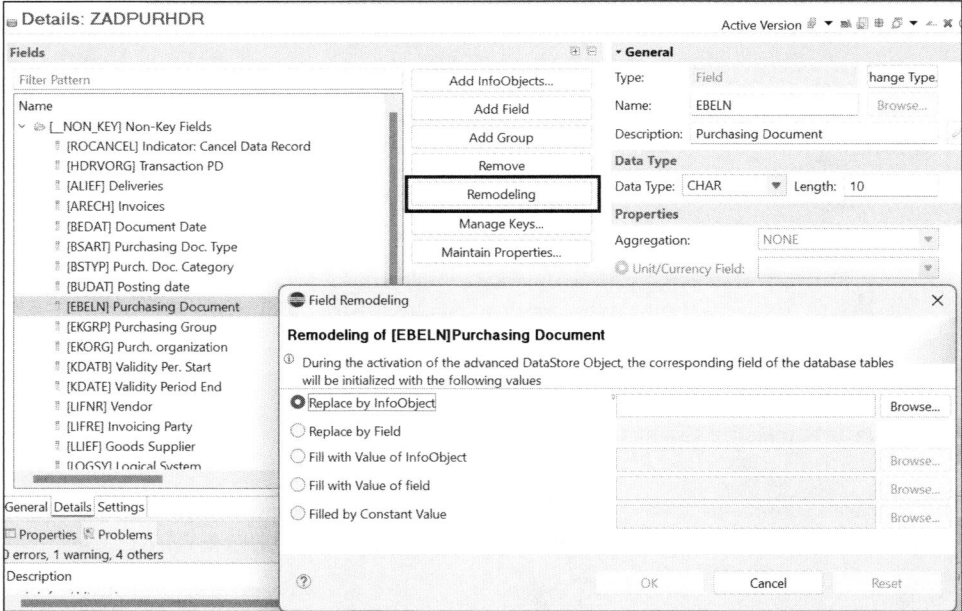

Figure 1.25 Remodeling in ADSO

This will get you into edit mode to adjust the structure, properties, and fields, including adding, removing, or modifying InfoObjects. Ensure that key figures, aggregations, and data types are appropriately adjusted. Update associated transformations and DTPs to align with the remodeled structure. After activating the changes, closely monitor the process for errors, conduct thorough testing, and, if applicable, transport the remodeled ADSO to the production environment. Following these steps ensures a systematic and validated remodeling process within the SAP BW/4HANA framework.

Real-Time Extraction Using Process Chains

In SAP BW/4HANA you can use the process chain with the **Streaming** option. Streaming process chains in SAP BW/4HANA offers the flexibility of frequent initiation and execution, enabling real-time updates to ADSOs. These process chains present a comprehensive solution catering to both periodic and highly frequent scheduling needs, including streaming scenarios. Consequently, they play a key role in streamlining the overall architecture of SAP BW/4HANA. The decision of whether to use a process chain for streaming can be determined by configuring a specific attribute within the process chain settings. There are two run modes—push and pull—and there is one push mode and two pull modes available. The key distinction lies in determining whether the initiation of data transfer is triggered by the consuming system or the source system.

In the push run mode, which is preferred in the case of ODP sources, SAP BW/HANA is triggered by an ODP daemon in the source system whenever new records are generated. This initiation prompts the immediate execution of a process chain, facilitating the seamless transfer of fresh data to the SAP BW InfoProvider. This mode is highly advantageous for ODP sources because the process chain operates solely when new data becomes available in the source, eliminating the need for defining periodic repetitions. To enable this, the daemon is scheduled as a regular job in the SAP source system. The frequency of check interval will be close to 15 seconds in this mode. Supported sources are `ODP_SAP`, `ODP_CDS_ODP_BW`, and `ODP_SLT`.

For the pull run mode, which is the preferred mode for SAP HANA sources, SAP BW/4HANA diligently monitors the source system every 60 seconds for the generation of new records. Upon a successful check, the process chain promptly commences to transfer the recently acquired data to the target InfoProvider. This mode, scheduled once in streaming mode, eliminates the need for defining periodic repetitions. Particularly suited for SAP HANA sources, this approach ensures that the process chain operates exclusively when fresh data becomes accessible in the source system. Supported sources are `ODP_HANA` and `ODP_SAP`.

SAP BW/4HANA Statistics

You can't use the technical content-based InfoProviders in SAP BW/4HANA. SAP BW/4HANA provides comprehensive application statistics accessible through ABAP CDS views. This new content serves as a replacement for the former technical content, offering insights into various aspects such as current query performance, process chain, and DTP efficiency. Additionally, it covers data volume statistics, DSO request details, and the valuable addition of data volume history statistics—all delivered with SAP BW/4HANA 2.0. This enhances the analytical capabilities of the platform, enabling users to gain deeper insights into the performance and historical trends of their SAP BW/4HANA applications.

> **Deployment Options (On-Premise and Cloud)**
>
> In your project, you may opt for SAP BW/4HANA as an on-premise model. However, it's important to note that SAP BW/4HANA is also designed for cloud deployment. Should you choose a cloud-based deployment, setting up the target SAP BW/4HANA system will involve configuring it according to your chosen cloud solution provider. It's crucial to plan resources adequately and determine the scaling of your architecture, whether it's scaling up or scaling out. As we embark on exploring the core focus of this book, which is SAP BW/4HANA migration, it's important to emphasize that the concepts and strategies discussed in the upcoming chapters are applicable to both on-premise and cloud deployment scenarios. Whether your SAP BW/4HANA instance resides on your own infrastructure or in the cloud, apart from the initial cloud-based system setup and configuration, the migration preparation and execution process entails similar considerations and best practices for a seamless transition to SAP BW/4HANA.

1.3 Summary

In this chapter, we gave you a high-level overview of SAP BW/4HANA and explored the history of SAP BW. You learned about the SAP BW 7.x objects, which were used in extraction and modeling; the layered scalable architecture (LSA); and the evolution of SAP BW/4HANA, which use LSA++. Later in the chapter, we discussed the SAP BW/4HANA design principles in detail and described the new modeling and reporting objects in SAP BW/4HANA. By completing this chapter, you should have a solid foundation for data modeling in SAP BW/4HANA. In the next chapter, we'll focus more on the SAP BW/4HANA migration overview.

Chapter 2
Overview of SAP BW/4HANA Migration

This chapter offers a comprehensive introduction to SAP BW/4HANA migration, providing an in-depth understanding of its core concepts and processes. By the end of this chapter, you'll understand the various conversion types available and recognize the key differences between them.

An organization on SAP Business Warehouse (SAP BW) 7.x on any database that needs to migrate to SAP BW/4HANA must first conduct a detailed analysis based on the current SAP BW version. This preliminary analysis of the existing SAP BW system with the available new versions is called an *analytical strategy workshop*. The outcome of an analytical strategy workshop is typically a clear road map for the implementation of the SAP BW/4HANA solution. This road map will guide the project team in building and configuring the system to meet the analytical needs of the organization effectively. It's an important step in the process of designing and implementing a successful SAP BW/4HANA solution that supports data analysis and reporting requirements. We'll cover this process in detail in Chapter 3. To migrate to SAP BW/4HANA, we need to categorize the customers based on the database size.

This chapter covers the overview of SAP BW/4HANA migration. We'll start exploring the multiple business scenarios in Section 2.1, followed by learning the conversion types in SAP BW/4HANA in Section 2.2. Then, we'll explore the conversion phases in SAP BW/4HANA in Section 2.3. Finally, we'll discuss why project management is important in the SAP BW/4HANA migration projects in Section 2.4.

2.1 Business Scenarios

Before getting into the details of SAP BW/4HANA migration, let's discuss a few business scenarios where the migration to SAP BW/4HANA is required.

2.1.1 Organizations without an SAP Business Warehouse Landscape

Consider this example: Organization A has used SAP S/4HANA for its operational reporting requirements, but they don't have an SAP BW system for the data warehouse requirement. The company plans to implement SAP BW/4HANA as the data warehouse solution for its analytical reporting requirements. Organization A has managed to fulfill its operational reporting requirements using SAP S/4HANA without the presence of a dedicated SAP BW system. SAP BW is the traditional stalwart of data warehousing, an indispensable repository for collecting and storing vast volumes of structured and unstructured data and preparing it for analytical purposes. However, the absence of this dedicated data warehousing solution has prompted Organization A to reevaluate its data management strategy, as their ambitions have outgrown the capabilities of their current setup.

With an eye on the future, Organization A plans to implement SAP BW/4HANA as its next-generation data warehouse solution to revolutionize their analytical reporting requirements. This strategic move is a significant step in their journey toward harnessing the full potential of their data. SAP BW/4HANA not only offers the traditional data warehousing functionalities but also incorporates in-memory computing, real-time data processing, and advanced data modeling capabilities.

By embracing SAP BW/4HANA, Organization A aims to elevate its analytical reporting capabilities. This transition will empower the company to extract actionable insights from the data more efficiently, enabling faster and more informed decision-making processes. By taking this step, the company seeks to not only keep pace with the changing landscape of BI but also make gains toward with data-driven excellence. These kinds of scenarios require implementing SAP BW/4HANA from scratch in a greenfield implementation.

2.1.2 Organization with an SAP Business Warehouse System Landscape

When the organization has an existing SAP BW system landscape, then the migration to SAP BW/4HANA comes with multiple options. The choice of migration path depends on factors such as the organization's existing landscape, data volume, customization level, and business requirements, as we'll discuss in the following sections.

Develop New Data Models

Imagine an organization with a rich historical dataset spanning 15 years that is currently managed within the confines of a dated SAP BW 7.x system. This aging setup runs on various databases such as Oracle or DB2. Over the years, the system has accumulated multidimensional data models that have become antiquated. However, recognizing the potential of leveraging the high-performance capabilities of SAP HANA, the organization has made a strategic decision to transition to the next-generation SAP BW/4HANA. In doing so, they aim to take full advantage of the SAP HANA database by revamping their data models, replacing the old structures with newer, more optimized counterparts such as Core Data Services (CDS) views and calculation views.

The existing SAP BW 7.x system, operating with historical data, has been functional but may have limitations in terms of data processing speed and agility. Transitioning to

SAP BW/4HANA signifies a forward-looking approach, with an emphasis on harnessing the in-memory processing capabilities of SAP HANA for optimal performance.

To fully embrace the capabilities of SAP BW/4HANA and the SAP HANA database, the organization intends to reconfigure its data models. The outdated multidimensional structures are due for a refresh, and this will be achieved through the creation of new, more efficient data models such as CDS views and calculation views. These modern models are not only more adaptable and user-friendly but are also better equipped to handle complex calculations and aggregations, making them the preferred choice for organizations seeking real-time analytics and robust data-driven decision-making. In essence, the move toward SAP BW/4HANA–optimized data models represents a strategic shift to a more agile, high-performance, and future-ready data management solution. In this scenario, the organization will prefer to go for a greenfield implementation approach.

Convert the Existing Data Models

When the organization has existing data models in the SAP BW landscape, there are multiple migration scenarios depending on the SAP BW version, the database for the SAP BW system, and the requirement to retain the historical data. You can see the options for the conversion based on the following scenarios:

- Scenario 1

    Within the same organizational setup operating SAP BW 7.x, which accommodates a significant dataset stored in a non-SAP HANA database, the imperative for migrating to SAP BW/4HANA is evident. However, the migration strategy isn't a uniform, universal solution, and it depends on whether data migration is required. This determination significantly shapes the chosen route and approach to facilitate a smooth transition to SAP BW/4HANA. The landscape requirements and the organization's data migration management requirements are key considerations when opting for this SAP BW/4HANA migration strategy, ensuring alignment with the goal of leveraging the advanced capabilities of the SAP HANA database. In this scenario, the organization wants to use the same landscape to convert to SAP BW/4HANA.

- Scenario 2

    When an organization aims to transition from any BW system operating on a non-SAP HANA database to SAP BW/4HANA within a completely new landscape, a methodical and step-by-step process is initiated. The first crucial phase involves setting up a fresh installation of the SAP BW/4HANA system landscape. Following this, the migration process centers on transferring data models from the older SAP BW 7.x system to the newly established SAP BW/4HANA system, guaranteeing a smooth transition to the modernized environment. Importantly, this approach doesn't involve the direct migration of data from the legacy system to the new SAP BW/4HANA system, ensuring that the organization capitalizes on the capabilities of the SAP HANA database.

- **Scenario 3**

    Let's envision an organization dealing with a substantial volume of data, aiming to retain historical records within its new SAP BW/4HANA landscape. In this scenario, the first step involves the installation of SAP BW/4HANA, followed by a sequence of actions that includes object conversion and the migration of historical data. This meticulous process ensures the seamless integration of historical data into the SAP BW/4HANA system while preserving its integrity and availability for future analytical purposes. This kind of migration approach is preferred when the database size is large and needs data to be migrated.

- **Scenario 4**

    Consider a scenario where a customer operates multiple SAP BW systems and needs to streamline these into a unified system. In such a case, a system consolidation approach is undertaken, where multiple SAP BW systems are carefully evaluated, and specific data models are extracted to create a new SAP BW/4HANA system. This method ensures a more efficient and consolidated system landscape, optimizing the organization's data management and reporting capabilities. This process requires guidelines from SAP and SAP tools for achieving this goal. The Data Management and Landscape Transformation group from SAP takes care of helping customers with these requirements.

Based on all of these scenarios, we can see that the migration from SAP BW 7.x to SAP BW/4HANA is based on multiple factors. First and foremost, the system landscape serves as a crucial determinant in shaping the migration strategy. Organizations often operate in diverse landscapes, with varying configurations and historical data structures. The existing landscape's architecture, including factors such as database systems, hardware, and data storage, plays a pivotal role in charting the migration path. It's essential to assess the compatibility of the current infrastructure with the requirements of SAP BW/4HANA, as this can impact the complexity of the migration and any potential infrastructure upgrades.

Second, data migration represents a fundamental consideration. Historical data, which holds significant value for many organizations, must be managed meticulously during the migration process. The decision to migrate, archive, or transform data models requires careful planning and execution. This stage is crucial not only for maintaining historical records but also for optimizing data quality and integrity in the new SAP BW/4HANA environment.

Furthermore, the conversion of classical objects into SAP HANA–optimized models stands as a pivotal undertaking during the migration journey. The transformation of data structures and objects is essential to harness the performance advantages of the SAP HANA database. This step requires meticulous planning, as it involves adapting existing data models and objects to align with the capabilities and architecture of SAP

BW/4HANA. The objective is to create an efficient and streamlined data model that leverages the in-memory processing capabilities of SAP HANA.

Considering these factors, the migration to SAP BW/4HANA is by no means a one-size-fits-all process. To effectively navigate this transition, organizations must adopt a tailored approach, customized to their unique circumstances. A detailed analysis of the existing system landscape, data requirements, and object conversion needs is fundamental to charting the most appropriate migration path. This strategic planning ensures that the migration aligns seamlessly with the organization's specific goals and data management needs, optimizing the transition to SAP BW/4HANA.

2.2 SAP BW/4HANA Conversion Types

To migrate or convert SAP BW or any BW system on to an SAP HANA system, there are multiple migration options available for SAP BW/4HANA, as listed here and shown in Figure 2.1:

- Greenfield implementation
- In-place conversion
- Remote conversion
- Shell conversion
- System consolidation

We'll discuss each of these in the following sections.

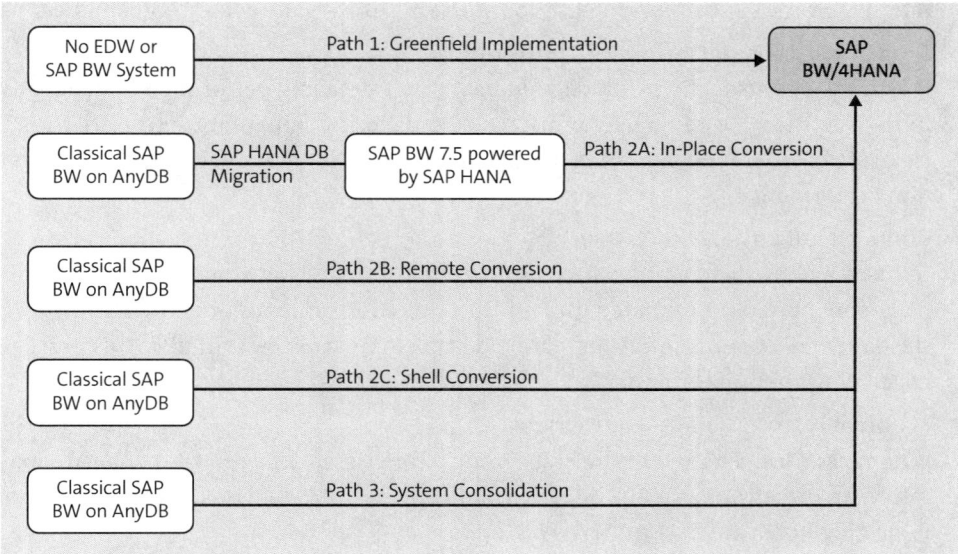

Figure 2.1 Migration Path to SAP BW/4HANA

2.2.1 Greenfield SAP BW/4HANA Implementation

As we've discussed in Scenario 1 in Section 2.1.2, an SAP BW/4HANA greenfield implementation is typically considered when an organization meets certain criteria or has specific business needs that align with the advantages of starting afresh with a new SAP BW/4HANA system. Here are some more situations in which opting for a greenfield implementation makes sense:

- **No existing SAP BW system**
  A greenfield implementation is the natural choice if an organization has never used SAP BW or any previous versions of it. This scenario occurs in cases where the organization is establishing a data warehousing solution for the first time.

- **Legacy system limitations**
  If the existing SAP BW system or other data warehousing solutions in use are outdated, inefficient, or unable to meet the organization's evolving data management and reporting requirements, a greenfield implementation can provide an opportunity to modernize and align with the latest best practices.

- **Significant changes in business needs**
  When the organization's business processes, data sources, or reporting requirements undergo substantial changes, it may be more practical to create a new SAP BW/4HANA system tailored to these evolving needs rather than attempting to retrofit the existing system.

- **Strategic shift toward SAP HANA**
  If the organization is making a strategic move toward adopting SAP HANA as the primary database platform for its data processing and analytics, a greenfield implementation allows for a seamless integration with SAP HANA's in-memory capabilities.

- **Data quality and performance optimization**
  Organizations struggling with data quality issues or facing performance challenges in their existing systems may opt for a greenfield implementation to start with a clean slate, redefining data models and structures to enhance data quality and system performance.

- **Complex data migration issues**
  In cases where the existing data models, structures, and data sources are too complex, convoluted, or outdated, the effort required for migrating and transforming the data may be substantial. A greenfield implementation can simplify this process by starting anew.

- **Minimal historical data requirement**
  When historical data retention isn't a critical requirement or can be managed separately, an organization may choose to build a new SAP BW/4HANA system without migrating historical data from the old system.

- **Strategic business expansion**
  Organizations embarking on new strategic business ventures or expanding into new markets may choose a greenfield implementation to design data models that specifically align with these new initiatives.

Undertaking a greenfield SAP BW/4HANA implementation represents a comprehensive endeavor, involving establishing the system from its foundational elements. This intricate process requires a considerable investment of both time and resources to ensure its successful development. Nevertheless, amid this complexity, customers have the opportunity to construct entirely new data models by leveraging CDS views and harnessing the power of SAP HANA calculations. By doing so, they unlock the full potential of the SAP HANA database, including its high-performance capabilities and data processing efficiency.

Leveraging CDS and SAP HANA calculation views in a greenfield SAP BW/4HANA implementation delivers a multitude of advantages. These tools optimize performance, enabling high-speed data processing and analytics, while providing flexible data modeling and simplified data integration. Real-time analytics become feasible, reducing data redundancy and supporting scalability as data volumes grow. Additionally, the tools streamline maintenance, lower resource overhead, and seamlessly integrate with the broader SAP ecosystem. In summary, using a greenfield implementation via CDS and SAP HANA calculation views empowers organizations to take advantage of the full potential of SAP HANA, enhancing their data-driven operations with efficiency, speed, and scalability and improving their analytics to derive valuable insights. All of this contributes to better decision-making and more efficient operations.

2.2.2 In-Place Conversion

In the context of Scenario 2 in Section 2.1.2, if your organization is interested in transitioning from the classical SAP BW running on any database to SAP BW/4HANA within the same system landscape, a two-step approach is required. The first step involves upgrading SAP BW 7.x to SAP BW 7.5 SP and migrating the classical database to SAP HANA, thus enabling your SAP BW 7.5 system to operate on the SAP HANA database. Once this supported version is in place, you can proceed with the in-place conversion method to seamlessly migrate from classical SAP BW to the more advanced SAP BW/4HANA. The process to migrate using in-place conversion is shown in Figure 2.2.

Figure 2.2 In-Place Conversion Process

2   Overview of SAP BW/4HANA Migration

You must have a sequence of activity before starting the in-place conversion. To upgrade from SAP BW 7.3 on any database to SAP BW 7.5 on SAP HANA, start with a system assessment and back up the existing database-based system. Use the maintenance planner to plan the upgrade, address Oracle-specific considerations, and download necessary software components. Execute the upgrade using Software Update Manager (SUM), closely monitor the upgrade, and address any issues. Following the upgrade, prepare for migration to SAP HANA by ensuring compliance with SAP's requirements, backing up the SAP BW 7.5 system on the specific database, and then using the database migration option (DMO) of SUM to transition to SAP BW 7.5 on the SAP HANA database. Validate the migration, conduct extensive testing, and communicate changes to stakeholders, providing training as needed. Monitor the system post-migration, adjusting for optimal performance and stability in the new SAP HANA environment. Figure 2.3 shows the first step for an SAP BW 7.5 upgrade and SAP HANA database migration using SUM.

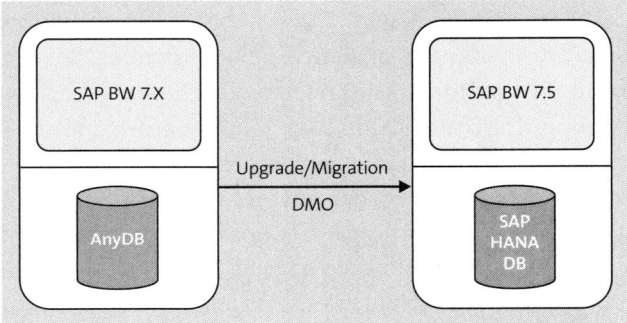

Figure 2.3 Step 1: DMO for SAP 7.5 Upgrade and SAP HANA Migration

The DMO can also be used to convert to Unicode when upgrading and migrating. The next step is to convert the SAP BW 7.5 SP 5 on SAP HANA system to SAP BW/4HANA, as shown in Figure 2.4.

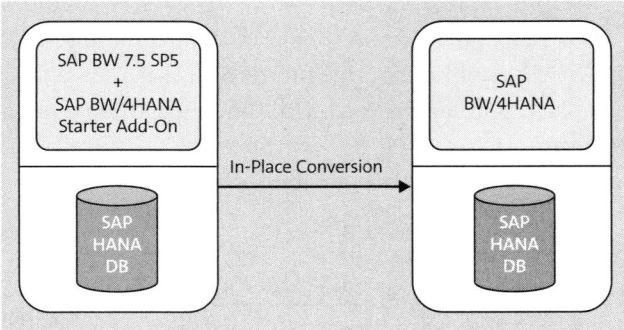

Figure 2.4 Step 2: SAP BW 7.5 to SAP BW/4HANA Using In-Place Conversion (Same Landscape)

2.2 SAP BW/4HANA Conversion Types

As depicted in Figure 2.5, the traditional SAP BW data flows, containing objects such as DataStore objects (DSOs), InfoCubes, and MultiProviders, undergo conversion in an in-place conversion scenario for SAP BW/4HANA. During this transformation, the former objects are transitioned into Advanced DSOs (ADSOs) and hybrid SAP HANA CompositeProviders (HCPRs) in the new SAP BW/4HANA framework. Notably, the in-place conversion involves not only the conversion of these objects but also the migration of data seamlessly within the same process. You need to note that the above process will he happening in the same system ID (SID), which is part of same system landscape, so there won't be a change of the system ID post-conversion to SAP BW/4HANA.

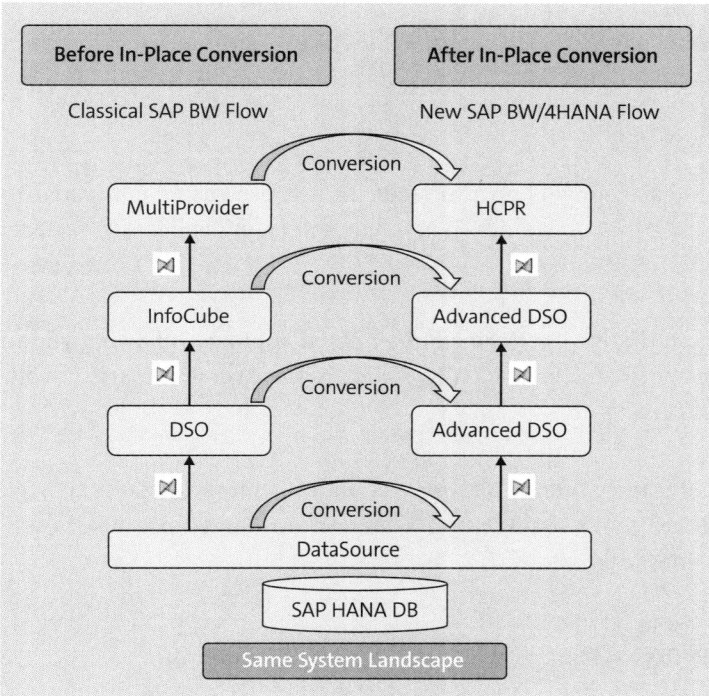

Figure 2.5 Data Flow Change before and after In-Place Conversion

These are the features when you go for an in-place conversion:

- **Simplified landscape**
  In-place conversion can help simplify your system landscape by transforming your existing SAP BW system into SAP BW/4HANA without the need to set up a new system, resulting in reduced complexity and maintenance.

- **Cost-efficiency**
  In-place conversion can be a cost-effective option as it eliminates the need to invest in new hardware, infrastructure, or licenses for a separate SAP BW/4HANA system. You'll be using your existing SAP BW 7.5 on SAP HANA system for the conversion.

- **Minimized data migration efforts**
  Migrating data from one system to another can be complex and time-consuming. In-place conversion reduces the need for extensive data migration efforts, which can be particularly beneficial for organizations with large and complex datasets. Because the database isn't changed, the complexity is reduced.

- **Preservation of historical data**
  In-place conversion allows you to retain historical data, reports, and business logic from your existing SAP BW system, which can be essential for ongoing business operations and reporting.

- **Leverage existing skills**
  If your organization has a team with expertise in SAP BW, an in-place conversion may be a more attractive option as it allows you to leverage existing skills and knowledge.

- **Reduced downtime**
  In some cases, an in-place conversion can result in less downtime compared to building a new SAP BW/4HANA system, as it involves fewer steps and less infrastructure setup.

- **Phased transition**
  In-place conversion can often be performed in a phased approach, allowing you to migrate specific areas or processes gradually, reducing the overall impact on your organization's operations.

- **Risk management**
  An in-place conversion can be seen as a lower-risk option, as it minimizes the need for significant changes to your existing SAP BW environment and can reduce potential disruption to business operations.

2.2.3 Remote Conversion

When an organization is currently using SAP BW 7.x on any database system and confronts several key circumstances, such as managing substantial volumes of data and seeking to transition to SAP BW/4HANA, the option of a remote conversion becomes particularly relevant. This process is employed when the organization opts for a fresh landscape for its SAP BW/4HANA deployment. The main driver for this choice is typically the desire to harness the advanced capabilities and performance enhancements that SAP BW/4HANA offers. However, it's important to note that this approach introduces a higher level of complexity compared to an in-place conversion.

In the case of remote conversion, the process often begins with the creation of an entirely new landscape dedicated to SAP BW/4HANA. This entails setting up new servers, databases, and infrastructure components that align with the requirements of SAP BW/4HANA. Subsequently, the data migration phase commences, which involves transferring the existing data, reports, and data models from the legacy SAP BW 7.x

2.2 SAP BW/4HANA Conversion Types

system to the newly created SAP BW/4HANA landscape. This migration process can be intricate, as it requires careful planning, data mapping, and the transformation of data structures to align with the new platform's requirements. The creation of this new landscape and the intricacies associated with data migration might require the involvement of specialized expertise and meticulous project management to ensure a smooth transition to SAP BW/4HANA. This flow is shown in Figure 2.6. Figure 2.7 shows the object-level flow.

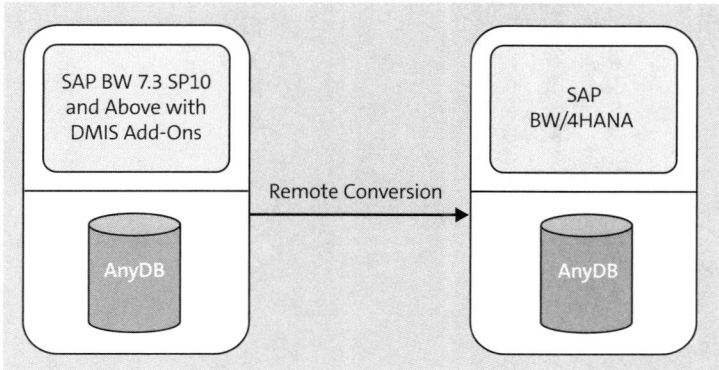

Figure 2.6 Remote Conversion Process

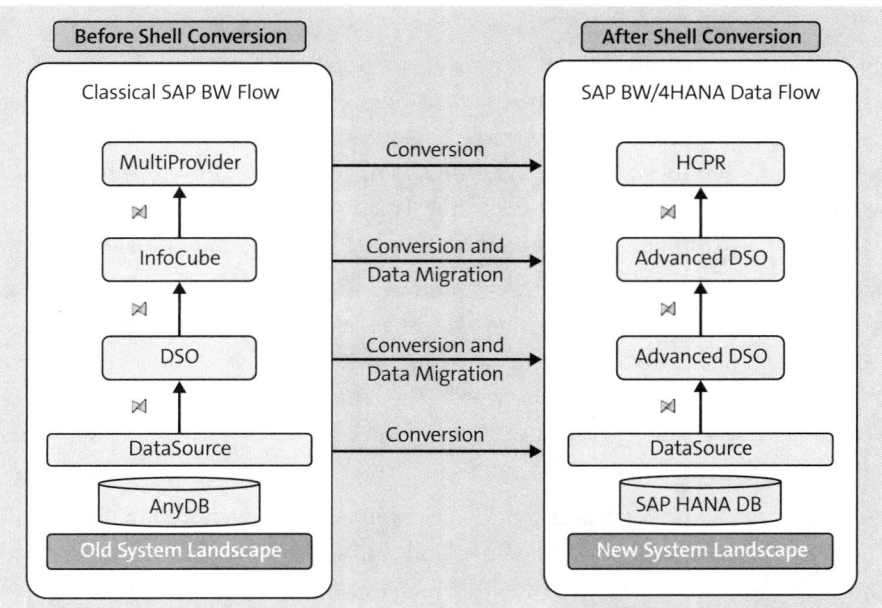

Figure 2.7 Remote Conversion Data Flow

In the context of remote conversion for data migration, the primary goal is to seamlessly transfer data from the existing system, referred to as the sender SAP BW system, to

59

the new SAP BW/4HANA system, known as the target SAP BW/4HANA system, throughout the course of the project. This migration process becomes particularly intricate when the data transfer occurs between systems in different landscapes, as shown in Figure 2.8.

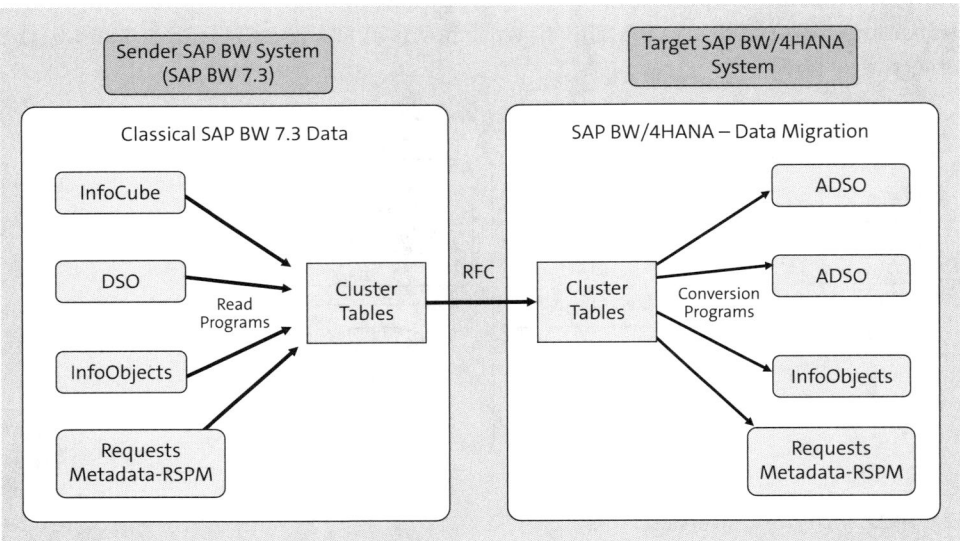

Figure 2.8 Remote Conversion: Data Migration

To facilitate this cross-landscape data transfer, a specialized migration tool is employed. The tool performs a crucial role by reading all relevant data from the sender system's objects. Subsequently, this extracted data is efficiently organized and stored in a cluster table. One noteworthy feature of this storage is the highly compressed format in which the data resides within the cluster. This compression not only optimizes storage space but also enhances the efficiency of the subsequent data transfer.

Once the data is securely housed in the cluster, the next step involves unpacking it within the target system. This unpacking process is instrumental in transforming the highly compressed data into a format compatible with the tables associated with the SAP BW/4HANA target objects. In other words, the cluster is meticulously unraveled, and the data is strategically placed into the appropriate tables, ensuring a seamless integration into the new system.

In essence, the migration tool acts as a bridge between the sender and target systems, orchestrating the extraction, compression, and subsequent unpacking of data to ensure a smooth and accurate transition during the data migration process in remote conversion scenarios. This meticulous approach is vital for maintaining data integrity and consistency in the transition to the advanced SAP BW/4HANA system.

These are the key points when going for a SAP BW/4HANA remote conversion:

- **Flexibility**
  Remote conversion offers remarkable flexibility, enabling organizations to seamlessly upgrade classical SAP BW systems, starting from version 7.3, regardless of the underlying database. This adaptability ensures a smooth transition to SAP BW/4HANA while accommodating diverse database environments and customizations, making it an ideal solution for businesses looking to modernize their data warehousing capabilities.

- **Parallel landscape requirement**
  Requires a parallel landscape setup, with both a sender (old SAP BW) system and target (SAP BW/4HANA) system, to facilitate a controlled and efficient migration process. The parallel landscape helps in running the project efficiently without impacting the production landscape.

- **Minimized disruption**
  Remote conversion often leads to reduced disruption in ongoing operations as it doesn't directly impact the existing SAP BW system. In contrast, in-place conversion may require significant changes to the current environment, potentially causing more interruptions.

- **Incremental approach**
  Enables an incremental approach to migration, allowing organizations to gradually adopt SAP BW/4HANA features and benefits while maintaining a level of continuity with the existing system. This helps the organization plan the conversion based on the criticality or phase by phase.

- **Highly compressed data storage**
  Stores data in a highly compressed format, reducing storage requirements and potentially improving overall system performance. This compressed data is stored in the cluster table of the sender and target systems.

- **Efficient data transfer**
  Efficiently transfers relevant data from sender system objects to the target system, streamlining the migration process and ensuring data integrity.

2.2.4 Shell Conversion

When the organization needs to convert the classical SAP BW objects to SAP BW/4HANA objects without the transfer of data, then the shell conversion needs to be used. This conversion is applicable from release 7.0 to 7.5. The primary goal of shell conversion is to update and adapt the structure and components of classical SAP BW objects to make them compatible with SAP BW/4HANA. This allows organizations to leverage the new capabilities of SAP BW/4HANA without the immediate need to migrate the data. It's particularly useful when a complete data migration isn't immediately feasible or necessary, but the organization still wants to benefit from the enhanced features of SAP BW/4HANA. To perform a shell conversion, a parallel landscape is required. This landscape should include both an SAP BW 7.x system (the sender system) and an SAP BW/4HANA system (the target system). The parallel landscape allows for the transformation and adaptation of objects in a controlled environment without affecting the production system. The process involves converting classical SAP BW objects, such as InfoProviders, transformations, and data transfer processes (DTPs), to their equivalent SAP BW/4HANA objects. This may include updating data models, transforming structures, and aligning configurations with SAP BW/4HANA standards. This flow is show in Figure 2.9. This is useful when the organization has less data volume.

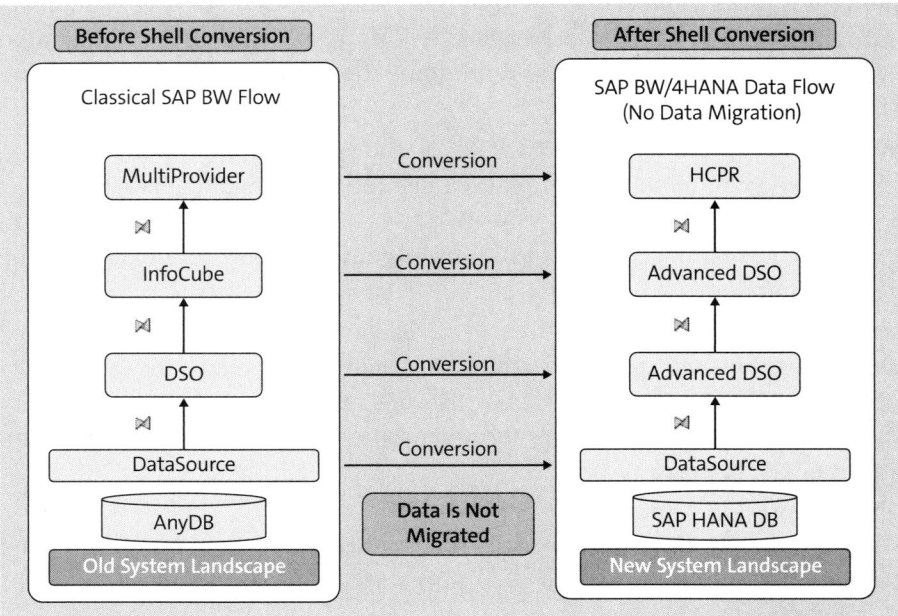

Figure 2.9 SAP BW/4HANA Shell Conversion Flow

There are key benefits when going for the shell conversion because there is no data migration to the target SAP BW/4HANA system. In this process, the data migration and preparation-based errors will be minimal if all the prerequisites are completed properly. This conversion is useful if you need to convert the SAP BW system to the SAP BW/4HANA system within a short period when compared to a remote conversion.

2.2.5 Landscape Transformation

In instances where an organization operates multiple SAP BW systems across diverse regions, each serving as a production environment for analytical reporting, complex scenarios may unfold. For example, when the organization undergoes new acquisitions, there might be a need to seamlessly integrate new SAP BW systems into the existing landscape. Alternatively, the organization might decide to carve out specific data flows from the current landscape, aiming to merge all the SAP BW systems into a unified global structure. These situations necessitate a detailed migration strategy, involving intricate analysis and planning. The Data Management and Landscape Transformation group from SAP plays a crucial role in assisting SAP customers globally by providing customized solutions tailored to their unique migration requirements. This is called system landscape transformation or system consolidation.

Consider a scenario where a multinational company acquires a new business unit that uses a separate SAP BW system. The challenge lies in harmonizing data reporting across the entire organization by seamlessly integrating the acquired SAP BW system into the existing landscape. This requires careful consideration of data structures, transformations, and business logic to ensure a unified analytical platform. Another example could be the need to streamline reporting processes by consolidating specific data flows from different regions into a centralized SAP BW system. This kind of migration involves custom approaches to align diverse data models and configurations.

It's important to note that the detailed custom migration scenarios described herein fall outside the scope of this book. As this topic revolves around custom approaches rather than tool-centric methodologies, you should seek additional resources or consulting expertise for a comprehensive understanding of SAP BW migration in complex, multi-landscape environments.

2.3 Phases in SAP BW/4HANA Conversion

When you choose any of the SAP BW/4HANA conversion approaches, there are multiple phases. Though there are many activities in each phase, the overall process can be divided into two phases: the conversion prepare phase, and the conversion realization phase, which we'll discuss in the following sections.

2.3.1 Conversion Prepare Phase

Figure 2.10 illustrates the intricate tasks associated with each phase of the SAP BW/4HANA migration. Later, in Chapter 3, we'll delve into a comprehensive discussion of the specific activities involved.

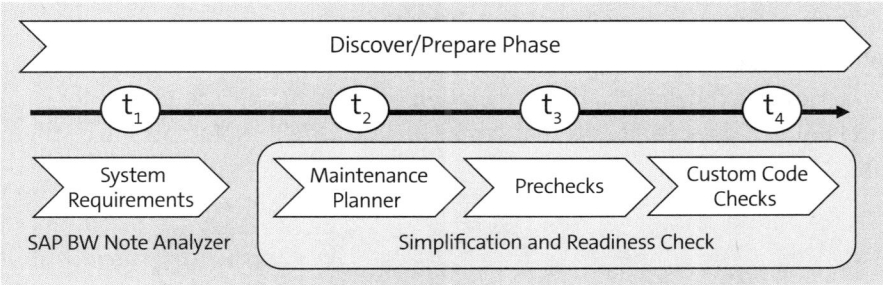

Figure 2.10 SAP BW/4HANA Conversion: Prepare Phase

The following list acts as a precursor, setting the stage for a detailed examination of the migration process and its integral components:

- **System requirements**
  Before staring any SAP BW/4HANA migration project, the first step is to understand the system requirements for SAP BW/4HANA. This will cover the sender system SAP BW version and target system SAP BW version and consider the database platform and release.

- **Maintenance planner**
  To perform the technical system conversion to SAP BW/4HANA, it's imperative to use the maintenance planner. As part of this process, it's recommended to execute the maintenance planner in the early stages of the SAP BW/4HANA system conversion project. The maintenance planner assesses and verifies the supportability of various items for the conversion, ensuring that all necessary prerequisites are met.

- **Prechecks**
  There are tools from SAP that facilitate your conversion project by furnishing prechecks designed to pinpoint crucial measures required to guarantee the compatibility of your system with the conversion process. It's advisable to execute these checks prior to commencing the realization phase to allow ample time for resolving any identified issues before initiating the conversion processes. These prechecks, delivered to customers intending to migrate to SAP BW/4HANA in the form of SAP Notes, serve as valuable resources. Customers can use these prechecks to discern the essential steps they must undertake before transitioning to SAP BW/4HANA. The outcome of these checks provides a comprehensive list of instances that necessitate attention before embarking on the conversion process.

- **Custom code**
  Before moving to SAP BW/4HANA, you check for custom code that needs to be adapted. To support you in detecting this code, SAP provides a code scan tool as part of the SAP BW/4HANA transfer cockpit. For example, you can verify whether custom code that has been developed long back in ABAP, and which is embedded in transformations and other objects, will be compatible with the SAP BW/4HANA data structures. This step needs the active involvement from the ABAP experts as there might be a requirement to retire and remove the obsolete custom code. Further details on this process will be discussed in depth in Chapter 3.

- **Simplification and readiness check**
  SAP has provided the detailed documentation on the simplification items for each classical SAP BW object and how that can be handled. We'll explore this in detail in Chapter 3.

2.3.2 Conversion Realization Phase

This is the phase where the actual SAP BW/4HANA migration will happen. Based on the detailed analysis, the organization can choose the best possible options for its SAP BW to SAP BW/4HANA conversion.

During the realization phase of the in-place conversion, as depicted in Figure 2.11, it's essential to have the necessary starter add-on and transfer cockpit. Once the custom code adjustments are completed, SUM is employed for the actual system conversion. Subsequently, post-conversion steps are undertaken, leading to the emergence of the SAP BW/4HANA system.

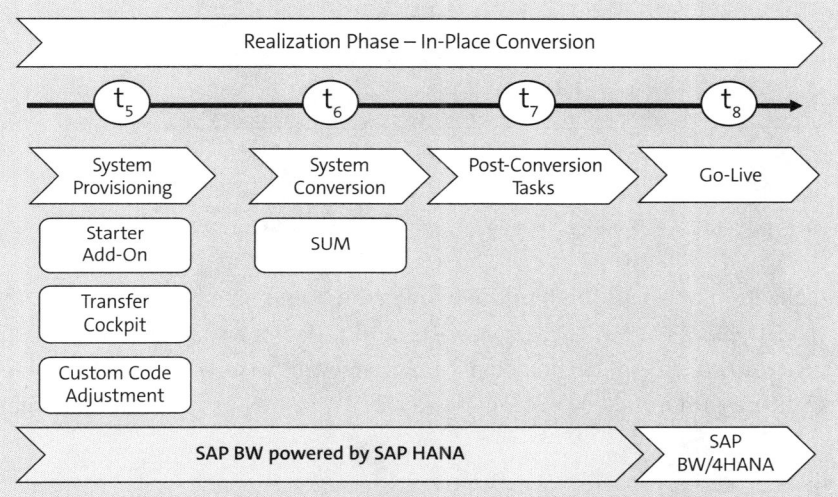

Figure 2.11 SAP BW/4HANA Conversion: Realization Phase for In-Place Conversion

In the realization phase of the remote conversion, depicted in Figure 2.12, a pivotal prerequisite is the installation of the SAP BW/4HANA system within the target landscape. Subsequently, the deployment of the Conversion Cockpit tool, an integral part of the Data Migration Integration Server (DMIS) add-on in both sender and target systems, becomes crucial. This tool serves as a cornerstone in orchestrating a smooth transition from the existing system to the advanced SAP BW/4HANA architecture. Prior to initiating the conversion, it's imperative to conclude the custom code adjustments, ensuring seamless compatibility with the new environment.

Within this process, the focus is on data migration, which is the important task preserving and transferring critical information accurately. Post-conversion, a series of steps are executed that culminate in the successful realization of the SAP BW/4HANA system. These post-conversion measures ensure the integration of advanced functionalities and complete the remote conversion.

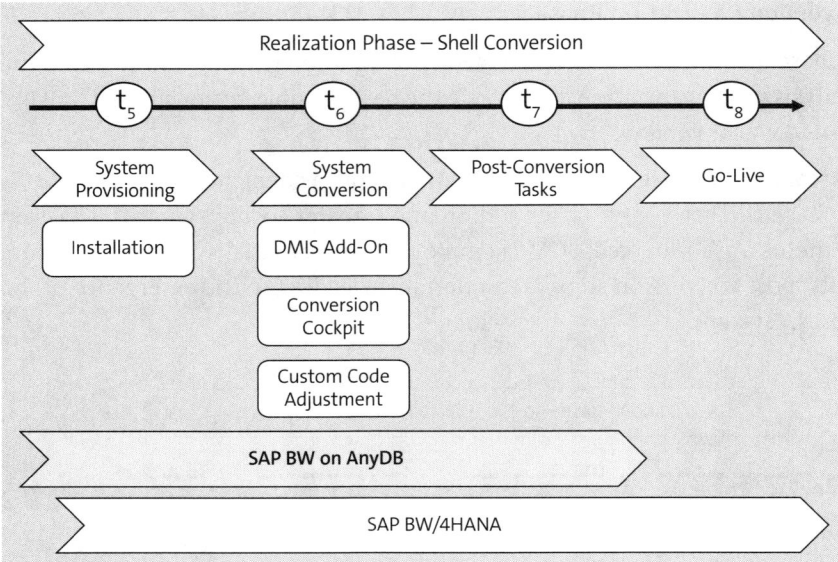

Figure 2.12 SAP BW/4HANA Conversion: Realization Phase in the Remote Conversion

During the realization phase of the in-place conversion, as depicted in Figure 2.13, it's essential to have the SAP BW/4HANA target system installed, the transfer cockpit, and the adjusted custom code. The data load isn't possible in this case, so the post-conversion steps are directly undertaken, leading to the SAP BW/4HANA system in the target landscape.

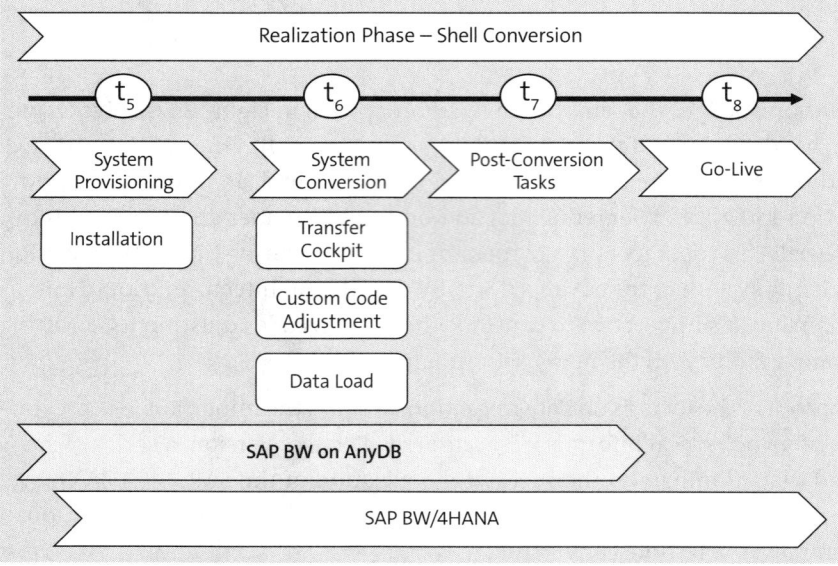

Figure 2.13 SAP BW/4HANA Conversion: Realization Phase in the Shell Conversion

2.4 Project Management

Project management plays a crucial role in an SAP BW/HANA migration project due to the complexity and strategic importance of the transition, as made clear in the central note for SAP BW/4HANA: SAP Note 2383530, "Conversion from SAP BW to SAP BW/4HANA." Here are the key reasons for effective project management in such endeavors:

- **Complexity of the migration process**
  Migrating from traditional SAP BW to SAP BW/4HANA involves a complex set of tasks, including data migration, object conversion, and system optimization. A well-defined project management process helps break down these complex tasks into manageable phases, ensuring a structured and efficient migration.

- **Resource coordination and allocation**
  Project management ensures effective coordination and allocation of resources, both human and technical. This involves managing the efforts of different teams, such as database administrators, ABAP developers, and SAP BW consultants and business analysts, to ensure a synchronized migration process.

- **Timeline management**
  SAP BW/HANA migration projects often have specific timelines and deadlines. Project management helps in creating realistic timelines for each conversion cycle, setting milestones for each task per cycle, and monitoring progress per cycle to ensure that the project stays on track and is completed within the scheduled time frame.

- **Risk management**
  Having a good project management team helps facilitate the identification, assessment, and mitigation of risks associated with the migration project. This proactive approach helps minimize potential disruptions and addresses challenges before they escalate.

- **Budget control**
  Managing costs is crucial in any migration project. Project management involves budget planning, tracking, and control to ensure that the migration is carried out within the allocated financial resources.

- **Change management**
  Migrating to SAP BW/4HANA often brings about changes in processes, data models, and workflows. Effective project management includes change management strategies to help teams adapt to the new environment, minimizing resistance and maximizing user acceptance.

- **Quality assurance**
  Project management includes quality assurance processes and rigorous testing to verify that the migration is meeting predefined standards and objectives. Sine the migration involves data transfer, there needs to be processes for reconciling the sender and receiver SAP BW/4HANA system. This involves testing the SAP S/4HANA

data load, checking the failed extraction, performing data validation, and ensuring that the new system operates as intended post-conversion.

- **Communication and stakeholder engagement**
  Project management facilitates clear and consistent communication with stakeholders, including management, end users, and IT teams. Keeping stakeholders informed and engaged throughout the migration process is critical for a successful transition.

- **Documentation and knowledge transfer**
  Whenever the conversion cycle is started, there must be a runbook that captures all the steps and issues, as comprehensive project management ensures proper documentation of the migration process. This runbook and issues/recommendations documentation are valuable for future reference, audits, and knowledge transfer within the organization.

2.5 Summary

You've now learned the basics of the SAP BW/4HANA in-place conversion, remote conversion, and shell conversion. You'll be able to analyze which conversion type to choose based on the organizational business scenario. You've also learned about the prepare and realization phases that are involved in the SAP BW/4HANA migration. In the next chapter, you'll see the detailed steps to prepare the system for the SAP BW/4HANA migration.

Chapter 3
System Preparation and Prerequisites

This chapter offers a thorough overview of an SAP BW system preparation and the prerequisites necessary for a successful SAP BW/4HANA conversion. The system preparation outlined here applies universally across all conversion types, including remote conversion, in-place conversion, shell conversion, and the SAP BW bridge. By establishing this groundwork, we pave the way for the subsequent chapters, which explore the practical execution of the conversion process.

The transition from SAP Business Warehouse (SAP BW) to SAP BW/4HANA involves extensive preparatory tasks, with various prerequisites that need to be fulfilled. While most preparation activities are consistent across all conversion scenarios, there may be some additional prerequisites depending on the chosen conversion approach. Let's examine this in detail. You'll start learning about the system requirements in Section 3.1 and then understand the impacts on the system landscape in Section 3.2. You'll learn how to prepare the system in Section 3.3. The supported source system is discussed in Section 3.4. You'll learn about using the maintenance planner in Section 3.5. We'll explore data volume management in Section 3.6. With the available details, you'll learn to do the sizing for SAP BW in Section 3.7. The steps to install the Note Analyzer report is discussed in Section 3.8. You'll learn about the SAP BW/4HANA readiness check in Section 3.9. The prechecks for the SAP BW/4HANA migration are discussed in Section 3.10. Finally, you'll learn about SAP BW/4HANA simplification in Section 3.11.

3.1 System Requirements for SAP BW/4HANA

Getting ready to switch to SAP BW/4HANA involves doing a lot of preparation tasks. It starts with the discover/prepare phase; you can see how they connect with the conversion process in Figure 3.1.

SAP strongly advises starting a conversion project with the latest support package (SP) for a release. This approach significantly minimizes the effort needed for implementing notes. In our case, we'll focus more on SAP BW/4HANA 2023, which was released and made generally available on October 30, 2023. The end of mainstream maintenance for SAP BW/4HANA 2023 is planned for December 31, 2030. If you're working with the previous version, SAP BW/4HANA 2021, keep in mind that the end of mainstream maintenance is planned for December 31, 2027.

3   System Preparation and Prerequisites

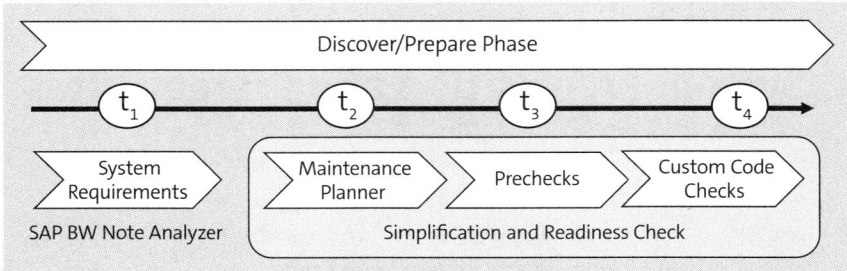

Figure 3.1 Prepare Phase in the SAP BW/4HANA Conversion

We'll now walk through a few important prerequisites:

- **Unicode**
  It's a prerequisite that your system needs to be a Unicode system before the conversion. If the system isn't Unicode, then you need to first start the Unicode conversion. You can see the status of your system from the selected SAP system by choosing **System • Status** and then looking for **Unicode System** under the **SAP System data** tab, as shown in Figure 3.2.

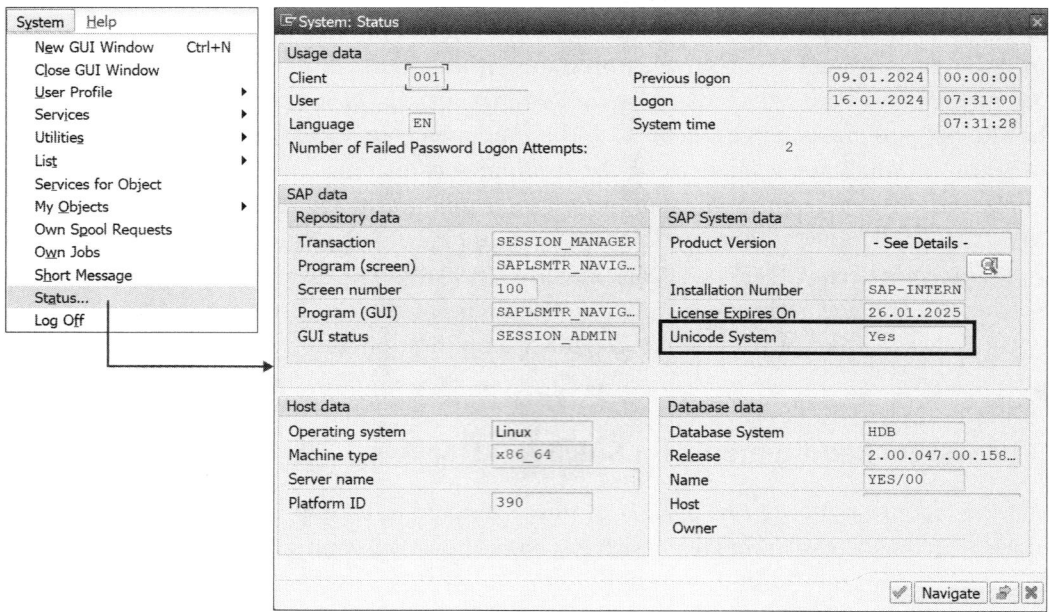

Figure 3.2 System Unicode Status

- **Database platform**
  If you're going for an SAP BW/4HANA conversion, then it's critical to have the right SAP HANA database platform. When you consider SAP BW/4HANA 2023, then you

must have the right SAP HANA database. If the system is on a non–SAP HANA database, then you need to do a database migration to SAP HANA.

The minimum SAP HANA release required for compatibility with SAP BW/4HANA 2023 is specified as SAP HANA 2.0 SP 07 Rev.71. This indicates that for a successful integration and optimal performance of SAP BW/4HANA 2023, the SAP HANA database must be at least at version 2.0 with SP 07 and have the revision level 71. This detailed specification ensures that the underlying SAP HANA infrastructure is aligned with the version and revision requirements, promoting a smooth and reliable operation of SAP BW/4HANA 2023. It's crucial to adhere to these minimum requirements to leverage the enhanced features and functionalities offered by the latest release of SAP BW/4HANA. If you plan to go for SAP BW/4HANA 2021, then the minimum SAP HANA database is SAP HANA 2.0 SP 05.

- **Dual stack split**
  In the context of SAP BW/4HANA conversion, a *dual stack* typically refers to the traditional dual-stack architecture of SAP NetWeaver, which includes both ABAP and Java stacks. In older SAP BW systems, the dual-stack architecture was common, but with SAP BW/4HANA, SAP recommends having a single-stack approach. During the conversion process from a traditional SAP BW system to SAP BW/4HANA, one key aspect is the transition from the dual-stack architecture to a single-stack architecture. The single-stack architecture involves only the ABAP stack, and the Javastack isn't used. This simplification is part of the modernization efforts in SAP BW/4HANA. Therefore, if you plan for the conversion, this step will be a prerequisite before the project.

3.2 Impacts on the System Landscape

When you migrate to SAP BW/4HANA, there can be many impacts to the existing system landscape that need to be considered before the project phase:

- **Application servers**
  It's general practice to reuse the existing application server in any upgrade projects. When considering the reuse of your application server(s) in conjunction with SAP BW/4HANA, it's crucial to ensure compatibility with the product. Achieving compatibility may necessitate updates to both the operating system (OS) and the SAP kernel. To find comprehensive details on these requirements, it's best practice to first check the Product Availability Matrix (PAM) to know the details related to SAP BW/4HANA. Ensuring alignment with the specified compatibility guidelines is vital for a seamless integration and optimal performance of your system. If you search for the target SAP BW version, for example, SAP BW/4HANA 2023, then you'll see the screen shown in Figure 3.3.

3   System Preparation and Prerequisites

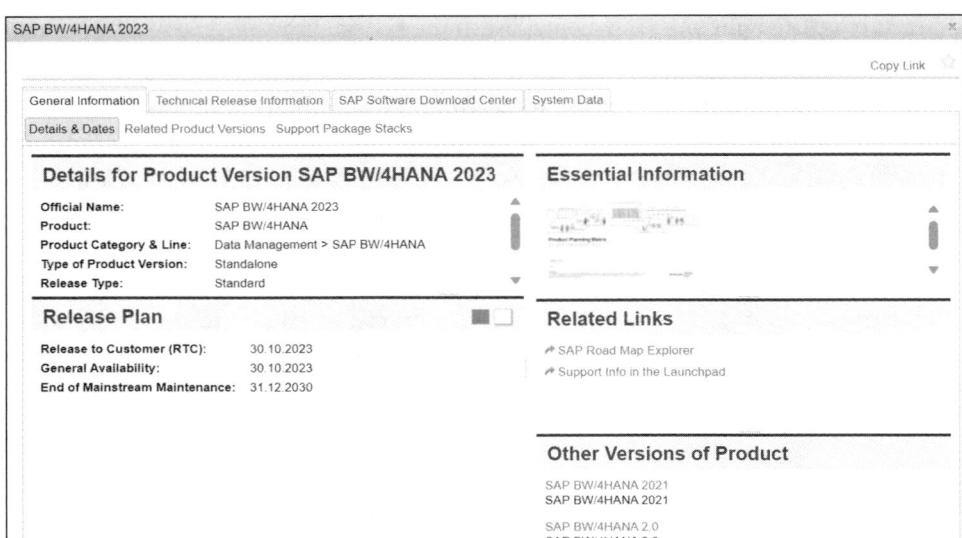

Figure 3.3 Product Availability Matrix for SAP BW/4HANA 2023

Once you choose **Essential Information**, which is a separate document, you'll find the latest details on the supported OS, SAP HANA database, and the browser. You can see the database/OS Product Planning Matrix, which lists the supported products or planned to be supported products by SAP. Similarly, you can see the details related to the language availability based on the target groups such as end users, administrators, and developers. Then, you'll have the information on SAPUI5 support.

If you plan to migrate to SAP BW/4HANA 2021, then you need to check the PAM for the same version and get the details of the supported OS and database. The detailed planning must start based on this information.

- **SAP BusinessObjects Business Intelligence (BI) Java components**
  With SAP BW/4HANA, the SAP BusinessObjects BI Java components are now considered obsolete. To align with this change, it's necessary to eliminate the technical Java system associated with these components from the landscape management database (LMDB) of the relevant productive system. This step ensures a seamless transition in line with the updated requirements of SAP BW/4HANA. Additionally, there is support for the temporary use of existing SAP Business Explorer (SAP BEx) web applications (SAP BusinessObjects BI Java) in SAP BW/4HANA through a dedicated pilot program. You can convert SAP BusinessObjects BI Java; the process involves transitioning BI Java to a supported platform within SAP BW/4HANA. SAP BusinessObjects BI Java will receive support until 2027 and can seamlessly connect to the SAP BW/4HANA system, ensuring a smooth upgrade without significant impact. You can also refer to SAP Note 2496706 "BEx Web Application Add-on for SAP BW/4HANA - Deployment and Limitations" for more details.

In alignment with the announced maintenance extensions, SAP also extended the mainstream maintenance for SAP Enterprise Portal 7.5 to the end of 2027, with extended maintenance to 2030.

- **SAP Fiori frontend server (FES) and SAP user interface (UI)**
  SAP BW/4HANA is co-installed with the SAP UI 7.5 component, encompassing the SAP Fiori and SAP UI. For streamlined integration with SAP solutions such as SAP S/4HANA, consider using a standalone FES. This approach enhances compatibility and simplifies the integration process. The SAP BW/4HANA Cockpit is a user-friendly web-based SAP Fiori UI. It features a collection of SAP Fiori apps grouped for tasks such as process chain modeling, monitoring, and administration, providing an intuitive experience. For SAP Fiori, frontend components are essential for SAP Fiori apps, serving to deliver the UI and establish the connection to the backend. Concurrently, backend components play a pivotal role by furnishing the necessary data for the optimal functioning of these applications. The deployment of newer SAP Fiori app frontend components necessitates the use of the FES), which is built on SAP NetWeaver AS ABAP. SAPUI5, SAP's JavaScript UI library, enables responsive web app development with cross-browser compatibility. Widely used for the SAP Fiori frontend, it allows developers to create custom apps, extending SAP Fiori libraries for specific business needs. Its versatility significantly enhances SAP Fiori app interfaces and functionality. Make sure that you refer to the maintenance window from the SAP Note 2217489, "Maintenance and Update Strategy for SAP Fiori Frontend Server."

- **SAP Analytics Cloud**
  When you use SAP Analytics Cloud in your current SAP BW version and after the migration, you need to make sure that everything is in sync. If you're unclear which SP or SAP Notes are needed to operate SAP Analytics Cloud together with the SAP BW backend, you can always update the system at least once per year to the latest SP. Refer to SAP Note 2541557, "SAP Analytics Cloud with BW Live Connection - Which BW Support Package Is Recommended?" for more information. You need to download the latest file, which will have the latest date along with the template such as *SAP_BW_Analytics_Cloud_Connection_XXXX_XX_XX.xml*, and implement all the notes in the same order. We'll explain how to install the SAP Note using report *Z_SAP_BW_NOTE_ANALYZER.txt* later. For this section, you need to remember to ensure that your new system has all the latest correction, for SAP Analytics Cloud.

- **Third-party tools**
  If the SAP BW system landscape uses third-party tools for extraction, transformation, and loading (ETL) purposes or any other reporting scenarios, then you need to check if those tools are supported by SAP BW/4HANA. This is an important factor because, if the tools isn't supported, then you need to plan for the alternate upon migrating to SAP BW/4HANA.

3 System Preparation and Prerequisites

3.3 Preparing the System

In this section, we'll walk through system preparation activities, such as the minimum supported system release and prechecks.

3.3.1 Minimum Supported Release for Original System

Before transitioning to SAP BW/4HANA, understanding the supported release details of SAP BW is crucial. The conversion method choice introduces a dependency, requiring verification of the supported release in both the SAP BW sender and SAP BW target systems. The remote conversion and shell conversion approaches involve the sender and target system concepts, which require thorough consideration. In contrast, the in-place conversion method streamlines the process, requiring only one system for the actual conversion. You must ensure compatibility across sender and target systems to guarantee a smooth transition to SAP BW/4HANA, considering the unique requirements of each conversion approach.

We'll discuss this for each conversion approach, but first we'll begin with the precheck minimum support releases, listed in Table 3.1.

Release	Minimum Support Package	Recommended Support Package
SAP BW 7.00	25	25 or higher
SAP BW 7.01	9	9 or higher
SAP BW 7.02	7	7 or higher
SAP BW 7.3	8	8 or higher
SAP BW 7.31	5	5 or higher
SAP BW 7.4	9	9 or higher
SAP BW 7.5	5	5 or higher

Table 3.1 Precheck Minimum Support Release

For an in-place conversion, we need only one system, as the conversion will be carried out directly in the same system. If you plan to go for the in-place conversion, then you need to have the following SP levels:

- SAP BW 7.5 powered by SAP HANA SP 05 or Higher. It's generally best practice to go for the latest SP. The minimum recommended SP level is SP 12.
- Direct in-place conversion from SAP BW 7.50 to SAP BW/4HANA 2023 SP 00 or higher (DW4CORE 4.00 SP 0 or higher) is supported.

As of SAP BW 7.50 SP 16, the in-place conversion is only supported with SAP BW/4HANA 2.0, SAP BW/4HANA 2021, and SAP BW/4HANA 2023 as a destination. Remote conversion involves two systems: sender, which is the original SAP BW 7.x system, and target, which is the SAP BW/4HANA system. The following list shows the SPs for the original system (keep in mind, it's recommended to go for the latest SAP BW/4HANA release):

- SAP BW 7.3 on any database, minimum SP from SP 10
- SAP BW 7.31 on any database, minimum SP from SP 10
- SAP BW 7.4 on any database, minimum SP from SP 12
- SAP BW 7.5 on any database, minimum SP from SP 05

Shell conversion involves two systems as well: the sender, which is the original SAP BW 7.x system, and the target SAP BW/4HANA system. Table 3.2 shows the SPs for the original system. It's recommended to go for the latest SAP BW/4HANA release.

Release	Minimum Support Release	Recommended Support Release
SAP BW 7.0	SP 28 and higher	Latest
SAP BW 7.01	SP 12 and higher	Latest
SAP BW 7.02	SP 10 and higher	Latest
SAP BW 7.3	SP 10 and higher	Latest
SAP BW 7.31	SP 10 and higher	Latest
SAP BW 7.4	SP 12 and higher	Latest
SAP BW 7.5	SP 12 and higher	Latest

Table 3.2 Supported Releases for Shell Conversion

3.3.2 Minimum Support Release for Target System

In this section, we'll cover the support release for the target system. Table 3.3 lists some key end of maintenance dates for SAP BW/4HANA. When you're planning for the SAP BW/4HANA conversion, you need to check the end of maintenance date to choose the right release.

Release	End of Maintenance	Next Release
SAP BW/4HANA 2.0	12/31/2024	SAP BW/4HANA 2021
SAP BW/4HANA 2021	12/31/2027	SAP BW/4HANA 2023
SAP BW/4HANA 2023	12/31/2030	SAP BW/4HANA 2026

Table 3.3 SAP BW/4HANA Key Maintenance Dates

3.4 Supported Source Systems

In any business warehouse system landscape, it's common for multiple source systems to be connected. One of the major sources of analytical reporting data will be from SAP ERP or SAP S/4HANA systems, and the SAP BW system fetches data from these systems using the SAP-delivered business content or extractors. There are also customer-generated extractors when there is no standard business content available. These systems are connected using the Remote Function Call (RFC) though the Service API (SAPI) framework. Similarly, here is the list of SAP Business Suite applications that are supported as source systems for SAP BW/4HANA.

- SAP ERP 6.0 or higher
- SAP Customer Relationship Management (SAP CRM) 5.0 or higher
- SAP Supply Chain Management (SAP SCM) 5.0 or higher

3.4.1 SAP Business Suite Applications

When you're working with SAP Business Suite applications, you need to have the minimum SAP NetWeaver support prior to the conversion, as listed in Table 3.4.

Release (SAP Source System)	Minimum Support Release	Recommended Support Release
SAP NetWeaver 7.0	14	Latest
SAP NetWeaver 7.0 EhP 1	0	Latest
SAP NetWeaver 7.0 EhP 2	0	Latest
SAP NetWeaver 7.1	19	Latest
SAP NetWeaver 7.1 EhP 1	13	Latest
SAP NetWeaver 7.3	1	Latest
SAP NetWeaver 7.3 EhP 1	1	Latest
SAP NetWeaver 7.4	5	Latest
SAP NetWeaver 7.5 or higher	0	Latest

Table 3.4 SAP Business Suite Applications: SAP NetWeaver Versions

3.4.2 SAP Business Warehouse

Data extraction from SAP BW systems is a standard practice, with SAP BW acting as the primary source system. Compatibility for conversion is guaranteed for systems running SAP BW 7.0 or higher. If your landscape features a SAP BW–based source system, it's crucial to verify that the sender SAP BW source system has the specified SAP

NetWeaver SP. Additionally, ensure seamless integration by maintaining up-to-date SPs for optimal performance and system functionality, as shown in Table 3.5.

Release (SAP BW Source System)	Minimum Support Release	Recommended Support Release
SAP BW 7.0	17	Latest
SAP BW 7.0 EhP 1	2	Latest
SAP BW 7.0 EhP 2	1	Latest
SAP BW 7.1	Not Supported	
SAP BW 7.1 EhP 1	2	Latest
SAP BW 7.2	Not Supported	
SAP BW 7.3	3	Latest
SAP BW 7.3 EhP 1	4	Latest
SAP BW 7.4	2	Latest
SAP BW 7.5 or higher	0	Latest

Table 3.5 Supported SAP BW Source Systems

3.4.3 SAP Source System and Operational Data Provisioning Readiness

In the realm of SAP BW/4HANA, operational data provisioning (ODP) takes center stage as the primary infrastructure for extracting and replicating data from SAP (ABAP) applications to the SAP BW/4HANA data warehouse. This pivotal role positions ODP as the key mechanism for seamless data transfer within the SAP environment, which is why the incorporation of SAP BW/4HANA with SAP NetWeaver systems, such as SAP Business Suite, SAP S/4HANA, or SAP BW, relies on ODP. Ensuring ODP readiness is imperative for both inbound and outbound connections across all interconnected SAP and SAP BW systems. ODP ensures a seamless, unified, and efficient integration process for managing data flow between these systems, contributing to the overall interoperability and functionality of the SAP environment. You need to verify the ODP compatibility of each system, including SAP Business Suite, SAP S/4HANA, and SAP BW, to facilitate smooth data provision. The mandatory ODP readiness guarantees optimal functionality and data flow between SAP BW/4HANA and associated systems. Regularly assess and update the ODP status to maintain the integrity of the integrated environment.

ODP Data Replication API 2.0 is a functional enhancement to the first version of this interface released and the recommended API for source systems. It facilitates internal connections between SAP BW/4HANA, SAP BW release 7.3 or higher, SAP Data Services 4.2, and SAP HANA smart data integration (SDI; ABAP adapter). It supports various data provider types, including DataSources/extractors (ODP-SAPI context), ABAP Core Data Services (CDS) views (ODP-CDS context), and InfoProviders of SAP BW/4HANA or SAP BW release 7.4 or higher (ODP-BW context). This API serves as a versatile tool for seamless integration across diverse data sources within the SAP ecosystem.

Table 3.6 lists the available SPs for ODP Replication 2.0.

Release (ODP-Based Extraction from an SAP Source System)	Minimum Support Release	Recommended Support Release
SAP NetWeaver 7.3	10	14
SAP NetWeaver 7.3 EhP 1	08	16
SAP NetWeaver 7.4	04	11
SAP NetWeaver 7.5	00	00
SAP BW/4HANA 1.0	00	00
SAP BW/4HANA 2.0	00	00

Table 3.6 SAP NetWeaver Support for ODP Data Replication AP1 2.0

3.4.4 Other Source Systems

In certain scenarios, an SAP BW 7.x system may require connectivity with systems that aren't reliant on ABAP application servers or are non-SAP systems. If these external systems need integration into your new SAP BW/4HANA environment, it's crucial to verify their compatibility with SDA or SDI. Ensure that the non-SAP source system aligns seamlessly with the specified integration technologies to guarantee a smooth and effective connection within the SAP BW/4HANA environment. You can refer to the source systems based on Database Connect (DB Connect) or Universal Data Connect (UD Connect) to get the list of connected systems.

3.5 Maintenance Planner

Before the migration to SAP BW/4HANA, you need to understand your existing system and get the plans for the target system. The maintenance planner, provided by SAP, serves as a hosted solution in SAP Support Portal to facilitate the planning, scoping, and maintenance of systems within your landscape. It enables the strategic planning of intricate tasks, such as installing a new system or updating existing ones. The scheduling flexibility allows for the deployment of all changes at a convenient time, minimizing downtime and optimizing system maintenance.

3.5 Maintenance Planner

For successful runtime execution, ensure that your system is equipped with the Google Chrome, Mozilla Firefox, or Microsoft Edge browser. Additionally, make certain that your SAP Solution Manager system is running on release 7.1 SP 05 or a later version to meet the required prerequisites. These specifications are essential for an optimal runtime environment and compatibility with the designated SAP Solution Manager system.

The LMDB in SAP Solution Manager serves as the primary repository for central landscape information. A comprehensive system landscape description forms the foundation for various SAP Solution Manager applications, including Monitoring and Alerting. For tasks such as calculating updates and upgrades through the maintenance planner in SAP Support Portal, an accurate landscape description becomes imperative. Detailed technical information about the system landscape is systematically collected and stored by SAP Solution Manager. This information comprises two integral components: software descriptions sourced from the SAP software catalog (SAP CR content) and data automatically transmitted by technical systems through the system landscape directory (SLD). Together, these elements contribute to a robust understanding of the technical system landscape within SAP Solution Manager. Figure 3.4 shows how the maintenance planner works.

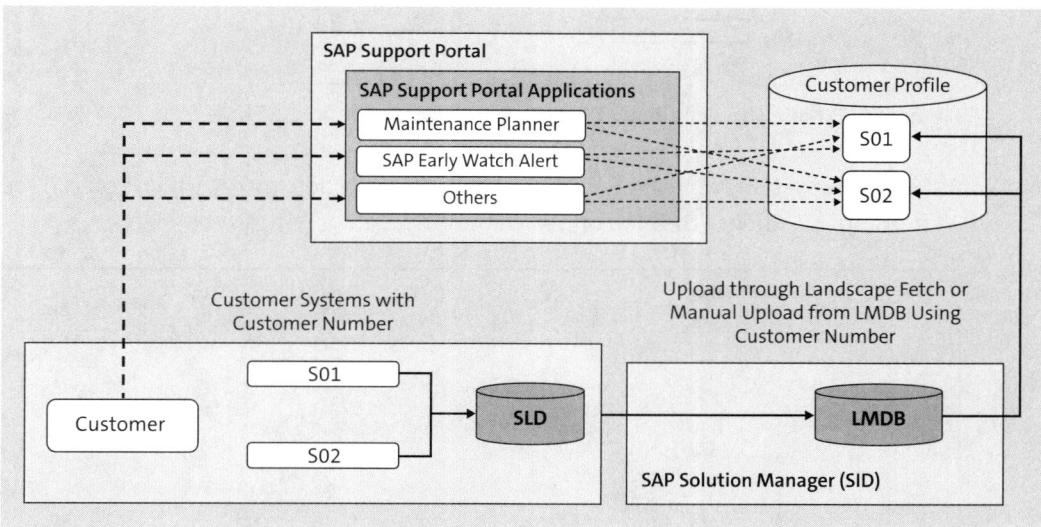

Figure 3.4 Maintenance Planner Architecture (Source: SAP)

Follow these steps to use the maintenance planner:

1. Go to *https://apps.support.sap.com/sap/support/mp* or navigate to the **Maintenance Planner** tile within the on-premise section on the SAP ONE Support launchpad, accessible through the SAP site. Log in using your SAP credentials (S user) to proceed with the maintenance planner. In the **Maintenance Planner** home screen, you can see the **Explore Systems** option, as shown in Figure 3.5.

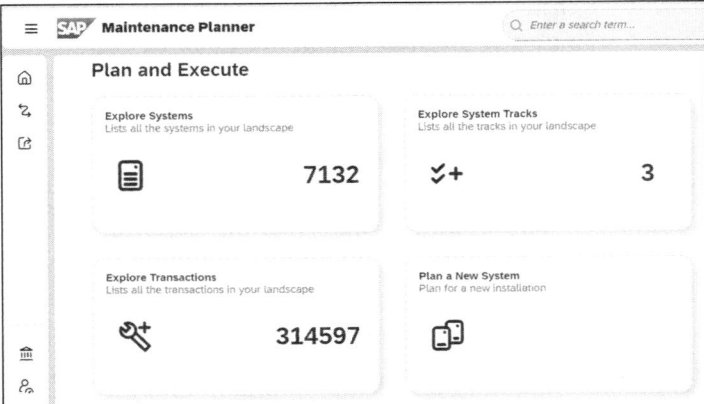

Figure 3.5 Maintenance Planner Home Screen

2. After you click on **Explore Systems**, filter on your specific system for the planning, as shown in Figure 3.6.

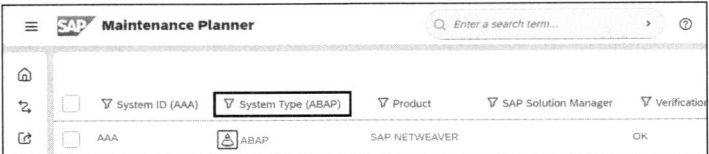

Figure 3.6 Explore the System in Maintenance Planner

3. Select the system you need to plan; in our case, we'll choose **AAA**, which you can see in Figure 3.7. Choose the **Plan** option in the circle.

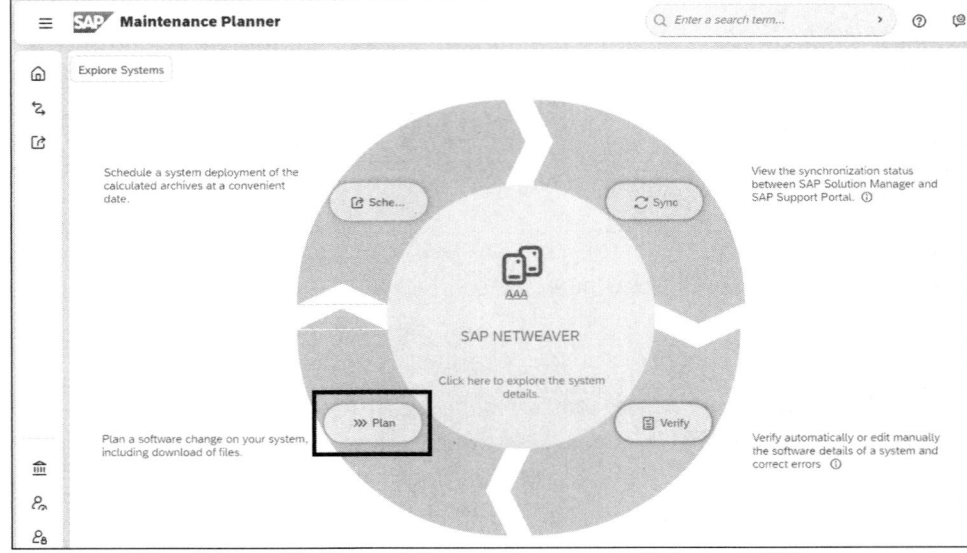

Figure 3.7 Maintenance Planner: Explore Options

3.5 Maintenance Planner

4. You'll arrive at the screen shown in Figure 3.8. Select the **Plan a Conversion to SAP BW/4HANA** option.

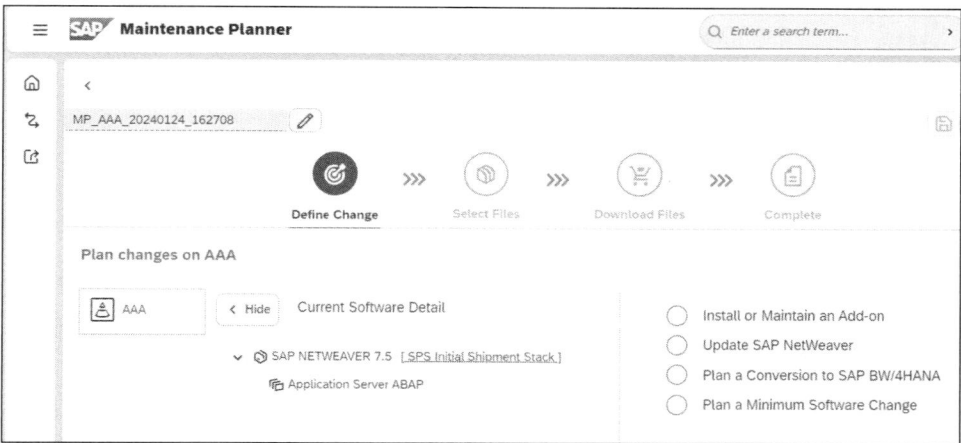

Figure 3.8 Maintenance Planner: Plan a Software Change Option

5. You'll arrive at the screen shown in Figure 3.9 in the **Define Change** step of the process. Follow these steps:

    ❶ Choose **Plan a Conversion to SAP BW/4HANA**.

    ❷ Choose the product version you want, in this case, **SAP BW/4HANA 2023**.

    ❸ Select the SP level that you want, in this case, **Initial Shipment Stack**.

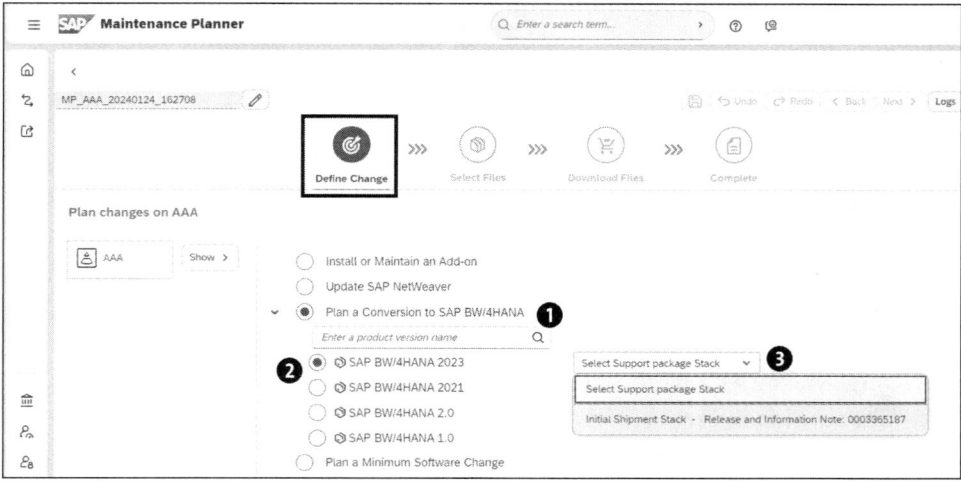

Figure 3.9 Plan the Conversion to SAP BW/4HANA 2023

In this stage, it's essential to understand that the system performs prechecks for add-ons. If any unsupported add-ons are detected, an error notification is generated, resulting in the halt of the SAP BW/4HANA conversion planning process. Subsequently,

81

you must select an alternate version and SP level or verify compatibility with the SAP BW version outlined in the preceding section.

6. A popup will appear regarding software changes. Opt to proceed, and then affirm your selection. Follow the on-screen instructions, and in the **Select Files** step, pick the files slated for installation.
7. Ultimately, in the **Download Files** step, opt for **Download Stack XML**, and choose to push them to the download basket from SAP Support Portal.
8. Select **Next**, and in the **Complete** step, you'll have the option to finalize and download the maintenance plan.

To execute the technical system conversion to SAP BW/4HANA, it's imperative to employ the maintenance planner. This tool, recommended for early use in the SAP BW/4HANA system conversion project, scrutinizes the following aspects for compatibility with the conversion:

- Any add-ons integrated into your system
- The status of active business functions within your system

As you've seen earlier, should any of the listed items lack a valid conversion path, such as an add-on not being released for conversion, the maintenance planner inhibits the conversion process. Following the verification, the maintenance planner generates download files encompassing add-ons, packages, DBD errors, and so on, and then formulates the stack configuration file (*stack.xml*). This configuration file plays a pivotal role in effecting the system conversion to SAP BW/4HANA.

When you use maintenance planner, you can plan based on the type of conversion approach you use; for example, you have the following options for the conversion types:

- For an in-place conversion, initiate the process by downloading the SAP BW/4HANA starter add-on along with the necessary software components using the maintenance planner. This step is crucial for ensuring a smooth execution of the technical system conversion to SAP BW/4HANA.
- In a remote or shell conversion scenario involving a sender and target system approach, initiate the planning process through the maintenance planner. Download the initial installation package for SAP BW/4HANA, excluding the SAP BW/4HANA starter add-on. It's important to highlight that the remote conversion requires the incorporation of the SAP Landscape Transformation package. To ensure optimal results, contemplate the integration of the latest SPs during the initial installation using the maintenance planner.

3.6 Data Volume Management

Despite the technological advancements of modern database systems, performance bottlenecks can still arise due to extensive data volumes. This can result in suboptimal system performance within applications and heightened resource usage from an administrative perspective. The impact of high data volumes extends beyond performance concerns, affecting the total cost of ownership (TCO) of a system, even as storage costs decline. To mitigate the adverse effects on costs, performance, and system availability caused by substantial data volumes, proactive measures need to be taken.

Data volume management (DVM) aims to minimize the data footprint, resulting in a shorter conversion duration by reducing the load size. With a focus on both pre- and post-conversion phases, DVM encompasses diverse capabilities. The data volume management work center in SAP Solution Manager serves as a central hub, featuring tools tailored for SAP HANA environments.

One notable tool in DVM is the guided self-service, allowing users to generate best practice documents for efficient data reduction before conversion. The same tool proves valuable post-conversion, aiding in the development of a blueprint for a strategic DVM approach. Another significant component is the Reorganization and Compression tool, applicable beyond SAP HANA contexts. This tool enables users to simulate savings achieved through reorganizing tables and databases, or through compressing the database. Furthermore, users can leverage the tool to simulate the future system size, offering insights for forecasting the impact of planned measures. For a comprehensive understanding and effective use of DVM, explore the data volume management work center and refer to the official documentation at *http://s-prs.co/v586701* for the latest guidelines and updates.

Apart from the Solution Manager tools, if you think from an organizational standpoint, the core in-house or SAP BW support teams play a pivotal role in executing manual DVM strategies in SAP BW. We'll discuss the options later in the book, but at a very high level. The team can adopt various approaches to optimize data handling. First, they actively engage in data avoidance, identifying and circumventing unnecessary data whenever feasible. This involves thorough data load analysis and the extraction patterns in the SAP BW system. Second, the data summarization is meticulously implemented through the data model, highlighting the significance of maintaining data granularity only when essential for reporting or legal requirements. The team systematically addresses data deletion, focusing on the removal of outdated information and establishing specific expiration times for temporary data tables such as persistent staging area (PSA) and change log. Data archiving is selectively employed for information that may be required in the future for reporting or legal compliance, aligning with predetermined retention times. When it comes to archiving, the service helps the team understand what data to archive, how much, and how to set up the archiving process. The team can take charge of developing a strategy for archiving that suits your specific needs. This ensures the entire team in the organization work together to handle data efficiently, improve system performance, and meet regulatory requirements.

3.7 SAP Business Warehouse Sizing

When you want to migrate to SAP BW/4HANA, it's important to properly size the system based on the available data. The proper sizing is very important to get the optimal SAP BW/4HANA–based servers as it can have an impact on the TCO. There is a standard report available for this feature.

The SAP BW/4HANA sizing report plays a crucial role in the implementation of SAP BW/4HANA by providing essential insights for resource planning, performance optimization, cost estimation, and scalability. This tool aids organizations in accurately determining the required hardware resources, including memory, CPU, and storage capacities, based on factors such as data volume, user concurrency, and query complexity. By leveraging the sizing report, organizations can make informed decisions about the configuration parameters and settings for their SAP BW/4HANA system, ensuring alignment with best practices for performance and stability. Moreover, the report facilitates proactive planning for scalability, allowing organizations to accommodate future growth seamlessly. Overall, the SAP BW/4HANA sizing report is an invaluable asset for effective decision-making, resource allocation, and successful implementation of SAP BW/4HANA solutions.

Sizing, a crucial aspect of system planning, involves the careful consideration of hardware requirements such as memory, CPU power, disk space, I/O capacity, and network bandwidth. This iterative process plays a vital role in translating business needs into tangible hardware specifications, setting the foundation for a well-configured system. Typically performed early in the project phase, sizing ensures that the chosen hardware aligns with the anticipated workload and user demands.

As part of the journey to upgrade or migrate your SAP BW system to SAP HANA, a pivotal step is estimating the memory size required for the new SAP HANA server. The availability of the sizing report (/SDF/HANA_BW_SIZING) from SAP BW version 7.3 onward facilitates this process. Leveraging this report is instrumental in providing insights into specific requirements, ensuring a seamless transition to SAP HANA, and optimizing the performance of your SAP BW system. The report aids in addressing questions related to system capabilities, offering valuable information on memory allocation, CPU usage, and other critical parameters. By using the sizing report, you can fine-tune your system architecture to align with SAP HANA's capabilities, thereby maximizing the benefits of the upgrade. This proactive approach ensures that your SAP BW system is adequately equipped to handle the enhanced capabilities and demands associated with the SAP HANA platform.

Follow these steps to execute the sizing report:

3.7 SAP Business Warehouse Sizing

1. In your SAP BW system, go to Transaction SE38, and choose **/SDF/HANA_BW_SIZ-ING**, as shown in Figure 3.10.

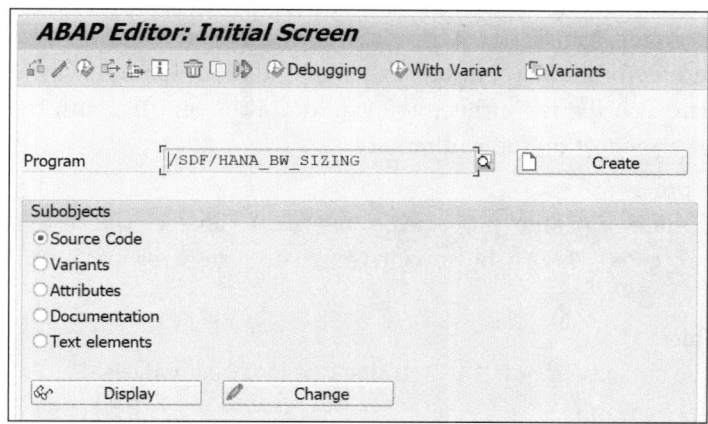

Figure 3.10 SAP HANA Sizing Report

2. Press [F8] or click **Execute** to see the selection screen for the report (see Figure 3.11) when you execute in the SAP BW on SAP HANA system.

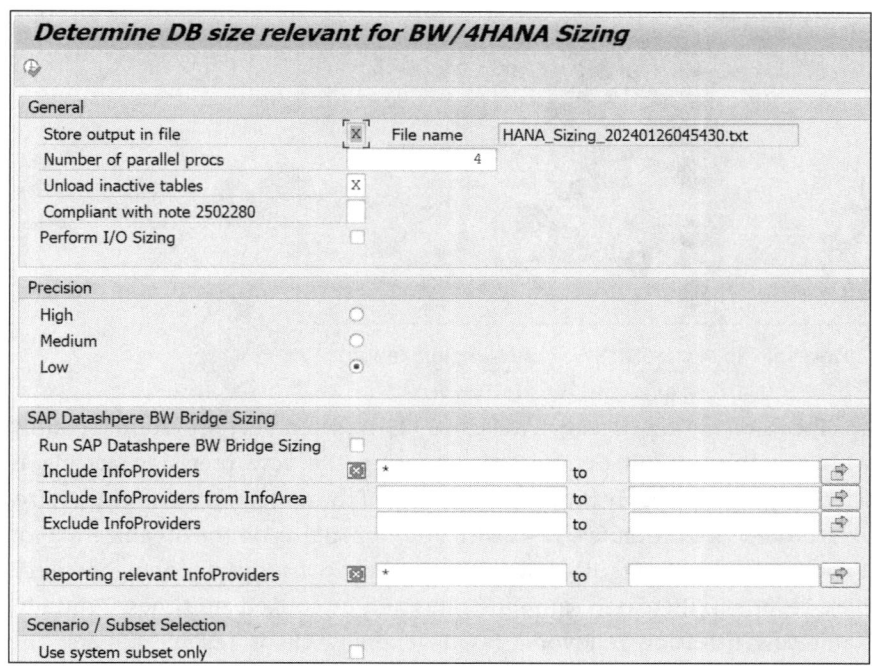

Figure 3.11 Sizing Input Area

3. Now, let's look at how this screen should be configured in a non-SAP HANA system. Fill out the **General** tab with the following selections, as shown in Figure 3.12:

85

❶ Store output in file

Select this checkbox if you want to store the output of the current sizing execution directly in a designated file on the application server. The file will be in *DIR_HOME*, accessible through Transaction AL11. When saving, ensure you provide a valid name for the file; for instance, in our example, the file is named *HANA_Sizing.txt*. After the execution is finished, you can locate the text file with the specified name in the application server directory.

❷ Number of parallel procs

Define the total number of parallel processes to be used to analyze the system when the report is executed. This can be provided based on system resource availability.

❸ Unload inactive tables

Use this option if the database is SAP HANA. Unload tables after analysis.

❹ Compliant with note 2502280

Determine the CPU and memory requirements for a given SAP BW system.

❺ Target Release BW/4HANA

Select this option when embarking on an SAP BW/4HANA migration. It's crucial and necessary to enable sizing calculations specifically tailored for the SAP BW/4HANA target system.

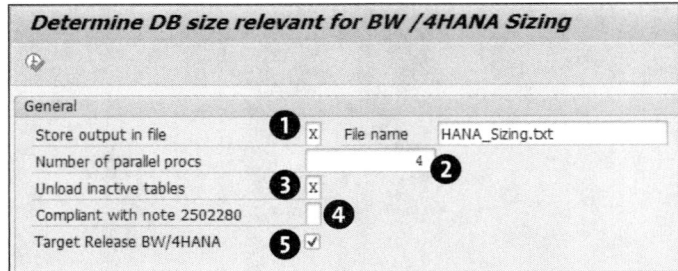

Figure 3.12 General Tab in the SAP BW/4HANA Sizing Report

4. Select the **Precision** option, as shown in Figure 3.13. As a default, the system is configured with the **Low** option preselected. Choosing the **Low** precision setting is designed to offer a reasonably thorough estimate of the anticipated SAP HANA size. However, the decision ultimately rests with you, depending on the organization- or project-specific requirements. It's essential to recognize that opting for higher precision levels results in increased sampling rates and prolonged runtimes. Typically, the default **Low** precision is favored, as it generally yields reliable results. The **Medium** and **High** precision alternatives are better suited for smaller systems characterized by lower row counts, where a more intricate analysis might be advantageous. You can weigh the trade-offs between precision and performance based on the scale and complexity of your system before selecting.

3.7 SAP Business Warehouse Sizing

Figure 3.13 Precision Options

5. Next, we'll see the options that the tool has for **SAP Datasphere BW Bridge Sizing**, as shown in Figure 3.14, which is used in SAP Datasphere:

   ❶ **Run SAP Datasphere BW Bridge Sizing**
   Check this flag if you need to do a sizing for SAP BW bridge or SAP Datasphere.

   ❷ **Include InfoProviders**
   Choose all or provide a specific list of InfoProviders that are to be loaded to the SAP BW bridge.

   ❸ **Include InfoProviders from InfoArea**
   Provide the list of InfoAreas that contain the Advanced DSO (ADSO) or InfoProviders to be loaded into the SAP BW bridge.

   ❹ **Exclude InfoProviders**
   Exclude certain InfoProviders or ADSOs by entering their technical names here.

   ❺ **Reporting relevant InfoProviders**
   Provide the list of technical names of ADSOs or InfoProviders that will be replicated in SAP Datasphere for reporting.

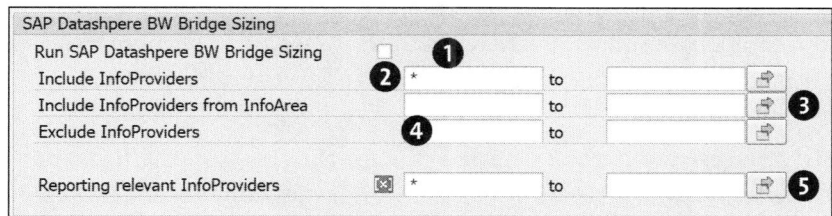

Figure 3.14 SAP Datasphere BW Bridge Sizing Screen

6. The next step is for **Scenario/Subset Selection**, as shown in Figure 3.15:

   ❶ Choose this option to activate the **Scenario/Subset Selection** functionality. This feature allows you to specify the SAP HANA size for specific segments of your source system by identifying the top-level InfoProviders that should be included or excluded in your scenario.

   ❷ If you aim for precise sizing of your target system, focusing solely on tables or InfoProviders, you can use this option. In this section, you can input a list of technical names for the top InfoProviders that are part of your scenario, with dependent objects being added automatically.

❸ The defined scenario can be either applied to a new SAP BW system or seamlessly integrated into an existing one. Additionally, you have the option to exclude the specified list from the sizing calculation if needed.

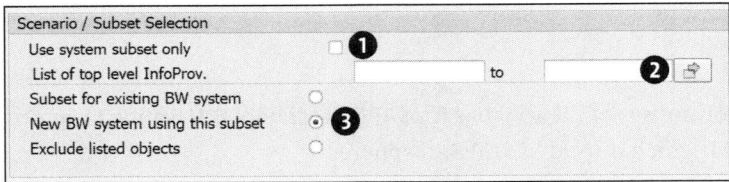

Figure 3.15 Scenario / Subset Selection

7. The next step is to consider the **Future Growth Simulation**, as shown in Figure 3.16:

   ❶ If you want the report to include projections for future growth as an integral part of its analysis, ensure that this flag is selected. By activating this option, the report will factor in anticipated growth scenarios, providing a more comprehensive and forward-looking assessment. This can be particularly valuable in strategic planning, allowing for a better understanding of the system's scalability and resource requirements over time.

   ❷ You have the flexibility to define the number of years to be considered for growth projections. This decision relies on your role as a system administrator, as you possess the knowledge to estimate the anticipated future growth rate, whether it's 5%, 10%, 15%, or any other specified percentage.

   ❸ You have the option to select either in percentage (%) or growth in gigabytes (GB). Make your choice based on your preference. Percentages are often more intuitive as they provide a proportional measure, making it easier to interpret compared to absolute growth numbers.

   ❹ In this field, you can input the anticipated growth value. For instance, if you project a 10% increase in the system size annually for the next three years, the report will then compute and furnish estimates for the expected SAP HANA size after the first year, second year, third year, and so forth.

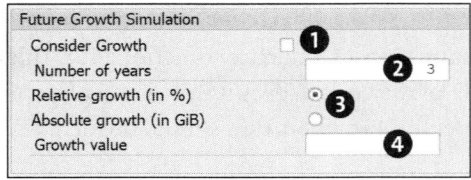

Figure 3.16 Future Growth Simulation

8. Next, fill out the **Planning Applications** section, as shown in Figure 3.17:

   ❶ To use the **Consider Planning Applications** option, select this flag.

   ❷ Use the **Calculate** buttons to fetch the size from the statistics tables in SAP BW.

❸ You can add the users in this input box.

❹ You can add the estimation of sizing here.

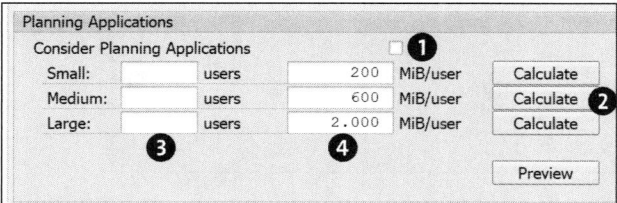

Figure 3.17 Planning Applications

9. Fill in the **Multi Temperature Data** section, as shown in Figure 3.18:

   ❶ In the specified field, input the technical names of any write-optimized (ADSO) entities you want categorized as **WARM DataStore Objects**. Only provide the technical names of the associated InfoProviders. Note that tables such as PSAs and change logs are automatically recognized as **WARM DataStore Objects** and don't require manual input. Recognizing that these database tables house warm data, it's crucial to understand that they will contribute partially to the overall sizing results. The inclusion of write-optimized objects in the **WARM DataStore Objects** category ensures a more nuanced assessment, accommodating data structures optimized for write-intensive operations and enhancing the accuracy of the sizing results to align with your specific data landscape.

   ❷ If your strategy involves the implementation of extension nodes and you have considerations for storing warm data, be sure to select **Consider Extension Nodes**. This choice is pivotal in configuring the system to accommodate the specific requirements associated with extension nodes and the storage of warm data, aligning your setup with optimal performance and data management practices.

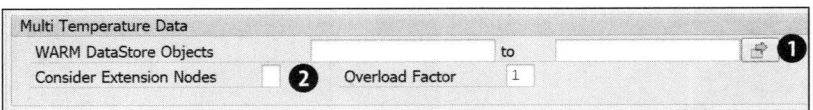

Figure 3.18 Multi Temperature Data Screen: Management in Sizing

10. Next is the **Memory Configuration** section, as shown in Figure 3.19:

    ❶ The **Predefined** radio button selection of memory configurations depends on the scale of your existing system. If uncertain, it's recommended to consult with your hardware partner for clarification. You should choose memory configurations that align with offerings from your hardware partner or that are pertinent to your specific considerations.

❷ If you prefer a custom size, use the highlighted custom option.

❸ The **GiB** field allows you to enter a specific size.

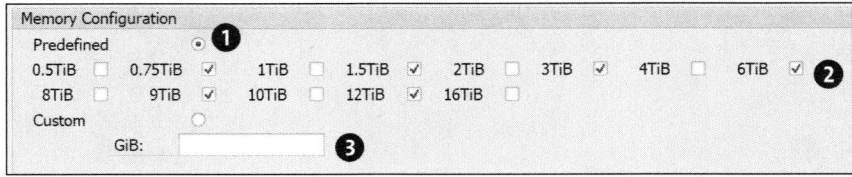

Figure 3.19 Memory Configuration

11. Select **Program • Execute in Background**, as shown in Figure 3.20.

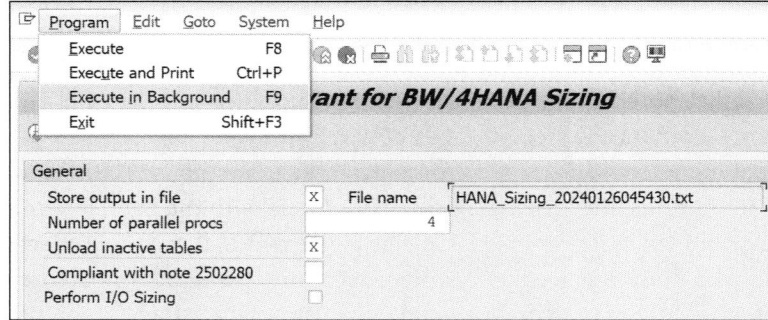

Figure 3.20 Sizing Report Execution

12. Using Transaction AL11, you can see this sample list of the directory (**DIR***), as shown in Figure 3.21. You can see a directory called **DIR_HOME**.

```
SAP Directories   26.01.2024 06:36:30 BW4

Name                Directory
DIR_ATRA            /usr/sap/BW4/DVEBMGS02/data
DIR_BINARY          /usr/sap/BW4/DVEBMGS02/exe
DIR_CCMS            /usr/sap/ccms
DIR_CT_LOGGING      /usr/sap/BW4/SYS/global
DIR_CT_RUN          /usr/sap/BW4/SYS/exe/uc/linuxx86_64
DIR_DATA            /usr/sap/BW4/DVEBMGS02/data
DIR_EXECUTABLE      /usr/sap/BW4/DVEBMGS02/exe
DIR_EXE_ROOT        /usr/sap/BW4/SYS/exe
DIR_GEN             /usr/sap/BW4/SYS/gen/dbg
DIR_GEN_ROOT        /usr/sap/BW4/SYS/gen
DIR_GLOBAL          /usr/sap/BW4/SYS/global
DIR_GRAPH_EXE       /usr/sap/BW4/DVEBMGS02/exe
DIR_GRAPH_LIB       /usr/sap/BW4/DVEBMGS02/exe
DIR_HOME            /usr/sap/BW4/DVEBMGS02/work
DIR_INSTALL         /usr/sap/BW4/SYS
DIR_INSTANCE        /usr/sap/BW4/DVEBMGS02
```

Figure 3.21 Home Directory

13. Double-click on the DIR_HOME row in Figure 3.21 to see the files loaded in Figure 3.22.

```
Directory: /usr/sap/BW4/DVEBMGS02/work

Usable Viewed Changed  Length  Owner   Lastchange  Lastchange  File Name
                               bw4adm  04.05.2023  19:08:54    HANA_Sizing_20230504230851.txt
                               bw4adm  04.05.2023  19:09:20    HANA_Sizing_20230504230920.txt
                               bw4adm  04.05.2023  19:31:39    HANA_Sizing_20230504233137.txt
X                      118503  bw4adm  04.05.2023  20:06:39    HANA_Sizing_20230504233201.txt
                               bw4adm  05.05.2023  14:05:00    HANA_Sizing_20230505180409.txt
                               bw4adm  05.05.2023  14:05:32    HANA_Sizing_20230505180518.txt
X                      898290  bw4adm  05.05.2023  14:37:47    HANA_Sizing_20230505180540.txt
                               bw4adm  24.08.2023  11:38:52    HANA_Sizing_20230824153718.txt
                               bw4adm  24.08.2023  11:49:39    HANA_Sizing_20230824154908.txt
X                      480816  bw4adm  24.08.2023  12:39:26    HANA_Sizing_20230824154947.txt
                               bw4adm  12.09.2023  04:21:15    HANA_Sizing_20230912082112.txt
                               bw4adm  21.09.2023  10:02:13    HANA_Sizing_20230921140034.txt
                               bw4adm  21.09.2023  10:23:53    HANA_Sizing_20230921142333.txt
X                      481707  bw4adm  21.09.2023  11:07:35    HANA_Sizing_20230921142420.txt
                               bw4adm  26.01.2024  09:01:26    HANA_Sizing_20240126045430.txt
```

Figure 3.22 Text File

14. You can now open the sizing output, as shown in Figure 3.23.

```
MINIMUM MEMORY SIZING RESULTS - CURRENT
=======================================

                                                  minimum          recomm.
                      Phys. memory per node:      768 GiB          768 GiB
Coord. / Worker Nodes:
  Memory Requirement (Minimum Total):             293    GiB       293 GiB
  Disk Space Requirement - data (Minimum Total):  190    GiB       190 GiB
  Disk Space Requirement - logs (Minimum Total):  147    GiB       147 GiB
  Number of Nodes  incl. coord. (Minimum Total):  1                1
  CPU requirements incl. coord. (Minimum Total):  72600  SAPS      72600 SAPS

NOTE:
 - Please carefully read documentation attached to SAP NOTE 2296290
   for a detailed description of the sizing procedure and its results!
 - You are confirming that you have read and understood the contents of SAP
   note 2502280 and that you fulfill the conditions described therein.

SIZING DETAILS
==============

(For   768  GiB nodes)  data [GiB]   total [GiB]   total [GiB]    util.
                                     incl. tmp.    (non-act.)
 Row Store                  3             5             5
 Coord. Column Store        23           46            46
 Worker Column Store        61          100            99
 Worker Planning                         93            93
 Caches / Services          50           50            50

 TOTAL (All Servers)       137          294           293          40 %
```

Figure 3.23 Sizing Output

Using the sizing data, the Basis team can make informed decisions regarding the SAP HANA server, including memory size, and determine whether a scale-up or scale-out architecture is preferable. For further insights on this topic, regularly review SAP Note

2610534 - "HANA BW Sizing Report (/SDF/HANA_BW_SIZING)" for the latest updates. For accurate results, execute this report in the live productive system.

3.8 Installing the Note Analyzer Report

SAP has provided a specialized Z report designed to install crucial SAP Notes, specifically crafted for distinct needs such as upgrades or migrations. These SAP Notes are presented in XML format, but to execute their implementation from XML, you use an ABAP-based tool to scan your system and analyze which SAP Notes need to be implemented. Consequently, the first phase of this process entails ensuring the availability of the Z report in the system.

Commence the process by creating a program named Z_SAP_BW_NOTE_ANALYZER. To gather essential information, refer to the central note for SAP BW/4HANA Conversion, SAP Note 2383530 - "Conversion from SAP BW to SAP BW/4HANA" on the SAP site. Open the **Attachment** section of the SAP Note, download the *SAP_BW4_Transfer_Note_Analyzer_YYYY-MM-DD.zip* attachment to the local workstation, and extract its contents. Inside the extracted ZIP file, locate the significant file named *Z_SAP_BW_NOTE_ANALYZER.txt* (see Figure 3.24).

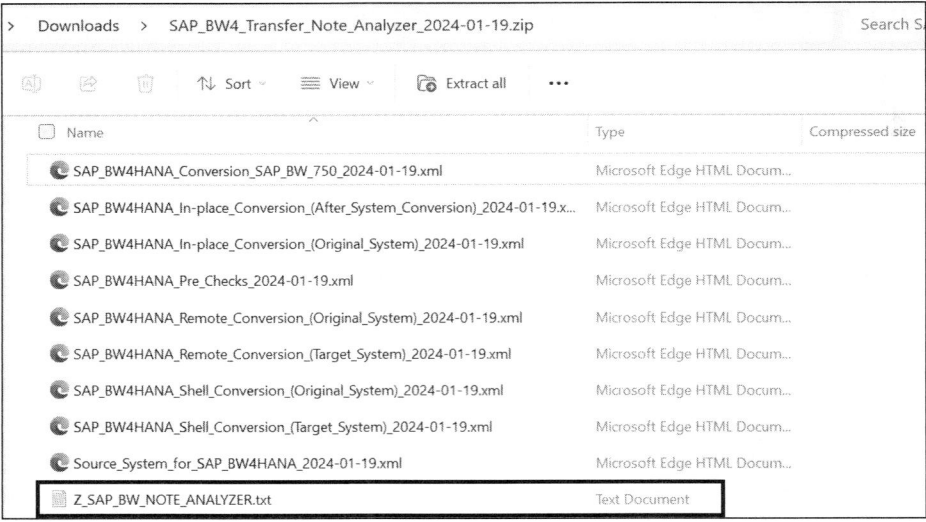

Figure 3.24 Note Analyzer Program

Moving on to program creation, access Transaction SE38 in the SAP BW system. Enter the report name as "Z_SAP_BW_NOTE_ANALYZER", give it the title of "SAP BW and SAP BW/4HANA Note Analyzer", and select the **Executable program** attribute key. Save the program as a local, temporary object. For content upload, navigate to the menu bar, and choose **Utilities** • **More Utilities** • **Upload/Download** • **Upload**. In the ensuing dialog

3.8 Installing the Note Analyzer Report

box, select the *Z_SAP_BW_NOTE_ANALYZER.txt* file, which you've copied from the ZIP file. Complete the process by saving and activating the program, ensuring that your Note Analyzer is now ready for effective use, as shown in Figure 3.25.

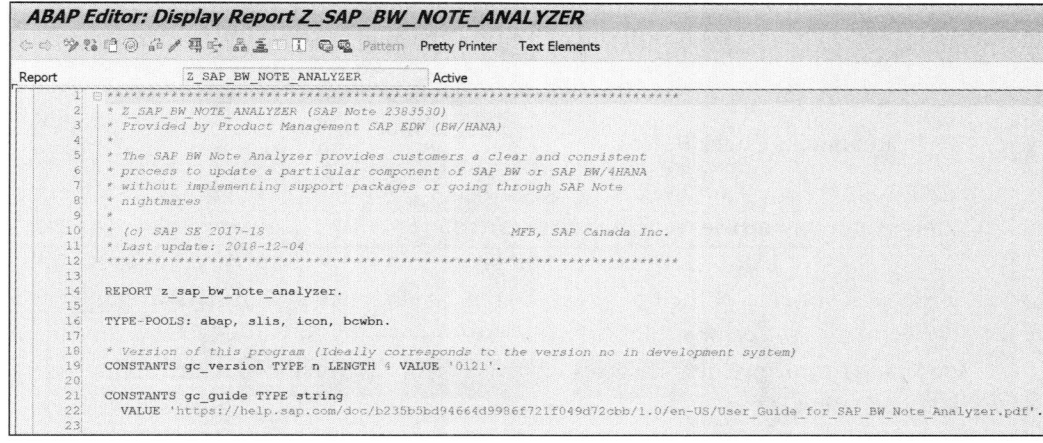

Figure 3.25 Note Analyzer Report

Once you execute this report (**Z_SAP_BW_NOTE_ANALYZER**), you'll see the screen shown in Figure 3.26. Selecting **Load XML file (No files loaded)** will trigger a prompt to include the XML files for the SAP Note scan in the system and list the implementation status.

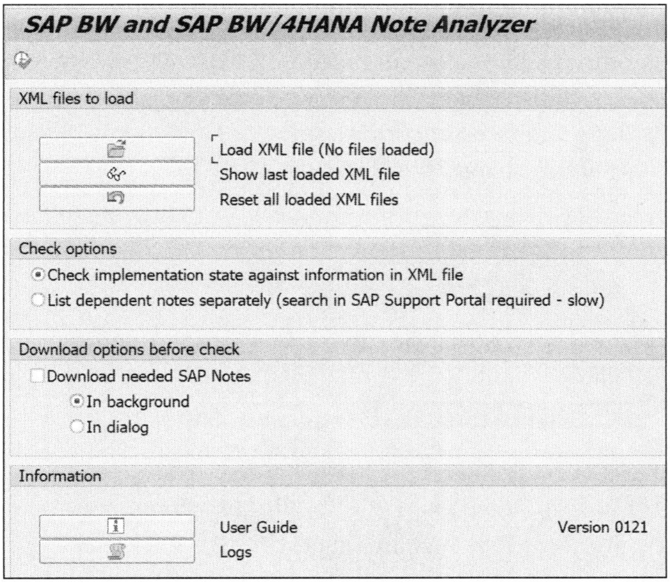

Figure 3.26 Note Analyze Report Execution

3.9 SAP Readiness Check for SAP BW/4HANA

In this section, you'll learn the details of SAP Readiness Check for SAP BW/4HANA and understand the ST-A/PI requirement. Then, you'll learn about the prerequisites and the process to execute the readiness check. You'll explore the functions available and finally learn how to view the SAP Readiness Check for SAP BW/4HANA.

3.9.1 Readiness Check Basics

SAP Readiness Check for SAP BW/4HANA serves as a valuable support system for customers undertaking the transition from existing SAP BW 7.x systems to the advanced SAP BW/4HANA platform. This comprehensive tool meticulously examines both the production systems, or their replicas, and the development systems. Its primary objective is to gauge the compatibility of these systems with the SAP BW/4HANA conversion process, scrutinizing the intricacies involved in the conversion and identifying essential preparation steps.

The results of this analysis are not only detailed but also intelligently consolidated into an interactive dashboard. This dashboard serves as a central hub for internal communication within an organization or for communication with SAP, providing stakeholders with a clear and concise understanding of the system's readiness for SAP BW/4HANA conversion.

Moreover, SAP Readiness Check for SAP BW/4HANA goes beyond being a mere diagnostic tool by offering a comprehensive pre-conversion overview. This overview proves instrumental in recognizing and addressing current system landscape preparation steps well in advance of the official commencement of the SAP BW/4HANA system conversion project. By providing this foresight, the tool empowers organizations to proactively plan and execute necessary steps, ensuring a smoother and more efficient transition to the enhanced capabilities of SAP BW/4HANA.

SAP Readiness Check for SAP BW/4HANA assesses specific areas, including the following:

- **Functional changes**
  - Evaluation of simplifications introduced in SAP BW/4HANA that are pertinent to your specific system installation
  - Explanation of their impact, detailed through corresponding SAP Notes
- **Installed add-ons**
  - Examination of both SAP and third-party add-ons installed in your system
  - Assessment of their compliance with the requirements of SAP BW/4HANA
- **System statistics**
  - Analysis of key statistical data pertaining to the system's performance and functionality

- **Technical system information**
  - In-depth insights into the technical aspects of the system, offering a detailed understanding of its configuration
- **System requirements**
  - Assessment of the system's adherence to the prerequisites and requirements for SAP BW/4HANA
- **Estimation of system size**
  - Evaluation of the anticipated size of the SAP BW/4HANA system
  - Recommendations for potential size reduction measures before initiating the conversion process
- **System configuration**
  - Information regarding the compatibility of configured SAP BW objects with SAP BW/4HANA
  - Insights into the availability and suitability of conversion tools for the configured objects

In essence, SAP Readiness Check for SAP BW/4HANA methodically examines and reports on these diverse facets, offering a comprehensive understanding of the system's readiness for the SAP BW/4HANA conversion, facilitating informed decision-making and proactive measures.

3.9.2 Basics of the ST-A/PI Add-On

The first step is to make sure the ST-A/PI add-on is the latest. The ST-PI and ST-A/PI tools, along with Transaction SDCCN, collect essential data for service sessions and SAP EarlyWatch Alert reports. Additionally, the AP/I tools feature analysis tools such as Transaction ST14 and Transaction ST12 for comprehensive analysis. Transaction ST13 provides specialized tools for evaluating SAP BW system performance. In Transaction SE38, choose **Report RTCCTOOL • Execute**. The results are shown in Figure 3.27.

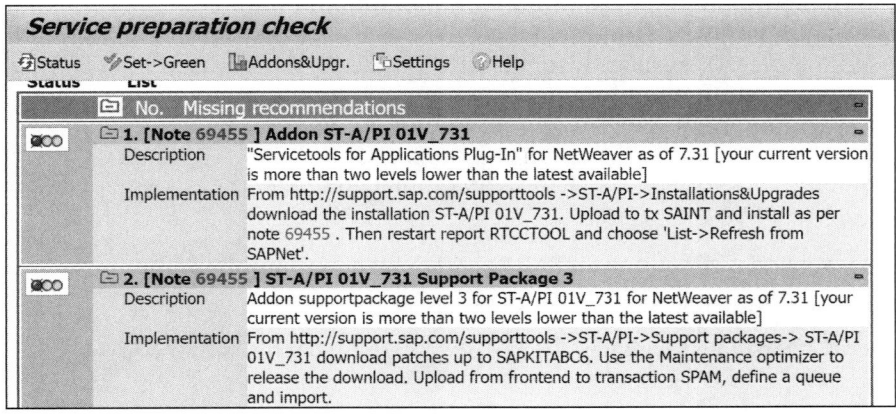

Figure 3.27 ST-A/PI Check

You can also see the version from Transaction RSA1 by choosing **System • Status • Choose System Data**. The data is shown in Figure 3.28.

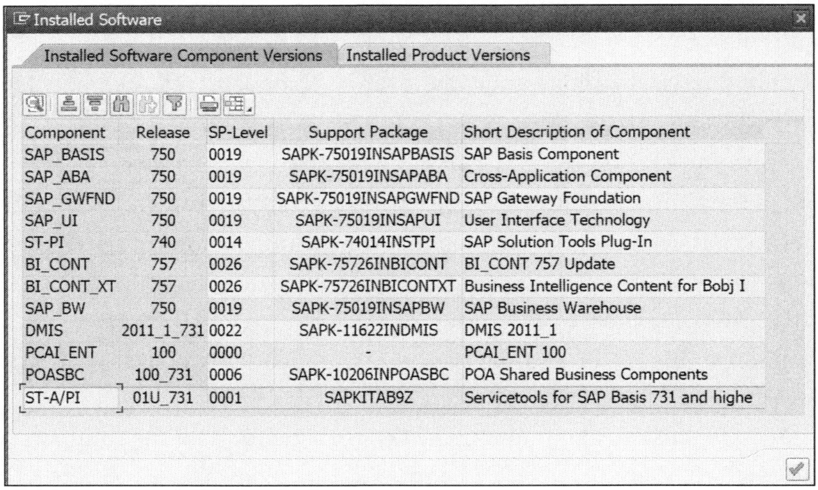

Component	Release	SP-Level	Support Package	Short Description of Component
SAP_BASIS	750	0019	SAPK-75019INSAPBASIS	SAP Basis Component
SAP_ABA	750	0019	SAPK-75019INSAPABA	Cross-Application Component
SAP_GWFND	750	0019	SAPK-75019INSAPGWFND	SAP Gateway Foundation
SAP_UI	750	0019	SAPK-75019INSAPUI	User Interface Technology
ST-PI	740	0014	SAPK-74014INSTPI	SAP Solution Tools Plug-In
BI_CONT	757	0026	SAPK-75726INBICONT	BI_CONT 757 Update
BI_CONT_XT	757	0026	SAPK-75726INBICONTXT	Business Intelligence Content for Bobj I
SAP_BW	750	0019	SAPK-75019INSAPBW	SAP Business Warehouse
DMIS	2011_1_731	0022	SAPK-11622INDMIS	DMIS 2011_1
PCAI_ENT	100	0000	-	PCAI_ENT 100
POASBC	100_731	0006	SAPK-10206INPOASBC	POA Shared Business Components
ST-A/PI	01U_731	0001	SAPKITAB9Z	Servicetools for SAP Basis 731 and highe

Figure 3.28 ST-A/PI Version

Obtain the latest **ST-PI** and **ST-A/PI** versions from SAP Service Marketplace. Transfer downloaded files to your server, and extract them using the SAPCAR tool at the SID admin level. Access Transaction SAINT to load the package, and install ST-AP. Then, update ST-PI via Transaction SPAM for system stability, security, and compatibility with the latest SAP technologies, ensuring optimal performance for your SAP environment.

3.9.3 Prerequisites

The first step for the readiness check is to refer to SAP Note 3061594 - "SAP Readiness Check for SAP BW/4HANA - Central Note." This SAP Note introduces the essential data collection framework managed by the RC_BW_COLLECT_ANALYSIS_DATA report. Efficient scheduling of the batch collection job hinges on the Activity 16 authorization assigned to the user ID, specifically for object S_DEVELOP. This authorization ensures seamless execution of the collection job, fostering a streamlined data analysis process within the SAP environment.

SAP Readiness Check for SAP BW/4HANA accommodates various eligible releases for conversion:

- Shell conversion is supported for SAP BW 7.0 or higher on any database.
- Remote conversion is available for SAP BW 7.30 or higher on any database.
- In-place conversion is facilitated for SAP BW 7.50 or higher on SAP HANA.

3.9 SAP Readiness Check for SAP BW/4HANA

Make sure you check the latest version of SAP Note 2575059 - "SAP BW Note Analyzer Files for SAP Readiness Check for SAP BW/4HANA and SAP Data Datasphere, SAP BW Bridge," and download the XML file from the attachment in this note, as shown in Figure 3.29.

File Name	File Size
SAP_BW4HANA_Readiness_Custom_Code_2023-02-18.zip	441 KB
SAP_BW4HANA_Readiness_Check_2023-12-01.zip	63 KB

Figure 3.29 XML Files for SAP BW/4HANA Readiness Check

Once you've downloaded the ZIP file, you can now execute the Note Analyzer report **Z_SAP_BW_NOTE_ANALYZER**, shown earlier in Figure 3.26. With the downloaded XML file, when you execute the report as shown in Figure 3.30, you have the following choices:

❶ Use this **Load XML file** button to choose your folder that has XML from the popup window.

❷ Choose the right XML file; in our case, we're now in the **SAP_BW/4HANA_Readiness_Check** XML.

❸ Once you see the filename is selected, click on **Open**, which will use this XML for the checks.

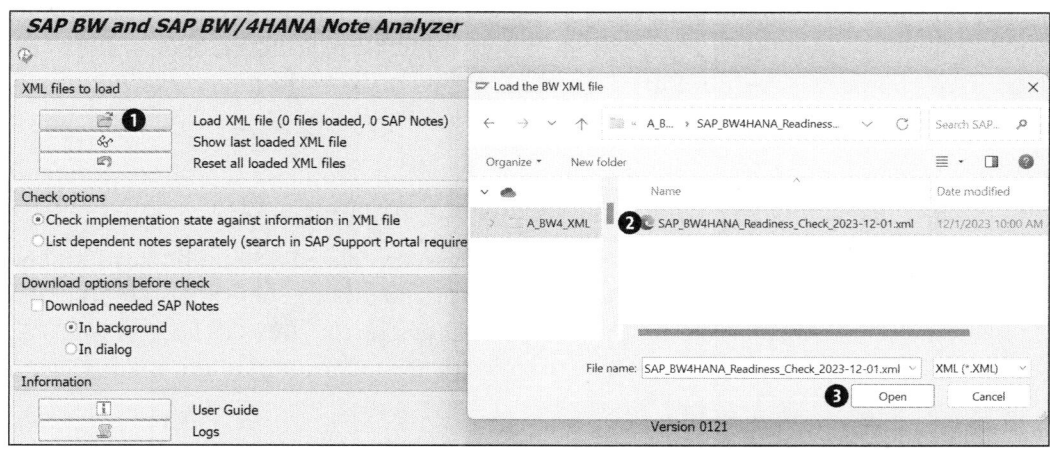

Figure 3.30 Readiness Check XML Upload

Run the report, bearing in mind that the process will require some time to validate the SAP Note. During this time, the system will verify the existence of the note and proceed to download it if it's accessible. Upon completion, you'll observe the note status displayed as green if it's available or red if it isn't accessible. You can see the result of the check in Figure 3.31. Install the notes one by one in the same order. After the notes are implemented, repeat the XML checks multiple times until all the note statuses are green.

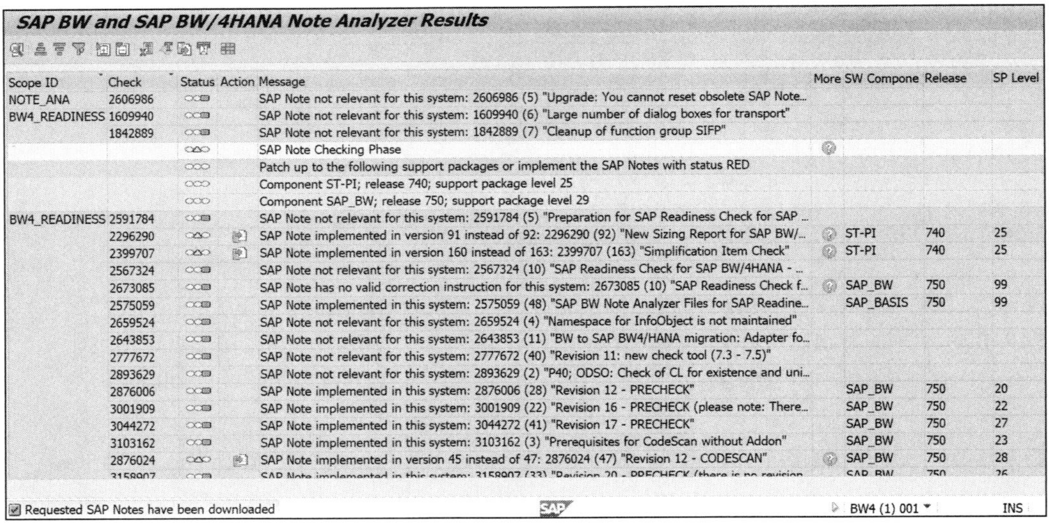

Figure 3.31 Results of the Readiness Check XML

You can see that the XML with code scan was available. You can now upload the XML and install the required SAP Notes.

For SAP BW versions 7.30 and above, the readiness check includes a code scan. If your SAP BW version is below 7.3, you must independently execute the code scan report. We'll now concentrate on SAP BW versions greater than 7.3. The following sections outline the process.

3.9.4 Executing SAP Readiness Check for SAP BW/4HANA

Once the steps in the previous section have been completed, you need to execute the SAP Readiness Check for SAP BW/4HANA in the production SAP BW system. The first step is data collection. Execute Transaction SE38 and run RC_BW_COLLECT_ANALYSIS_DATA in the productive client of the system scheduled for analysis, as shown in Figure 3.32:

❶ The transformation target product needs to be selected as **SAP BW/4HANA**.

❷ Choose the SAP BW/4HANA target version.

❸ You need all the details in **Scope Selection**, so check all the check boxes in this section.

❹ This is used to start the analysis in background.

Click on **Schedule Analysis** to see the screen shown in Figure 3.32.

Goto System Help
SAP Readiness Check
Schedule Analysis
Refresh Execution status
My Jobs
Download Analysis Data
Help
Transformation Target
SAP BW/4HANA
SAP Datasphere, SAP BW Bridge
SAP BW/4HANA Target Release
Target BW/4HANA Version
SAP BW/4HANA 2023 [Initial Shipment Stack]
Scope Selection
HANA Sizing Data
Code analysis
Object analysis
Source system readiness
Analysis Execution Status
Analysis Finished
Analysis Data Status
Anaylsis Data from 07:32:49 26.01.2024
Delete Buffered Data for Selected Items

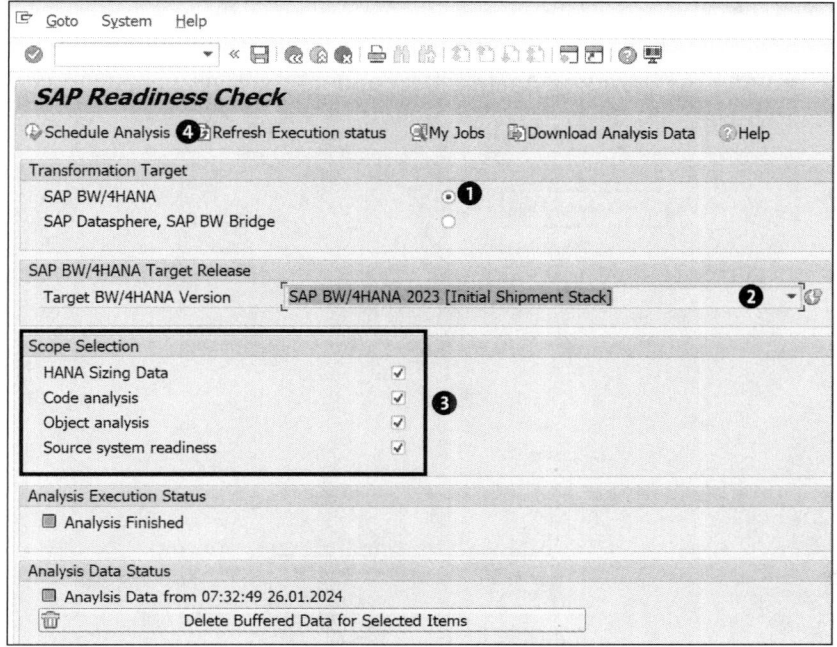

Figure 3.32 SAP Readiness Check for SAP BW/4HANA

Click the button in Figure 3.33 to see the jobs status as shown in Figure 3.34.

Information
Job RC_BW_4_HANA:MONITOR (17112100) is scheduled.
Check the status in transaction SM37 and download
the analysis file once it is finished.

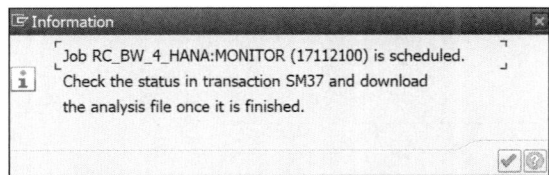

Figure 3.33 Background Job Popup

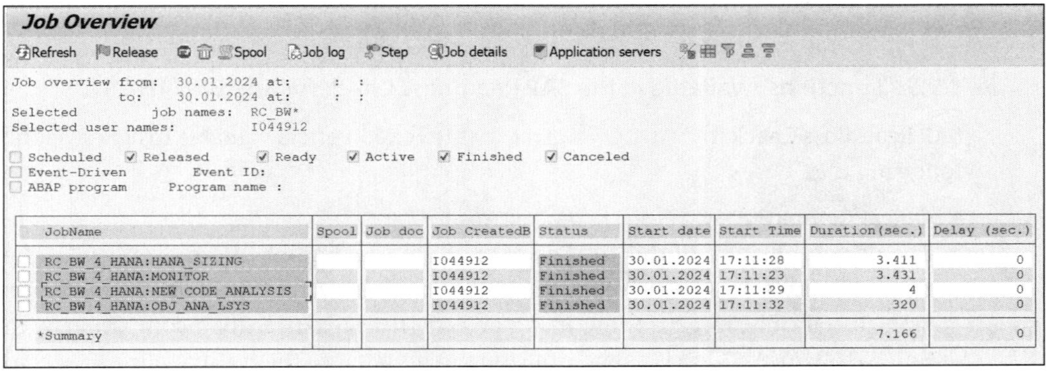

Job Overview
Refresh
Release
Spool
Job log
Step
Job details
Application servers
Job overview from: 30.01.2024 at: : :
to: 30.01.2024 at: : :
Selected job names: RC_BW*
Selected user names: I044912
Scheduled
Released
Ready
Active
Finished
Canceled
Event-Driven
Event ID:
ABAP program
Program name :
JobName
Spool
Job doc
Job CreatedB
Status
Start date
Start Time
Duration(sec.)
Delay (sec.)
RC_BW_4_HANA:HANA_SIZING I044912 Finished 30.01.2024 17:11:28 3.411 0
RC_BW_4_HANA:MONITOR I044912 Finished 30.01.2024 17:11:23 3.431 0
RC_BW_4_HANA:NEW_CODE_ANALYSIS I044912 Finished 30.01.2024 17:11:29 4 0
RC_BW_4_HANA:OBJ_ANA_LSYS I044912 Finished 30.01.2024 17:11:32 320 0
*Summary 7.166 0

Figure 3.34 Jobs Overview

3   System Preparation and Prerequisites

Once you've seen the list of active jobs, you can monitor when all the jobs are finished. If all the jobs are completed, go back to the readiness check report selections, as shown in Figure 3.32, and click the **Download Analysis Data** option. You'll get a popup to save the file (see Figure 3.35).

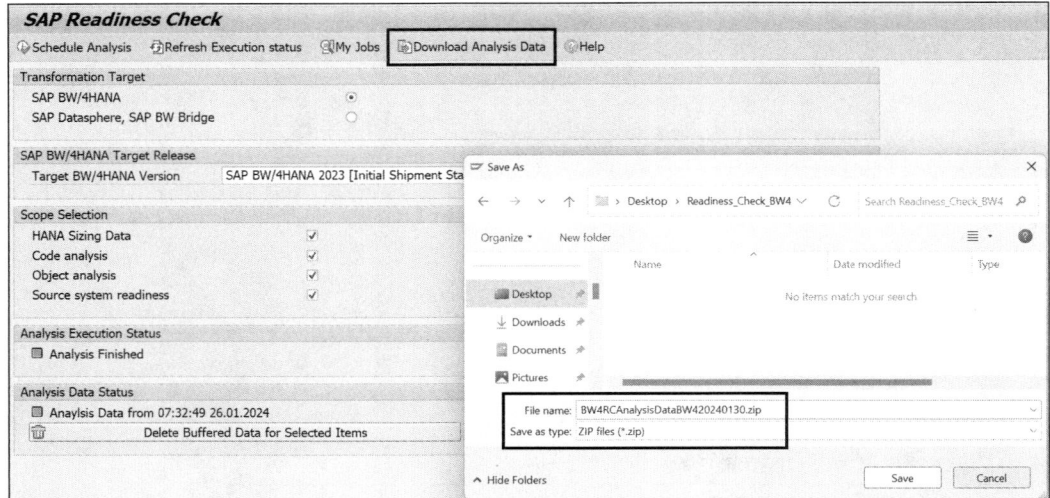

Figure 3.35 Downloading the Data

After the file has been downloaded, you can see the details (see Figure 3.36).

Figure 3.36 The Downloaded File

Once you've downloaded the ZIP file, the next step is to use this file and understand the details that will help to check for SAP BW/4HANA readiness.

3.9.5   Functions Available in the SAP Readiness Check for SAP BW/4HANA

SAP Readiness Check for SAP BW/4HANA in this version offers valuable insights via the following tiles:

- **Object Compatibility**
  The primary feature of this tile lies in providing a comprehensive overview of the compatibility status of active SAP BW objects in the system earmarked for conversion. Specifically, it focuses on listing objects that are currently incompatible with SAP BW/4HANA. There is a toggle button integrated into the tile that allows users to seamlessly switch between in-place and remote/shell conversion options, with a

100

note on the potential impact on findings. Compatibility statuses are categorized into three main types:

- **Convertible**: Objects are automatically converted during the conversion process.
- **Adjustment Required**: Objects need manual intervention to ensure compatibility with SAP BW/4HANA.
- **Nonconvertible**: Objects are either skipped during the conversion process or require replacement.

Findings are logically grouped based on when they should be addressed in the conversion process, providing a structured approach. References to SAP Notes (mostly simplification items) are given whenever applicable to offer additional guidance in addressing compatibility issues. (For more details about **Object Compatibility** see the left pane of Figure 3.41 later in this section.)

- **Source System Readiness**

    This tile furnishes an overview of the compatibility status of source systems linked to the system slated for conversion. It exclusively lists systems that are currently incompatible with SAP BW/4HANA. Key features include the following:

    - **Ready for ODP 1.0/2.0**: Systems imply that the system type can transition to ODP without additional efforts. If the system solely supports ODP 1.0, upgrading may be recommended.
    - **Adjustments Required**: Systems indicate that the current connection type is no longer supported and requires a change. Examples include replacing a web service connection with an SAP HANA connection or SDA.
    - **System Failure**: Systems indicate that no information could be collected about the source system, often stemming from an issue with the RFC connection. If the system is still essential, addressing this problem becomes imperative.

    References to SAP Notes (primarily simplification items) are provided whenever applicable, serving as valuable resources for further guidance and resolution.

- **Custom Code Compatibility**

    As a result of modifications made to SAP BW objects and the streamlining of various ABAP function modules, classes, reports, exits, interfaces, variables, and other elements in SAP BW/4HANA, it's essential for customers to gain insight into the necessary adaptations for their custom code developments. This overview is crucial for ensuring compatibility between their system and solutions with SAP BW/4HANA. A comprehensive understanding of the specific implications on the analyzed system plays a pivotal role in the success of a conversion project.

    The **Custom Code Compatibility** tile delivers a comprehensive assessment of custom code in the analyzed system, addressing potential incompatibility with SAP BW/4HANA. Given the varied use of custom code, which may depend on deprecated ABAP entities or SAP BW objects, this information is pivotal. The tile covers diverse custom code locations in all possible areas of SAP BW mentioned earlier.

- **Add-On Compatibility**
 This tile presents a comprehensive overview of the compatibility status of implemented add-ons within the SAP BW system scheduled for conversion. Notably, users have the flexibility to include comments for individual findings, offering utility for third-party add-ons. It's essential to recognize that the analysis relies on maintenance planner, underscoring the importance of ensuring that the system's data is current within this tool for accurate and reliable compatibility assessments.

- **Database Sizing**
 This tile summarizes current database size and distribution, recommends target system size (e.g., memory, nodes, SAPS), and lists the largest SAP BW objects and tables based in the system on their **Estimated Size on SAP HANA**. This information is crucial for strategic planning and optimizing system performance. The analysis relies on the SAP HANA BW Sizing report (/SDF/HANA_BW_SIZING).

3.9.6 Viewing the SAP Readiness Check for SAP BW/4HANA

You can launch SAP Readiness Check in public mode from SAP for Me via these steps:

1. Go to *https://me.sap.com/readinesscheck* to reach the **SAP Readiness Check** screen (see Figure 3.37).

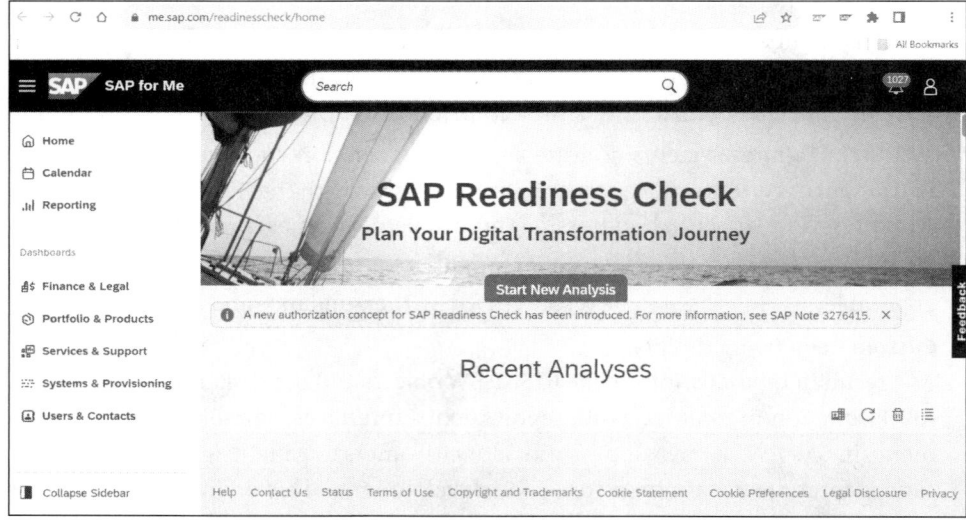

Figure 3.37 SAP Readiness Check Landing Page

2. Click the **Start New Analysis** button to reach the **Create Analysis** screen (see Figure 3.38).

3. Enter a name in the **Analysis Name** field, and then choose the **Browse** option next to the **Analysis File** field to select the file from your local workstation which you've downloaded already. The files appear in the popup, as shown in Figure 3.38.

3.9 SAP Readiness Check for SAP BW/4HANA

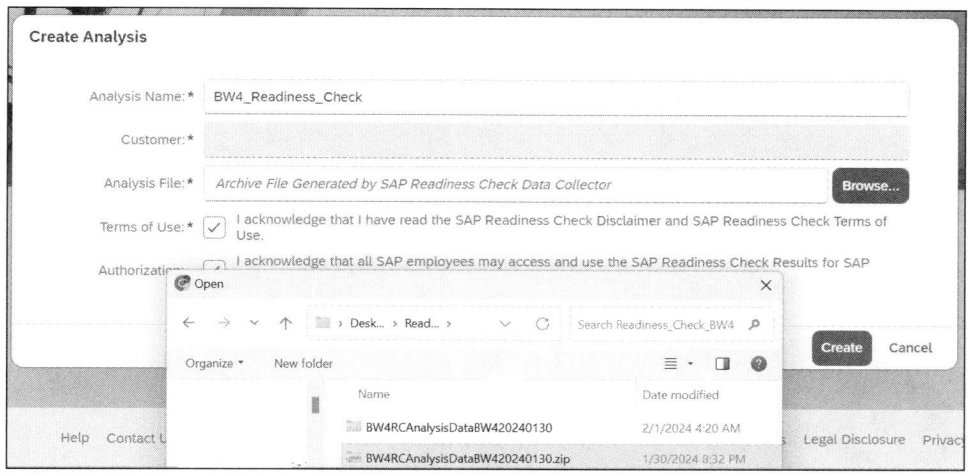

Figure 3.38 Choose the File

4. Choose the relevant *.zip* file, and then you'll see the file name in the **Analysis File** input screen. Click **Create** (see Figure 3.39).

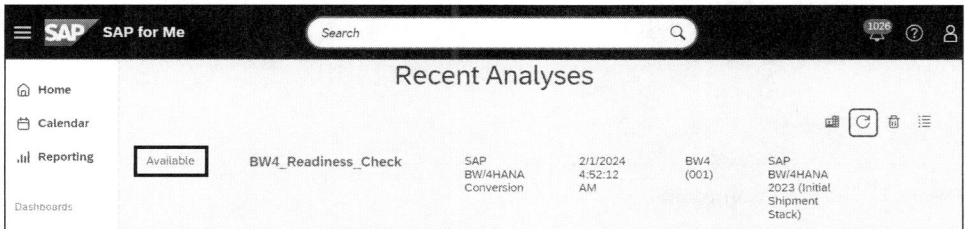

Figure 3.39 Create Analysis

Now you'll see its status as **Available** in Figure 3.40.

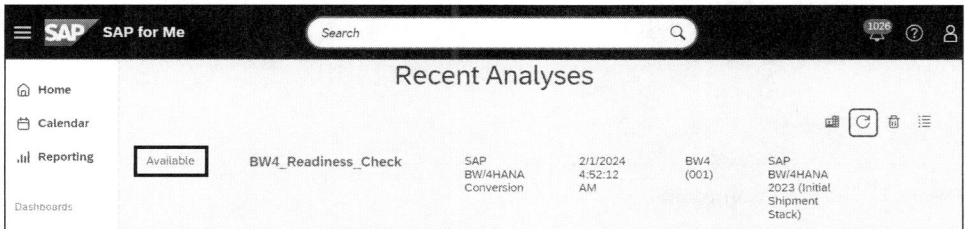

Figure 3.40 Available Status

103

5. Click on the description (in our case, **BW4_Readiness_Check**), and you'll see the screen shown in Figure 3.41, which includes a toggle button for **Remote/Shell Conversion** and **In-Place Conversion** in the **Object Compatibility** tile.

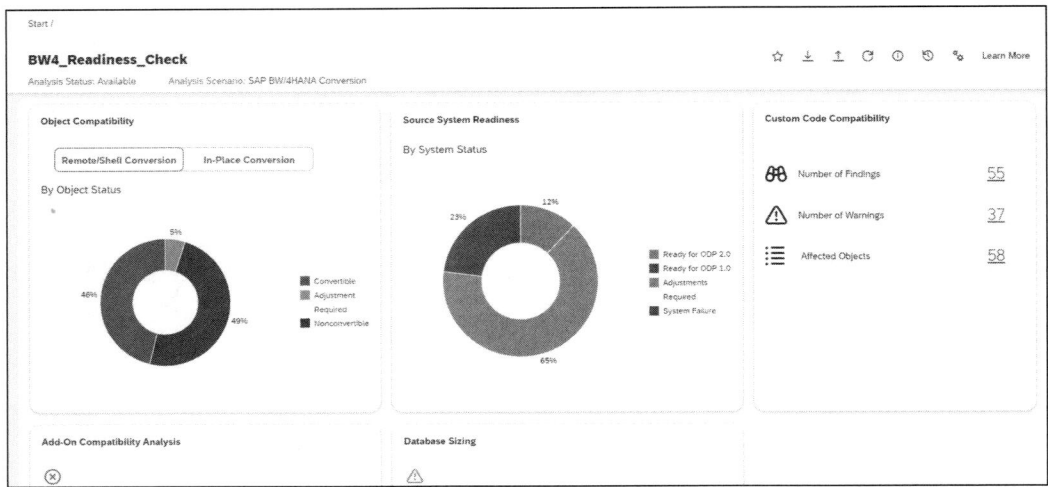

Figure 3.41 Readiness Check Tiles

6. Click on the **Nonconvertible** section of chart in **Object Compatibility**, and you'll see the data shown in Figure 3.42.

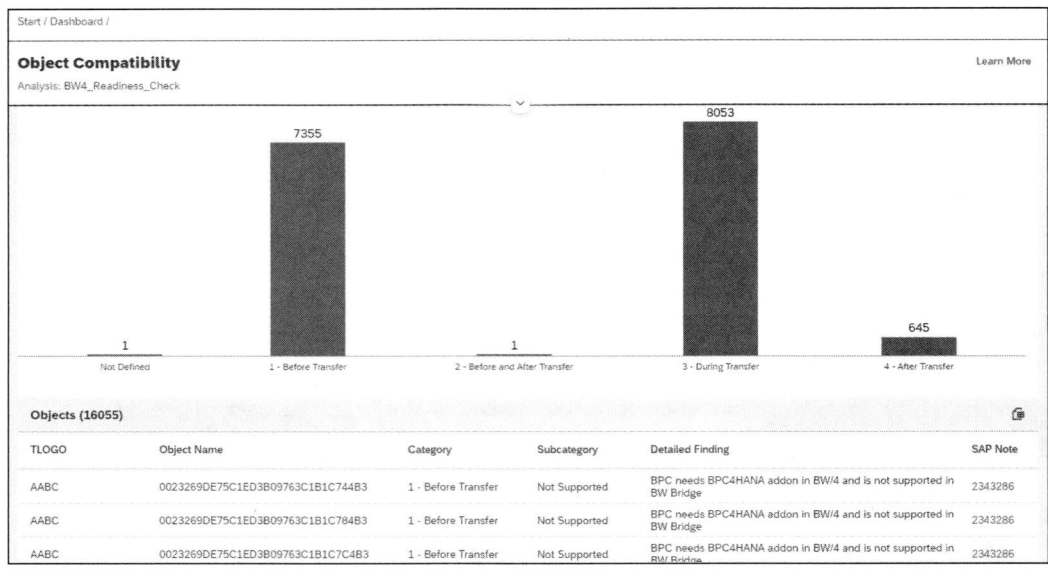

Figure 3.42 Object Compatibility

7. If you choose **Affected Objects** on the right side of Figure 3.41, shown previously, you'll see the screen shown in Figure 3.43. Similarly, you can check the **Add-On Compatibility Analysis** and **Database Sizing** in the respective tiles on that same screen.

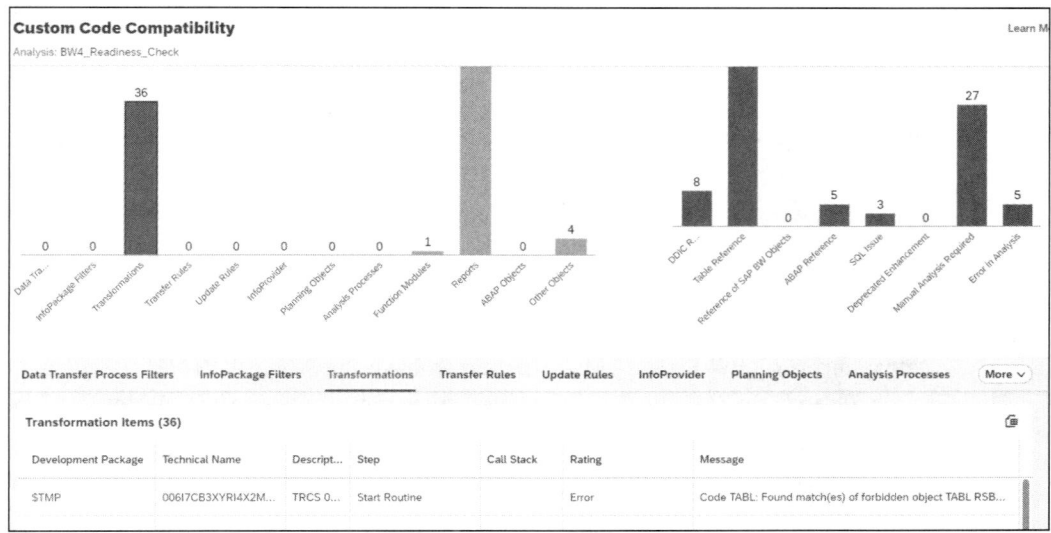

Figure 3.43 Custom Code Compatibility

3.10 Prechecks for SAP BW/4HANA Migration

In this section, you'll learn about the essential prechecks for the SAP BW/4HANA migration, starting with the steps to install the precheck tools. You'll also see the process to execute the precheck tools and understand the options to fix the issues observed during prechecks based on the issues. We'll then discuss the sizing report followed by using the code scan tool to find the unsupported codes and fix them. Finally, you'll learn about the cleanup reports that can be used before the migration.

SAP facilitates your conversion project through prechecks, which serve as essential tools in identifying crucial steps to ensure the compatibility of your system with the conversion process. These checks should be executed before entering the realization phase, allowing sufficient time to address any issues that may arise before initiating the conversion processes. Delivered to customers in the form of SAP Notes, these prechecks provide actionable insights, outlining mandatory steps that must be undertaken prior to converting to SAP BW/4HANA. The results generated from these prechecks highlight specific instances that require attention and resolution before proceeding with the actual conversion.

By incorporating these prechecks into your conversion planning, you gain a proactive approach to system compatibility, enabling timely resolution of potential challenges.

Leveraging SAP Notes as the medium for delivering these checks ensures that customers have access to comprehensive guidance and necessary steps for a successful transition to SAP BW/4HANA. As a result, the prechecks contribute significantly to the overall efficiency and smooth execution of the conversion project.

3.10.1 Installing Precheck Tools

To get the precheck tools, you need to have certain SAP Notes installed in your SAP BW system, as follows:

1. Install SAP Note 2777672 - "Revision 11: New Check Tool (7.3 - 7.5)."
2. Ensure that you've successfully installed the SAP Readiness Check notes, as discussed in the previous section. The XML file can be obtained from SAP Note 2575059 - "SAP BW Note Analyzer Files for SAP Readiness Check for SAP BW/4HANA and SAP Datasphere, SAP BW Bridge." Make certain to have the *SAP_BW4HANA_Readiness_Check_[Latest].zip* XML file. Do the manual steps mentioned in the SAP Note.
3. Check SAP Note 2383530 - "Conversion from SAP BW to SAP BW/4HANA," download *SAP_BW4_Transfer_Note_Analyzer_2024-01-30.zip* from the SAP Note, and check for these XML files, as shown in Figure 3.44.

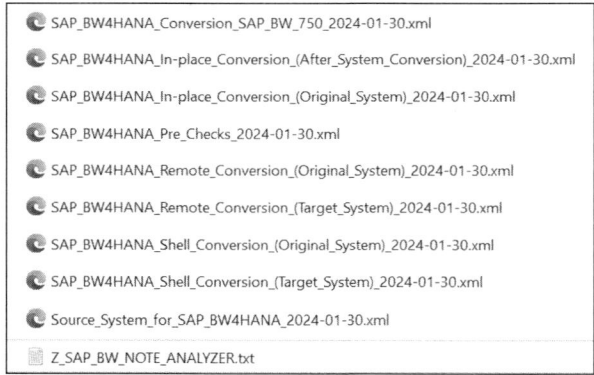

Figure 3.44 XML Files after Extract

4. Use the following XML for prechecks:
   - For SAP BW from 7.00 to 7.40, use *SAP_BW4HANA_Pre_Checks_[last_update].xml*.
   - For SAP BW 7.50, the prechecks can't be installed in isolation but are bundled with the transfer toolbox. Use *SAP_BW4HANA_In-place_Conversion_(Original_System)_[last_update].xml*.

Once you've installed all the notes based on these XML files, you'll have the tools you need for precheck. Make sure that report RS_B4HANA_CONVERSION_CONTROL is now available in the system by checking Transaction SE38 or SA38. This report can also be accessed via Transaction RSB4HCONV.

3.10 Prechecks for SAP BW/4HANA Migration

3.10.2 Executing Prechecks

In this section, we'll walk through the steps to execute a precheck, including the individual steps for in-place conversions and remote conversions.

Initial Steps

Follow these initial steps to execute the precheck, no matter your conversion type:

1. In your SAP BW system, execute Transaction RSB4HCONV to open the screen shown in Figure 3.45. On the **Prepare Phase** tab, click the **Execute Pre-Checks** button.

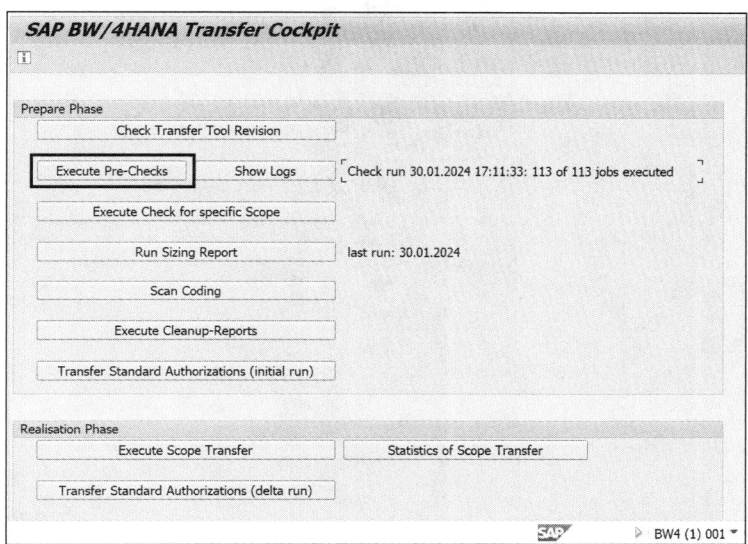

Figure 3.45 SAP BW/4HANA Transfer Cockpit

2. When he popup appears, as shown in Figure 3.46, select **Execute in Batch** and **Collect Statistics**. Then, click the execute icon .

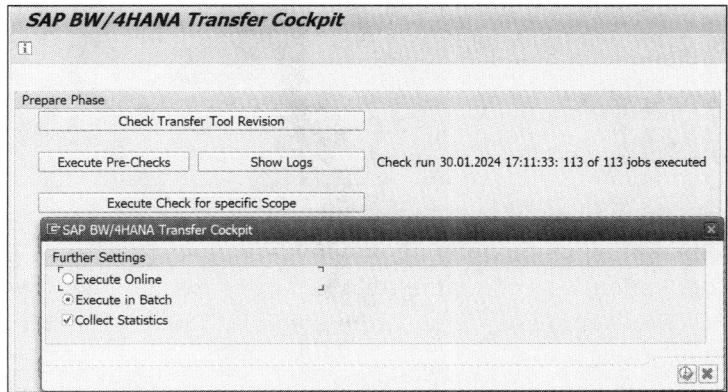

Figure 3.46 Popup for Prechecks

3. The execution will take some time to complete. When it's finished, you can see that jobs **RS_RC*** are **Active** in the upper screen and all jobs are in status **Finished** in the lower screen (Figure 3.47).

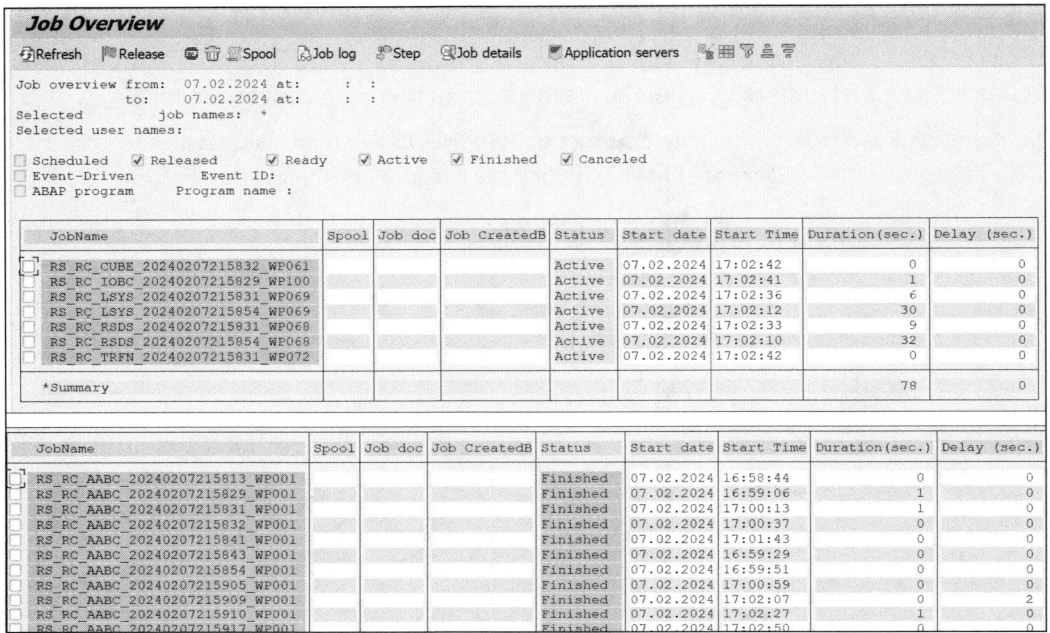

Figure 3.47 Job Status for Precheck Analysis

4. In Transaction SM37, wait for all **Active** jobs to be completed that start with **RS_RC***. Once all jobs are in the **Finished** status, you'll see the **Check Run** statement updated with the date, time, and job count, as shown in Figure 3.48.

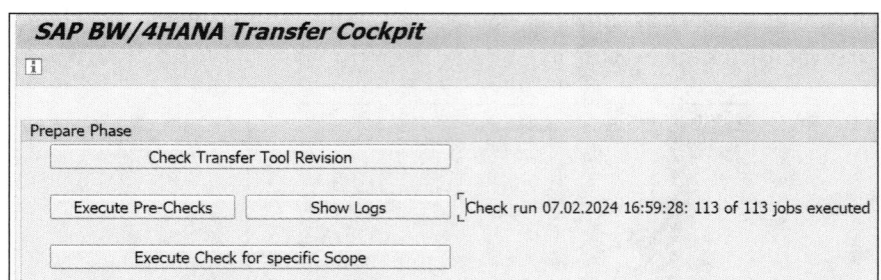

Figure 3.48 Precheck Job Status

5. Choose the **Show Logs** option. From here, the steps will differ based on whether you're doing an in-place conversion or a remote conversion. Let's look at the steps for an in-place conversion first.

3.10 Prechecks for SAP BW/4HANA Migration

Precheck for In-Place Conversion

From the screen shown in Figure 3.49, follow these steps:

1. Choose **Inplace** in the popup window.

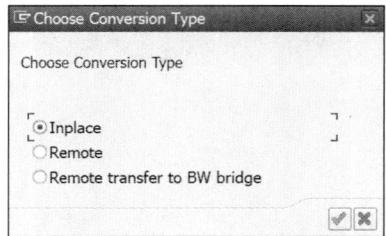

Figure 3.49 Show Logs for In-Place

2. Click the execute button, and you'll see the result shown in Figure 3.50:
   - ❶ **Pre-check Findings** and tool numbers, such as **1** for **Before Transfer Tool**
   - ❷ **Check Message** showing the details of the findings

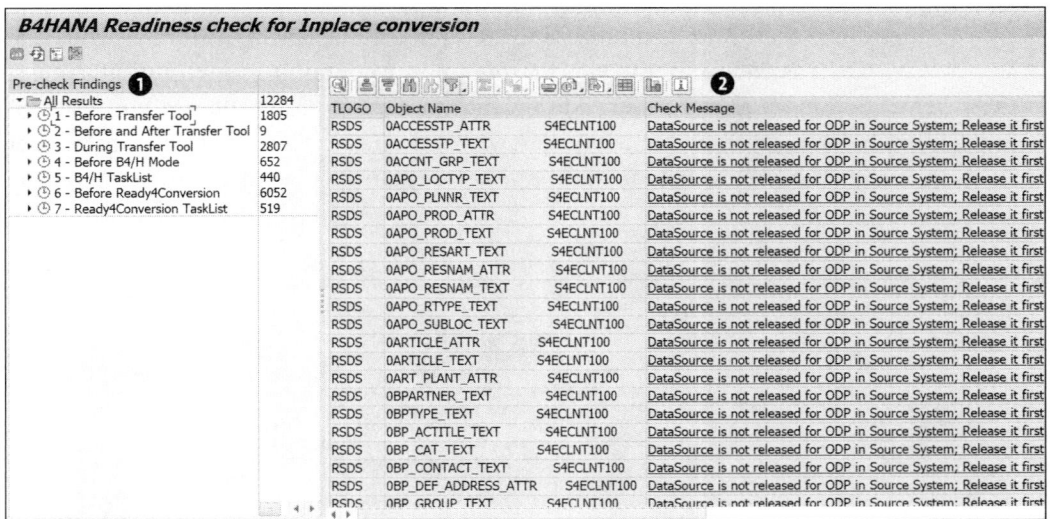

Figure 3.50 Precheck Findings

3. Now you can see the **Pre-check Findings** in a different order. Follow these steps, shown in Figure 3.51:
   - ❶ Select this icon to change the layout.
   - ❷ Preview the **Selected Groups** to see if you need any fields from the **Available Groups**. We'll stick with **Handling Phase**, **TLOGO**, and **Infoarea**.
   - ❸ Press [Enter].

3   System Preparation and Prerequisites

❹ Choose **TLOGO**, which will have the details of the Transportable Objects types in SAP BW, to see the details.

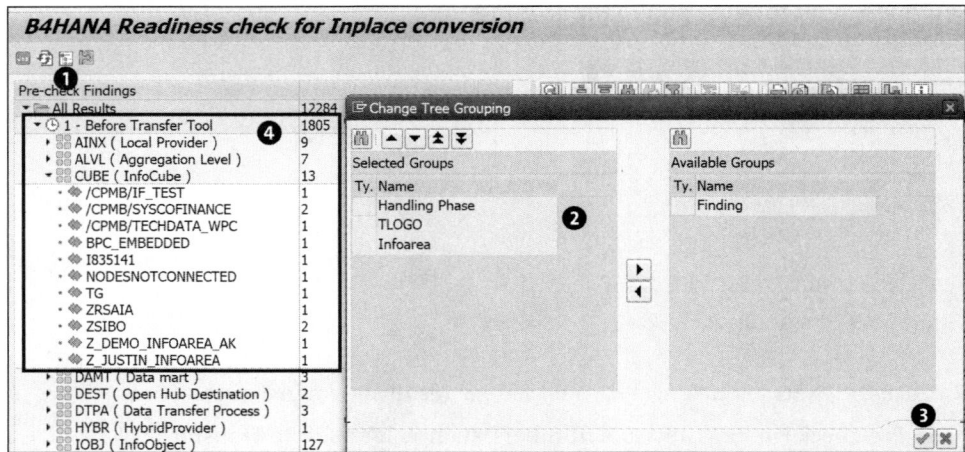

Figure 3.51 Change Layout

4. Click the execute icon [✓] in Figure 3.51 to see the right side of the screen, which has the **TLOGO** as **Cube** and the **Object Name** column containing the technical name of the InfoCube (Figure 3.52). In addition, in the **Check Message** column, you'll see the details of the issue, and in the **Processed at** column, you'll see **Before Transfer Tool**, which means you need to fix this issue before starting the SAP BW/4HANA transfer tool.

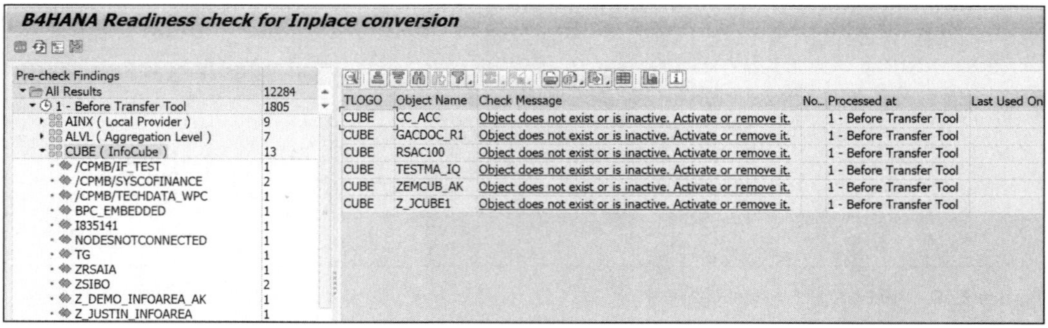

Figure 3.52 Status of Cube

5. Now focus on all the issues shown in the right window and fix them. You can re-execute this check multiple times and ensure the issues are fixed.

6. In the **All Results** folder, you can see many objects, such as TLOGO (see Figure 3.53). Ensure that you click on each TLOGO name under **Before Transfer Tool** on the left side of the screen and try to fix them based on the suggestion. The precheck report can be run multiple times to ensure this error list is reduced as much as possible. For

example, clicking on **ODSO,** which is listed under the **Before Transfer Tool** option in the **All Results** folder, will show the results on the right side of the screen shown in Figure 3.53.

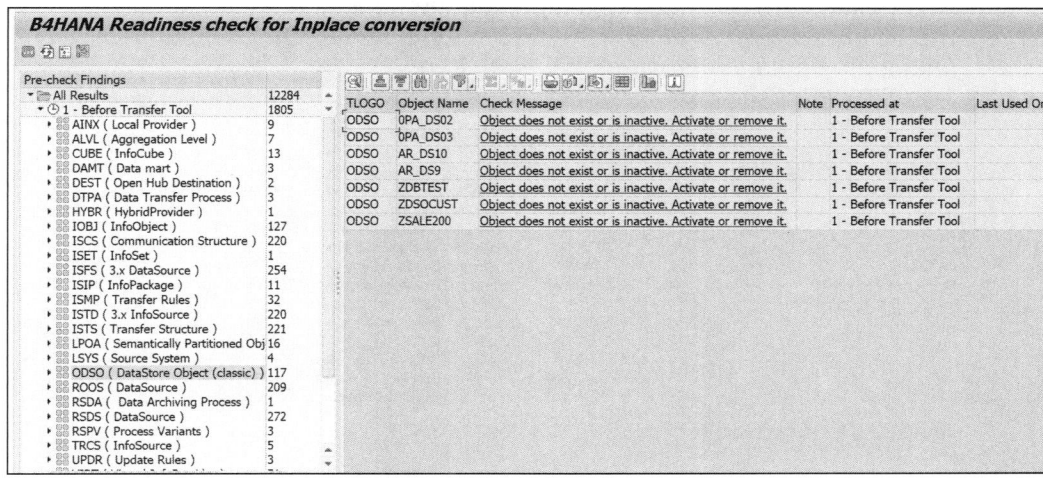

Figure 3.53 Before Transfer Tool Option

7. Collapse the **Before Transfer Tool** option, and you can see other options such as **Before and After Transfer Tool** (see Figure 3.54). Double-click on **RSDS** to see that you need to manually modify a few DataSources using SDI. However, this isn't a show-stopper for your conversion, so you can add this effort to the overall project plan.

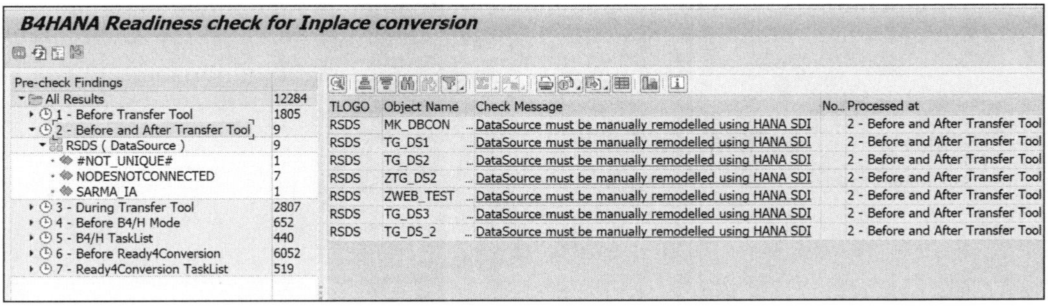

Figure 3.54 Before and After Transfer Tool

8. The next findings for the **During Transfer Tool** (see Figure 3.55) will give you a glimpse of how the tool will handle these TLOGO objects during conversion. This is just informational, and there aren't many activities if you see objects in this section. All the **CUBE**s will be converted to ADSOs, as shown in the right part of the window. Per the SAP BW/4HANA conversion cockpit tool, the classic optimized DSO (ODSO) is converted to an ADSO, as shown in Figure 3.56.

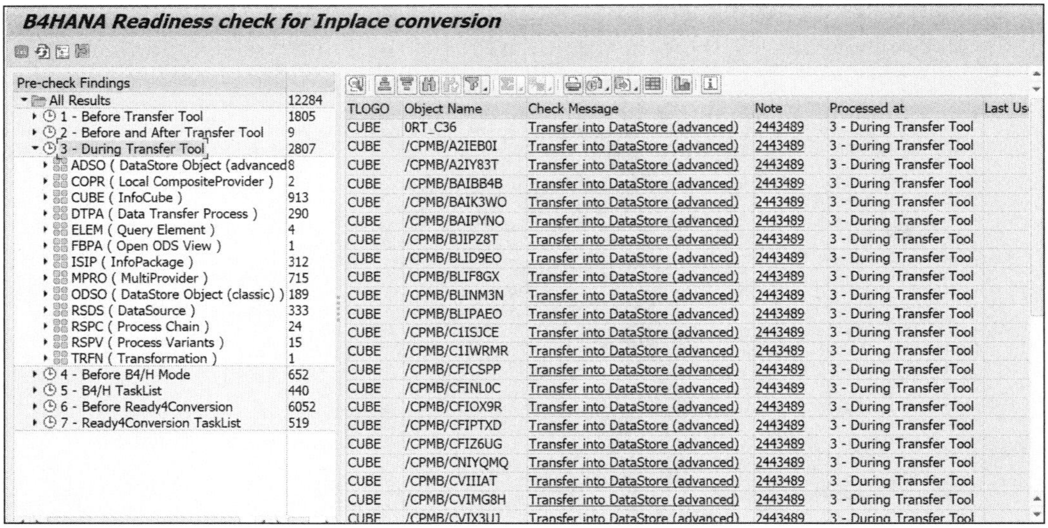

3   System Preparation and Prerequisites

Figure 3.55 During Transfer Tool

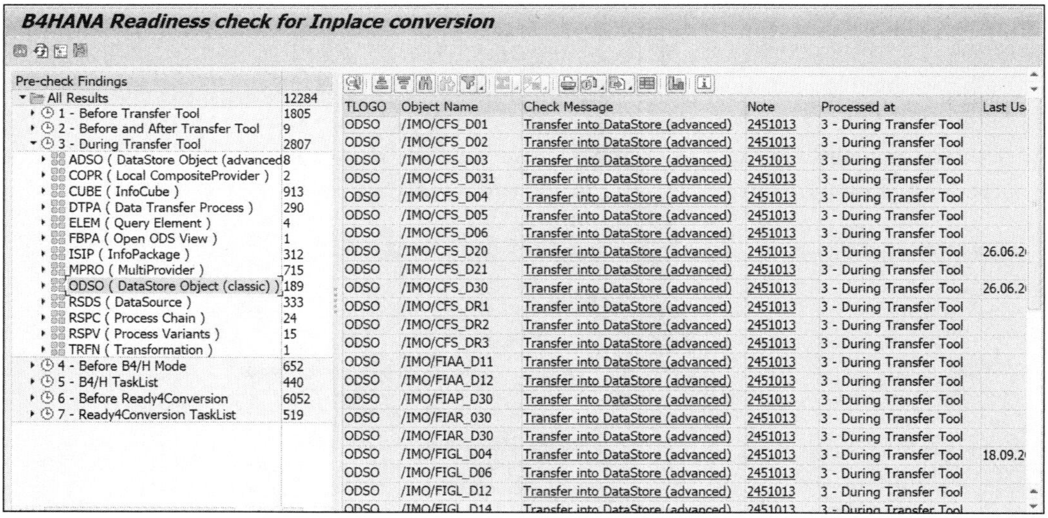

Figure 3.56 ODSO Conversion

You also have other options for the in-place conversion:

- **Before B4/H Mode**
  You can see the issues that are related to the Before BW/4HANA conversion mode, if there are any errors, then you need to fix them, which is covered in this section.

- **B4/H TaskList**
  Here, you can see the results based on the SAP BW/4HANA task list mode.

3.10 Prechecks for SAP BW/4HANA Migration

- **Before Ready4Conversion**
  This will list all the results before the Ready4Conversion phase. Check the findings, and fix the issues shown in this section.

- **Ready4Conversion TaskList**
  If you see the list of SAP BW objects in Ready4Conversion mode with any errors in the results, those need be fixed. The details of the analysis result are shown in Figure 3.57.

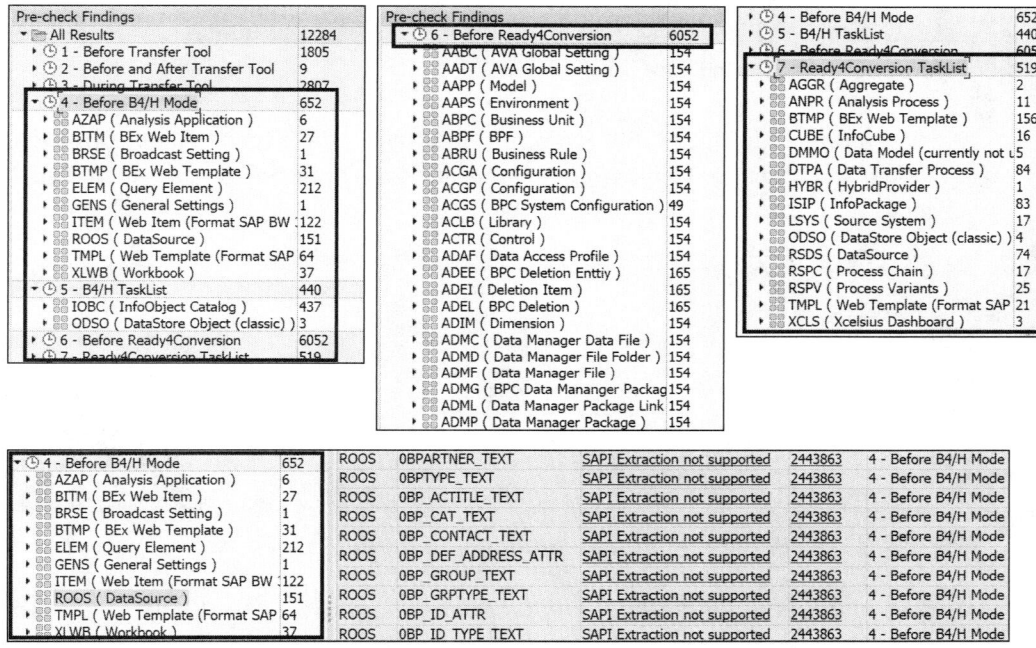

Figure 3.57 Other Modes for In-Place Conversion

Precheck for Remote Conversion

We'll now walk through the precheck steps for a remote conversion. From the screen shown earlier in Figure 3.49, follow these steps:

1. Choose the **Show Logs** option, and select **Remote** in the popup window, as shown in Figure 3.58. The resulting screen is shown in Figure 3.59.

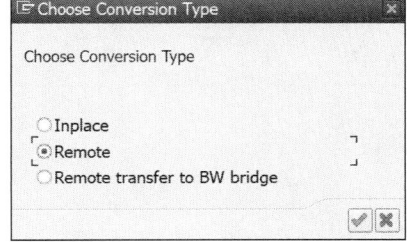

Figure 3.58 Remote Conversion

113

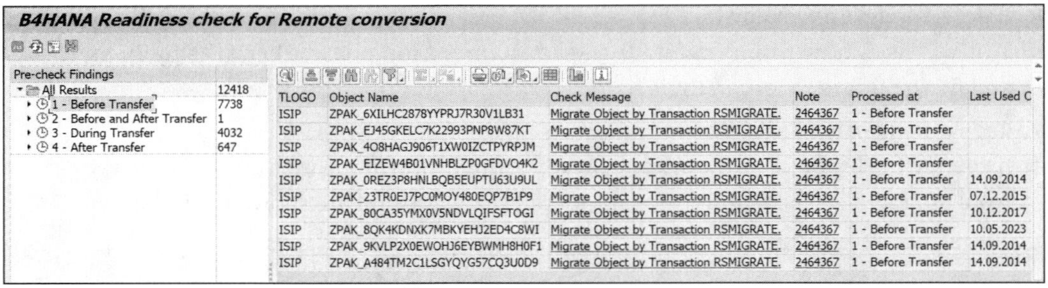

3   System Preparation and Prerequisites

Figure 3.59 Precheck Findings

2. As you can see in Figure 3.59, you'll have multiple options such as **Before Transfer**, **Before and After Transfer**, **During Transfer**, and **After Transfer**. Choose **Before Transfer** to see the findings shown in Figure 3.60.

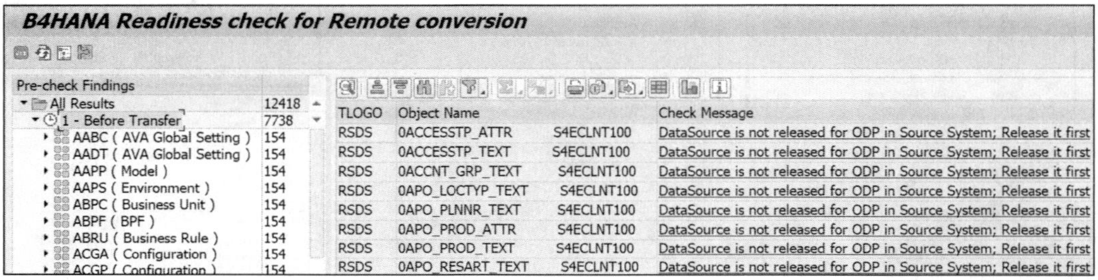

Figure 3.60 Before Transfer

For example, when you choose **During Transfer**, the details based on TLOGO are shown in Figure 3.61. You can then review them to ensure they are correct.

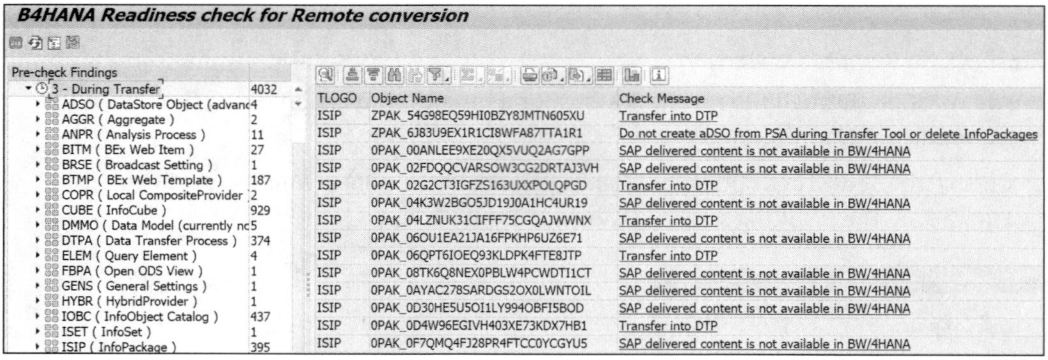

Figure 3.61 During Transfer

3.10.3   Fixing Issues in the Precheck Report

The most important part of the precheck is the part before conversion. You can change the format of findings as shown in Figure 3.62. Click on the Change Layout icon in

3.10 Prechecks for SAP BW/4HANA Migration

Figure 3.61. In the **Change Tree Grouping** popup, you can use the icon to bring the fields from **Available Groups** to **Selected Groups**, and vice versa. This will change the way the **Precheck Findings** are displayed; for example, if you send **TLOGO** from **Selected Groups** to **Available Groups**, you won't see data related to TLOGO.

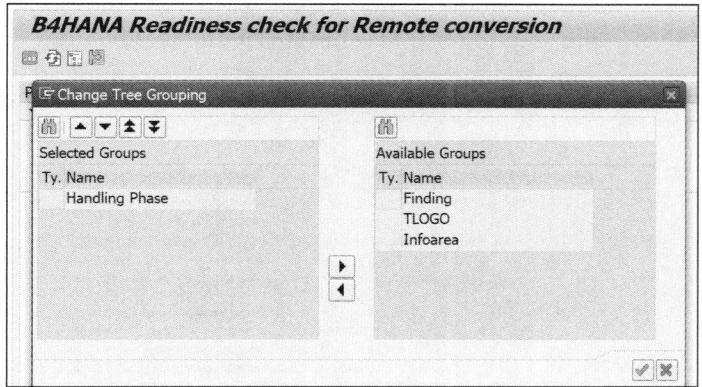

Figure 3.62 Format the Findings

Issues can occur for any SAP BW object. Once you click, you'll see the options under **Pre-check Findings**, as shown earlier in Figure 3.59. If necessary, you can expand the **Before Transfer Tool** option to see the issues and findings, as shown in Figure 3.63. You can see that the tool expects certain **Before Transfer** activities to be completed if you see results in **ISMP** that are related to the transfer rule. In the **Check Message** column, you can see **Migrate Object by Transaction RSMIGRATE**. To fix this issue, you need to migrate this transfer rule to transformation using Transaction RSMIGRATE.

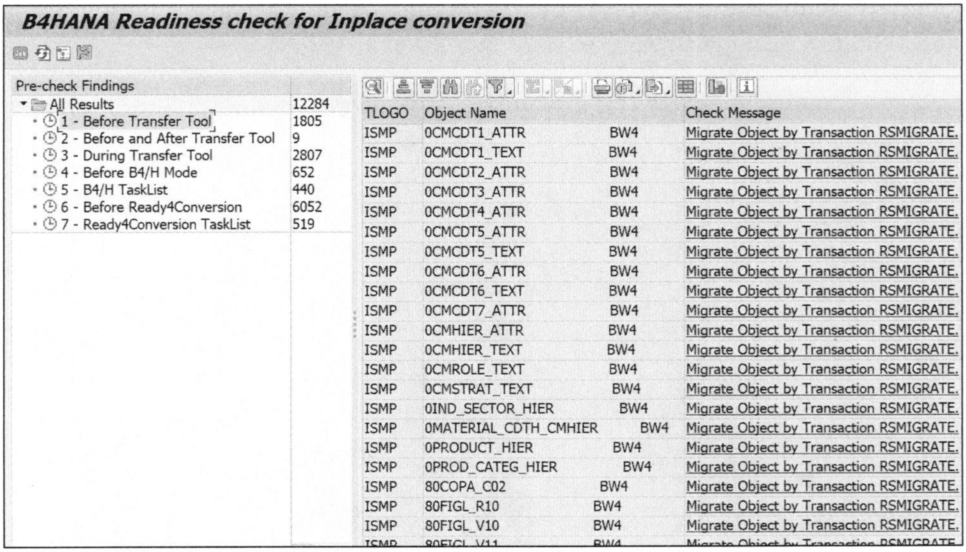

Figure 3.63 In-Place Conversion

115

3 System Preparation and Prerequisites

After you've fixed all the objects, then you won't see such errors. You need to migrate all SAP BW 3.x objects in the same way. You need to check the findings related to each TLOGO and repair them accordingly, You'll see references to specific SAP Notes for how to fix certain objects.

When you check the logs, you'll have the option to choose the remote conversion. Once you've selected that option, the screen will appear as shown in Figure 3.64.

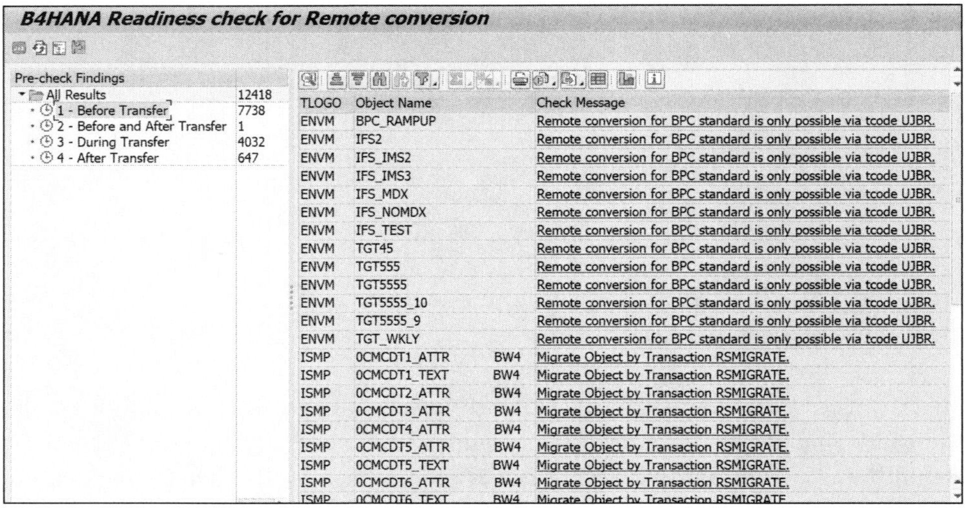

Figure 3.64 Remote Conversion

3.10.4 Sizing Report

You can execute the sizing report from the precheck report by choosing the **Run Sizing Report** button, as shown in Figure 3.65.

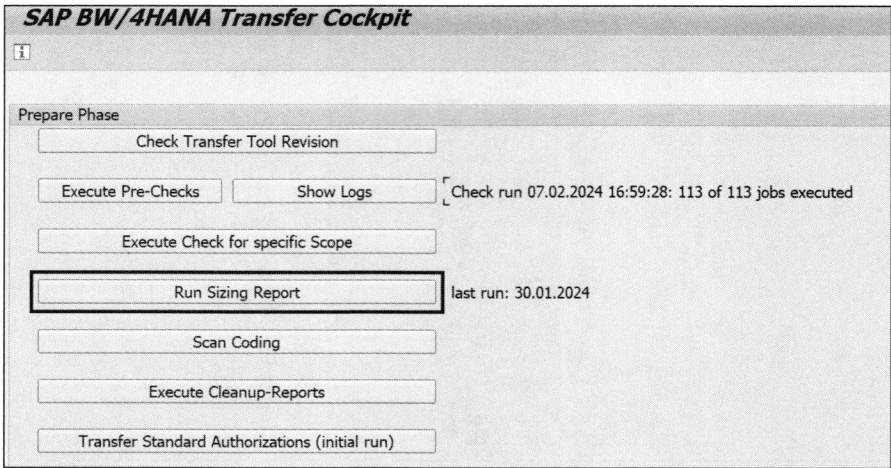

Figure 3.65 Sizing Report

3.10 Prechecks for SAP BW/4HANA Migration

This opens the screen shown earlier in Figure 3.11. Next, you can also see the option for **Scan Coding** in the precheck tool, which we'll discuss next.

3.10.5 Code Scan

With the **Scan Coding** option in Figure 3.65, you can assess the compatibility of ABAP code in the current system for SAP BW/4HANA, reducing potential errors during the actual conversion. The details of the code scan are given in Figure 3.66:

❶ Choose this if you need to find about **All Blacklisted ABAP objects**.

❷ Choose the staging objects that have ABAP code.

❸ Choose other objects.

❹ Provide the restrictions here.

❺ Delete the XREF entries and write XREF entries in remote conversions.

❻ View the logs based on the previous execution.

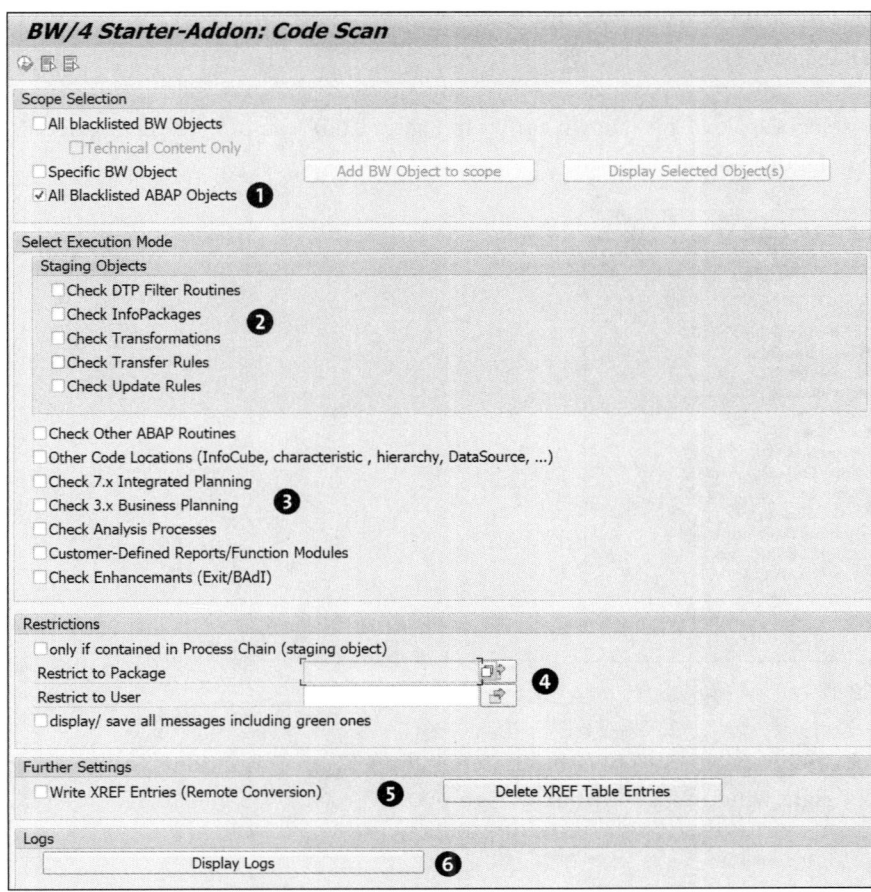

Figure 3.66 Code Scan Options

117

Once you choose **Display Logs**, you'll see the screen show in Figure 3.67.

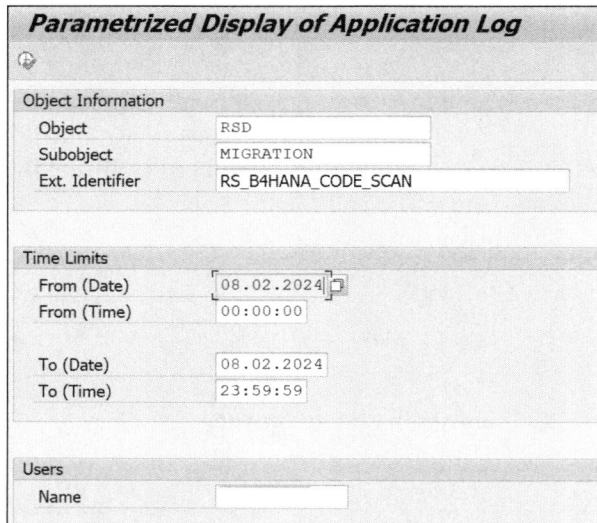

Figure 3.67 Display of Application Logs

After you select **Display Logs** (shown earlier in Figure 3.66), you can see the log details, as shown in Figure 3.68.

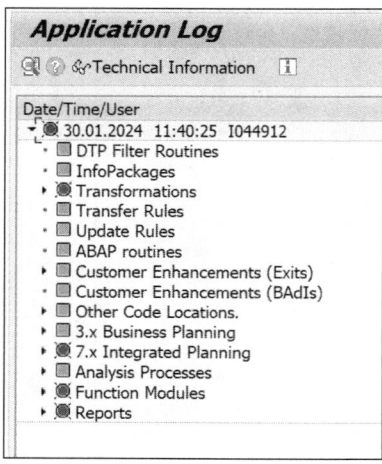

Figure 3.68 Details of Logs

If you click on the specific SAP BW objects, for example, **Transformations**, you'll see the error in the right window, as shown in in Figure 3.69.

3.11 SAP BW/4HANA Simplification List

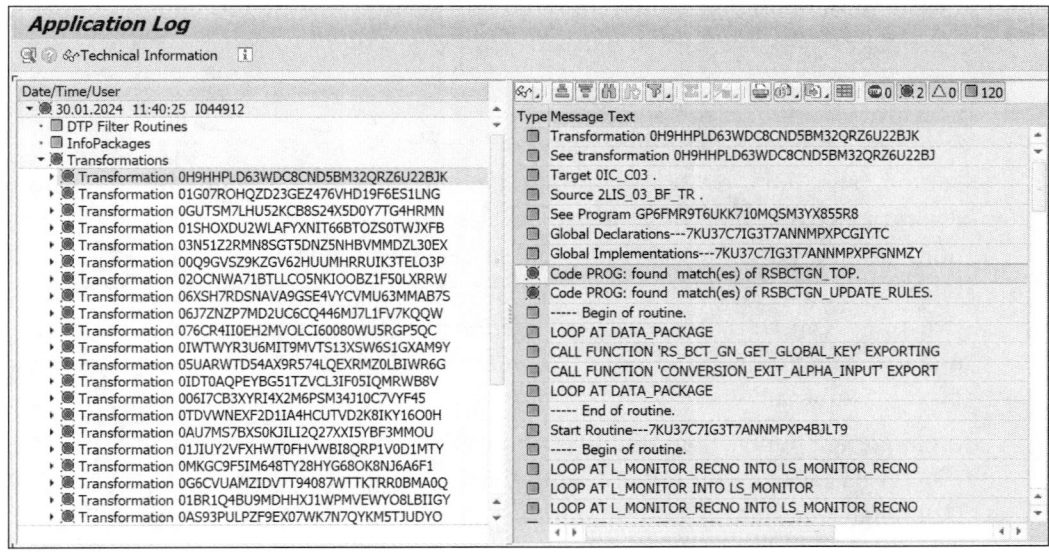

Figure 3.69 Detailed Logs of the Code Scan Tool

3.10.6 Cleanup Reports

From the precheck tool cockpit, you have the **Execute Cleanup-Reports** option. When you click this, you'll see the task list shown in Figure 3.70.

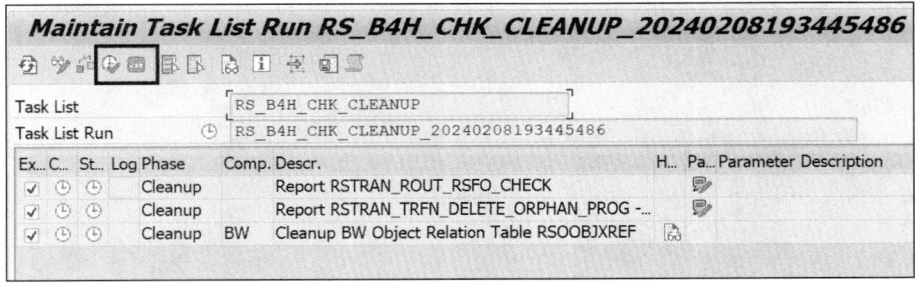

Figure 3.70 Cleanup Report

You can use the icon 🔽 to execute the task list and the 🔲 icon to start or resume the job. The other options shown here will be used in the realization phase, which we'll cover in later chapters.

3.11 SAP BW/4HANA Simplification List

When you execute the precheck tool using Transaction RSB4HCONV or using **ABAP Report** • **RS_B4HANA_CONVERSION_CONTROL**, you can see the list of logs by choosing the **Before Conversion** option. Those result can provide detailed insights on how the

119

specific object will be converted after conversion. In the following sections, we'll see the essential simplification list for SAP BW/4HANA.

3.11.1 SAP Business Warehouse 3.X Data Flows

For the migration to SAP BW/4HANA or SAP Datasphere via the SAP BW bridge from object types associated with SAP BW 3.x data flows, certain replacements and adaptations are necessary. Objects requiring replacement include update rules, SAP BW 3.x InfoSources, transfer rules, SAP BW 3.x DataSources, and transfer and communication structures. Objects necessitating adaptation include InfoPackages, process chains, process chain variants, and VirtualProviders based on SAP BW 3.x InfoSources (Cube Type = V, Cube Subtype = S).

It's important to note that the SAP BW 3.x data flow migration tool (Transaction RSMIGRATE) isn't available in SAP BW/4HANA. Initiate migration to SAP BW/4HANA or SAP Datasphere using SAP BW bridge starting with Transaction RSMIGRATE for SAP BW 3.x data flows. Manual migration is required for releases prior to SAP BW 7.3, excluding hierarchy DataSources and transfer rules. Complete migration from SAP BW 3.x to SAP BW 7.x data flows before using the SAP BW transfer cockpit for conversion. Ensure compatibility with SAP BW/4HANA or SAP Datasphere through the SAP BW bridge, as listed in Table 3.7.

SAP BW Object	Status in SAP BW/4HANA	Required Action	Reference SAP Note
- DataSource (SAP BW 3.x) - Transfer structure (SAP BW 3.x) - Communication structure (SAP BW 3.x) - InfoSource (SAP BW 3.x) - Transfer rules (SAP BW 3.x) - Update rules (SAP BW 3.x)	Not compatible with SAP BW/4HANA	- Migrate to SAP BW 7.x - Use Transaction RSMIGRATE tool	2470352
InfoPackage (SAP BW 3.x)	Not compatible with SAP BW/4HANA	Migration of SAP BW 3.x to SAP BW 7.x required	2464367
InfoPackage Real-Time Data Acquisition (RDA)	Not compatible with SAP BW/4HANA	Manual redesign	2447916

Table 3.7 Data Flow Simplification

SAP BW Object	Status in SAP BW/4HANA	Required Action	Reference SAP Note
Push InfoPackage for web service	Not compatible with SAP BW/4HANA	■ Manual redesign ■ **Write-Interface enable**d option for ADSO alternative in SAP BW/4HANA	2441826

Table 3.7 Data Flow Simplification (Cont.)

3.11.2 Core Data Warehouse Modeling Objects

The next simplification is related to the core data warehouse modeling objects, as shown in Table 3.8.

SAP BW Object	Status in SAP BW/4HANA	Required Action	Reference SAP Note
InfoObject Catalog	Not supported	Report available, handled during conversion	2442621
InfoObject	Supported	Handled in SAP BW/4HANA conversion cockpit	2464367
DataStore Object (DSO)	Not supported	SAP BW/4HANA conversion cockpit will change this to ADSO	2447916
InfoCube	Not supported	SAP BW/4HANA conversion cockpit will change this to ADSO	2443489
InfoSet	Not supported	Converted to HCPR	2444912
MultiProvider	Not supported	Converted to HCPR	2444718
VirtualProvider	Not supported	Manual redesign required	2444913
HybridProvider	Not supported	Manual redesign required	2442730
Aggregates	Not supported	Manual change required, drop aggregate	247050
Semantic groups	Not supported	Converted to ADSO by tool	2472609

Table 3.8 SAP BW Objects Simplifications

SAP BW Object	Status in SAP BW/4HANA	Required Action	Reference SAP Note
ADSO	Supported	Handled by tool	2487023
CompositeProvider	Not supported	Manual redesign required	2442062
InfoSpoke	Not supported	Manual redesign required	2437637
SAP HANA views (attribute and analytic)	Not supported	Manual change required	2442775

Table 3.8 SAP BW Objects Simplifications (Cont.)

Let's look at these in more detail:

- **InfoObject catalogs**
  SAP BW/4HANA no longer supports InfoObject catalogs; instead, they have been replaced by InfoAreas (AREA). To transition, use report RSDG_IOBJ_IOBC_MIGRATE_TO_AREA to migrate InfoObject assignments to InfoAreas. During the report execution, the system operates in an InfoArea-only mode. It's important to be aware that once the system is in this mode, the authorization object S_RS_IOBJA is evaluated instead of S_RS_IOBJ.

- **InfoObject**
  In SAP BW/4HANA, certain types of InfoObjects aren't supported, including characteristics with ORECORDMODE as an attribute, those enabled for real-time data acquisition, and characteristics using class CL_RSR_REMOTE_MASTERDATA as their master data read class. Moreover, key figures with Date or Time data types and aggregation SUM aren't available. It's important to note that the extraction of data from InfoObjects in SAP BW/4HANA to target systems can only be accomplished using ODP-BW, as export DataSources are no longer applicable.

- **DSO**
  For write-optimized DSOs incorporated in queries or MultiProviders, the transition involves transferring them into ADSOs with the **Keep inbound data, extract from inbound queue** option. It's crucial to note that these write-optimized DSOs must possess a semantic key; otherwise, the SAP BW/4HANA transfer tool will halt, necessitating the removal of the ODSO from the MultiProvider or the deletion of associated queries for the process to proceed. Additionally, the activation of data post-transfer doesn't occur automatically in this scenario. In the case of write-optimized DSOs not used in queries or MultiProviders, the migration results in their transfer to ADSOs without activation. This approach is designed to offer greater flexibility for write-optimized applications, particularly in corporate memory scenarios.

Standard ODSOs will be transferred into ADSOs with activation and change logs, ensuring compatibility with the updated structure. For ODSOs designed for direct update, the transfer tool facilitates their migration into ADSOs tailored for direct updates.

In the context of ADSOs, a noteworthy change involves the key composition, especially for compounding characteristics. Unlike in ODSOs, where compounding characteristics didn't have to be part of the key, the transfer tool will automatically check and adapt the key during the transfer, ensuring inclusion of compounding characteristics in the key.

Furthermore, the transfer tool streamlines the process by automatically replacing process chain variants related to classic ODSOs with advanced counterparts, ensuring a smooth transition and adherence to updated procedures. Specifically, it replaces variants such as Activate Data in DataStore Objects (Classic) and Deletion of Requests from Change Log of DSO (classic) with advanced variants such as Activate Data in ADSOs and Clean Up Old Requests in ADSOs.

There are certain restriction in DSOs that you need to maintain in your project process. In-memory ODSOs with the SAP HANA–optimized option can't be directly converted into ADSOs. Use report RSDRI_RECONVERT_DATASTORE to revoke this option before conversion. Note that field-based DSOs aren't supported in the SAP BW/4HANA conversion; manual remodeling is required for such scenarios.

For write-optimized ADSOs, it's essential to enable the **Activate Data** option to define a semantic key. If transferring a write-optimized DSO to such an ADSO without the semantic key, unexpected behavior may occur, especially if the semantic key was used in transformations such as currency conversion. Remodel this scenario manually by creating an InfoSource, defining the ODSO's semantic key as the InfoSource key, and establishing transformations accordingly.

When using a process variant for deleting old requests from changelogs or write-optimized DSOs (ADSODEL), be aware of restrictions on the corresponding variant for ADSOs. It's replaced by process type ADSOREM.

Additionally, the transfer of ODSOs with key figure aggregation type NO2 is only supported for direct update ODSOs. For other DSO types, manual redesign is necessary before initiating the transfer.

- **InfoCube**

  The SAP BW InfoCube isn't supported in SAP BW/4HANA, SAP Data Warehouse Cloud, and SAP BW bridge. Basic InfoCubes can be transformed into an ADSO with the **All characteristics are key, reporting on union of inbound and active table** option using the SAP BW/4HANA transfer tool.

  During the transfer, the tool automatically replaces the Compression of InfoCubes process chain variant with Clean Up Old Requests in ADSOs. Navigation attributes of ADSOs are exposed for staging processes only, and the tool generates a new HCPR to enable their usage in queries, as detailed in SAP Note 2185212.

For InfoCubes with noncumulative key figures and NLS, refer to SAP Note 2618005. Conversion of InfoCubes with an audit dimension is possible, but it's a technical conversion, resulting in the deletion of the audit InfoObject field.

However, there are restrictions; VirtualProvider or HybridProvider type InfoCubes can't be converted, as stated in SAP Notes 2444913 and 2442730. Zero elimination support for ADSOs starts with SAP BW/4 2.0 SP 04. Conversion of InfoCubes with key figure aggregation type NO2 isn't supported; manual redesign is necessary. SID materialization for ADSOs from InfoCubes might affect performance and can be enabled post-transfer. Additionally, partitioning methods differ between InfoCubes and ADSOs, with fields automatically determined but changeable after transfer. Unoptimized InfoCubes on SAP HANA must be SAP HANA–optimized using Transaction RSMIGRHANADB before conversion.

- **InfoSet**
  The SAP BW InfoSet functionality isn't available in SAP BW/4HANA, but a conversion path exists through the SAP BW/4HANA transfer tool, allowing transformation into a CompositeProvider. During this conversion process, it's important to note that InfoSet names of InfoObjects aren't directly supported in the resulting CompositeProvider. Consequently, an InfoObject mapping is required, and adjustments to queries, query views, and dependent objects defined on top of InfoSets are necessary. These adaptations can be efficiently handled using the SAP BW/4HANA transfer tool.

  However, several restrictions apply to the conversion process. InfoSets featuring Most Recent Reporting for InfoObjects or Left Outer: Add Filter Value On-Condition aren't compatible with the SAP BW/4HANA Transfer Tool. Similarly, InfoSets lacking Additional Grouping Before Join and Join Is Time-Dependent aren't supported. Additionally, InfoSets with only one PartProvider, those using navigational attributes from an InfoCube in combination with a temporal join, and those with fields based on InfoObjects compounded to referenced key figures face limitations in the conversion process and may not be supported by the SAP BW/4HANA transfer tool.

- **MultiProvider**
  The SAP BW MultiProvider isn't supported in SAP BW/4HANA; however, it can be transformed into a CompositeProvider using the SAP BW/4HANA transfer tool. There are specific cases where conversion may not be possible, such as when the **Accept Inconsistency During Compounding** setting is enabled or when navigation attributes from InfoObjects based on SAP HANA Model are mapped in a way that renders them unusable as navigation attributes.

  During the conversion process, an InfoObject mapping may be necessary. This is crucial because the CompositeProvider requires a consistent mapping of navigation attributes. It's essential to ensure that navigation attributes are mapped consistently with their corresponding characteristics. A new Compatibility Mode for the CompositeProvider is introduced, enabling the association of fields to navigation attributes, which significantly reduces the need for name changes related to inconsistent navigational attribute mappings in MultiProviders. However, it's important to note that the transferred CompositeProvider in Compatibility Mode has limitations—it's not reusable in other CompositeProviders and must not include joins.

Furthermore, InfoSet InfoObjects are no longer supported in the CompositeProvider, necessitating a new mapping if InfoSets are part of a MultiProvider. Any required InfoObject mapping automatically triggers adaptations to queries, query views, aggregation levels, and other dependent objects defined on top, and this adjustment process is facilitated using the SAP BW/4HANA transfer tool.

- **VirtualProvider**
  The SAP BW VirtualProvider isn't supported in SAP BW/4HANA, and there's no direct tool for replacing it. A manual redesign is needed. For VirtualProviders based on data transfer process (DTP) or business application programming interface (BAPI), it's recommended to remodel them as Open ODS views. In the case of VirtualProviders using a function module, consider remodeling the logic in an SAP HANA calculation view and embedding it into a CompositeProvider. For those based on SAP HANA Model, add the model directly as a PartProvider to a CompositeProvider, or use it as a source for an Open ODS view with necessary conversions.

- **HybridProvider**
  The SAP BW HybridProvider isn't supported in SAP BW/4HANA. Instead, consider using the ADSO as an alternative for the HybridProvider based on a classic DSO. For HybridProviders based on direct access, opt for the ADSO paired with a streaming process chain.

- **Aggregates**
  SAP BW/4HANA doesn't support aggregates. In-place conversions, starting with SAP BW powered by SAP HANA, should ideally have no aggregates in the system. However, if aggregates persist (refer to the Checklist Tool), they must be deleted before converting to SAP BW/4HANA. Use program RSDU_AGGR_GARBAGE_DEL_HDB to remove any remaining aggregates in SAP BW powered by SAP HANA. It's important to note that remote conversions simply disregard aggregates.

- **Semantic Partitioned Objects**
  SAP BW/4HANA doesn't support Semantically Partitioned Objects (SPOs). They have been allowed to be automatically converted to ADSOs and semantic groups since Q1/2018. Initially designed for performance in large SAP BW InfoProviders, SPOs are less crucial with SAP HANA. Still, using SPOs or semantic groups may have benefits for administration or operations.

- **ADSO**
  DTPs generated prior to SAP BW release 7.5 SP 4 store request IDs as metadata for each data load into ADSOs. However, SAP BW/4HANA ADSOs don't include request IDs in their metadata. For ADSOs that retain request IDs in the metadata due to old request management, it's necessary to perform a one-time processing using the scope transfer tool for metadata cleanup. This process is expedient as it solely addresses metadata without impacting transaction data.

- **CompositeProviders and analytical index**
  In SAP BW/4HANA, certain components from earlier versions, such as the central CompositeProvider from Transaction RSLIMO and the analytical index (AINX) created in the Analysis Process Designer (APD) or in Transaction RSDD_LTIP, are no longer supported. To adapt, users can manually create alternatives using the SAP BW modeling tools. For replacing an analytic index, a new field-based ADSO can be crafted. Meanwhile, to substitute a central CompositeProvider, users can manually create a new SAP HANA CompositeProvider (TLOGO HCPR) using the SP BW modeling tools. Additionally, there is support for automatic conversion using the SAP BW/4HANA transfer tool for central CompositeProvider migration.

- **SAP HANA attribute and analytical views**
  If your organization has an SAP HANA database with attribute and analytical views for data modeling, then you need to know that in SAP BW/4HANA, external SAP HANA views of the attribute or analytic view type aren't supported, leading to potential loss of references from other manually created SAP HANA views. To address this, implement SAP Note 2555138 and SAP Note 2659784. Cease activities in the SAP BW system that could trigger SAP BW object activation, and avoid using Transaction RS2HANA_ADMIN. Migrate generated external SAP HANA views of the attribute or analytic view type to calculation views using SAP HANA Studio. The Quick View provides a Migrate function that automatically adjusts references for other affected content objects (see SAP Note 2325817 for details). Activate the **Calculation View-Only** setting in the SAP BW backend, and use Transaction RS2HANA_ADMIN to repair manually migrated SAP HANA views, overwriting them while retaining references. For the step-by-step migration process, you can use SAP Note 2236064 as a reference. After migration, ensure that all external SAP HANA views are calculation views by adding an entry in table RS2HANA_VIEW_SET with OBJECT = RS2HANA_CALCVIEW_ONLY and VALUE = X. Execute the /$TAB command in the ABAP backend's command field to reset table buffers, applying the new setting. Alternatively, use Transaction SE24 to execute method SET_CALCVIEW_ONLY(X) if available in class CL_RS2HANA_VIEW_SETTINGS. You need to start the SAP BW/4HANA migration only after completing this activity.

3.11.3 Data Staging

Next, take a look at the simplifications in data staging, as summarized in Table 3.9.

SAP BW Object	Status in SAP BW/4HANA	Required Action	Reference SAP Note
InfoPackage and PSA	Not supported	Converted to DTP	2464367
InfoPackage group	Not supported	Manual conversion to process chain	2487598

Table 3.9 Data Staging Simplifications

3.11 SAP BW/4HANA Simplification List

SAP BW Object	Status in SAP BW/4HANA	Required Action	Reference SAP Note
DTP	Supported	Handled in the SAP BW/4HANA conversion cockpit	2464541
Transformation	Supported	Handled in the SAP BW/4HANA conversion cockpit	2479567
Real-time data acquisition (RDA)	Not supported	Manual change required	2447916
Generic and export DataSources	Not supported	Manual change required	2470315
InfoSource (7.x)	Supported	Manual pre- and post-activity required	2755139
Hierarchy DataSources	Partially supported	Manual changes required due to ODP	2480284
Open hub destination	Supported	Manual activity required	2469516
DAP and NLS	Not supported	Alternate: DTO in SAP BW/4HANA	2441922

Table 3.9 Data Staging Simplifications (Cont.)

Let's look at these further:

- **InfoPackage and PSA**
  In SAP BW/4HANA, the traditional components, such as the PSA service and InfoPackage, have been phased out. Consequently, the conventional method of loading data through InfoPackages is no longer supported. Instead, the primary mechanism for data loading is the use of DTPs, offering direct access to a variety of source system types without relying on a PSA.

  If PSA was previously employed in SAP BW for tasks such as data distribution, sorted access, or SAP HANA transformations, the transition involves introducing a write-optimized ADSO between the DataSource and the target ADSO. Additionally, the creation of an (ADSO) from the PSA becomes essential, especially when dealing with a Full-InfoPackage featuring distinct filter conditions or file names.

  The transfer tool provides flexibility by allowing users to either bypass the InfoPackage and directly create the DTP on the DataSource or transfer the PSA to an ADSO, thereby replacing the InfoPackage with a DTP. For scenarios where no ADSO is created from the PSA, the Full-InfoPackage is seamlessly merged into the Delta-DTP. In cases where an ADSO is derived from the PSA, different InfoPackages are systematically transferred to distinct Full-DTPs. Furthermore, Delta extraction from the

PSA-ADSO is facilitated through a dedicated Delta-DTP. The Full DTPI_s ID is generated by substituting the initial five characters of the InfoPackage ID with the prefix "DTPI_". The conversion process also generates a Delta DTPI for each data source featuring active Delta (initialization) ISIPs. The delta DTPI's ID incorporates a hash derived from the original Data Source and Source System names. The filter conditions of this delta DTPI align with all active InfoPackage (ISIP) delta initializations. During metadata conversion, the DTP is dynamically generated, incorporating this selection condition as a filter in an in-place transfer. In the case of a remote transfer, this filter setting is implemented during the request conversion of the DTP's target.

Replacing PSA with ADSO isn't applicable for hierarchy data sources. For one InfoPackage, its settings will transfer automatically to existing DTPs, which is extracting from the source. For multiple InfoPackages, implement SAP Note 3254923, and use report RSB4H_CONV_HIER_ISIPS_TO_DTP for in-place conversion. Adjust process chains and run the transfer tool. In remote conversion, apply SAP Note 3044301, and run the report after importing DataSources, transformations, and InfoObjects. Direct InfoPackage display isn't available in the target system.

The transfer tool faces limitations in handling specific InfoPackages, including those associated with SAP BW 3.x DataSources, Real-Time Data Acquisition (RDA), Push InfoPackages for Web Service Source Systems, and scenarios involving multiple Delta InfoPackages for File Source Systems reading different files or employing different routines to determine file names. For SAP BW 3.x DataSources, migration to SAP BW 7.x DataSources using Transaction RSMIGRATE is recommended. RDA considerations are covered in SAP Note 2447916, while Web Service Source Systems are addressed in SAP Note 2441826. Situations involving multiple Delta InfoPackages with variations in file sources require special attention as they're not supported due to their unique characteristics.

Note that InfoPackage groups aren't supported in SAP BW/4HANA, so if you have any in your landscape, they must be manually replaced with process chains.

- DTP

  With SAP BW/4HANA, the PSA is no longer supported, which impacts DTPs with error handling. The new Data Transfer Intermediate Storage (DTIS) technology, introduced in SP 04, covers DTP error stack functionality (excluding maintenance). Before system conversion, clear existing error stack tables by deleting requests or using an error DTP. The transfer tool identifies and removes error DTPs without a standard DTP or with error handling turned off. Refer to SAP KB Article 2494151 for more details on error DTP display and editing.

  Real-Time Data Acquisition DTPs and Direct Access DTPs aren't supported in SAP BW/4HANA. In these cases, an ADSO serves as a replacement for the PSA, enabling "extract once, deploy many" scenarios. The conversion tools in SAP BW/4HANA support this transition, as outlined in SAP Note 2464367 "BW4SL - InfoPackages and Persistent Staging Areas".

DTPs involving error handling and using SAP HANA runtime for transformations aren't supported in SAP BW/4HANA 1.0. To address this, SAP Note 2497506 introduces data packaging for DTPs based on ODP DataSources using semantic groups. This correction helps mitigate out-of-memory situations on the source system level. However, memory shortages may still occur in the SAP BW/4HANA target system. Alternatively, using an ADSO as the source object for DTPs employing semantic groups can replace the PSA, with the conversion tools facilitating this process (see SAP Note 2464367 "BW4SL - InfoPackages and Persistent Staging Areas").

- **Transformations**
  Transformations with the Read from DataStore (Classic) rule type are unsupported in SAP BW/4HANA. After conversion, these transformations can't be activated due to unspecified DataStore keys. Transformations with non-ABAP object-oriented (OO) compatible coding have limited functionality, accessible only in SAP GUI and the SAP BW modeling tools. There's no automated analysis for rule type usage. During migration, the transfer tool converts classic DSOs to ADSOs, adjusting transformation rules accordingly. Activating the Read from DataStore rules requires specifying all keys; otherwise, manual adjustment is needed. Recommended updates for noncompatible coding are advised to leverage new features in the SAP BW modeling tools for SAP BW/4HANA.

- **Real-Time Data Acquisition (RDA)**
  RDA is unavailable in SAP BW/4HANA. Instead, streaming process chains must be employed. There is currently no automated transfer process to replace RDA settings with streaming process chains. Manual steps are required. During this process, data can't be loaded in RDA mode. Initially, disable and remove RDA settings in the original SAP BW system, replacing them with standard SAP BW loading. Once completed, a standard process chain for delta data loading is established. Although this chain can be used, high-frequency loading is discouraged. To transition streaming, run the transfer tool, either in in-place or remote scenarios, for your process chain. This transfers the InfoPackage to DTP (if needed) and SAPI DataSources to ODP DataSources. The resulting chain can then be switched to streaming mode and scheduled accordingly.

- **Generic and Export DataSources**
  The SAPI for DataSources isn't supported in SAP BW/4HANA, leading to the unavailability of generic DataSources (maintained in Transaction RSO2), export DataSources (that starts with 8, generated from Transaction RSA1), and technical content DataSources (which has the prefix as OTCT, activated via Transaction RSA1). For generic extraction from tables or views, we recommend creating corresponding ABAP CDS views and using the source system type ODP with ABAP-CDS context in the target system. All relevant objects are accessible through the ODP context BW. When loading data from one SAP BW/4HANA system to another, or to a SAP BW system, choose the source system type ODP with BW context in the target system. Over

time, DataSources from technical content are gradually replaced with ABAP CDS views (namespace RV*), offering capabilities for real-time monitoring or extraction of system statistics.

- **InfoSource**
  When migrating from SAP BW to SAP BW/4HANA or SAP Datasphere using the SAP BW bridge, the transition involves reorganizing and reauthorizing InfoSources. In SAP BW, they are structured by application component, while in SAP BW/4HANA and SAP Datasphere, InfoSources are organized by InfoArea. Existing InfoSources are placed in the Unassigned Nodes InfoArea during the conversion, and they should be assigned to specific InfoAreas for security. The assignment is recommended pre-conversion for SAP BW 7.5 and post-conversion for earlier versions. Use the SAP BW modeling tools for efficient multiple InfoSource assignment to a new InfoArea.

- **Hierarchy DataSource**
  Hierarchy DataSources via ODP aren't supported in SAP NetWeaver 7.0x to 7.3x Source systems. The use of ODP-BW or ODP-SAPI requires specific SAP Notes based on the system version (SAP Note 2418250 for ODP-BW, SAP Note 2418250, and SAP Note 2768527 for ODP-SAPI). Alternative options include using the SAP BW source system type for loading hierarchies in SAP BW target systems (7.3x, 7.4x, or 7.5x) or loading hierarchies to a hierarchy InfoObject in SAP BW >= 740 SP 5 and using ODP-BW in SAP BW/4HANA. For SAP BW/4HANA remote conversion, note the specific restrictions on hierarchy DataSources with the IDoc transfer method in target releases >= 7.00 but < 7.30 (SAP Note 2469120). After conversion, reconnect hierarchy DataSources to InfoObjects in SAP BW/4HANA using a transformation and the ODP-SAPI source system.

- **Open Hub Destinations**
  Certain Open Hub Destinations are unavailable in SAP BW/4HANA, including SAP BW 3.x destinations (InfoSpokes) and those of the Third-Party Tool type. For database table destinations with a technical key, data deletion is necessary due to differences in storing data by Request ID in SAP BW/4HANA. Recommendations include deleting or replacing SAP BW 3.x or third-party tool destinations with SAP BW/4HANA-compatible features such as SDI. For database table destinations, either automatic conversion or manual truncation is required based on the target table's location in the same or another database schema. Truncating tables before switching modes is advised, especially for larger data volumes.

- **Data archiving and NLS**
  For organizations using SAP BW 3.x data archiving, that is, ADK-Based Archiving, archiving processes linked to SPOs, and write-optimized DSOs with keys in reporting, the migration to SAP BW/4HANA, SAP Datasphere, or SAP BW bridge, presents specific challenges. There is no dedicated tool for migrating the mentioned cases to ADSO-based archiving processes. The recommended approach is to reload all data

from the archives to the InfoProviders and then convert the InfoProviders to ADSOs. In SAP BW/4HANA, data archiving can be accomplished using the new concept, along with NLS or data tiering features such as Data Tiering Optimization (DTO). To minimize expenses, data archiving processes with a nonkey characteristic for time slice can be converted, preserving its use in the archive as the primary partitioning characteristic. However, creating new processes with this approach in SAP BW/4HANA isn't supported. Refer to SAP Note 2767207 for additional information on this scenario.

For SAP Datasphere and SAP BW bridge, the archiving functionality is currently unavailable, and there is no support for tiering options such as DTO or NLS. In SAP BW bridge, data is stored in ADSOs using the SAP HANA Native Storage Extension (NSE) concept. To prepare for transferring objects using archiving or other tiering options, cold data must be dropped or loaded back as hot/warm data, as the objects may otherwise be blocked from transfer.

3.11.4 Source Systems

Source system simplifications are listed in Table 3.10.

SAP BW Object	Status in SAP BW/4HANA	Required Action	Reference SAP Note
Myself source system	Not supported	Manual creation of ODP-BW source system	2479674
SAP source system	SAPI not supported	Manual creation of ODP-SAP source system	2473145
DB Connect	Supported	Use SAP HANA smart data integration (SDI)	2447932
UD Connect	Supported	Use SAP HANA smart data integration (SDI)	2441884
Data services	Not supported	Manually create ODP-HANA or data services	2441836
Web services	Not supported	Use SAP Cloud Integration for write interface DSOs	2441826
SAP BW source system	Partially supported	Manual creation of ODP-BW source system	2473145
Flat file	Supported	Create flat file source system manually	

Table 3.10 Source System Simplifications

Let's look at these in more detail:

- **Myself source system**
  The Myself source system is no longer supported in SAP BW/4HANA. To achieve similar functionality, it should be replaced with an ODP source system with SAP BW context. This replacement enables the transfer of data between InfoProviders within the SAP BW/4HANA system. During the switch to Ready for Conversion mode, any remaining Myself source systems will be automatically deleted.

- **SAP source system**
  The SAP and SAP BW source systems based on SAPI aren't supported in SAP BW/4HANA. To transfer data from SAP systems using extractors/DataSources (SAPI) to SAP BW/4HANA, configure an ODP source system with the SAP (Extractors) Context. Ensure ODP compatibility of DataSources by referring to SAP Note 2232584. Additionally, data can be transferred using other ODP contexts such as BW, SLT queue (for real-time replication), or ABAP CDS views (for SAP S/4HANA source systems). The SAP BW/4HANA transfer tool seamlessly copies DataSources from SAPI or SAP BW source systems to corresponding ODP source systems, ensuring delta queues' continuity. Implementation of specific notes in the source system is required, as advised by the Note Analyzer in SAP BW/4HANA. In cases where note installation isn't feasible, alternative measures may be necessary, such as stopping operations in the source system and performing an Init Simulation in SAP BW. Any remaining SAP source systems during the Ready for Conversion Mode switch will be automatically deleted.

- **DB Connect source system**
  DB Connect source systems aren't supported in SAP BW/4HANA. For systems using connection types One Logical System per DB Schema or SAP HANA Smart Data Access, replace them with SAP HANA source systems Local SAP HANA Database Schema or SAP HANA Smart Data Access. Execute program `RSDS_HANA_MIGRATE_DBCONNECT` for the default database connection. Manually replace source systems with connection type **Source System = DB Connection** using SAP HANA Smart Data Access as your SAP HANA source systems. Any remaining DB Connect source systems during the **Ready for Conversion Mode** switch will be automatically deleted. In shell and remote conversions, if the DataSource has the connection type **Source System = DB Connection**, you need to manually create an SAP HANA DataSource of an identical name and with identical fields manually in the target SAP BW/4HANA system before you transfer the corresponding dataflow.

- **UD Connect source system**
  UD Connect source systems aren't supported in SAP BW/4HANA. For primary use cases involving Java Database Connectivity (JDBC) connector types, manually replace UD Connect source systems with source systems of type SAP HANA. However, for other documented use cases of UD Connect in SAP BW versions below 7.30,

such as BI ODBO Connector, BI SAP Query Connector, BI XMLA Connector, or SDK_ JAVA, no replacement is available currently. Any remaining UD Connect source systems during the **Ready for Conversion** mode switch will be automatically deleted.

- **Data services source system**
  SAP Data Services and external system (partner ETL) source systems aren't supported in SAP BW/4HANA. In SAP BW/4HANA 2.0, a new write interface for DSOs replaces the PSA table push capability of DataSources in SAP Data Services. SAP Data Services versions 4.2 SP 12 and higher can use this interface, while support for SAP Integration Suite is planned. The Write Interface isn't externally accessible, limiting its use for ETL by SAP Partners. There's no automated conversion from DataSources in SAP Data Services to DSOs in SAP BW/4HANA 2.0. For manual conversion, create a new DSO based on the existing DataSource, ensuring data distribution from the PSA table before conversion. Any remaining SAP Data Services or External System source systems will be deleted during the Ready for Conversion Mode switch. After conversion, enable the **Write Interface** property in the DSO for use in SAP Data Services. In SAP BW/4HANA 1.0, consider using SAP Data Services or external systems to write to a native SAP HANA table and then use an SAP HANA source system to bring the data into SAP BW/4HANA.

- **Web Service source system**
  Web Service source systems aren't supported in SAP BW/4HANA. In version 2.0, the Write Interface for DSOs replaces the PSA table push for Web Service source systems. SAP Integration Suite and SAP Process Integration (7.50 SP 15 or higher) can use the new REST interface. No automated conversion is available, so you must manually create a DSO from the Web Service data source, distribute the PSA data, and switch to Ready for Conversion Mode. Any remaining Web Service systems are deleted. Post-conversion, enable the **Write Interface** property, and find URIs in the SAP BW modeling tools under **Template URIs**.

3.12 Summary

With the completion of this chapter, you're now aware of the system preparation and prerequisites when you're migrating to SAP BW/4HANA. You can now understand the impacts to your system conversion during the migration project. You're now aware of the supported product versions and required SP level in the sender SAP BW system and the target SAP BW/4HANA system. You've learned about data volume management and know how to perform SAP BW/4HANA sizing. You can now execute the SAP Readiness Check for SAP BW/4HANA and understand the findings. You've learned about the simplification in SAP BW/4HANA as well. In the next chapter, we'll focus on remote conversion to SAP BW/4HANA.

Chapter 4
Remote Conversion

This chapter outlines the detailed procedure for implementing SAP BW/4HANA remote conversion. Building upon the system preparation and prerequisites discussed in Chapter 3, this section emphasizes using remote conversion to convert objects and migrate data to the target SAP BW/4HANA system. Upon completing this chapter, you'll be proficient in executing SAP BW/4HANA remote conversion and gain a comprehensive understanding of task lists and associated activities in the system landscape.

SAP BW/4HANA remote conversion involves the conversion of the SAP Business Warehouse (SAP BW) system to SAP BW/4HANA along with data. In this chapter, we'll start with an overview of remote conversion in Section 4.1, followed by the SAP BW/4HANA target system provisioning steps in Section 4.2. Then, you'll learn about the SAP BW sender system provision for the conversion in Section 4.3. Preparing the SAP ERP or SAP S/4HANA source system is discussed in Section 4.4. Once all the systems are provisioned, the steps to prepare for the remote conversion are explained in Section 4.5. You'll learn about the basics of SAP BW/4HANA conversion cockpit in Section 4.6. Then, we'll discuss the SAP BW data model in detail, which will be used for the conversion demo in Section 4.7. Once the system landscape is ready, we'll create the SAP BW/4HANA remote conversion package in Section 4.8. Next, you'll learn the detailed process to execute the SAP BW/4HANA remote conversion with the step-by-step process in Section 4.9. The SAP BW system will be ready for go-live if you verify all details that have been mentioned in post-conversion in Section 4.10. Finally, Section 4.11 provides some useful reports and tables.

4.1 Overview of Remote Conversion

In this section, we delve into the fundamental aspects of remote conversion, building upon the groundwork laid in the previous chapter's preparation activities. Once the system is adequately primed for the SAP BW/4HANA conversion, the focus shifts to the realization phase where the actual conversion activities are initiated for SAP BW/4HANA; this phase is critical to the project's overall outcome. The management of system availability and downtime is paramount, so meticulous planning and execution are required throughout this phase. Figure 4.1 shows the realization phase activities in an SAP BW/4HANA remote conversion. We'll explore each activity in detail throughout this chapter.

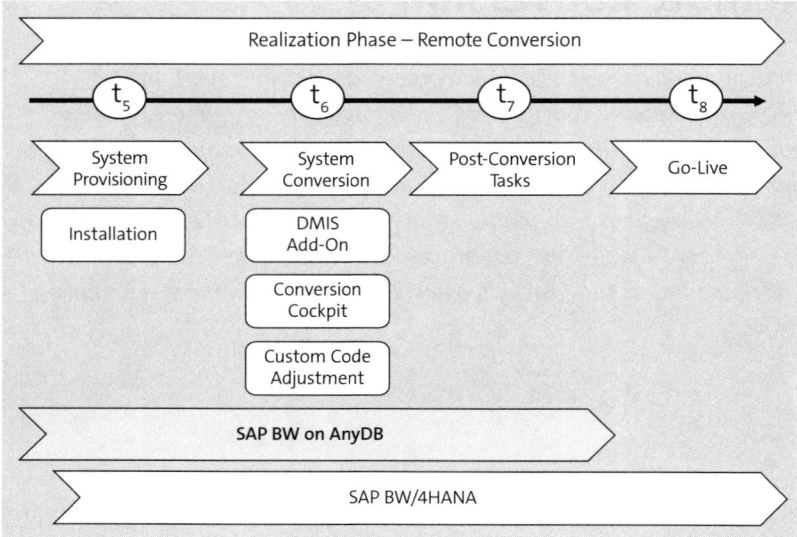

Figure 4.1 Remote Conversion: Realization Phase

4.2 SAP BW/4HANA Target System Provisioning

In the system landscape you're provisioning, you need to make sure you have a production landscape, which you can call N, and a project landscape, which you can call N + 1. In the following sections, we'll walk through the steps involved for SAP BW/4HANA target system provisioning.

4.2.1 Installing the SAP HANA Database

The initial stage of SAP BW/4HANA system provisioning involves the installation of the SAP HANA database. The sizing of the SAP HANA database is a critical consideration, and it's determined through a sizing exercise conducted in the prepare phase. Properly sizing the SAP HANA database is essential, considering the calculated year-over-year growth rate for accurate project planning and resource allocation. Once the size of the target SAP BW/4HANA based system is finalized, the organization will have the details for the scale-up or scale-out architecture. Scaling up enlarges your machine's RAM for processing, while scaling out combines multiple computers into a unified system, overcoming hardware limitations. In SAP HANA, distributing workloads across servers enhances performance, with each index server assigned to a dedicated host. Tables can be allocated to different hosts (database partitioning), and a single table can be split across multiple hosts (table partitioning).

In contrast to the appliance delivery method, SAP HANA tailored data center integration offers a more open and flexible approach. This approach is specifically designed to cater to the diverse needs of customers looking to integrate SAP HANA into their data center infrastructure. It provides a tailored solution that can be adapted to the unique requirements and preferences of each organization, offering greater flexibility in the deployment process.

When venturing into the cloud, SAP HANA presents itself as a comprehensive solution that includes both infrastructure and managed services. This means that organizations can leverage the power of SAP HANA without the burden of managing the underlying infrastructure. In the cloud space, SAP HANA provides various deployment options, including SAP HANA One, SAP Business Technology Platform, and SAP HANA Enterprise Cloud, allowing organizations to choose the solution that best aligns with their business objectives and IT strategies. Whether through a preconfigured instance with SAP HANA One, a more customizable platform with SAP HANA Cloud Platform, or a fully managed cloud service with SAP HANA Enterprise Cloud, SAP HANA offers a range of options to suit different business requirements and preferences in the dynamic landscape of digital transformation.

The SAP HANA, platform edition, is the technical foundation of the SAP HANA platform and various SAP HANA editions. The SAP HANA platform edition comprises SAP HANA database, SAP HANA Client, SAP HANA Studio, SAP HANA, extended application services engine, and others. SAP Note 2000003 "FAQ: SAP HANA" has detailed information related to SAP HANA.

To install SAP HANA, platform edition, follow these steps:

1. Go to *http://s-prs.co/v586702*, and choose **INSTALLATIONS & UPGRADES**, as shown in Figure 4.2.

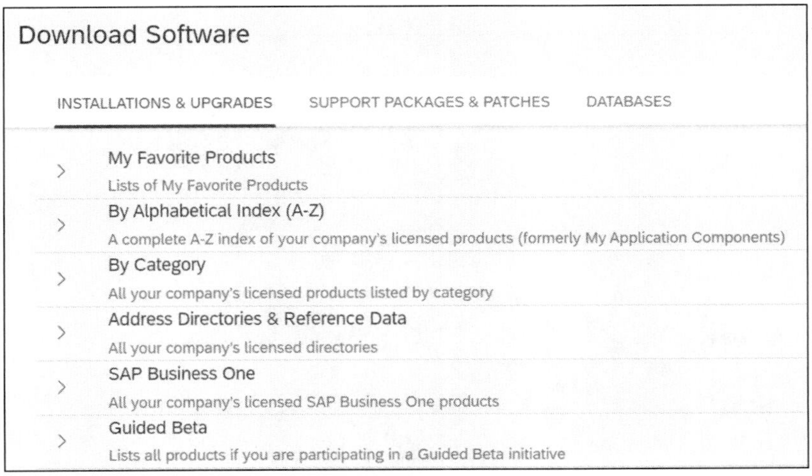

Figure 4.2 Software Center

2. Choose **By Alphabetical Index (A-Z)**, and select **H** to list the SAP HANA edition options, as shown in Figure 4.3.

Figure 4.3 SAP HANA, Platform Edition

3. Choose **SAP HANA PLATFORM EDITION**, and you'll see the options shown in Figure 4.4.

Figure 4.4 SAP HANA, Platform Edition 2.0

4. Choose **INSTALLATION** in Figure 4.4. Then, you'll see the list and have the option to choose the items available to download, as shown in Figure 4.5.

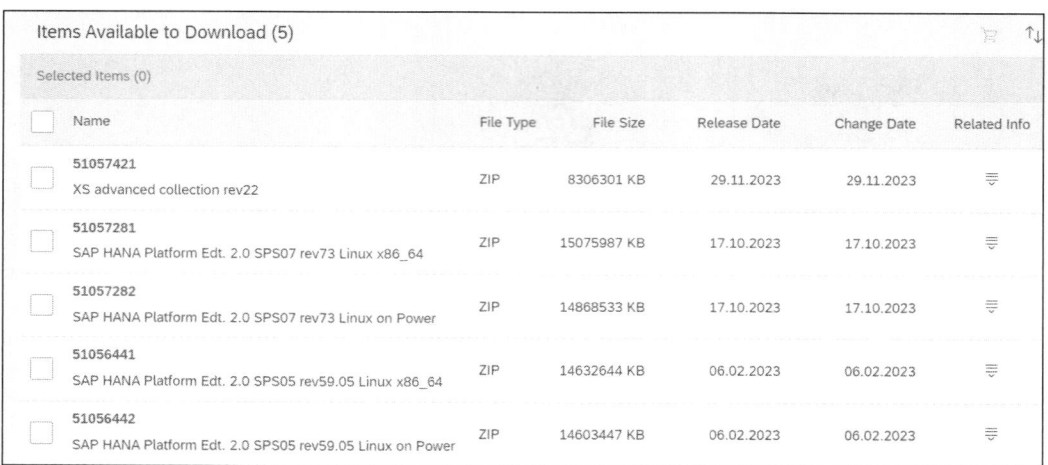

Figure 4.5 Download Option

5. Choose the SAP HANA Platform versions you want, tick the boxes, and add them to your download basket by clicking the cart icon. Open the basket to see your selected software. Install the download manager, enter your SAP credentials, and continue. Download the items individually or all together.

The responsibility for acquiring and installing SAP HANA depends on the chosen deployment model:

- SAP HANA tailored data center integration requires installation on validated hardware by a certified administrator or SAP HANA hardware partner. SAP HANA appliance installation is exclusive to certified hardware partners using validated hardware and a specific operating system (OS). SAP doesn't support systems or content developed with different systems. Support Package Stacks (SPS) can be downloaded and applied based on agreements with the hardware partner.
- As of SAP HANA 2.0 SPS 03, the SAP HANA database lifecycle manager tool (HDBLCM) can automatically verify that the software used to update the system is authentic and not tampered with. This can prevent various security attacks, for example, man-in-the-middle attacks, or detect changes to software packages stored on a common share, where more people have written access. The SAP HANA database lifecycle manager tool verifies the authenticity and integrity of the software packages prior to the installation or update.

To verify the authenticity of a SAP archive (SAR), use the following command:

```
/usr/sap/hostctrl/exe/SAPCAR -dVf <archive name> /usr/sap/hostctrl/exe/libsapcrypto.so
```

Because this deals with SAP HANA administration, we recommend you go through SAP Note 2577617 "Verifying Software Authenticity and Integrity during SAP HANA Installation and Update."

SAP HANA provides flexibility in the installation and updating of its components by offering two varieties of the HDBLCM tool. First, the SAP HANA *resident* HDBLCM, integral to the SAP HANA system, allows users to conveniently install and update components. Second, users can leverage the SAP HANA database lifecycle manager (HDBLCM), which comes bundled with the SAP HANA database installation kit. This versatile toolkit enables users to seamlessly manage the deployment and updates of SAP HANA components, providing a comprehensive and user-friendly solution tailored to the specific needs of the installation or updating process. With these options, users can choose the method that best aligns with their preferences and requirements, ensuring a smooth and efficient management of SAP HANA components throughout their lifecycle.

Detailed information about the supported OSs for SAP HANA can be found in SAP Note 2235581. To run SAP HANA, it's necessary to have one of the specified versions of the following Enterprise Linux distribution products: Red Hat Enterprise Linux for SAP Solutions, Red Hat Enterprise Linux for SAP HANA, SUSE Linux Enterprise Server for SAP Applications, or SUSE Linux Enterprise Server.

SAP strongly recommends using Red Hat Enterprise Linux for SAP Solutions or SUSE Linux Enterprise Server for SAP Applications due to their robust features and extended support cycles. For a more in-depth understanding of the Linux product flavors, their unique feature sets, and associated benefits, get in touch with your respective Linux sales representative. This ensures a well-informed decision in selecting the most suitable OS for SAP HANA. The details from PAM are shown in Figure 4.6.

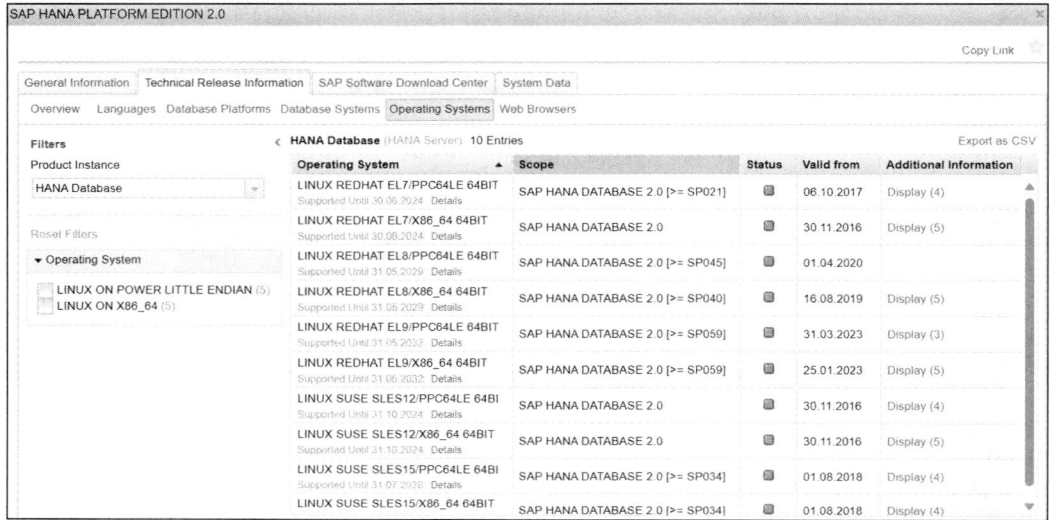

Figure 4.6 SAP HANA Supported OSs

Selecting the correct support package is crucial during the installation of SAP HANA. Consistently opt for the latest SP level as it aids in preventing known issues, as shown in Figure 4.7.

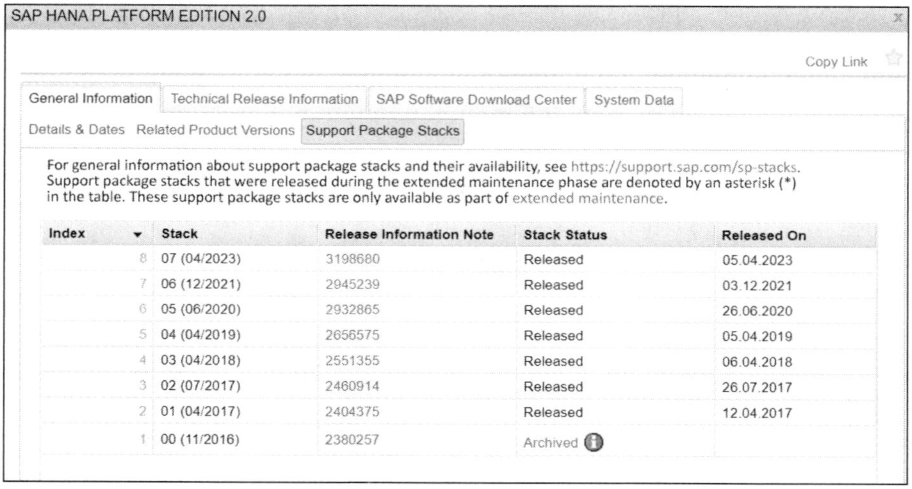

Figure 4.7 Support Packages Available as of February, 2024

4.2.2 Installing SAP BW/4HANA

Once you've installed the SAP HANA database, the next step is to install SAP BW/4HANA. You need to use maintenance planner for this activity. To generate the *stack.xml* file and download all the necessary software packages for SAP BW/4HANA, you must complete a maintenance task transaction. Follow these steps to install SAP BW/4HANA:

1. Open the maintenance planner from SAP Portal at *http://s-prs.co/v586703*.
2. Choose **Access Maintenance Planner**, and you'll be redirected to the homepage of maintenance planner, as shown in Figure 4.8. Click **Plan a New System**.

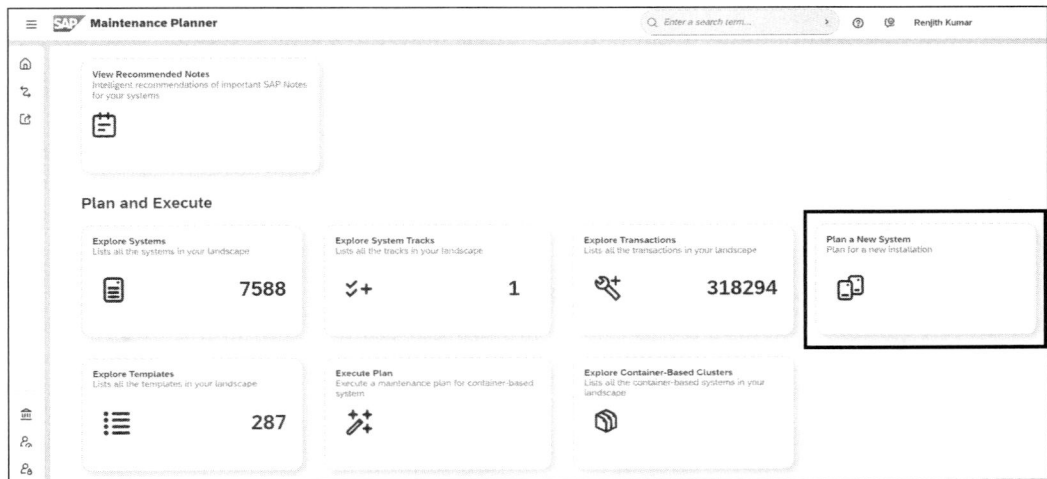

Figure 4.8 Maintenance Planner Home Screen

3. On the screen that appears, select **Plan**, as shown in Figure 4.9.

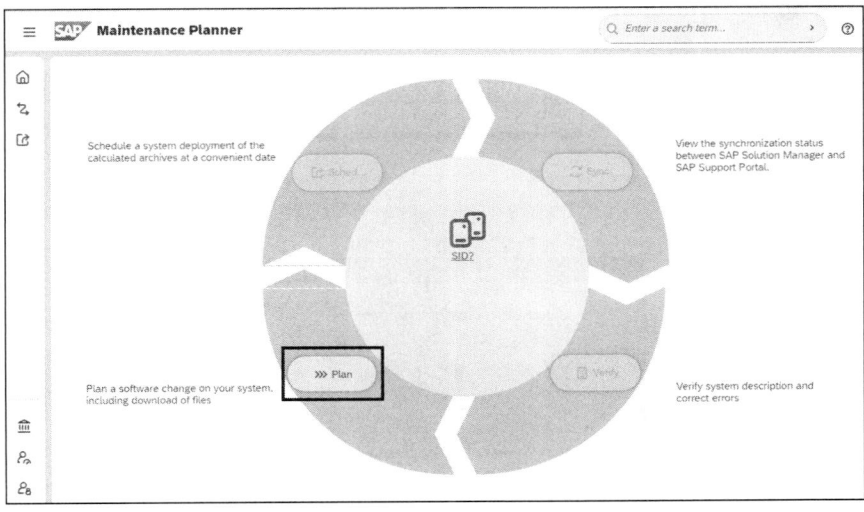

Figure 4.9 Planning the System

4. In the next screen, enter your system ID in the **SID** input box, as shown in Figure 4.10.

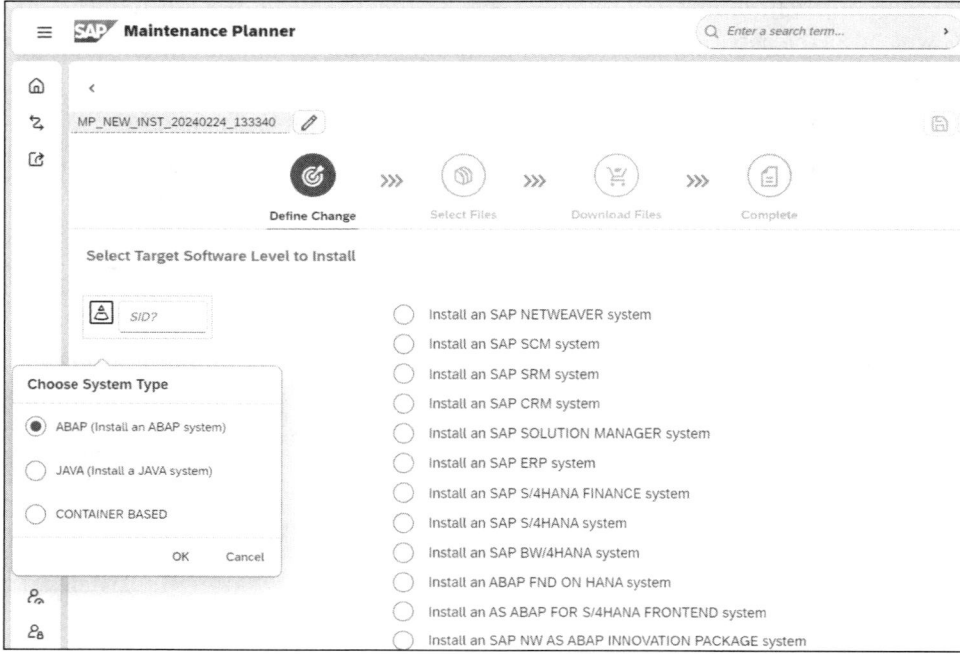

Figure 4.10 System ID Selection

5. For our demo, we'll enter the SID as "B4H", as shown in Figure 4.11. Next, choose the **Install an SAP BW/4HANA System** radio button for the SAP BW/4HANA versions.

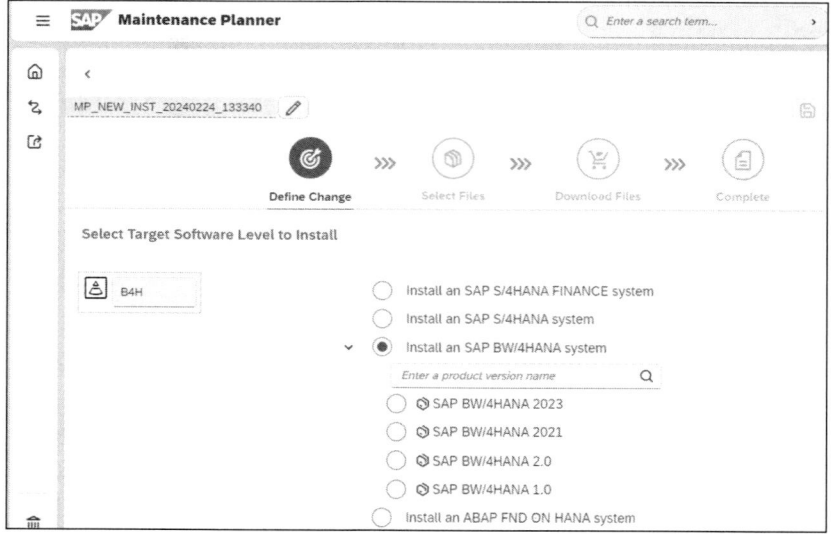

Figure 4.11 System ID and Installation Details

4.2 SAP BW/4HANA Target System Provisioning

6. The target SAP BW/4HANA version is determined according to your business plan, so choose the relevant version. In our case, we'll choose **SAP BW/4HANA 2023**, as shown in Figure 4.12.

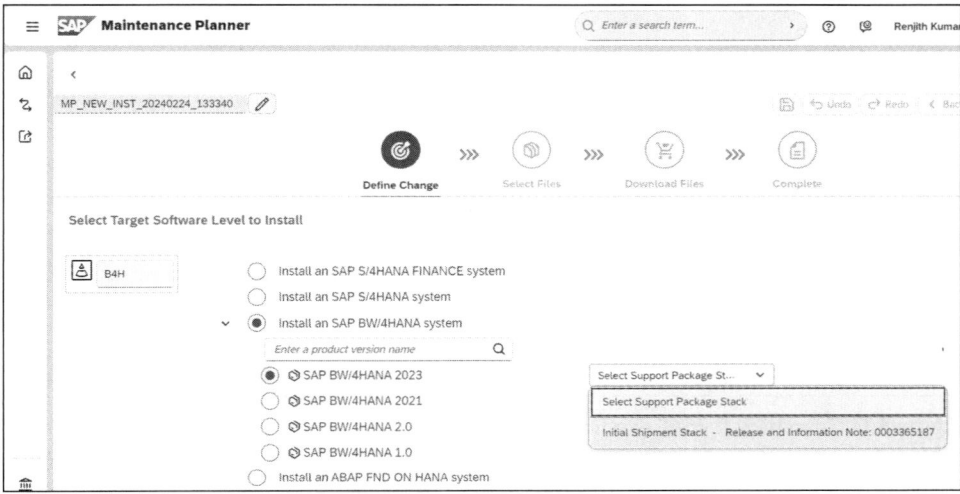

Figure 4.12 Target SAP BW/4HANA Version

7. Select the option based on your target SAP BW/4HANA version. In our demo, we chose **SAP BW/4HANA 2023**. You can also other relevant checkboxes such as **BW/4HANA Content** and **BW/4HANA Content Basis** based on your business needs. Once you have made the required selection then click **Confirm Selection**, as shown in Figure 4.13.

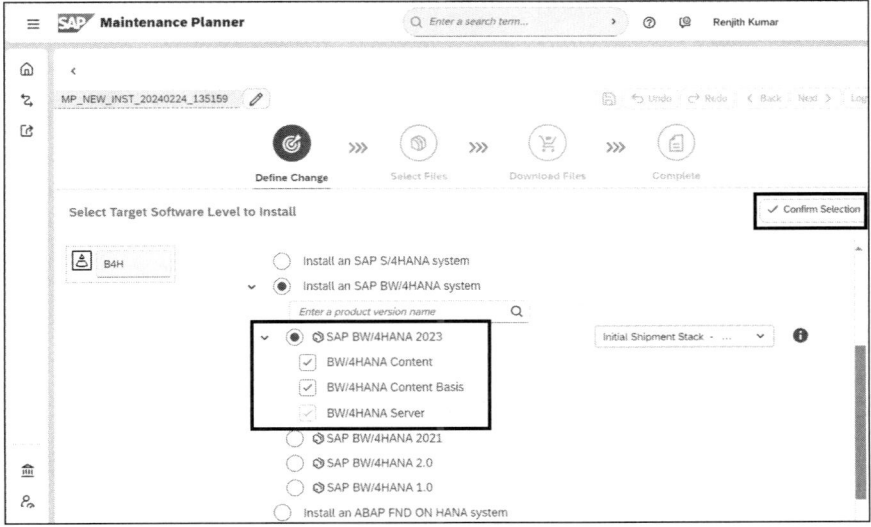

Figure 4.13 Confirm Selection

8. You'll then move to the next step, **Select Files**. Choose the **Install or Maintain an Add-on** option, and search for "LT", as shown in Figure 4.14.

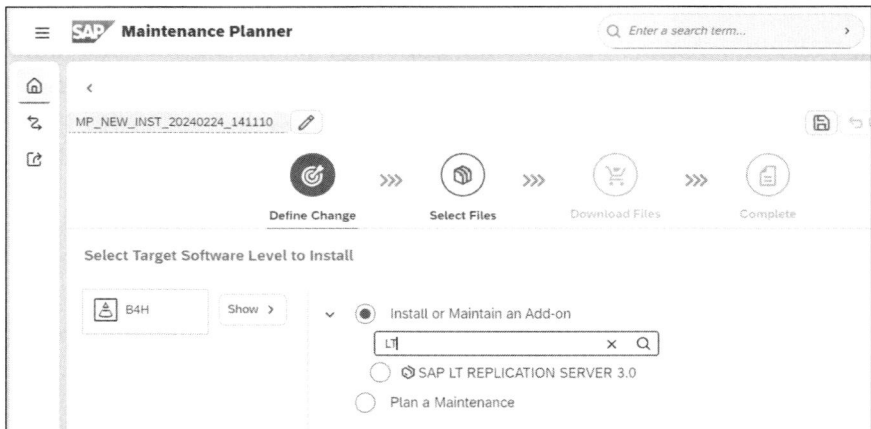

Figure 4.14 Searching for SAP Landscape Transformation

9. Select the **SAP LT REPLICATION SERVER 3.0** radio button, as shown in Figure 4.15. You'll see the list under **Select Support package Stack**; it's best practice to go with the latest **SPS**.

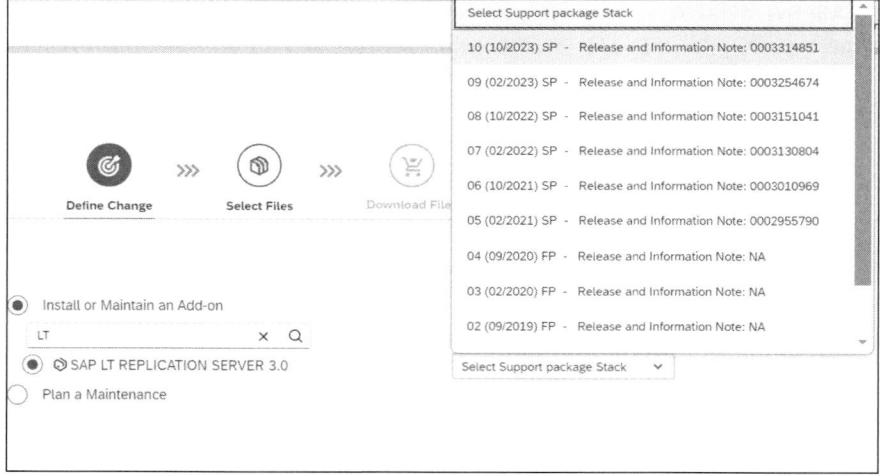

Figure 4.15 Support Packages

10. Choose **10 (10/2023) SP** to see the details, as shown in Figure 4.16. You can see that only SP **10** is now visible
11. Click **Confirm Selection**, as shown in Figure 4.17.
12. You'll then be able to see the **Target Software Detail**, as shown in Figure 4.18. Once you've reviewed all the details, choose **Next**.

4.2 SAP BW/4HANA Target System Provisioning

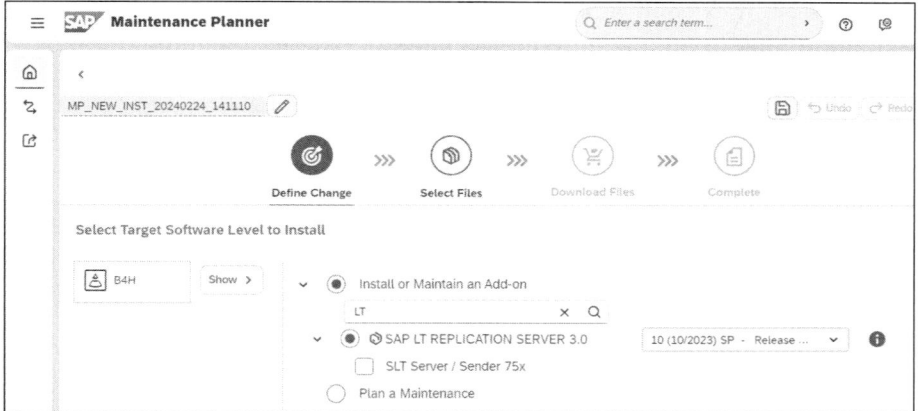

Figure 4.16 SAP Landscape Transformation Server and Sender

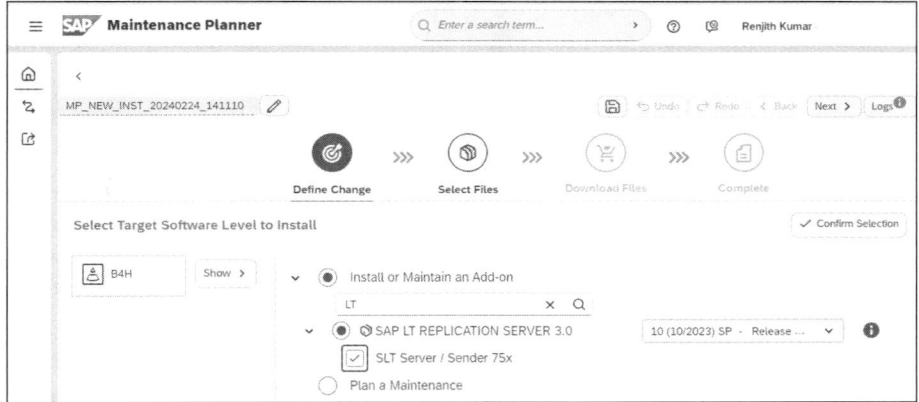

Figure 4.17 Confirm Selection

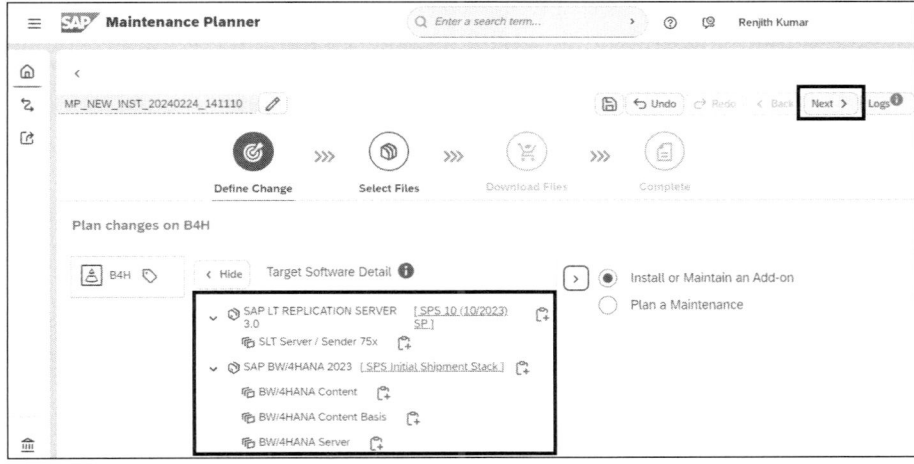

Figure 4.18 Target Software Details

13. You'll now see the option to **Select OS/DB dependent files**. Choose the desired OS and database based on your system landscape and project planning, as shown in Figure 4.19, and then click **Confirm Selection**.

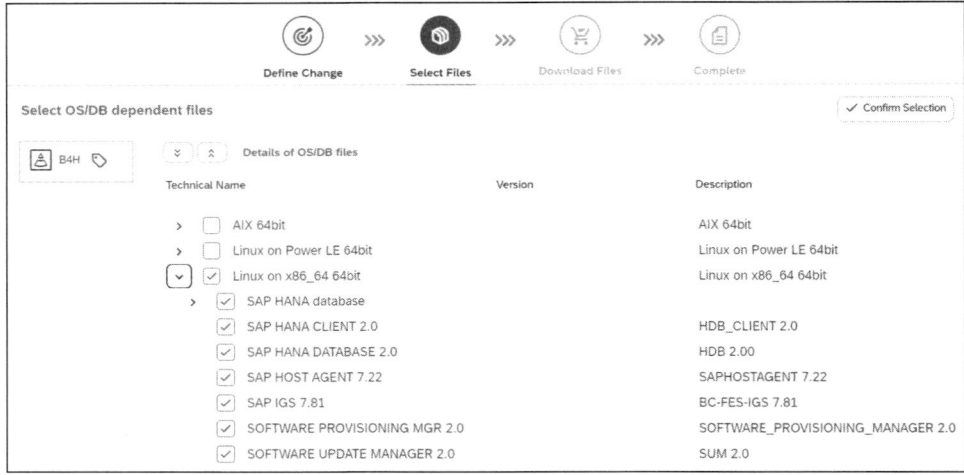

Figure 4.19 OS Files Selection

14. You'll be presented with the options shown in Figure 4.20. Choose the relevant files under **Select Stack Dependent and Independent Files**.

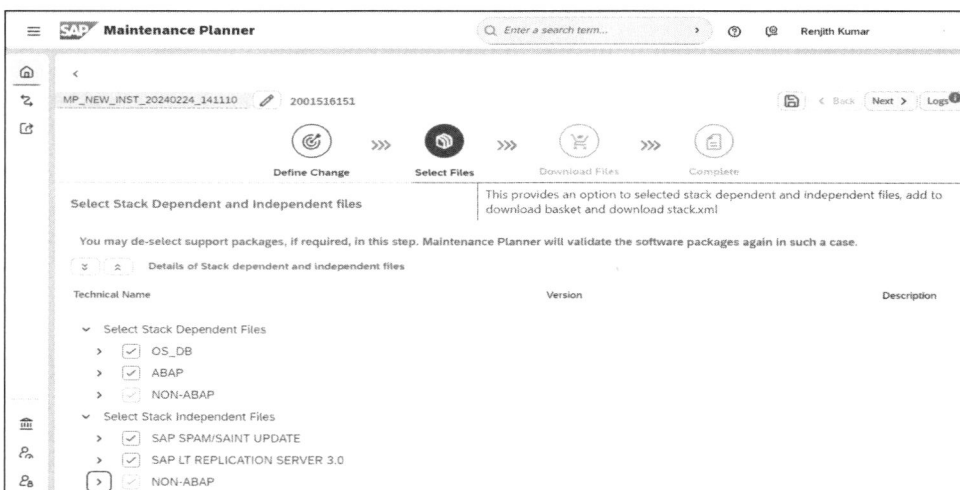

Figure 4.20 Confirm Selection

15. We'll use **SAP LT Replication Server 3.0**, which has **DMIS** underneath (see Figure 4.21). Choose **Next**.

4.2 SAP BW/4HANA Target System Provisioning

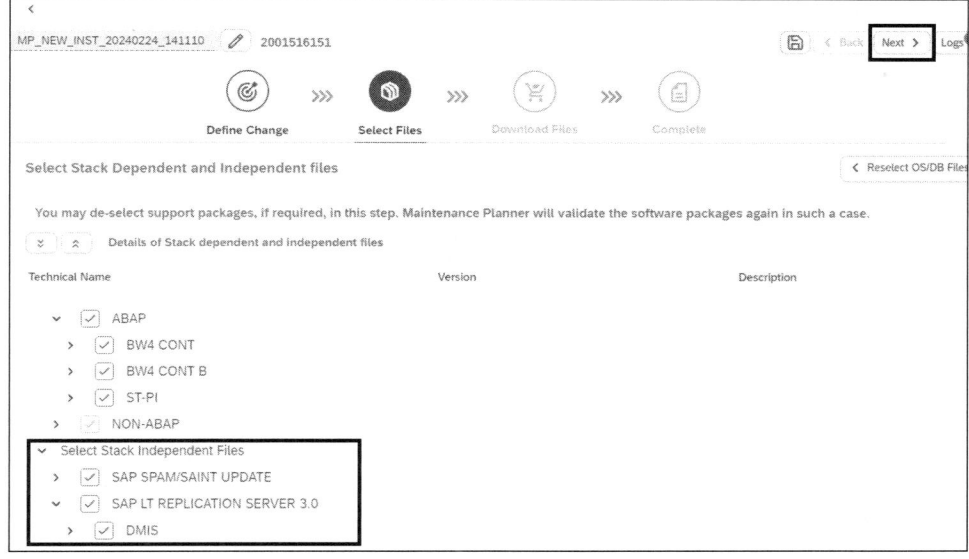

Figure 4.21 DMIS Available under SAP LT Replication Server 3.0

16. On the next screen, as shown in Figure 4.22, click **Download Stack XML**, and download the selected packages using SAP Download Manager. You can then use SAP Download Manager again to export the files to your system landscape. Unpack the SAR file to your EPS inbox.

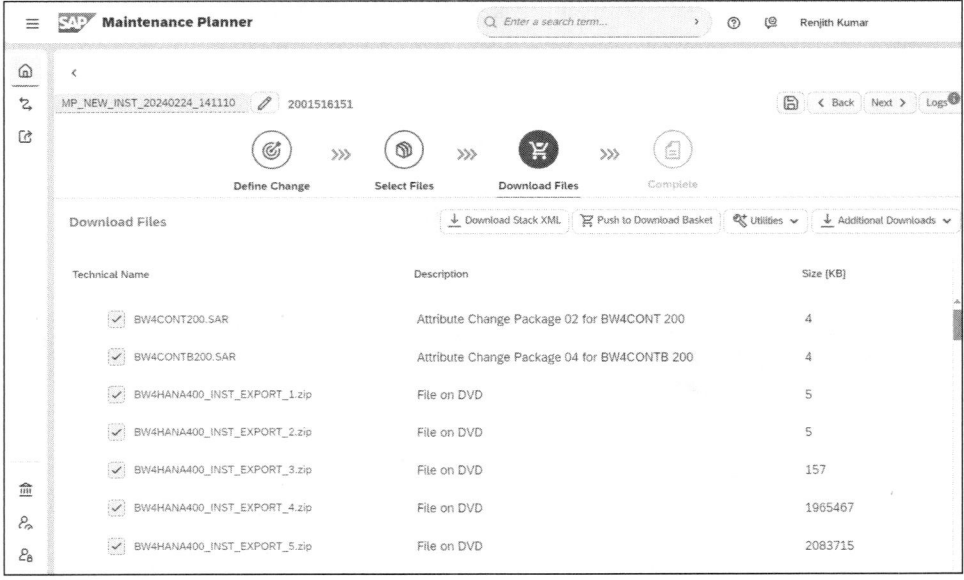

Figure 4.22 Download Stack XML

17. You can see **DMIS** listed on this screen as well, as shown in Figure 4.23. Once you've checked all the download files and completed the **Download Stack XML** step, you can choose **Next** to complete the maintenance transaction.

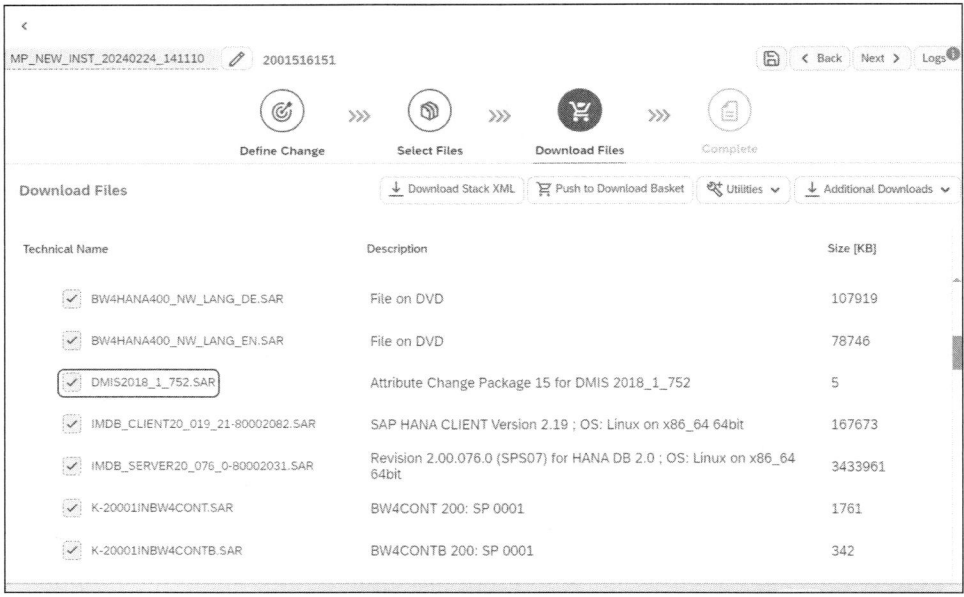

Figure 4.23 Locating DMIS

18. You'll see the screen shown in Figure 4.24, where you can download the file as a PDF by clicking **Download PDF**.

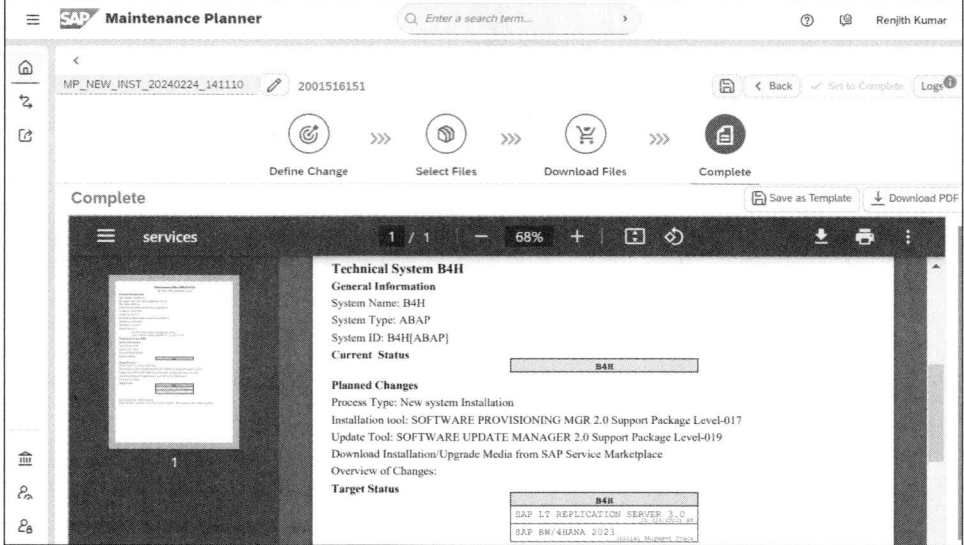

Figure 4.24 The Complete Screen

19. After you have the downloads from *Stack.XML* and validated the information per your project plan, we're ready for the installation. You need to have the Basis team ready to install SAP BW/4HANA using Software Provisioning Manager (SWPM 2.0). Install the required components for SAP BW/4HANA version. Install the SAP BW/4HANA content add-on, and ensure you have the latest kernel. Then, perform the SAP HANA Server installation. Finally, you can apply the latest SPs on top of SAP BW/4HANA using Software Update Manager (SUM).

The SAP BW/4HANA 2021 OS (application server) has a minimum requirement of SLES 15 or RHEL 8. Similarly, the minimum requirement for the database server is SLES 15 or RHEL 8. The minimal supported SAP HANA release for SAP BW/4HANA 2021 is SAP HANA 2.0 SP 05 Rev. 56. For installation and upgrades, SAP BW/4HANA 2021 uses SUM 2.0 and SWPM 2.0, both optimized for the SAP HANA database. Remote conversion to SAP BW/4HANA 2021 is supported with SAP BW/4HANA 2021 SP 0 or higher (DW4CORE 3.00 SP 0 or higher). SAP BW/4HANA 2021 contains component UIBASO01 700.

The essential requirements for the SAP BW/4HANA 2023 OS (application server) include a minimum of SLES 15 or RHEL 8. For additional details on supported OSs, refer to the Product Availability Matrix (PAM) for SAP BW/4HANA 2023 under **Essential Information** and the **Product Planning Matrix**. The minimum requirement for the database server is SLES 15 or RHEL 8, and the SAP HANA release supported by SAP BW/4HANA 2023 is a minimum of SAP HANA 2.0 SP 07 Rev. 71. Installation and upgrades for SAP BW/4HANA 2023 involve the use of SUM 2.0 and SWPM 2.0, both optimized for the SAP HANA database. Remote conversion to SAP BW/4HANA 2023 is supported with SAP BW/4HANA 2023 SP 0 or higher (DW4CORE 4.00 SP 0 or higher). SAP BW/4HANA 2023 contains component UIBASO01 758.

In the SAP BW/4HANA installation process, SWPM is a vital tool for the seamless installation and management of SAP software components. Embedded within SWPM is the robust SAPInst framework and a user-friendly browser-based interface known as SL-UI (Software Provisioning Manager User Interface). SL-UI leverages SAPUI5, a client-side HTML5 rendering library built on JavaScript, creating a link between the web browser and the SAPInst executable on the installation host through HTTPS. Notably, SL-UI can initiate automatically if the host has a compatible web browser installed, adhering to the default browser settings. This integration ensures an efficient and user-friendly experience throughout the deployment and maintenance processes of SAP BW/4HANA.

SWPM automates critical tasks such as system landscape configuration, software component installation, and post-installation procedures. This automation ensures a seamless and efficient deployment of SAP BW/4HANA. When working in conjunction with tools such as SUM, SWPM becomes instrumental in managing the entire lifecycle of SAP software. It encompasses not only the initial installation but also subsequent upgrades and ongoing maintenance, making the process comprehensive and effective. Post-installation, you'll see the components listed in Table 4.1.

Component	Description
SAP_BASIS	SAP Basis component
SAP_ABA	Cross-application component
SAP_GWFND	SAP Gateway foundation
SAP_UI	User interface technology
ST-PI	Solution tools plug-in
BW4CONT	SAP BW4/HANA content add-on
BW4CONTB	SAP BW4/HANA content Basis add-on
DW4CORE	SAP Data Warehouse
UIBAS001	UI for Basis applications
BPC4HANA	SAP Business Planning and Consolidation for SAP S/4HANA
DMIS	Data migration integration server
ST-A/PI	Service tools for SAP Basis 731 and higher

Table 4.1 Components of SAP BW/4HANA

20. The next step is to install standalone engines such as SAP Web Dispatcher and the standalone gateway. SAP Web Dispatcher plays a role as an intermediary between your web browser and the SAP system, managing requests for a web application. It forwards incoming requests (HTTP, HTTPS) to the appropriate application server (AS) in the SAP system. The capacity of SAP NetWeaver AS ABAP depends on configured dialog work processes, while an SAP NetWeaver AS Java's capacity depends on server processes. SAP Web Dispatcher ensures that, for stateful applications, your requests are directed to the server managing your application using a session cookie for HTTP connections and the client's IP address for Secure Sockets Layer (SSL) connections.

Moreover, SAP Web Dispatcher makes decisions about whether incoming requests should be sent to an ABAP or Java server. Additionally, it ensures proper forwarding by using session cookies and client IP addresses. It's important to note that the BUCHANAN SWPM GUI, which facilitates these processes, requires a compatible web browser. Ensure you have at least one of the following web browsers installed on the host: Microsoft Internet Explorer 11 or higher, Microsoft Edge, Mozilla Firefox, or Google Chrome. Always use the latest version of these web browsers for optimal performance and compatibility with SAP Web Dispatcher.

4.2 SAP BW/4HANA Target System Provisioning

21. The final step is to install the GUI and frontend tools. Then, you need to have SAP HANA Studio and Eclipse for the ADTs (ADT) and SAP BW modeling tools. Once the SAP BW modeling tools is installed, you need to add the SAP BW system to the SAP BW modeling tools. You can connect your SAP BW/4HANA system to the SAP BW modeling tools using your logon credentials.

4.2.3 Post-Installation Steps

Before starting your tasks within SAP BW/4HANA, it's imperative to undergo critical preparatory steps in SAP BW/4HANA system after installation. We'll walk through these steps in the following sections.

SAP_BW4_SETUP_SIMPLE Initial Setup

Following the installation process, initiate the SAP_BW4_SETUP_SIMPLE task list through Transaction STC01. This task list is a comprehensive mechanism for executing essential system setup activities.

These tasks include the creation and configuration of the background user dedicated to SAP BW/4HANA. Additionally, the task list involves specifying the SAP BW/4HANA client, which is necessary for seamless system operation. Furthermore, it encompasses the installation of technical content, predominantly consisting of InfoObjects and variables. These components are important parts of SAP BW/4HANA functionality and ensure a well-structured foundation for your subsequent endeavors within the platform.

In the SAP BW/4HANA system, open Transaction STC01, as shown in Figure 4.25.

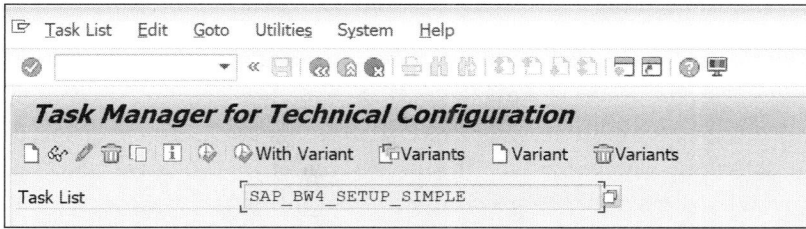

Figure 4.25 SAP BW/4HANA Initial Setup

You can see the following details of the task list in Figure 4.26:

❶ Task list run number
❷ Task list item and description
❸ Parameters for the task list
❹ **Display/Change** mode, where you can execute the task list and resume the job

4 Remote Conversion

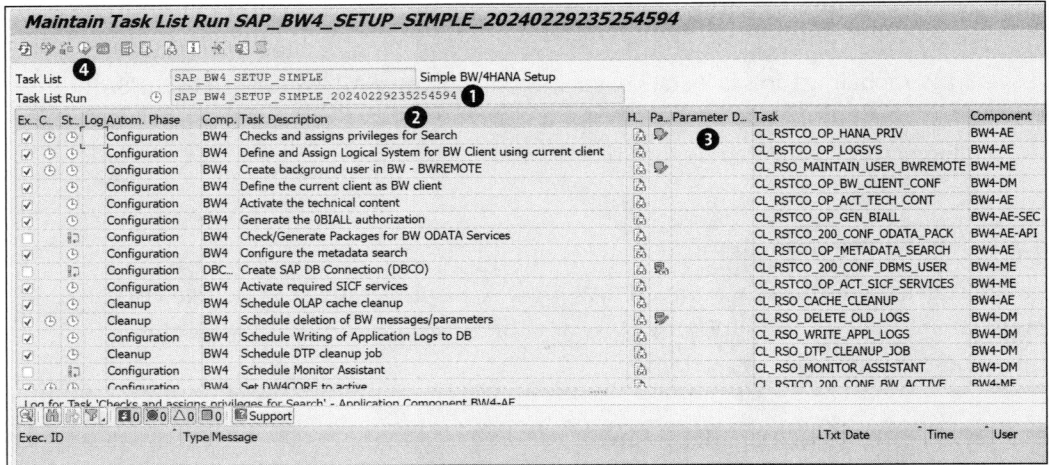

Figure 4.26 Setup Task List Execution

Additionally, you'll see the following tasks on this screen:

- **Create background user in BW - BWREMOTE**
  Use this procedure to generate a background user within SAP BW, denoted generally as BWREMOTE. Assign the S_BI-WHM_RFC and S_BI-WX_RFC profiles to this user, and allocate the SAP_RO_BCTRA role.

  The background user, BWREMOTE, is indispensable for SAP BW background processes, especially in data extraction. It's specifically entered in the destination within the source system. When establishing a connection to the source system, the system prompts the user to input a password for the background user.

- **Create RFC destination for after-import handling of client dependent BW objects (RSTPRFC)**
  Employ this procedure to establish an RFC destination facilitating the import of client-dependent objects into a designated target client.

  The subsequent import post-processing occurs within client 000. To ensure accurate creation, activation, and potential scheduling of objects in the target client, an RFC logon involving a user change becomes essential in the target client. Consequently, a specific RFC destination must be set up within the target system for this purpose. During the import post-processing phase, this destination is used for the logon process, transitioning from client 000 to the specified target client.

- **Create virtual time hierarchies for BW4**
  This task triggers the activation of various virtual time hierarchies designed for SAP BW/4HANA reporting. The activated hierarchies include the following:
  - Year - Quarter - Month - Day: 0YEA_QUA_MON_DAY
  - Year - Month - Day: 0YEA_MON_DAY
  - Week - Day: 0WEEK_DAY
  - Year - Quarter - Month: 0YEA_QUA_MON

152

4.2 SAP BW/4HANA Target System Provisioning

It's noteworthy that the time interval's validity spans from −7 years (7 years in the past) to +10 years (10 years in the future), calculated from the time of task execution.

- **Activate BW/4 launchpad**

  This task encompasses a series of SAP BW-specific configuration steps aimed at enabling the SAP BW/4HANA cockpit within your system.

  This inclusive process involves activating the gateway if not done previously, ensuring all necessary web services are activated through Transaction SICF, activating essential OData services, registering SAP BW/4HANA as an OData source system, replicating the SAPUI5 catalog, and invalidating the SAPUI5 cache. Through the systematic execution of these steps, the SAP BW/4HANA cockpit is effectively configured, ensuring seamless integration and optimal functionality within your system.

SAP BW/4HANA Web Cockpit Setup

For the SAP BW/4HANA web cockpit setup, use the task list **SAP_BW4_LAUNCHPAD_SETUP** via Transaction STC01, as shown in Figure 4.27. (This task list is deprecated from SAP BW/4HANA 2.0 SP 07 onward, so you'll get an error when you execute.)

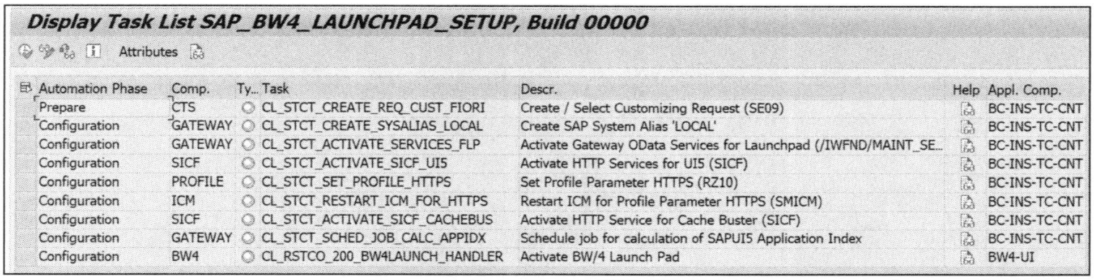

Figure 4.27 SAP_BW4_LAUNCHPAD_SETUP

For SAP BW/4HANA 2.0 SP 07 and later, use task list **SAP_FIORI_LAUNCHPAD_INIT_SETUP** via Transaction STC01 for initial SAP Fiori launchpad configuration, as shown in Figure 4.28.

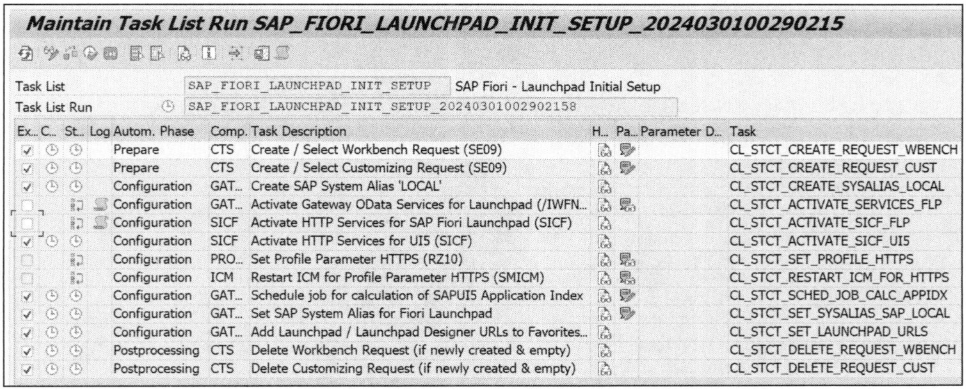

Figure 4.28 SAP_FIORI_LAUNCHPAD_INIT_SETUP

4   Remote Conversion

When you execute the task list, exclude the following tasks:

- **Activate Gateway OData Services for Launchpad (/IWFND/MAINT_SERVICE)** (technical name `CL_STCT_ACTIVATE_SERVICES_FLP`)
- **Activate HTTP Services for SAP Fiori Launchpad (SICF)** (technical name `CL_STCT_ACTIVATE_SICF_FLP`)

Subsequently, run `SAP_BW4_SETUP_SIMPLE` for SAP BW/4HANA–specific SAP Fiori launchpad configuration.

Once you've completed the configuration of SAP BW/4HANA web cockpit you can access it from the SAP BW GUI using Transaction BW4WEB, and it will prompt you for the logon screen. Once you've provided your credentials, you'll be taken to the screen shown in Figure 4.29.

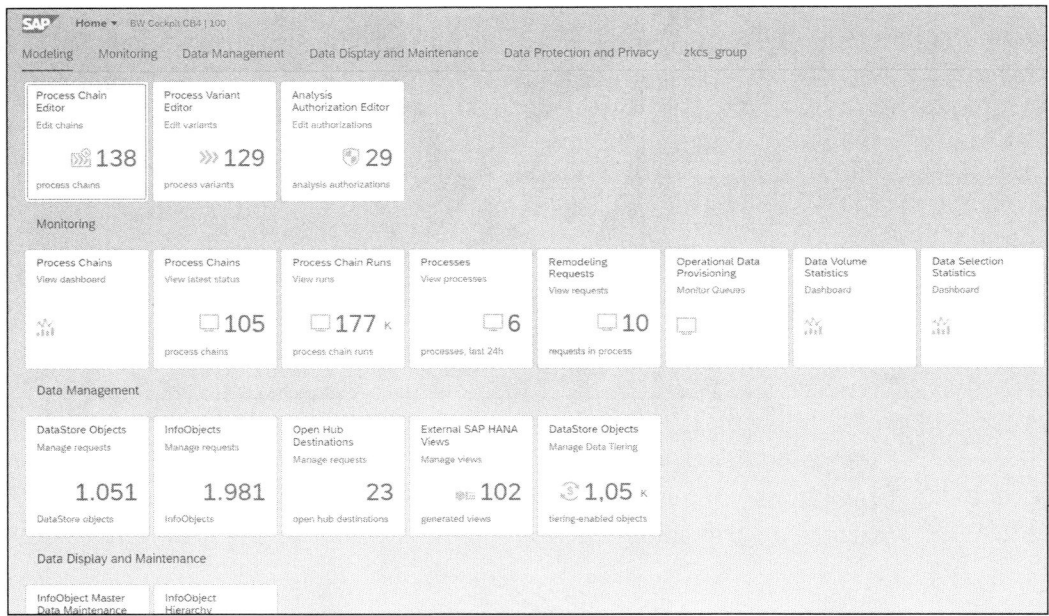

Figure 4.29 SAP BW/4HANA Cockpit

If you're in Eclipse, you can access the SAP BW/4HANA web cockpit by clicking the icon shown in Figure 4.30.

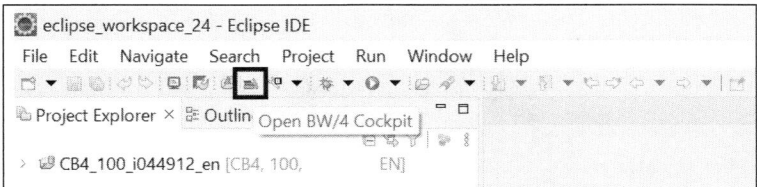

Figure 4.30 SAP BW/4HANA Web Cockpit: Eclipse

Roles in SAP BW/4HANA

User roles in SAP BW/4HANA are important in defining authorizations for various tasks within the system. SAP provides predelivered user roles containing authorizations for roles and user tasks specifically designed for SAP BW. Each user role includes the necessary authorizations for basic user tasks.

In Role Maintenance (Transaction PFCG), consider duplicating SAP-delivered roles before assigning them to users. This prevents potential overwrites by new standard roles during upgrades or release changes. SAP BW/4HANA roles start with SAP_BW4, as follow:

- **BW Modeler (SAP_BW4_MODELER)**
  In charge of connecting source systems, modeling data flows, performing DTPs, and scheduling process chains in the development system.

- **Reporting Developer (SAP_BW4_REPORTING_DEVELOPER)**
  Responsible for creating and executing queries, administrating analysis authorizations, and managing currency and unit conversion types.

- **BW Administrator (production system) (SAP_BW4_ADMINISTRATOR)**
  Responsible for handling data loading, process monitoring, and error analysis, as well as executing data archiving processes in the production system.

- **BW Operator (production system) (SAP_BW4_OPERATOR)**
  Involved in data loading, process chain scheduling, process monitoring, and basic troubleshooting in the production system.

- **BW Authorization Administrator (development system) (SAP_BW4_AUTHORIZATION_ADMIN)**
  Responsible for creating and managing analysis authorizations in the development system.

- **Reporting User (SAP_BW4_REPORTING_USER)**
  In charge of executing queries in SAP Analysis for Microsoft Office; SAP Analytics Cloud; or any generic BI clients from third-party providers.

4.2.4 Creating Source Systems

After completing the installation of SAP BW/4HANA and configuring the initial Basis settings, including the installation of Eclipse alongside ADTs and SAP BW modeling tools, the subsequent task involves establishing connections between all source systems and SAP BW/4HANA. Please be aware of a significant alteration in SAP BW/4HANA regarding source systems: the source system based on Service API (SAPI) is no longer supported. You can only create source systems of type operational data provisioning (ODP). Typically, the Basis team is responsible for source system creation. Ensure the creation of the ODP source system for the SAP BW on SAP HANA system, establishing RFC connections between systems. Connect all SAP ERP and SAP S/4HANA

4   Remote Conversion

systems accordingly. If you have data in the current SAP BW on SAP HANA system, create flat file–based source systems. For DB Connect–based source systems, establish SAP HANA smart data integration (SDI) source systems, ensuring consistent names for both SDI and DB Connect.

In this section, you'll learn how to create the ODP-SAP source system, the ODP-BW source system, and the flat file source system.

Creating an ODP-SAP Source System

You can create a new SAP-ODP system from the **DataSource** folder in **Project Explorer**, as shown in Figure 4.31.

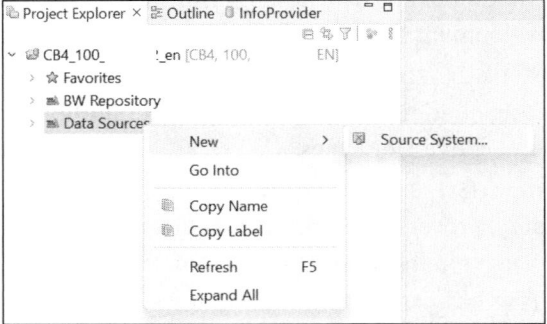

Figure 4.31 Creating a New Source System

Then, follow these steps:

1. Provide the **Name** of the **BW Project** and the **Description**, as shown in Figure 4.32.

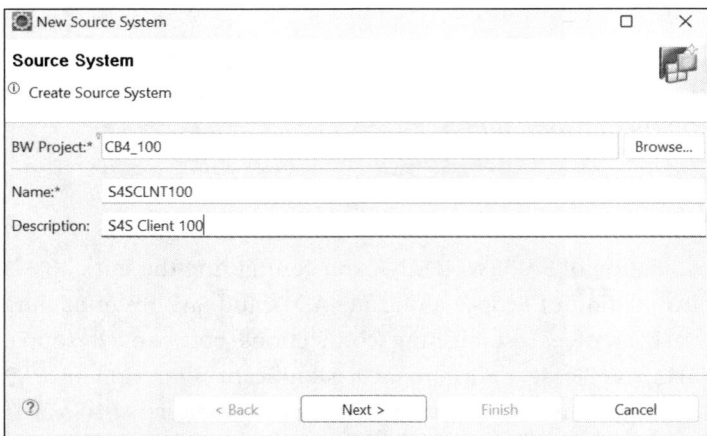

Figure 4.32 Source System Creation Wizard

2. Once you choose **Next**, there will be an option to choose the **Connection Type**. Choose **ODP** as the type for this demo, as shown in Figure 4.33. Click **Next**.

4.2 SAP BW/4HANA Target System Provisioning

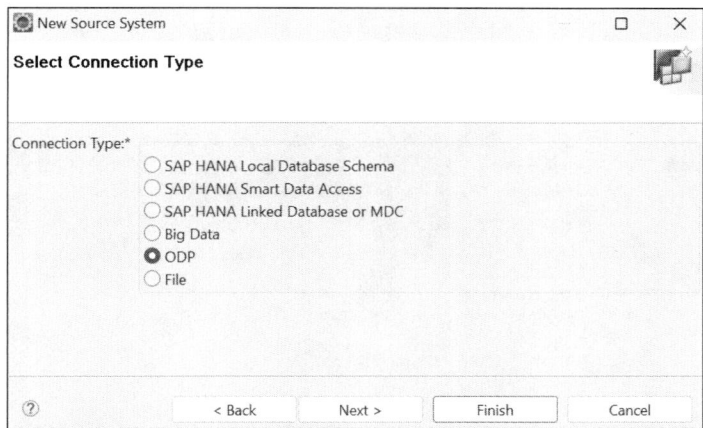

Figure 4.33 Connection Type Selection

3. In the next step, you need to choose the **ODP Page Selection**. For our scenario, we need to connect using **RFC with Remote System**, as shown in Figure 4.34.

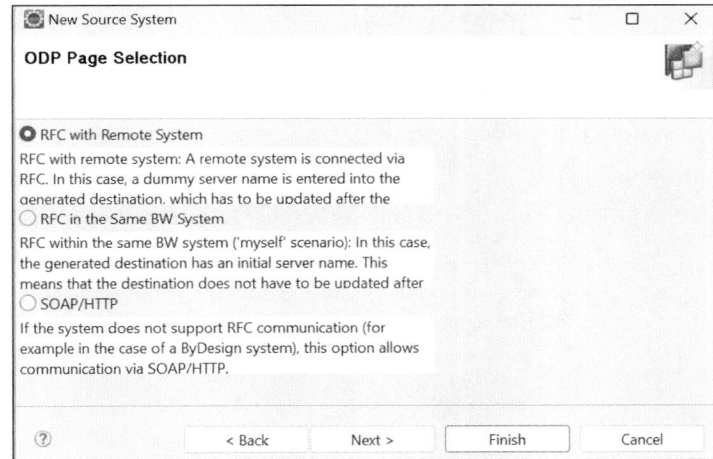

Figure 4.34 ODP Page Selection

4. Choose **Finish**, and you can see the system has been created with the following details, as shown in Figure 4.35:

   ❶ SAP BW/4HANA system details are given in the directory.

   ❷ **DataSources** shows the source system.

   ❸ The **ODP_SAP** folder will show the SAP ERP or SAP S/4HANA system.

   ❹ The name of the system that you've connected is shown in the directory.

   ❺ **Specific Properties** lists the details of the RFC for the source and SAP BW/4HANA.

   ❻ If you're connecting the SAP ERP or SAP S/4HANA system, then your ODP context will be SAPI.

4 Remote Conversion

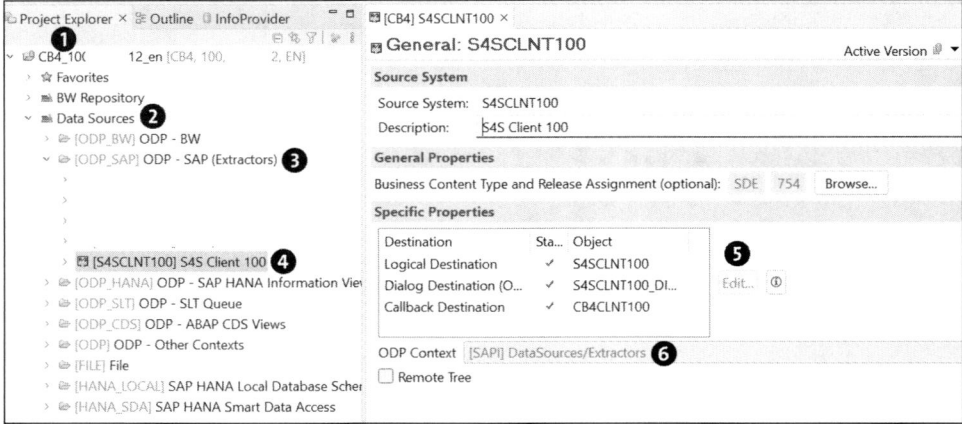

Figure 4.35 Created ODP System Details

Creating an ODP-BW Source System

You'll need to connect your sender SAP BW system to the target SAP BW/4HANA system. You need to use the ODP-BW for this connectivity. Ensure that you have the **ODP Context** set to **SAP NetWeaver Business Warehouse** as shown in Figure 4.36:

❶ **ODP_BW** is for connecting the SAP BW system to SAP BW/4HANA.

❷ The name of the logical system is provided here.

❸ Logical destination is provided here.

❹ For the SAP BW system, the **ODP Context** is set to **SAP NetWeaver Business Warehouse**.

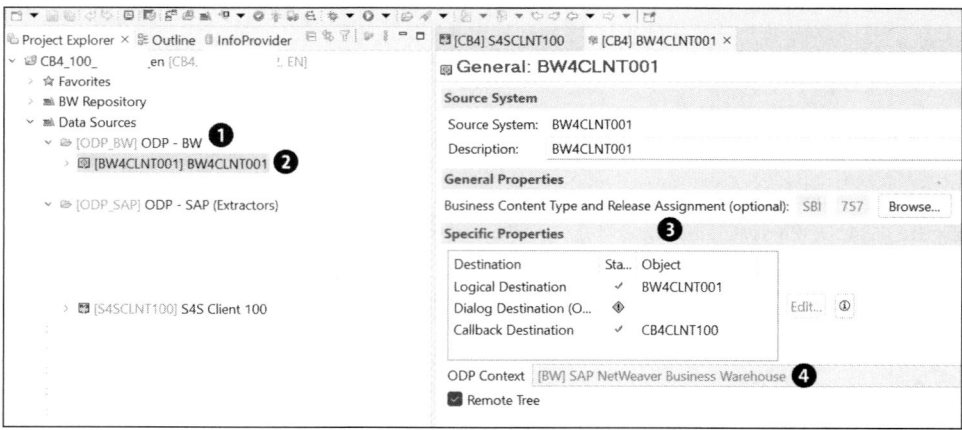

Figure 4.36 ODP-BW Source System

Creating Flat File Source Systems

You can see the details of the flat file system in Figure 4.37. Under the **Data Sources** folder, choose **[FILE] File**. Right-click to open the context menu and choose **New • Source**

158

4.2 SAP BW/4HANA Target System Provisioning

System. You need to provide the **Name** and **Description** of the source system, and choose **Next**. In the wizard, then choose **File** • **Next**, and continue to finish the flat file source system setup.

Figure 4.37 Source System of Type File

Once the flat file source system (**ZRKFFILE**) is created, you can see it in the screen shown in Figure 4.38.

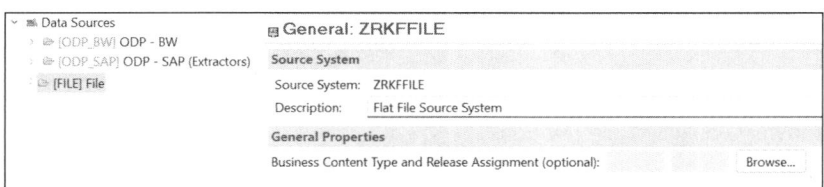

Figure 4.38 Flat File Source System

4.2.5 Settings in the SAP BW/4HANA System

In this section, you'll learn about the basic initial settings in SAP BW/4HANA. We'll start with the transfer of global settings and then you'll learn how to transfer exchange rates. We'll discuss maintaining the special characters and the material length settings. You'll learn about the transport path, RFC and how to expose the DataSource to ODP.

Transfer Global Settings

Let's begin by transferring the global settings. This action can be performed from any source system you're responsible for, but prioritize the system with the most up-to-date entries. Right-click on the source system, and select **Transfer Global Settings**, as shown in Figure 4.39.

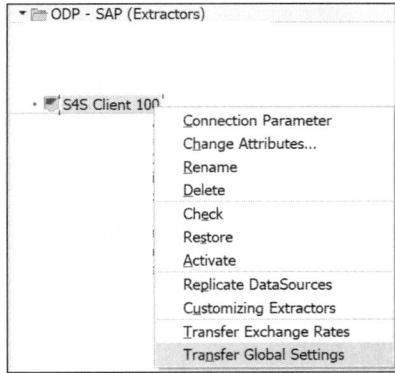

Figure 4.39 Transfer Global Settings

You also have the option to conduct the global settings transfer, as shown in Figure 4.40.

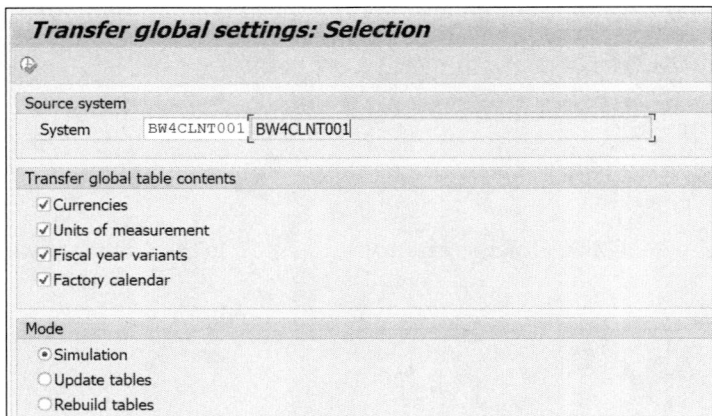

Figure 4.40 Transfer Global Settings: Options

There are three modes for transferring global settings:

- **Simulation**
  Provides a preview of updates, showing tables and record counts.
- **Update tables**
  Actively updates related tables, overwriting with source system changes.
- **Rebuild tables**
  Reconstructs tables by deleting old entries and writing new ones.

Transfer Exchange Rates

You can also transfer the exchange rates. Right-click on the source system and select **Transfer Exchange Rates**, as shown in Figure 4.41.

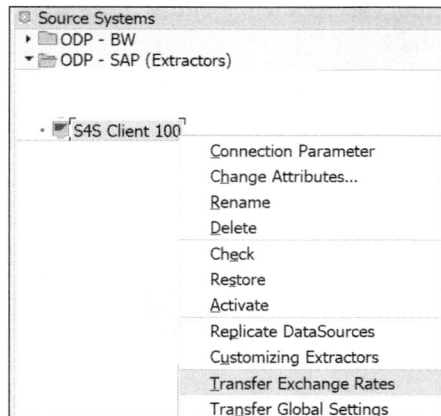

Figure 4.41 Transfer Exchange Rates

The **Transfer Exchange Rates** option takes you to the screen with the input selection, as shown in Figure 4.42. Choose the relevant source system (here, **S4SCLNT100**).

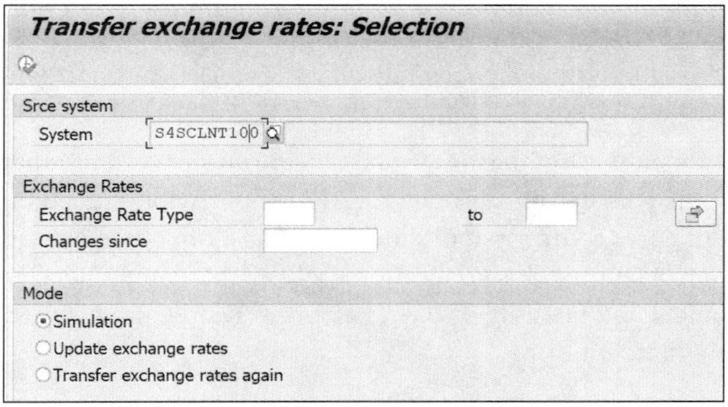

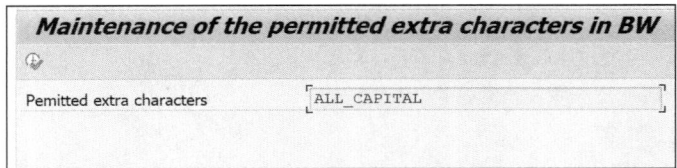

Figure 4.42 Selections for Transfer Exchange Rates

Maintain Transaction RSKC Similar to Your Sender System

You need to maintain the **Permitted extra characters** in Transaction RSKC. This will allow the SAP BW system to load the data from the source with all the allowed characters, as shown in Figure 4.43.

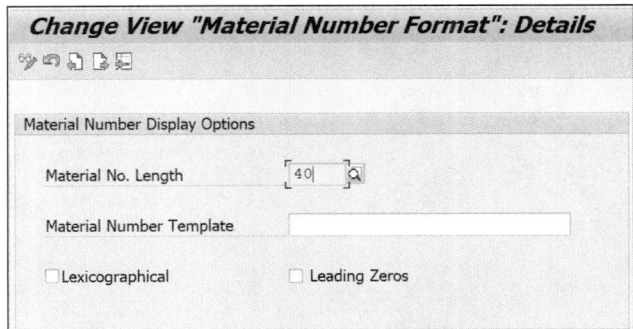

Figure 4.43 Transaction RSKC Maintenance

Transaction OMSL Settings

Maintain the **Material No. Length** in Transaction OMSL, as shown in Figure 4.44.

Figure 4.44 Material Number Maintenance

4   Remote Conversion

Transport Path and Remote Function Call

Verify the seamless flow of object transport based on conversion by confirming the presence of essential transport paths in both your SAP BW sender and target systems. It's crucial to establish a comprehensive transport path not only within the system itself but also with other relevant systems to facilitate effective object transport and accommodate any required fixes.

Furthermore, conduct a thorough check of the RFC destination in Transaction SM59 that connects to the sender system. This step ensures that the RFC connectivity is functioning correctly. A properly configured source system should result in a smooth operation without encountering any errors during this verification process. This comprehensive approach ensures the reliability and integrity of the transport infrastructure between systems in your SAP BW environment.

Replicate DataSources

In the SAP BW modeling tools, select and right-click on the relevant source system (here, **S4SCLNT100**), and choose **Replicate** to replicate all the DataSources from all the sender systems (see Figure 4.45).

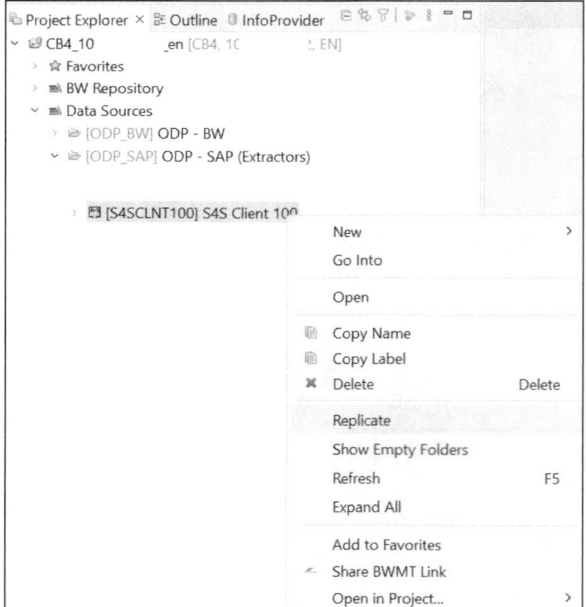

Figure 4.45 Replication of DataSources

Expose to Operational Data Provisioning

You need to expose all the replicated and activated DataSources in SAP BW/4HANA to ODP. SAP has delivered a report for this purpose. The functionality provided by report BS_ANLY_DS_RELEASE_ODP, shown in Figure 4.46, is integral for exposing standard

DataSources to ODP in your system. This report streamlines the process of making standard DataSources available for consumption through ODP, enhancing flexibility and accessibility for downstream applications and processes.

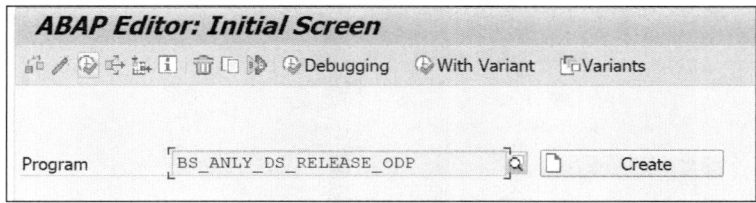

Figure 4.46 Expose Standard DataSources

In comparison, report RODPS_OS_EXPOSE serves a similar purpose but is specifically designed for custom DataSources. This tool facilitates the exposure of custom DataSources, enabling them to be seamlessly integrated into the ODP framework, as shown in Figure 4.47.

Figure 4.47 Expose Generic DataSource

4.2.6 Installing Essential Add-Ons

In this section, you'll learn about the essential add-ons: Data Migration Integration Server (DMIS), SAP BW/4HANA content, and SAP Business Planning and Consolidation, as well as third-party add-ons.

DMIS Add-On

A remote conversion requires the installation of the SAP Landscape Transformation Replication Server, which is shipped as the DMIS add-on, on both the original SAP BW as well as the target SAP BW/4HANA systems (SAP Note 2513088) The DMIS add-on contains the SAP BW/4HANA conversion cockpit. If you're on SAP BW/4HANA 2021, then you need to install the SAP LT Replication Server 3.0 add-on. To use DMIS 2011 on both the sender and target systems, the DMIS minimal requirement is SP 16, but best practice is to use the latest released SP. Make sure that both sender and target systems are on the same DMIS SP level if DMIS 2011 is used in both systems. For example, you can see the DMIS add-on for the sender ❶ and target systems ❷ in Figure 4.48.

Installed Software ❶

Component	Release	SP-Level	Support Package	Short Description of Component
SAP_BASIS	750	0019	SAPK-75019INSAPBASIS	SAP Basis Component
SAP_ABA	750	0019	SAPK-75019INSAPABA	Cross-Application Component
SAP_GWFND	750	0019	SAPK-75019INSAPGWFND	SAP Gateway Foundation
SAP_UI	754	0005	SAPK-75405INSAPUI	User Interface Technology
ST-PI	740	0014	SAPK-74014INSTPI	SAP Solution Tools Plug-In
BI_CONT	757	0026	SAPK-75726INBICONT	BI_CONT 757 Update
BI_CONT_XT	757	0026	SAPK-75726INBICONTXT	Business Intelligence Content for Bobj I
BW4HANA	100	0001	SAPK-10001INBW4HANA	SAP BW/4HANA Starter Add-On
SAP_BW	**750**	**0019**	**SAPK-75019INSAPBW**	**SAP Business Warehouse**
DMIS	**2011_1_731**	**0022**	**SAPK-11622INDMIS**	**DMIS 2011_1**
PCAI_ENT	100	0000	-	PCAI_ENT 100
ST-A/PI	01U_731	0001	SAPKITAB9Z	Servicetools for SAP Basis 731 and highe

Installed Software ❷

Component	Release	SP-Level	Support Package	Short Description of Component
SAP_BASIS	753	0007	SAPK-75307INSAPBASIS	SAP Basis Component
SAP_ABA	75D	0007	SAPK-75D07INSAPABA	Cross-Application Component
SAP_GWFND	753	0008	SAPK-75308INSAPGWFND	SAP Gateway Foundation
SAP_UI	754	0010	SAPK-75410INSAPUI	User Interface Technology
ST-PI	740	0014	SAPK-74014INSTPI	SAP Solution Tools Plug-In
BW4CONT	200	0011	SAPK-20011INBW4CONT	SAP BW4 HANA Content Addon
BW4CONTB	200	0011	SAPK-20011INBW4CONTB	SAP BW4 HANA Content Basis Addon
DW4CORE	**200**	**0007**	**SAPK-20007INDW4CORE**	**DATA Warehouse**
UIBAS001	400	0008	SAPK-40008INUIBAS001	UI for Basis Applications
BPC4HANA	200	0006	SAPK-20006INBPC4HANA	BPC/4HANA
DMIS	**2011_1_731**	**0022**	**SAPK-11622INDMIS**	**DMIS 2011_1**
ST-A/PI	01U_731	0001	SAPKITAB9Z	Servicetools for SAP Basis 731 and highe

Figure 4.48 SAP BW Sender and SAP BW/4HANA Target System Components

SAP BW/4HANA Content

SAP BW/4HANA Business Content integrates the data warehouse capabilities of SAP BW/4HANA with real-time analytics using SAP HANA. It adheres to the LSA++ architecture, using optimized modeling objects such as advanced DataStore objects (ADSOs) and CompositeProviders. This preconfigured set of information models, based on consistent metadata, enables quick and cost-effective implementation, as well as flexibility for use without adjustment, customization, or as a template during implementation on SAP BW/4HANA 2.0 SPS 01 or higher. As shown in Figure 4.49, SAP BW/4HANA content has two add-ons:

- **SAP BW4 HANA Content Addon (BW4CONT)** has SAP HANA-optimized business warehouse content and objects such as InfoSources and transformations for data provisioning.

- **SAP BW4 HANA Content Basis Addon (BW4CONTB)** has InfoObjects, InfoAreas, variables, conversion exits, master data read classes, and so on.

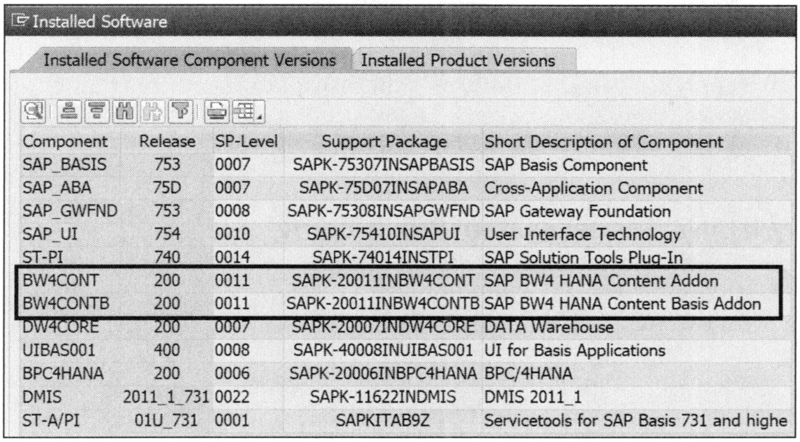

4.2 SAP BW/4HANA Target System Provisioning

Component	Release	SP-Level	Support Package	Short Description of Component
SAP_BASIS	753	0007	SAPK-75307INSAPBASIS	SAP Basis Component
SAP_ABA	75D	0007	SAPK-75D07INSAPABA	Cross-Application Component
SAP_GWFND	753	0008	SAPK-75308INSAPGWFND	SAP Gateway Foundation
SAP_UI	754	0010	SAPK-75410INSAPUI	User Interface Technology
ST-PI	740	0014	SAPK-74014INSTPI	SAP Solution Tools Plug-In
BW4CONT	**200**	**0011**	**SAPK-20011INBW4CONT**	**SAP BW4 HANA Content Addon**
BW4CONTB	**200**	**0011**	**SAPK-20011INBW4CONTB**	**SAP BW4 HANA Content Basis Addon**
DW4CORE	200	0007	SAPK-20007INDW4CORE	DATA Warehouse
UIBAS001	400	0008	SAPK-40008INUIBAS001	UI for Basis Applications
BPC4HANA	200	0006	SAPK-20006INBPC4HANA	BPC/4HANA
DMIS	2011_1_731	0022	SAPK-11622INDMIS	DMIS 2011_1
ST-A/PI	01U_731	0001	SAPKITAB9Z	Servicetools for SAP Basis 731 and highe

Figure 4.49 SAP BW Content Add-Ons

SAP Business Planning and Consolidation for SAP BW/4HANA

SAP Business Planning and Consolidation for SAP BW/4HANA (SAP BPC for SAP BW/4HANA) is a finely tuned solution for the SAP HANA platform. It offers a unified application and user interface to fulfill both bottom-up and top-down financial and operational planning needs. Administrators can set up installations in either a standard configuration, offering flexibility for end-user report building, or an embedded configuration seamlessly integrated into SAP BW/4HANA's Analytic Manager and Data Warehouse Management. In context to remote conversion with embedded models, ensure that your SAP BPC in the sender system is upgraded to the latest SP 7.5 with the supported version for SAP BW/4HANA, and switch all InfoCubes to Loading mode before initiating a remote conversion in your current solution. Turn off the audit feature for InfoCubes where it's enabled in your existing setup. Following the model transfer, transition all converted ADSOs to planning mode to ensure the proper functioning of SAP BPC features.

Third-Party Add-Ons

In situations where your sender SAP BW system incorporates multiple third-party add-ons, it's crucial to verify their seamless compatibility with the SAP BW/4HANA environment. To ensure a smooth transition, reach out to the respective third-party vendors and inquire about the specific support for SAP BW/4HANA, including details on supported releases and compatibility information.

If your reporting environment heavily relies on a particular third-party add-on, it's imperative to conduct a thorough examination before proceeding with the conversion activity. This involves assessing not only the general compatibility but also the critical functionalities provided by the add-on within the context of the SAP BW/4HANA framework. A comprehensive evaluation helps mitigate potential issues and ensures that the third-party add-ons perform optimally in the updated environment, safeguarding the integrity of your reporting processes during and after the conversion process. This proactive approach can save time, prevent disruptions, and contribute to a successful migration to SAP BW/4HANA.

4.3 SAP BW/4HANA Sender System Provisioning

You need to make sure the sender system is properly provisioned for the remote conversion. This involves two steps, which we'll discuss in the following sections.

4.3.1 Data Migration Integration Server Checks

As discussed in Chapter 3, if you need to install the latest DMIS add-on based on the target SAP BW/4HANA system, you can go for DMIS 2011 for SAP Landscape Transformation Replication Server 2.0 and DMIS 2018 for SAP Landscape Transformation Replication Server 3.0. Because remote conversion generally involves sender systems with lower SAP BW versions, you'll mostly end up using DMIS 2011_1_7XX with the minimum SP 14. If the DMIS 2011_1_731 add-on (SP 14 or above) isn't part of the SAP BW/4HANA installation, use the add-on installation tool (Transaction SAINT) to install it on your SAP BW/4HANA system. In general, before installing DMIS, ensure that your system has the latest Transaction SPAM/SAINT update installed. If a newer version is accessible on the SAP Support Portal, import the updated Transaction SPAM/SAINT release. DMIS 2011_1_731 is compatible with the following SAP_BASIS versions: 731, 740, 750, 751, 752, 753, and 754. This means that you can use DMIS 2011_1_731 seamlessly with systems running any of these specified SAP_BASIS versions.

> **DMIS Compatibility**
>
> Make sure that the sender system DMIS version and target SAP BW/4HANA system DMIS is the same. This is a prerequisite for the SAP BW/4HANA remote conversion.

4.3.2 Sender System Landscape Preparation with Note Analyzer

System landscape preparation is a foundational step in the remote conversion process. After successfully installing SAP BW/4HANA and the system: landscape DMIS add-on, it's essential to maintain the system's integrity by updating it to the latest version of the SAP BW/4HANA transfer cockpit and SAP BW/4HANA conversion cockpit. This update is accomplished via the Note Analyzer tool in SAP BW, which uses the corresponding XML file and implements all required SAP Notes. This comprehensive approach guarantees a smooth transition and optimal performance, setting the stage for a successful remote conversion journey. We've already seen the Note Analyzer process in detail in Chapter 3. When you download the Note Analyzer, you can see the XML files shown in Figure 4.50.

```
SAP_BW4HANA_Conversion_SAP_BW_750_2024-01-30.xml
SAP_BW4HANA_In-place_Conversion_(After_System_Conversion)_2024-01-30.xml
SAP_BW4HANA_In-place_Conversion_(Original_System)_2024-01-30.xml
SAP_BW4HANA_Pre_Checks_2024-01-30.xml
SAP_BW4HANA_Remote_Conversion_(Original_System)_2024-01-30.xml
SAP_BW4HANA_Remote_Conversion_(Target_System)_2024-01-30.xml
SAP_BW4HANA_Shell_Conversion_(Original_System)_2024-01-30.xml
SAP_BW4HANA_Shell_Conversion_(Target_System)_2024-01-30.xml
Source_System_for_SAP_BW4HANA_2024-01-30.xml
Z_SAP_BW_NOTE_ANALYZER.txt
```

Figure 4.50 Note Analyzer: XML File List

Let's explore the landscape preparation based on these XML files. Following the instructions outlined in previous sections, start by installing *Z_SAP_BW_NOTE_ANALYZER.txt* across all sender, target, and source systems. Download the specified XML files, as shown in Figure 4.50, and then run ABAP report Z_SAP_BW_NOTE_ANALYZER in the sender SAP BW, target SAP BW/4HANA, and connected SAP source systems (e.g., SAP ERP, SAP Customer Relationship Management [SAP CRM], SAP Supplier Relationship Management [SAP SRM], etc.). If SAP S/4HANA is part of the landscape, make sure to run the report in that system too. Based on the outcomes from the XML, implement SAP Notes with red and yellow statuses in sequential order. Run the report multiple times with the selected XML files until all SAP Notes display a green status across all systems. Don't skip any manual activity that is part of the note; if there is any SAP Note with yellow status try to fix that and make it green. When the note status isn't green based on the XML, then you'll have issues during the conversion, so make this an essential preparation activity during the remote conversion, carry out this same activity diligently across the entire system landscape, including sandbox, development, quality, production, and pre-production environments. It's advisable to maintain the latest XML in the production system, aligning it with the configuration in the quality system. You need to analyze with different XML files in different systems, as shown in Figure 4.51:

❶ **Install Precheck in the original SAP BW system.**
  – For SAP BW from 7.3 to 7.4, use *SAP_BW4HANA_Pre_Checks_[last_update].xml*.
  – For SAP BW 7.5, the prechecks can't be installed isolated but are bundled with the transfer cockpit, so use *SAP_BW4HANA_Conversion_SAP_BW_750_[last_update].xml*.

❷ **Install the transfer cockpit and DMIS add-on in the original SAP BW system.**
For SAP BW from 7.3 to 7.4 acting as the sending system for a remote conversion, use *SAP_BW4HANA_Remote_Conversion_(Original_System)_[last_update].xml*. For SAP BW 7.5, use *SAP_BW4HANA_Conversion_SAP_BW_750_[last_update].xml*.

❸ **Install ODP updates in each system.**
For the source systems connected to SAP BW such as SAP ERP, SAP S/4HANA, SAP CRM, and SAP SRM, use *Source_System_for_SAP_BW4HANA_[last_update].xml*.

❹ **Install the transfer cockpit and DMIS add-on in the target SAP BW/4HANA system.**
For SAP BW/4HANA acting as the receiving system for a remote conversion, use *SAP_BW4HANA_Remote_Conversion_(Target_System)_[last_update].xml*.

❺ **Install all the SAP Notes for the transformation based on the XML files from the SAP Note.**
This needs to be done in the target SAP BW/4HANA system. Refer to SAP Note 2603241, "Overview and summary of the most important SAP Notes in the context of BW." Download the notes from *SAP_BW_Transformation_[Last_Update].xml* in your target SAP BW/4HANA system.

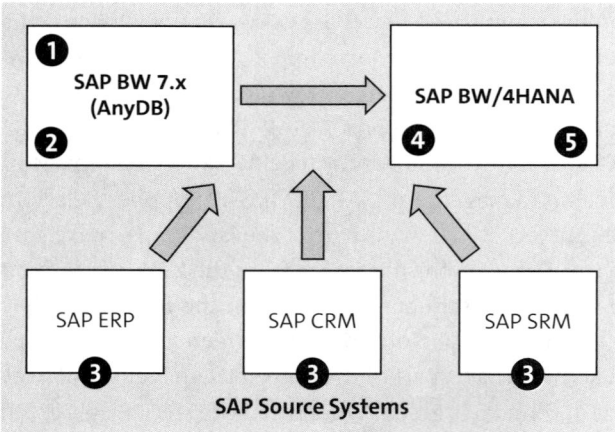

Figure 4.51 Sender System Landscape

4.4 Source System Provisioning

Based on the list shown earlier in Figure 4.50, make sure you execute the Note Analyzer for all the source systems connected to the SAP BW system, such as SAP ERP, SAP S/4HANA, SAP CRM, or SAP SRM. You can use *Source_System_for_SAP_BW4HANA_[last_update].xml* for the test. This will update all the related ODP-based corrections in the system. Make sure you check for the compatibility of certain SAP ERP system-based solutions (e.g., IS-retail) with SAP BW/4HANA during system preparation.

You then need to make sure to expose all the SAP-delivered DataSources and generic DataSources in all connected SAP source system to ODP using the following reports see Figure 4.52:

- SAP-delivered DataSources can be exposed using ABAP reports. Execute Transaction SE38, enter "BS_ANLY_DS_RELEASE_ODP", and click **Execute**.
- Generic DataSources can be exposed to ODP using report RODPS_OS_EXPOSE. You need to expose one DataSource at a time.

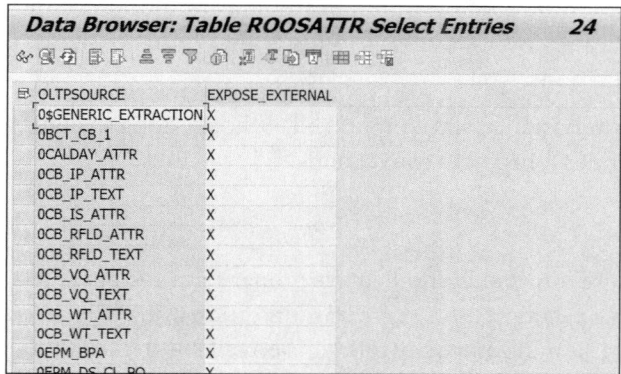

Figure 4.52 Table for ODP Expose Status

4.5 Preparing for Remote Conversion

For a successful execution of a remote conversion project, efficient project planning is very important. We'll walk through essential planning steps in this section.

4.5.1 Project Planning and Understanding

As you commence the remote conversion process, begin by focusing on high-level project planning. Ensure that you establish clear boundaries for all major milestones, outlining key objectives and timelines. This proactive approach fosters alignment and clarity among team members, paving the way for smooth project execution. By delineating these milestones up front, you create a road map that guides the project's progression and facilitates effective communication and coordination.

It's also crucial for the project team to possess a comprehensive understanding of the project landscape. This includes ensuring clarity on both the current (N) and upcoming (N + 1) landscapes. By having a clear grasp of these environments, the team can navigate potential challenges more effectively and make informed decisions. Assessing both landscapes allows for a holistic view of the project's evolution, enabling proactive planning and resource allocation. Additionally, it fosters a proactive approach to addressing any discrepancies or gaps between the current and future states, ensuring smoother transitions and successful project outcomes.

Finally, the organization must have a precise understanding of the required downtime for the SAP BW sender system. This information should be communicated clearly to end users to manage expectations and minimize disruptions. By proactively outlining the expected downtime, stakeholders can plan accordingly and mitigate any potential impacts on operations. Transparent communication fosters trust and cooperation among all parties involved, ensuring a smoother transition process. Additionally, it allows for effective coordination of activities and facilitates a more efficient use of resources during the downtime period.

4.5.2 User Roles and Authorizations

In this section, we'll discuss the importance of user roles and authorizations, including authorization objects and user roles for remote conversions.

User Roles

The user who executes the remote conversion should have proper roles and authorization. There are two kind of users, execution users and communication users. When you use the DMIS add-on, the single role and composite role will be available in the system. Both kinds of users are described here:

- **Execution user**
  This role serves as a composite role specifically tailored for execution expert users with extensive project administration authorizations. It consolidates and incorporates the authorization settings from various roles, listed in Table 4.2, providing users with a holistic set of permissions required for efficient execution and project management. The included authorizations cover a spectrum of functionalities, ensuring that individuals assigned to this role have the necessary access and control over project-related activities.

Role Name	Role Description
SAP_DMIS_SLOP_BASIS	SAP DMIS SLOP Basis authorizations for CWB functionality
SAP_DMIS_SLOP_BASIS_700	SAP SLOP additional authorizations for Basis release 700
SAP_SLOP_MASTER SAP_SLOP	Master user
SAP_SLOP_RFC_USER	SAP SLOP standard RFC user
SAP_DMIS_SLOP_APPL1	SAP SLOP additional authorizations 1
SAP_DMIS_SLOP_APPL2	SAP SLOP additional authorizations 2
SAP_DMIS_SLOP_APPL3	SAP SLOP additional authorizations 3

Table 4.2 Execution User Roles

Role Name	Role Description
SAP_DMIS_SLOP_APPL4	SAP SLOP additional authorizations 4
SAP_LTRL_MASTER	SAP SLOP specific authorizations for SAP BW/4HANA remote conversion

Table 4.2 Execution User Roles (Cont.)

- **Communication user**
  The SAP BW/4HANA conversion cockpit offers flexibility by not having preset destinations or user names. For seamless communication between the original SAP BW system and the target SAP BW/4HANA system, it's necessary to set up an RFC destination using the **Define RFC Destination** process. Users within these RFC destinations should be configured as dialog type users and must have the necessary authorizations, which are specified in certain composite roles, listed in Table 4.3. This ensures a smooth and secure transition process during the conversion, allowing users to effectively manage and execute tasks in both systems.

Role Name	Role Description
SAP_DMIS_SLOP_BASIS	SAP DMIS SLOP Basis authorizations for CWB functionality
SAP_DMIS_SLOP_BASIS_700	SAP SLOP additional authorizations for Basis release 700
SAP_SLOP_RFC_USER	SAP SLOP standard RFC user
SAP_DMIS_SLOP_APPL1	SAP SLOP additional authorizations 1
SAP_DMIS_SLOP_APPL2	SAP SLOP additional authorizations 2
SAP_DMIS_SLOP_APPL3	SAP SLOP additional authorizations 3
SAP_DMIS_SLOP_APPL4	SAP SLOP additional authorizations 4
SAP_LTRL_MASTER	SAP SLOP specific authorizations for SAP BW/4HANA remote conversion

Table 4.3 Communication User Roles

User Role for Remote Conversion Execution

SAP Note 228033, "SAP BW Landscape Transformation Solutions - Authorization Role Required in Customer System," has an attachment with all the required objects and details such as OBJNAME, OBJTYPE, DEVCLASS, ACTVT, TABLE, and so on.

To define the roles, follow these steps:

1. Download locally the file *SAP_LTRL_MASTER.TXT*, which is attached to SAP Note 2280336.

4 Remote Conversion

2. Use Transaction PFCG to create SAP_LTRL_MASTER roles in your SAP BW sender system.
3. Create them by uploading the file *SAP_LTRL_MASTER.TXT*.
4. Generate an authorization profile for the authorizations. When prompted for an authorization profile name, a valid name in the customer namespace is proposed.
5. Assign the role SAP_LTRL_MASTER to the user who will execute the SAP BW landscape transformation solutions, as shown in Figure 4.53.

Figure 4.53 SAP_LTRL_MASTER

Authorization Objects and Roles

To run the SAP BW landscape transformation solutions in customer systems, the end user needs enough permissions. Role SAP_LTRL_MASTER contains the necessary authorizations required to execute the SAP BW landscape transformation solutions. It will provide access to the following authorization objects:

- LTRL_EXE: Authorization for LT Business Warehouse Transformation
- S_ADMI_FCD: System Authorizations
- S_ALV_LAYO: ALV Standard Layout
- S_APPL_LOG: Application Log
- S_BDS_D (BC-SRV-KPR-BDS: Authorizations for Accessing Documents
- S_BTCH_ADM: Background Processing: Background Administrator
- S_BTCH_JOB: Background Processing: Operations on Background Jobs
- S_DEVELOP: ABAP Workbench
- S_GUI: Authorization for GUI Activities
- S_RFC : Authorization Check for RFC Access
- S_RS_ADMWB: Data Warehousing Workbench - Objects
- S_RS_AUTH: BI Analysis Authorization in Role
- S_RS_ICUBE : Data Warehousing Workbench - InfoCube
- S_RS_ODSO : Data Warehousing Workbench - DataStore Object
- S_RZL_ADM: CCMS: System Administration
- S_TABU_CLI : Cross-Client Table Maintenance

4.5 Preparing for Remote Conversion

- S_TABU_DIS : Table Maintenance (via standard tools, e.g., Transaction SM30)
- S_TABU_NAM: Table Access with Generic Standard Tools
- S_TCODE: Transaction Code Check at Transaction Start

The authorization roles for various activities in the SAP system are defined as follows:

- **LTRL_EXE**
  This role allows activities such as display, change, and delete (ACTVT = 01, 02, 03, 06, 07, 16).
- **RSANPR**
  Authorization for activity 03 is granted in this role.
- **S_ADMI_FCD**
  Permissions for function codes PADM and SUM are provided in this role.
- **S_ALV_LAYO**
  Activities related to SAP List Viewer (ALV) layouts are authorized (ACTVT = 23).
- **S_ALV_LAYR**
  Like S_ALV_LAYO, this role authorizes ALV layout activities (ACTVT = 23).
- **S_APPL_LOG**
  This role grants authorization for activity 03, related to application logs.
- **S_BDS_D**
  Authorization for activity 03 concerning Business Document Service (BDS) documents is provided in this role.
- **S_BTCH_ADM**
  Authorization is given to execute batch administration with BTCADMIN set to Y.
- **S_BTCH_JOB**
  Permissions for various batch job actions such as delete, list, plan, protect, release, and show are included in this role.
- **S_CTS_ADMI**
  This role authorizes transport-related activities, with CTS_ADMFCT set to TADD and TABL.
- **S_DATASET**
  Permissions for activities 33 and 34 related to datasets are granted in this role.
- **S_DEVELOP**
  Activities encompassing display, change, delete, and others are authorized in this role (ACTVT = 01, 02, 03, 07, 16).
- **S_DMIS**
  Authorization is given for activities 01, 02, and 03 related to DMIS.
- **S_GUI**
  Permissions are given for GUI activities with ACTVT values 60 and 61.

- S_RFC

  Authorization is given for RFC-related activities (ACTVT = 16).

- S_RFC_ADM

  This role authorizes activities 03, 36, and 39 concerning RFC administration.

- S_RS_ADMWB

  Authorization is given for activity 03 in the Administration Workbench.

- S_RS_ADSO

  Permissions are given for activities 03 and 23 related to ADSO.

- S_RS_AUTH

  Authorization is given for BIAUTH with a wildcard (*).

- S_RS_B4H

  Authorization is given for activity 16 related to SAP BW/4HANA.

- S_RS_COMP

  Authorization for various composite-related activities, including display, change, delete, and execute (ACTVT = 01, 03, 06, 16) based on queries.

- S_RS_COMP1

  Like S_RS_COMP, this role authorizes composite-related activities (ACTVT = 01, 03, 06, 16).

- S_RS_DTP

  Authorization is given for DTP activities.

- S_RS_HCPR

  Permissions are given for activities 03 and 23 related to SAP HANA CompositeProviders (HCPR).

- S_RS_ICUBE

  Authorization is given for activities 03 and 23 related to InfoCubes.

- S_RS_ODSO

  Permissions are given for activities 03 and 23 related to ODSOs.

- S_RS_OHDST

  Authorization is given for activities 03 and 23 concerning open hub destinations (OHDST).

- S_RS_PC

  Authorization is given for activities 03 and 23 related to process chains.

- S_RS_TR

  Permissions is given for activities 03 and 23 related to transformations.

- S_RZL_ADM

  Authorization is given for activity 03 in the Runtime Repository administration.

- S_TABU_CLI

  Authorization is given for client maintenance (CLIIDMAINT = X).

- S_TABU_DIS

  Permissions are given for display, change, and deletion activities (ACTVT = 02, 03, BD) in table maintenance.

- S_TABU_NAM

  Authorization is given for activity 02 in table maintenance and specification for tables starting with CNVLTRL*.

- S_TC

  Authorization is given for Transaction STC01 with activities 03 and 16.

- S_TCODE

  Permissions are given for various transactions, including Transaction RSA1, Transaction RSD1, Transaction RSDCUBE, Transaction RSRV, Transaction CNVLTRL*, Transaction SE16, Transaction SM21, Transaction SM50, Transaction SM51, Transaction SM66, Transaction ST06, Transaction ST04, Transaction ST22, Transaction SU53, Transaction RSOADSO, Transaction RSPC, Transaction CNV_MBT_SLOP, Transaction STC01, and Transaction RSB4HCONV.

- S_TRANSLAT

  Authorization is given for translation activities with ACTVT = 02.

- S_TRANSPRT

  Permissions are given for transport-related activities with ACTVT values 03, 05, and 43.

- PLOG

  Authorization is given for PLOG activities.

- S_USER_AGR

  Permissions are given for user administration activities with ACTVT values 01 and 02.

- S_RS_TTOOL

  Authorization is given for activities 03 and 23 related to transformation tools.

4.6　Basics of the SAP BW/4HANA Conversion Cockpit

In this section, you'll learn about the SAP BW/4HANA conversion cockpit, the essential transaction to open the task list, and the steps for making the essential checks before the remote conversion. You'll also learn how to evaluate the scope objects by using the simulation task list. Once you've installed all the necessary SAP Notes, you'll have access to all the required transactions. Let's start from the beginning.

The conversion cockpit is an important in the migration process from the legacy SAP BW to the modern SAP BW/4HANA systems. It will be available once you've completed all the preliminary steps such as DMIS add-on installation and SAP Notes implementation based on the required XML. The conversion cockpit's functionalities are designed to enhance the efficiency and effectiveness of the conversion process.

4.6.1 Overview

In this section, we'll provide a general overview of the SAP BW/4HANA conversion cockpit, including its key characteristics, package types, activity types, and phase and status management.

Key Characteristics

Following is a detailed exploration of the characteristics of the SAP BW/4HANA conversion cockpit, shown in Figure 4.54:

- **Status management and progress tracking**
  The SAP BW/4HANA conversion cockpit manages the status of each conversion activity and tracks their progress. This ensures a comprehensive and well-documented record of all executed tasks, allowing stakeholders to monitor the evolution of the conversion process.

- **Centralized execution platform**
  Acting as a centralized hub, the SAP BW/4HANA conversion cockpit consolidates various conversion activities into a single location. This not only streamlines the execution process but also provides a unified interface for users to initiate, monitor, and manage different tasks related to the migration.

- **Documentation and guided step-by-step activities (process tree)**
  A key feature of the SAP BW/4HANA conversion cockpit is its emphasis on documentation and guidance. The tool employs a process tree format, offering users a clear and structured overview of the conversion activities. This guided step-by-step approach ensures that users can navigate through the process seamlessly, reducing the likelihood of errors and enhancing overall efficiency.

- **Restart capability after interruptions**
  Recognizing the dynamic nature of conversion processes, the SAP BW/4HANA conversion cockpit incorporates a restart mechanism. In the event of interruptions, users can resume the DTP without having to restart the entire set of activities. This feature not only saves time but also contributes to a more resilient conversion process.

- **Selective re-execution of activities in case of issues**
  The SAP BW/4HANA conversion cockpit introduces a level of flexibility by allowing the selective re-execution of activities. In situations where issues arise, users can pinpoint and rerun specific tasks without the need to repeat the entire conversion process. This targeted approach enhances troubleshooting efficiency and minimizes potential downtime.

4.6 Basics of the SAP BW/4HANA Conversion Cockpit

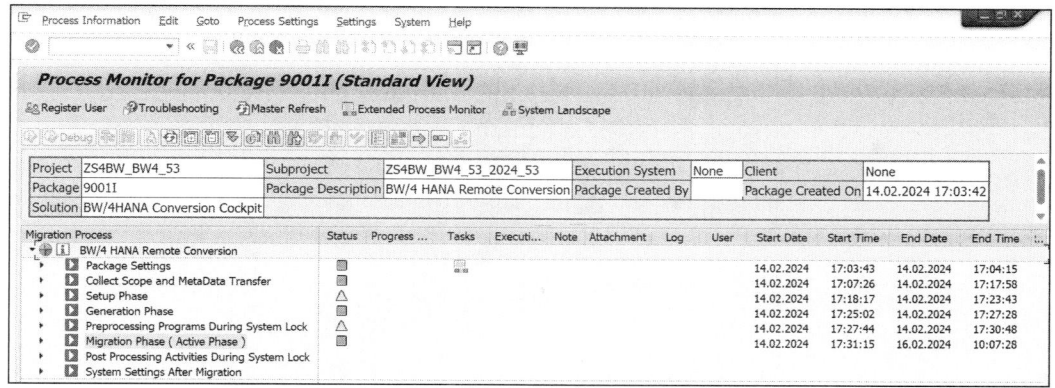

Figure 4.54 Process Tree in the SAP BW/4HANA Conversion Cockpit

The conversion process tree embedded within the SAP BW/4HANA conversion cockpit serves as a comprehensive road map for remote conversion. It outlines a series of activities and steps, ensuring a guided journey through each critical stage of the conversion process. This structured approach provides users with clear, step-by-step guidance for a seamless, easy transition from classic SAP BW to the advanced SAP BW/4HANA system.

Moreover, the process tree isn't merely a linear sequence; it incorporates well-defined dependencies between individual steps. These dependencies are strategically designed to prevent manual errors by establishing a logical order and ensuring prerequisites are met before progressing to subsequent stages. The interlinking of steps enhances the overall reliability and accuracy of the conversion process, allowing users to navigate confidently through the conversion journey while optimizing efficiency and reducing the risk of oversights.

Process Tree

The utilization of the process tree is integral for carrying out activities in a remote conversion scenario. These activities are executed within the designated execution systems, encompassing both the original system and the target system. Within the process tree interface, all essential conversion activities are conveniently displayed, offering an overview of the entire conversion process.

To fortify the security and consistency of the remote conversion process, SAP Landscape Transformation introduces multiple measures. These measures are designed to uphold a secure and seamless conversion experience, ensuring the integrity and reliability of the data transformation from the original system to the target system.

Package Type in the SAP BW/4HANA Conversion Cockpit

The SAP BW/4HANA conversion cockpit employs packages to systematically structure the remote conversion process. Within an execution system, be it the original system or the target system, packages may assume an active or inactive status, with only one active package permitted at any given time within a client. Consequently, conversion activities are only executable for the active package.

If you need to create a new package, it's imperative to first conclude or deactivate the existing active package. This ensures a methodical and orderly progression of the conversion process, allowing for the seamless initiation of subsequent activities within the designated package framework. Figure 4.55 shows the **Package ID 9001I** as the **Active Package**.

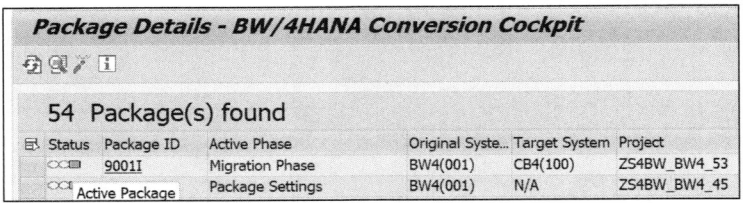

Figure 4.55 Package List

Phase Management

Every activity within the package is allocated to a distinct phase in the process tree. The execution of an activity is contingent upon the active status of the corresponding phase within the package. Upon initiating a package, the initial phase is set to active. Subsequently, the activation of the next phase occurs once the first activity in that phase is commenced. This progression is conditional on the successful completion (either with "ok" or a warning) of all programs from the preceding phase. This careful sequencing ensures a controlled and systematic advancement through the phases, maintaining a prerequisite-based flow in the execution of activities within the package. Figure 4.56 shows the details of phases in the migration process, such as **Package Settings**, **Collect Scope and Metadata Transfer**, **Setup Phase**, **Generation Phase**, **Preprocessing Programs During System Lock**, **Migration Phase**, **Post Processing Activities During System Lock**, and **System Settings after Migration**. Each of these phases will have its own subline items as well.

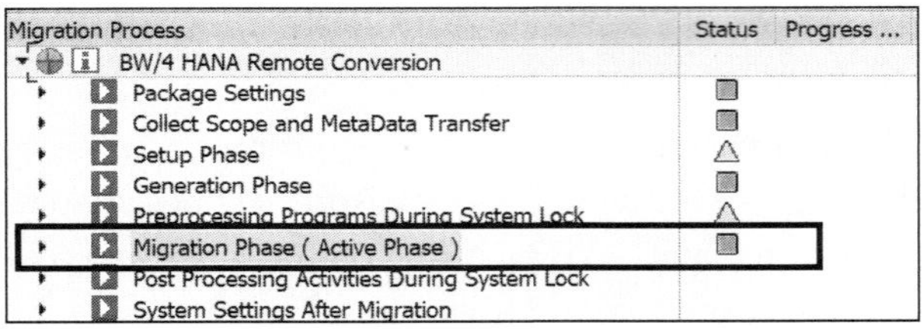

Figure 4.56 Phase Management

Status Management

Within each phase, specific predecessor-successor relationships govern the sequence of activities. Upon completion, each activity is assigned a status (**OK**, **Warning**, or **Error**).

In cases where an activity remains unexecuted or carries an **Error** status, the commencement of its successor activity is prohibited. Furthermore, if an activity within a phase is marked with an **Error** status, the initiation of activities in the subsequent phase is also restrained. This interdependency ensures a structured and error-aware execution flow, preventing the progression to subsequent tasks until the prerequisite activities have been successfully executed and validated. Figure 4.57 shows the **Status** as green for success ▣, yellow for completed with a warning △, or red for an error ⦿.

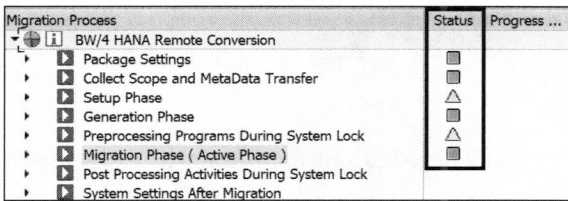

Figure 4.57 Status Management

Activity Type

Within the process tree, there is a mix of individual activities executed sequentially and group activities. When initiating a group activity, a series of background activities is triggered automatically. As each activity initiates its successor, user intervention becomes unnecessary.

In the event of errors or warnings after the group activity, activating the extended view of the process tree becomes crucial. This expanded view displays all activities affiliated with the group activity, along with their respective statuses. Within this extended process tree, users can independently execute each activity, facilitating targeted actions to address issues and ensure a smooth progression through the conversion process. Alternatively, you can individually execute each activity in the extended process tree, as shown in Figure 4.58.

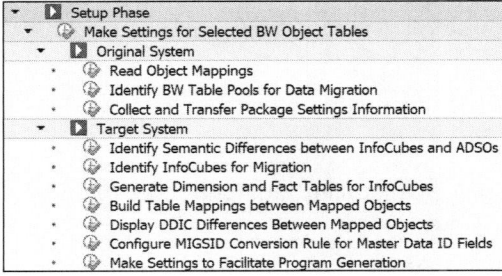

Figure 4.58 Activity Type

4.6.2 Scope Check Tool

If you've outlined a specific scope for your remote conversion project, make sure you conduct prechecks and address most errors beforehand. Additionally, consider running a simulation before initiating the conversion tool. Once you've finalized the scope, such as process chain or InfoCube, proceed with the following steps:

1. Access Transaction STC01 in the SAP BW sender system, and enter the **Task List** name as "SAP_BW4_TRANSFER_CHECK_REMOTE", as shown in Figure 4.59.

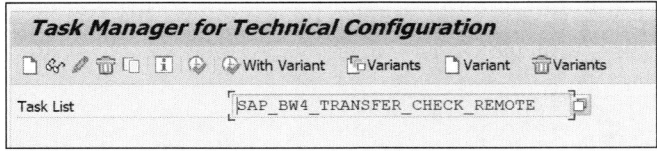

Figure 4.59 Transfer Check Remote Tool

2. Click the execute icon, and then you'll have options for two activities, as shown in Figure 4.60.

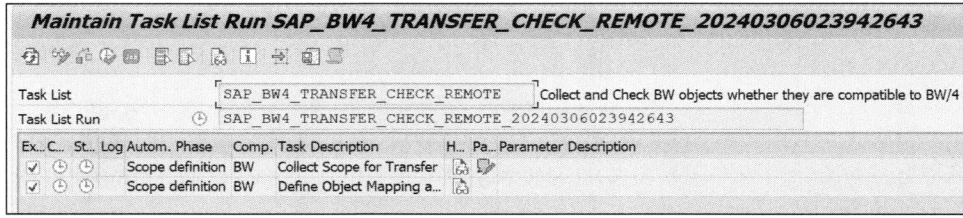

Figure 4.60 Collect Scope

3. In the list of activities, first get into edit mode by clicking, and then choose the parameter icon for the **Scope definition** row. The next screen, **Select Start Objects**, will have the options for selecting the objects, as shown in Figure 4.61.

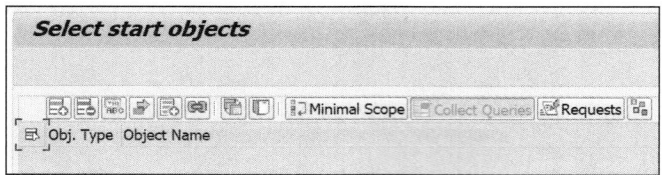

Figure 4.61 Select Start Objects

4. Choose the icon to enter multiple objects manually, and you'll see the popup window shown in Figure 4.62.

5. For our demo, we're going to collect the InfoCube (ZPRKC1) for the conversion, so select the TLOGO (**Object Type**) as **CUBE**, and enter the cube name as "ZPRKC1" in the **Object Name** field, as shown in Figure 4.63. Then, choose the execute icon.

4.6 Basics of the SAP BW/4HANA Conversion Cockpit

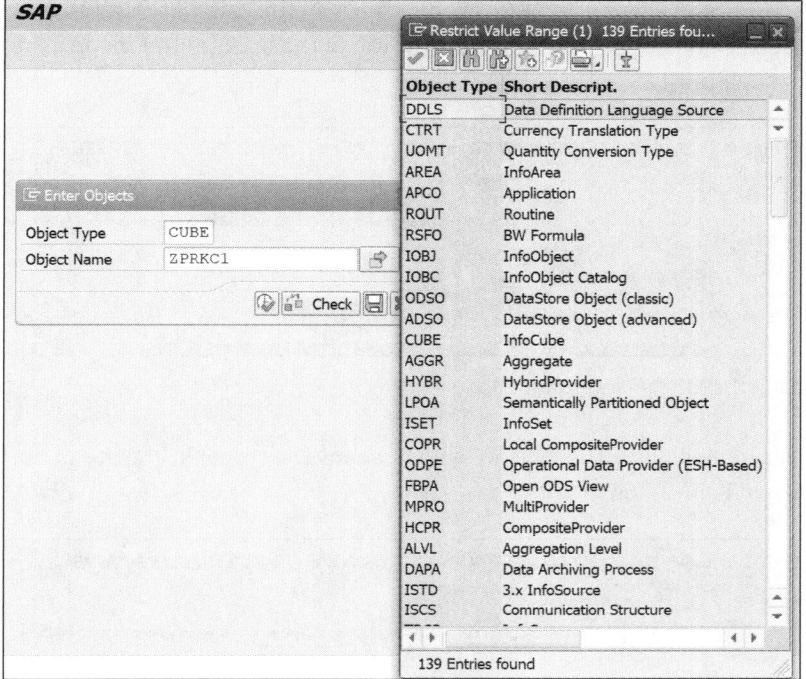

Figure 4.62 Choose Your Object

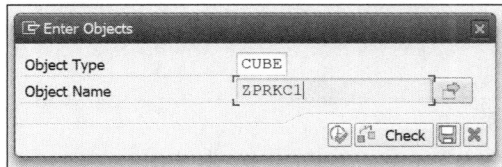

Figure 4.63 Start Object Details

You'll then see the list of selected objects, as shown in Figure 4.64.

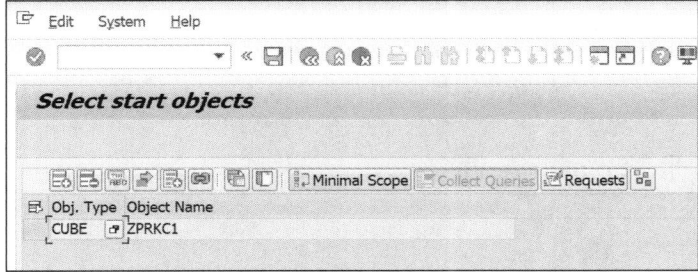

Figure 4.64 Start Object Confirmation

181

6. You can see the InfoCube name (**ZPRKC1**) is now shown in the **Parameter Description** column. In the display mode, choose to start the job in the background mode using ▥, as shown in Figure 4.65.

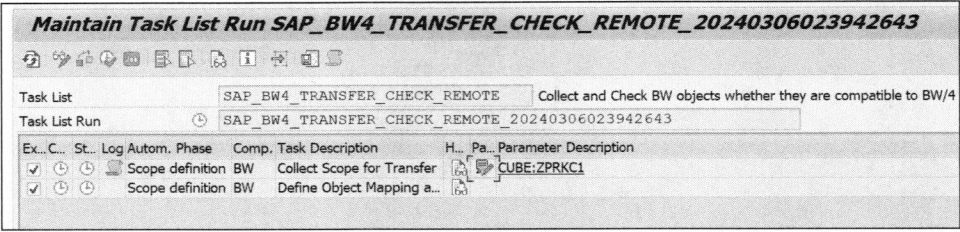

Figure 4.65 Starting the Job in Background Mode

7. The **Status** column will have the ✱ icon, which means that specific activity is running, as shown in Figure 4.66.

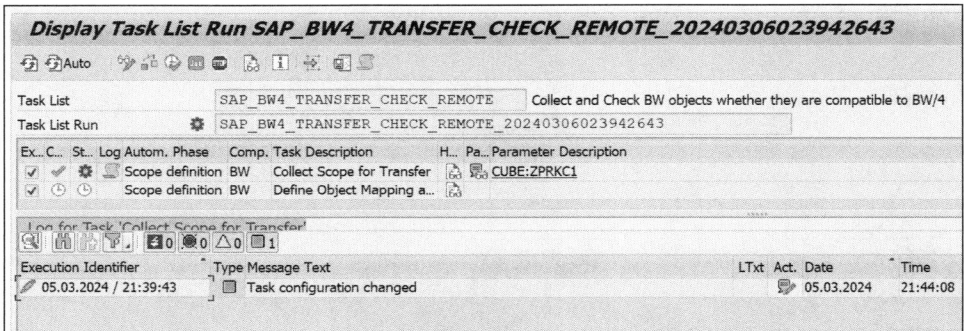

Figure 4.66 Task List Status

8. Once the scope definition is completed without any errors, it will have the green status icon, and the tool will by default move into the next activity, which is **Define Object Mapping**. Because that activity needs manual input, the tool will have the **Stop** status, as shown in Figure 4.67.

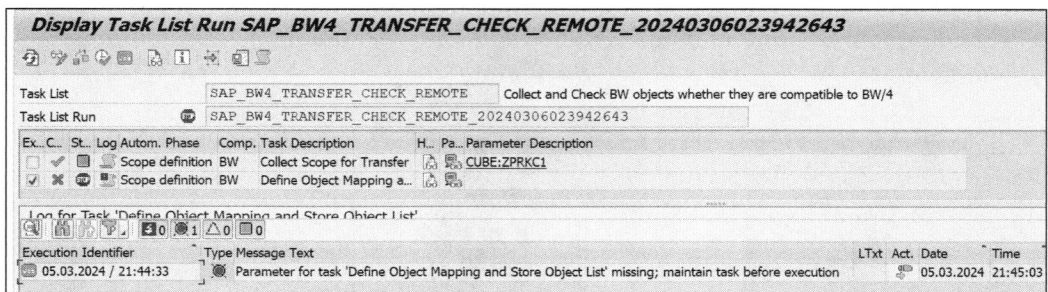

Figure 4.67 Object Mapping Status

4.6 Basics of the SAP BW/4HANA Conversion Cockpit

9. Get into edit mode by clicking . Choose the **Edit Parameter** option, which will take you to the **Maintain Object List** window, as shown in Figure 4.68. You'll see the details of the source object and its associated original TLOGO in the **Original** column. You can the target TLOGO in the **New TLOGO** column, along with the system-generated object name in the **New Object Name** column. (For example, **CUBE** is shown as **ADSO** in row 1 with the same name **ZIPC1**.) The details of the transfer status of the object are given in the **Transfer As** checkbox.

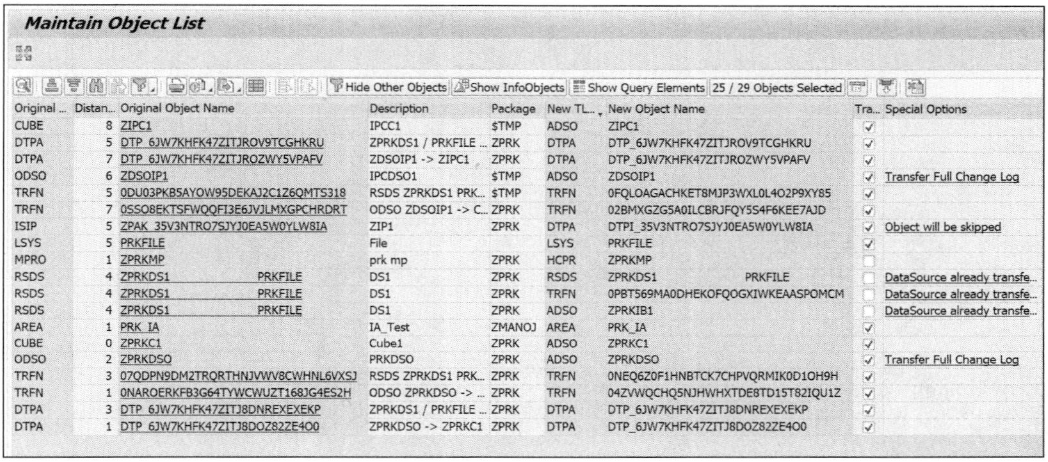

![Maintain Object List screen showing object mapping details with columns for Original, Distan, Original Object Name, Description, Package, New TL, New Object Name, Tra, and Special Options]

Figure 4.68 Object Mapping Details

4.6.3 Task List Based on Transaction STC01 and Transaction STC02

You can use Transaction STC01 to execute the task list for the first time. Once you execute the task list, it will create the task list run based on that task list. Note the task list run number for further execution. Follow these steps:

1. Open Transaction STC01.
2. Provide your desired **Task List** to be executed (e.g., "SAP_BW4_TRANSFER_CHECK_REMOTE").
3. Click **Execute**.
4. The task list and task list run name are shown in Figure 4.69. The task list run is **SAP_BW4_TRANSFER_CHECK_REMOTE_20240306023942643**.

Figure 4.69 Task List Run Name

183

4 Remote Conversion

Transaction STC02 (Task List Run Monitor) allows you to reopen previously saved task list runs for execution or status checking. Throughout the project duration, you'll use this transaction multiple times to access saved task list runs. Upon accessing the initial screen of the transaction (see Figure 4.70), you'll see options to search based on either **Created By**, which is the username, or based on the **Task List** name.

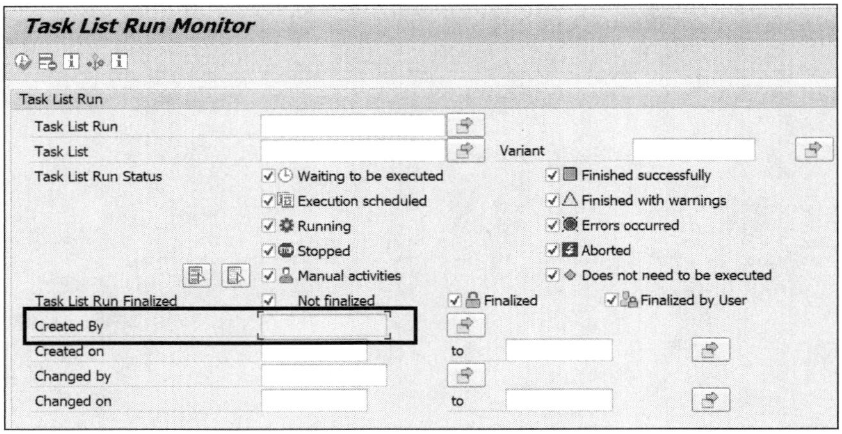

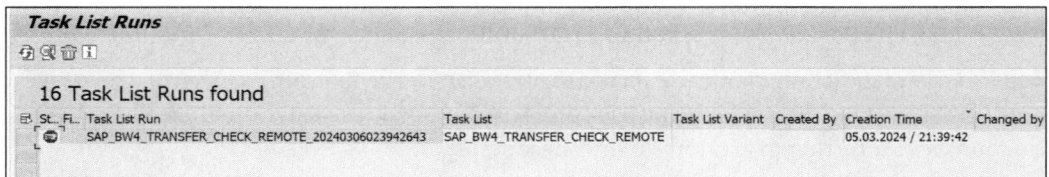

Figure 4.70 Task List Monitor

Based on the **Created By** user name, you'll see a task list with the **Stop** status, as shown in Figure 4.71.

Figure 4.71 Task List Status

Clicking or selecting the designated row within the task list runs will take you to the maintenance screen of that task list. Here, you'll find all the activities available for execution or analysis, as shown in Figure 4.72.

Figure 4.72 Object Mapping Status

Following is a summary of the icons used in the task list:

- To refresh the task list run, use the ⟳ icon.
- To get into edit/display mode, use the ✏ icon.
- To check the task list, use the 🔍 icon.
- To execute the task list in a dialog, use the ▶ icon.
- To execute the task list in background mode, use the [Job] icon. The same can be used to resume a failed task list run.
- To save the task list run at the current state, use the 💾 icon.
- To display the parameter in the task list item, use the 🗔 icon.
- To edit, insert, or change the parameter in the task list item, use the ✎ icon.

4.6.4 Essential Checks before Remote Conversion

There are two essential checks to perform before a remote conversion: one for the sender system and one for the target system. We'll explore both in this section.

Sender System

Make sure the prerequisite is completed and users have enough authorization, and then ensure the table space is 30% more than the current database size. You can use Transaction DB02, and then choose **Current Status • Overview**. Check the **Database Memory and CPU Details** area, as shown in Figure 4.73.

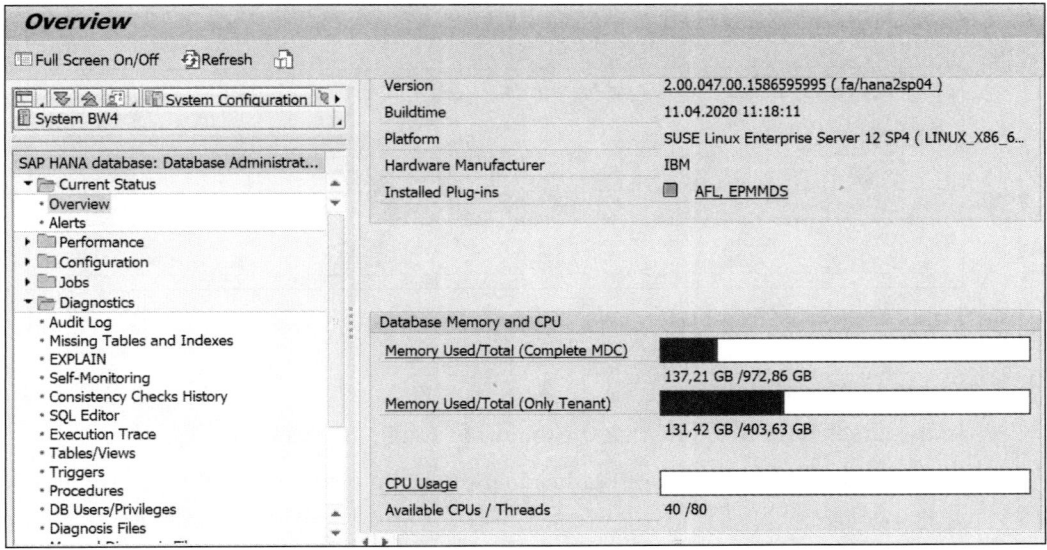

Figure 4.73 System Overview

Select the **Diagnostics** folder, and navigate to **Tables/Views**. Enter the table name "CNVLTL_BWCLU" into the designated field, and execute using icon. The relevant schema name for your system will be displayed. In our demo system, the schema used is **SAPBW4**, as shown in Figure 4.74. Allocate 30% of the overall database size to this schema, as the cluster table is located within it exclusively.

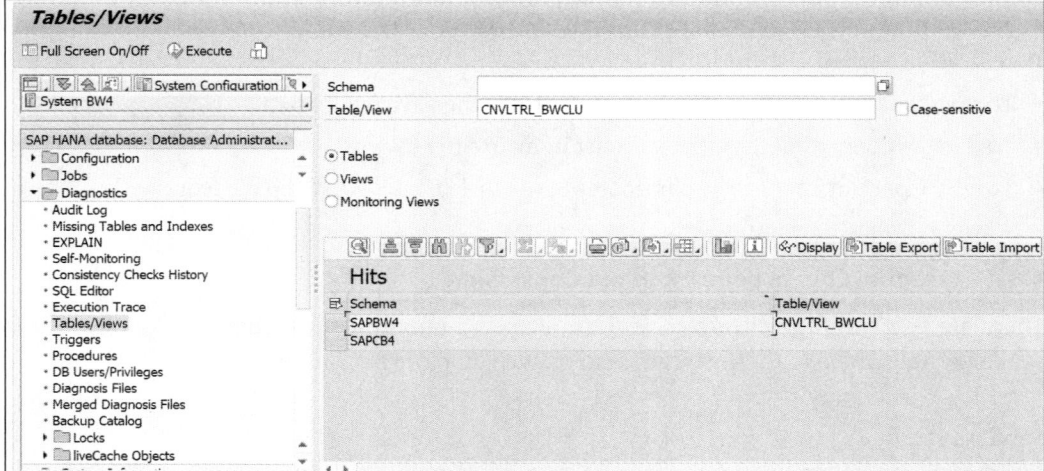

Figure 4.74 Schema Name for the Cluster

You can check Transaction RZ04 in the SAP BW system to know the details of the available work process in the system, such as the dialog work process (**Dia**). Ensure that you have a greater number of background work processes (**BP**) (see Figure 4.75).

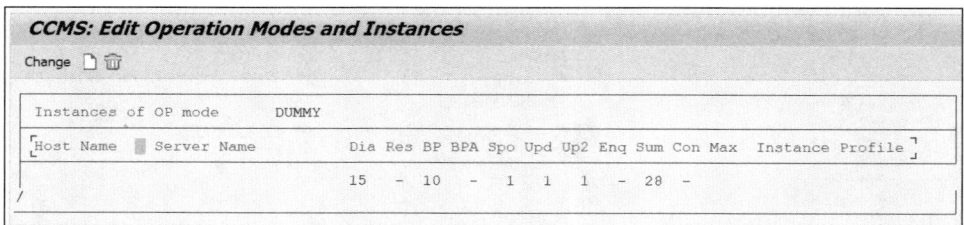

Figure 4.75 Work Process Overview

Target System

In the target system, ensure that the following details are available:

- Target SAP BW/4HANA system installation is completed with all post-installation activities.
- The target system has valid DMIS per the prerequisite.
- The target system is connected to all SAP ERP systems using SAP–ODP connectivity.
- The target system has the latest XML files for the transformation.

4.7 SAP Business Warehouse Data Model

- The sender SAP BW and target SAP BW/4HANA systems are connected using ODP–BW.
- The SDI has been created for all DB Connect source systems.
- The global settings, currencies, and units based on the sender system have been transferred.

4.6.5 Evaluating Scope Objects

Once you have all the prerequisites met, then you need to evaluate the scope objects and make sure that they're ready for the conversion using task list SAP_BW4_TRANSFER_CHECK_REMOTE in Transaction STC01.

4.7 SAP Business Warehouse Data Model

For our remote conversion demo, we'll use the following SAP BW data flow to be converted to the SAP BW/4HANA system:

- **Delta Queue (RSA7) – SAPI for the DataSource in the SAP ERP system**
  In the sender SAP ERP or SAP S/4HANA system, check Transaction RSA7 (SAP BW Delta Queue Maintenance) to see the status of delta. You can see that our DataSource **ZPRRTDS** has the status based on the last loaded sales document number. You can also see the delta-relevant fields, in this case, the sales document number (**VBELN**), as shown in Figure 4.76.

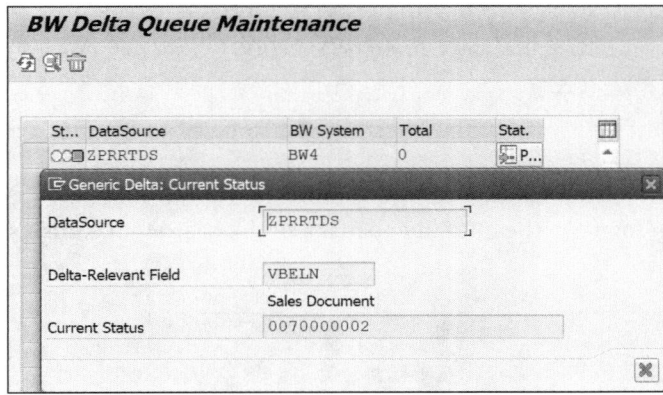

Figure 4.76 SAPI Delta Queue: Transaction RSA7 in SAP ERP

- **SAP BW sender system objects**
  You can view the details of system objects, as shown in Figure 4.77. The InfoArea **ZPRKRMTCV** has SAP BW objects such as InfoCube **ZPRKRCUB** and **ZPRKRDSO** that will be used for the remote conversion (see Figure 4.77).

4   Remote Conversion

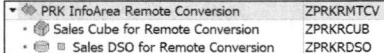

Figure 4.77 SAP BW InfoProviders

- **Sender system InfoCube flow**
  The data model for the sender system InfoCube (**ZPRKRCUB**) is shown in Figure 4.78.

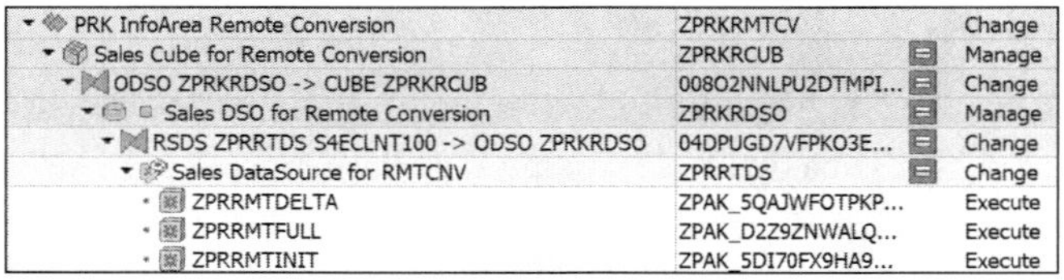

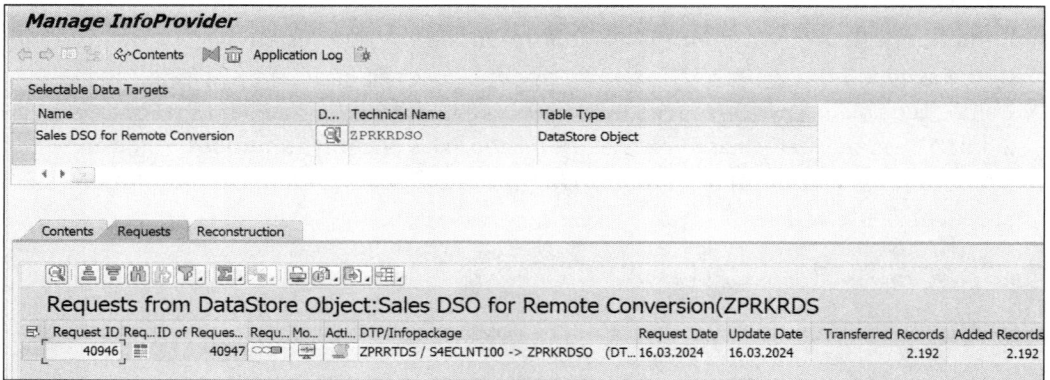

Figure 4.78 SAP BW Sender InfoCube Flow

Check the request in the source DSO (**ZPRKDDSO**) for InfoCube (**ZPRKRCUB**), as shown in Figure 4.79.

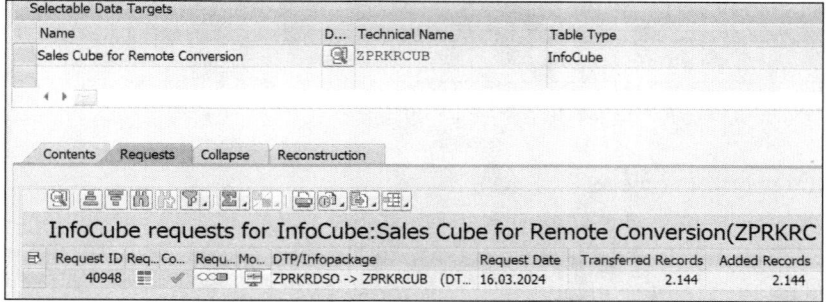

Figure 4.79 SAP BW Sender: DSO Request

Check the request management for the InfoCube to make sure it has **2144** records, as shown in Figure 4.80.

Figure 4.80 InfoCube Request Status

188

4.7 SAP Business Warehouse Data Model

- **Data flow in SAP BW sender**
  We'll use the InfoCube **ZPRKRCUB** for our scope selection, as shown in Figure 4.81.

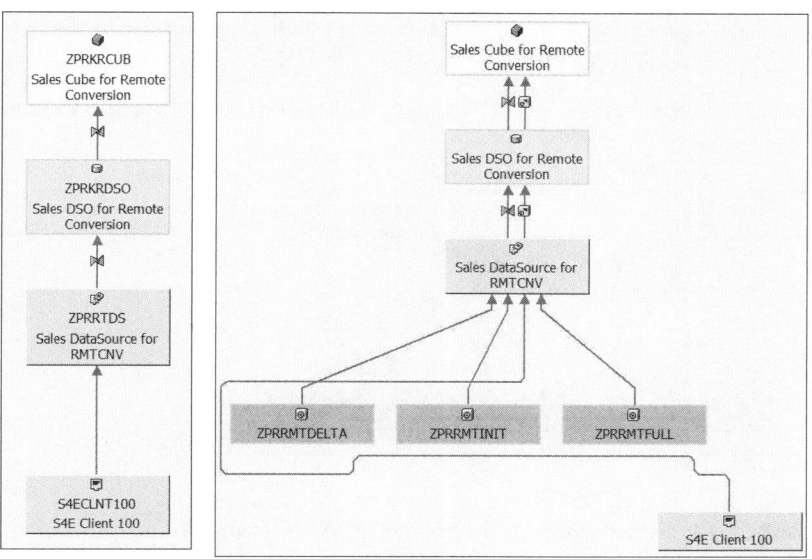

Figure 4.81 SAP BW Sender System Data Flow

You can find the details for the InfoCube and DSO of the sender system tables in Figure 4.82. You can access these tables using Transaction SE16.

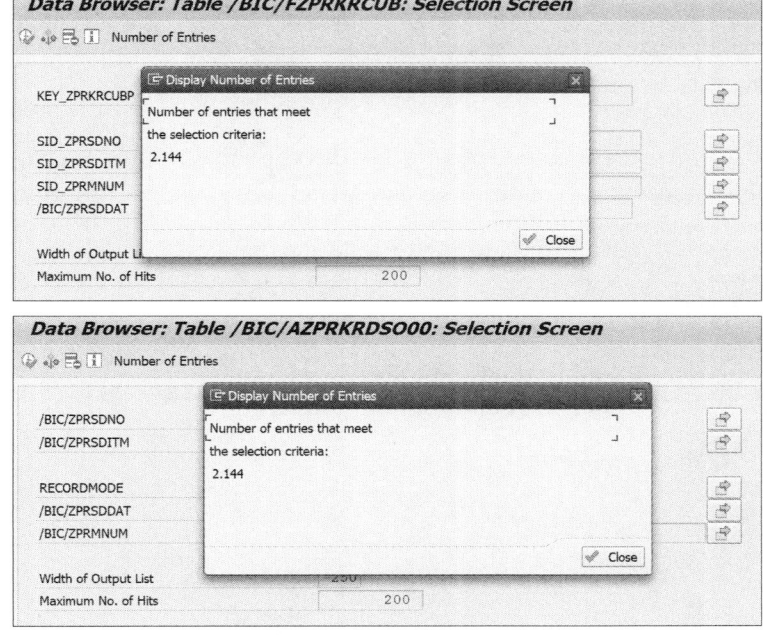

Figure 4.82 Sender System Tables

189

4.8 Package Creation

If you need to start a remote conversion, you use Transaction RSB4HCONV. The first step is to create a new package as the conversion requires an active package in the system. To create a new package, follow these steps:

1. Choose the **Execute Scope Transfer** in the **Realization Phase** tab of the transfer cockpit, as shown in Figure 4.83.

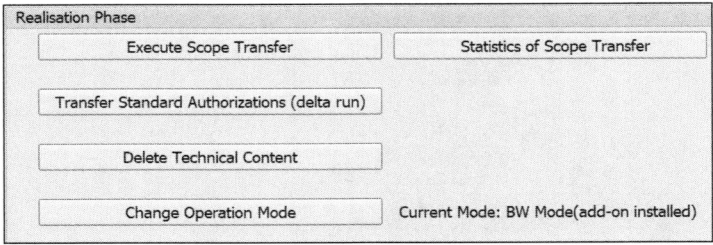

Figure 4.83 Scope Transfer

2. In the **Type of Receiver System** popup, choose **BW/4HANA**, as shown in Figure 4.84.

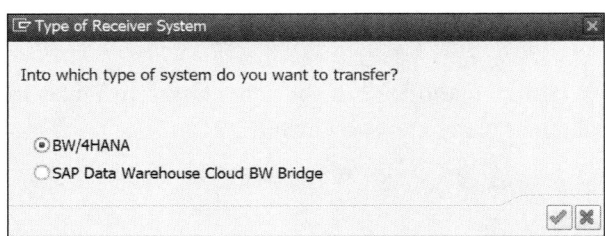

Figure 4.84 Type of Receiver System

3. The next window provides the **Transfer Scenario Selection** option. To transfer using remote conversion, choose the **Remote (Data and Metadata)**, as shown in Figure 4.85.

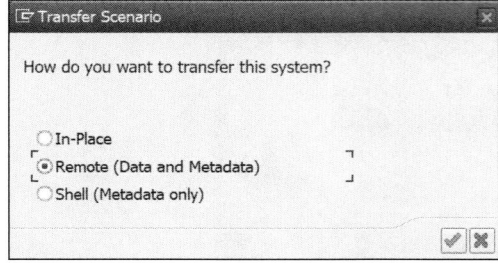

Figure 4.85 Transfer Scenario

4. In the **Create or Show?** popup, you'll have the options to create a new package or show the existing package. Because we're going to create a new conversion package, select **New**, as shown in Figure 4.86.

Figure 4.86 Package Selection

5. The transfer cockpit swiftly generates a fresh Package (**9001K**) encapsulating vital details such as the project name, description, designated solution (in this case, the **BW/4HANA Conversion Cockpit**), creator's user name, and the time stamp of creation. This comprehensive package also includes a process tree, offering a top-level overview of the activities within the **Migration Process**. By default, the **Package** is presented in the **Standard View**, though users have the option to switch to the extended view through icons available in the process monitor. This extended view enhances accessibility of activities used in multiple phases, and there are a few buttons available that help with registering users and monitoring activities seamlessly. The initial screen of the process monitor, as shown in Figure 4.87, serves as a centralized hub for tracking the progression of tasks within the migration process. You can see the following details:

❶ Package number

❷ Phases in the process tree, including the following:
- **Package Settings** (the active phase)
- **Collect Scope and MetaData Transfer**
- **Setup Phase**
- **Generation Phase**
- **Preprocessing Programs during System Lock**
- **Migration Phase**
- **Post Processing Activities during System Lock**
- **System Settings after Migration**

❸ Other phases

❹ Icons for execution and refresh of the process tree

❺ Icons for user management and process tree nodes management

4 Remote Conversion

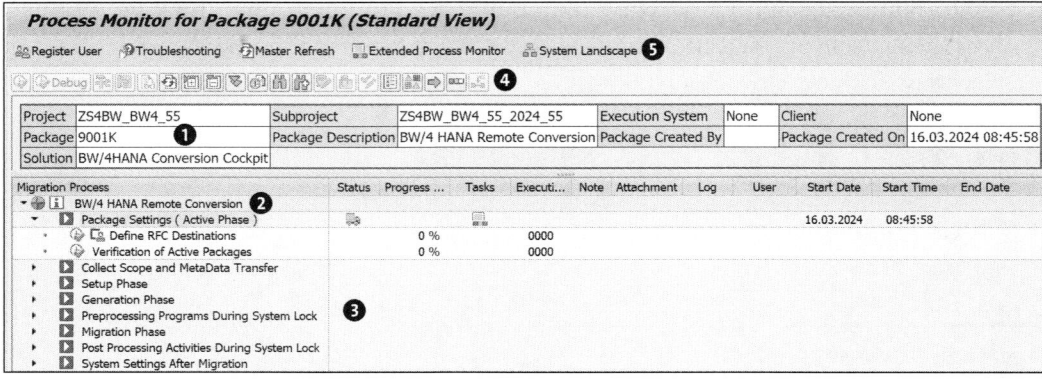

Figure 4.87 Process Tree: Initial View after Creation

4.9 Execution

In this section, we'll execute the actual remote conversion for the specific scope objects. You'll learn about each phase in detail and know how to execute the task list item. You'll also see how the remote conversion framework converts the SAP BW objects and migrates the data.

4.9.1 Package Settings

In the process monitor shown in Figure 4.87, you can see that **Package Settings** is the initial phase in **Migration Process**. The package number is **9001K**, so it won't be the active package, but for remote conversion execution, we need the package to be in active state with no other package active. So, the first step is to make your package the active package so that no other package will be active in the system because you can't have more than one active package in the system. Follow these steps:

1. Select the **Define RFC Destinations** option shown in Figure 4.87. Note that the **Package Settings** has **(Active Phase)** in green and an icon that depicts its active running phase, as shown in Figure 4.88.

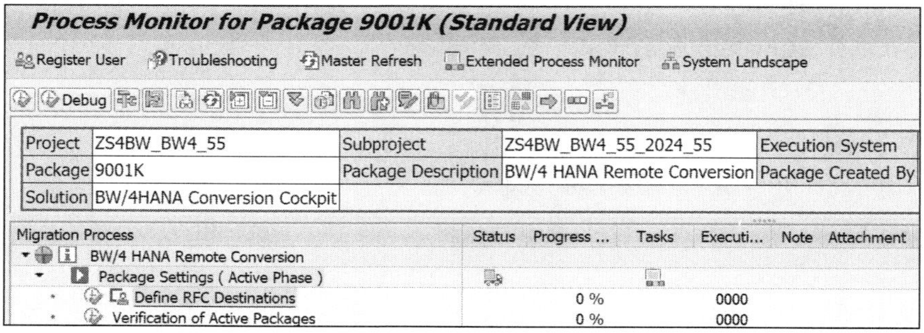

Figure 4.88 RFC Destination

192

2. Now choose ![icon], which will take you to the next screen. Provide the **RFC Destination** for the target SAP BW/4HANA system, and make sure test connection and other tables are successful. Enter the RFC destination name in the **Receiver system** field as "CB4CLNT100", as shown in Figure 4.89.

Figure 4.89 RFC Destination Maintenance

3. Once you choose **Test connection**, you'll get a message of ✅ Successfully checked, and you can do an authorization check and a remote logon too to check that everything is fine. Click the 💾 icon to save, and go back by using 🔙 icon. You can see the status is green, as shown in Figure 4.90.

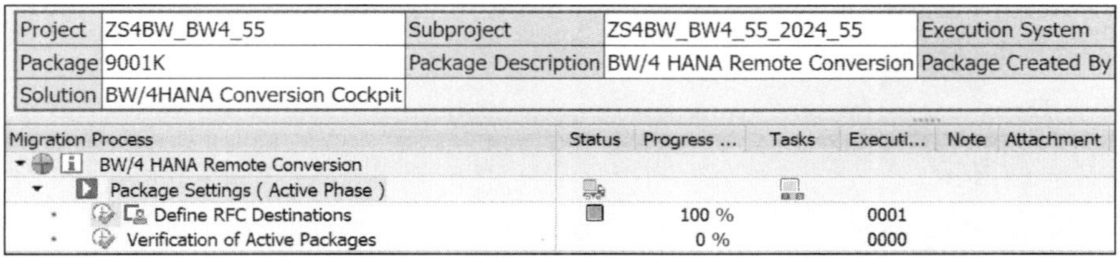

Project	ZS4BW_BW4_55	Subproject	ZS4BW_BW4_55_2024_55	Execution System			
Package	9001K	Package Description	BW/4 HANA Remote Conversion	Package Created By			
Solution	BW/4HANA Conversion Cockpit						
Migration Process			Status	Progress ...	Tasks	Executi...	Note Attachment
▼ BW/4 HANA Remote Conversion							
▼ Package Settings (Active Phase)							
· Define RFC Destinations				100 %		0001	
· Verification of Active Packages				0 %		0000	

Figure 4.90 RFC Destination Execution Status

The **Verification of Active Packages** option is used to check if there are any active packages in the system. If not, package 9001K becomes the active package. If it's green, then no other package can be executed further until this package is closed. Click on the ![icon] beside **Verification of Active Packages**, to go to the screen shown in Figure 4.91.

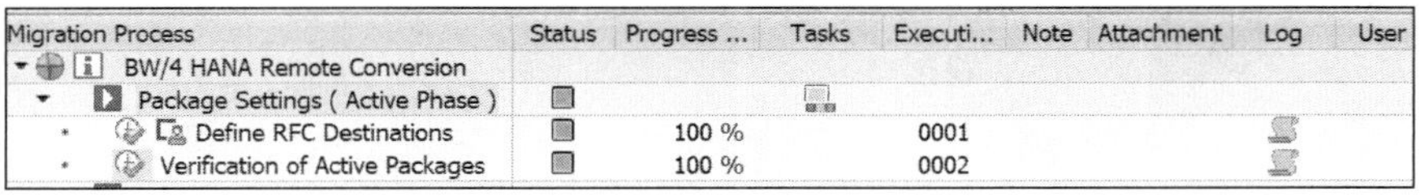

Migration Process	Status	Progress ...	Tasks	Executi...	Note	Attachment	Log	User
▼ BW/4 HANA Remote Conversion								
▼ Package Settings (Active Phase)								
· Define RFC Destinations		100 %		0001				
· Verification of Active Packages		100 %		0002				

Figure 4.91 Active Packages

You'll notice the status is green and **Progress** is at **100%**. If any errors occur, the status will turn red. You can address these errors using the logs accessible via the 📋 icon and then attempt to re-execute by clicking on another ![icon] icon. Each re-execution will be numbered, for instance, as 0002 if it's the second attempt, and this number will increment with each subsequent execution. Multiple executions are common during tasks such as scope selection. This is just an overview of how the execution process appears.

193

4 Remote Conversion

You can now see that the 🚚 icon isn't there because all the underlying items in this phase are completed, but you still see **(Active Phase)** because you didn't execute the next phase.

4.9.2 Collect Scope and Metadata Transfer

Once you've completed the **Package Settings** phase, you'll be ready to start the next phase, which is **Collect Scope and MetaData Transfer**, as shown in Figure 4.92.

Project	ZS4BW_BW4_55	Subproject	ZS4BW_BW4_55_2024_55	Execution System	
Package	9001K		Package Description	BW/4 HANA Remote Conversion	Package Created By
Solution	BW/4HANA Conversion Cockpit				

Migration Process	Status	Progress …	Tasks	Executi…	Note	Attachment
▾ ⓘ BW/4 HANA Remote Conversion						
▾ ▶ Package Settings (Active Phase)						
· 📋 Define RFC Destinations		100 %		0001		
· 📋 Verification of Active Packages		100 %		0002		
▾ ▶ Collect Scope and MetaData Transfer						
· 📋 Collect Scope for Transfer (Data flow)		0 %		0000		
· 📋 Define Object Mapping and Store Object List		0 %		0000		
· 📋 Prepare Propagate Requests		0 %		0000		
· 📋 Map InfoObjects for InfoProvider		0 %		0000		
· 📋 Select Main Partition Criteria for Semantically Par		0 %		0000		
· 📋 Determine Usage of Involved Objects		0 %		0000		
· 📋 Checklist for Usage of Involved Objects		0 %		0000		
· 📋 Confirm Metadata in Target System with Data Tr		0 %		0000		
▸ ▶ Setup Phase						

Figure 4.92 Collect Scope and MetaData Transfer

> **Task List Name**
>
> Until you initiate and execute the **Collect Scope for Transfer (Data flow)** step, the task list won't be generated. To verify this, you can access Transaction STC02, provide the user name in the **Created By** field, and click **Execute**. You'll observe that no task list has been generated at this stage based on the date.

As shown in Figure 4.93, the search for task list using Transaction STC02 won't have any name for the task list run.

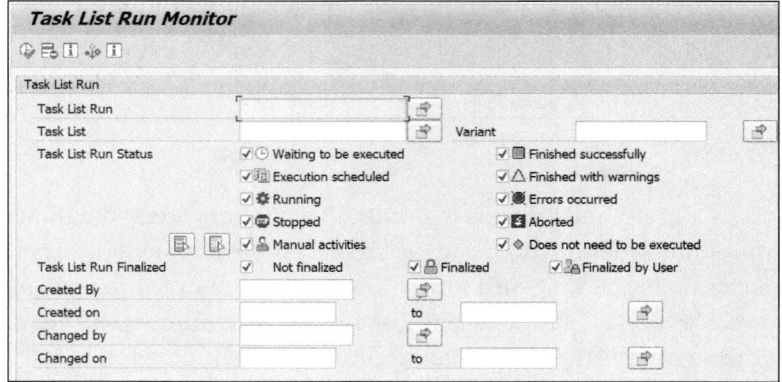

Figure 4.93 Transaction STC02: Selection Fields

194

In Figure 4.93, we can now execute using the icon without any selections. You'll see that there is no running task list run for the remote conversion, as shown in Figure 4.94.

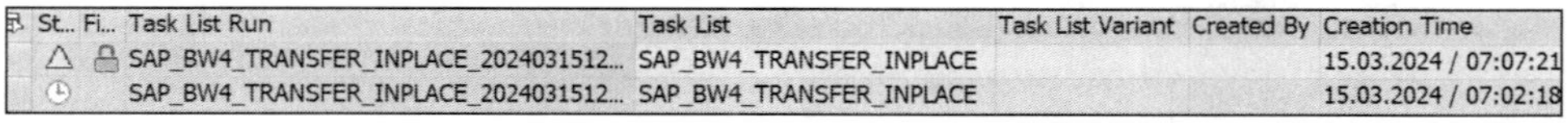

St... Fi... Task List Run
Task List
Task List Variant
Created By
Creation Time
SAP_BW4_TRANSFER_INPLACE_2024031512...
SAP_BW4_TRANSFER_INPLACE
15.03.2024 / 07:07:21
SAP_BW4_TRANSFER_INPLACE_2024031512...
SAP_BW4_TRANSFER_INPLACE
15.03.2024 / 07:02:18

Figure 4.94 Current Task List Based on User Selection

Now you can come back to the process tree and execute the task list item for **Collect Scope and MetaData Transfer**, as shown in Figure 4.95.

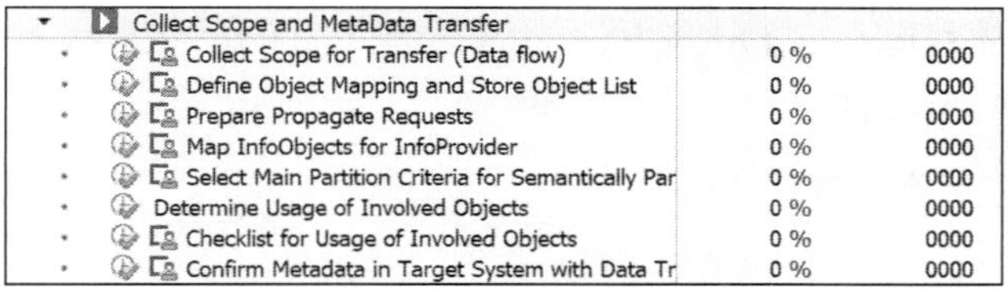

Collect Scope and MetaData Transfer
Collect Scope for Transfer (Data flow) 0 % 0000
Define Object Mapping and Store Object List 0 % 0000
Prepare Propagate Requests 0 % 0000
Map InfoObjects for InfoProvider 0 % 0000
Select Main Partition Criteria for Semantically Par 0 % 0000
Determine Usage of Involved Objects 0 % 0000
Checklist for Usage of Involved Objects 0 % 0000
Confirm Metadata in Target System with Data Tr 0 % 0000

Figure 4.95 Collect Scope for Transfer

Once you click the execute icon, you'll see the options to select the start objects for the conversion, as shown in Figure 4.96. You can use the following icons:

- : Add or remove the object.
- : Add a single object based on TLOGO.
- : Add multiple objects such as InfoCubes, InfoSets, and process chains based on TLOGO.
- : Add all data warehouse objects.
- : Add top line process chains.
- : Add from the remote system.
- Minimal Scope: Collect only limited objects.
- Requests: Turn on new request management.

Edit System Help
Select start objects
Minimal Scope
Collect Queries
Requests
Obj. Type Object Name

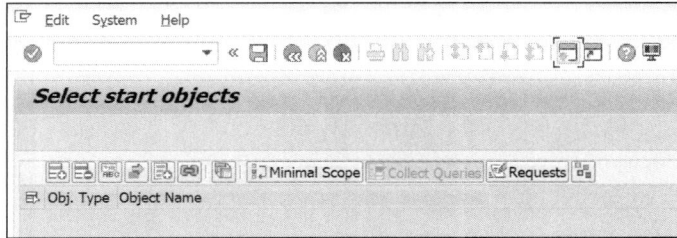

Figure 4.96 Select Start Objects

Now let's click the icon to add our top-level InfoCube, which will be able to collect all the dependent objects so there is no need to add any other objects. However, you can add a process chain if you need to, but make sure you add other objects such as InfoSets and MultiProviders as well that won't be part of the process chain. For our demo, we'll add the InfoCube name in the **Object Name** field and set the TLOGO (**Object Type**) as **CUBE** for InfoCube, as shown in Figure 4.97.

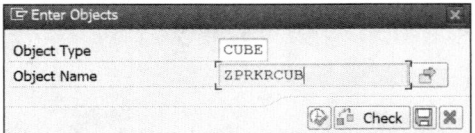

Figure 4.97 Enter Object: InfoCube Selection

Once you choose to execute the **Select start objects** window, then you'll see that **CUBE** and the associated **Object Name** are added as a row, as shown in Figure 4.98.

Figure 4.98 Addition of Start Objects

If you need to add more SAP BW start objects, then you need to choose and add objects based on your conversion requirement. You need to make sure to turn on Requests, which will use new request management and help with performance. In the popup that appears, choose **OK**. Read the provided SAP Note for the behavior, and then press [Enter]. Once turned on, you'll see the **Requests** button is highlighted (see Figure 4.99).

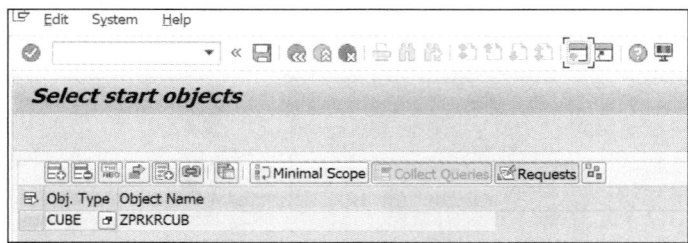

Figure 4.99 Enabling Optimized Request Selection

If you don't need more objects, such as associated queries, choose **Minimal Scope**, which will bring only the required objects for the conversion. If you need all objects leave, it as is. Now choose **Save** and go back, and you'll see the popup for the processing option, as shown in Figure 4.100.

4.9 Execution

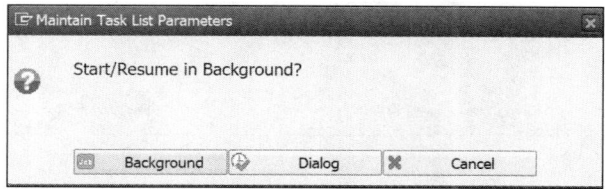

Figure 4.100 Job Execution Option

Choose **Background**, and then you'll go back to the process tree and see that job in running status with the icon. The active phase is now changed to the task list item **Collect Scope and Metadata Transfer**, as shown in Figure 4.101.

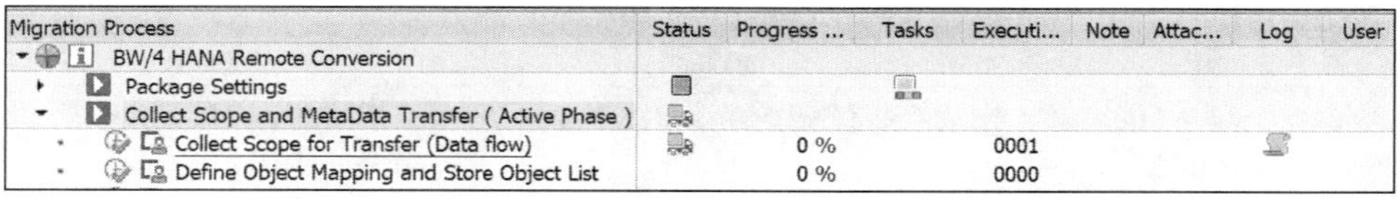

Figure 4.101 Collect Scope for Transfer: Active Phase

While the job is running, which will take time based on the object count, you can click refresh to see the status. In our demo, the job is completed, so you'll see a green icon on your screen (Figure 4.102).

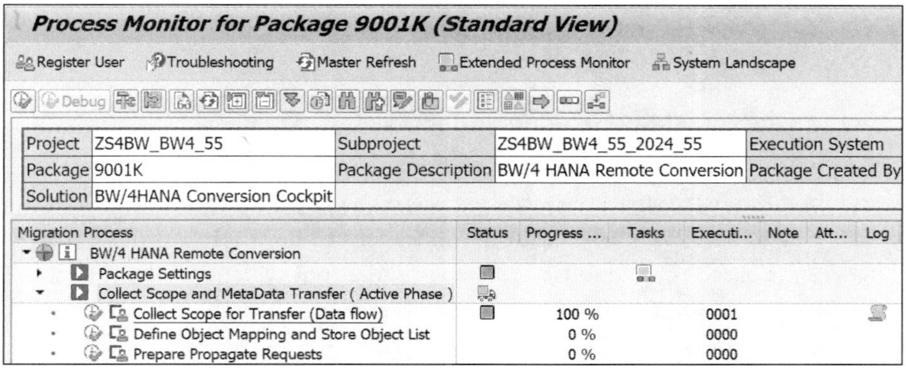

Figure 4.102 Collect Scope for Transfer: Final Status

After observing the status displayed as green, yellow, or red, you'll notice the corresponding task list name in Transaction STC02. The job's status aligns with the process tree; it's important to recognize that the cockpit-based process tree initiates the Transaction STC01 job in the background. This job endeavors to gather all associated objects. Therefore, it's crucial to understand that this task generates Transaction STC01 job in the background, which you can then observe in Transaction SCT02. Figure 4.103 shows that the job is stopped and waiting for next steps.

197

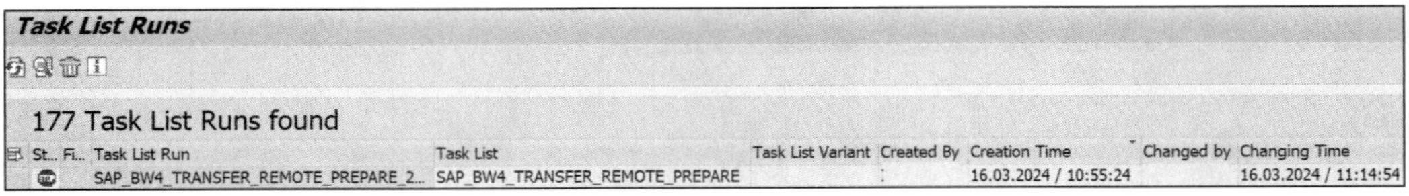

Figure 4.103 Latest Task List Run

Once you can see the task list run name in Transaction STC02, you can double-click on it to see the status in Transaction STC02. The job is green in **Collect Scope for Transfer**, and it's currently stopped in the **Define Object Mapping** step waiting for the input parameter, as shown in Figure 4.104. You can use ![icon] to add parameters here and try to execute and check logs until its green. But you can also do that from the process tree, as the same logs from Transaction STC02 will be shown there.

Figure 4.104 Task List in Transaction STC02

Now go back to the process tree to move to the next step.

4.9.3 Define Object Mapping and Store Object List

You'll find yourself in the **Define Object Mapping and Store Object List** step, which is a crucial and time-intensive stage. Here, the focus is on gathering information about the old object type and specifying how the tool will transform it into the new object type. Using algorithms based on previous scope objects, the tool systematically collects all necessary objects essential for conversion and ensures that they are accurately mapped to the appropriate SAP BW/4HANA object.

For instance, it attempts to map InfoCubes to ADSOs and ODSOs to ADSOs. In cases where a persistent staging area (PSA) table exists, it asks whether you want to replace it with an inbound ADSO. Additionally, new names for these ADSOs in the target system might be necessary. Typically, in extensive SAP BW/4HANA conversion projects, this activity is iterated multiple times after addressing numerous errors. Therefore, it's essential to have a proficient project management team with an error-tracking process to mitigate prolonged runtimes. With a higher number of objects, the execution time for this activity increases, potentially extending beyond several hours, as shown in Figure 4.105.

4.9 Execution

Migration Process	Status	Progress ...	Tasks	Executi...	Note
▼ ⓘ BW/4 HANA Remote Conversion					
▶ ▶ Package Settings	☐		▣		
▼ ▶ Collect Scope and MetaData Transfer (Active Phase)	▣				
• ⊕ 🗐 Collect Scope for Transfer (Data flow)	☐	100 %		0001	
• ⊕ 🗐 Define Object Mapping and Store Object List		0 %		0000	
• ⊕ 🗐 Prepare Propagate Requests		0 %		0000	
• ⊕ 🗐 Map InfoObjects for InfoProvider		0 %		0000	
• ⊕ 🗐 Select Main Partition Criteria for Semantically Partitioned		0 %		0000	
• ⊕ Determine Usage of Involved Objects		0 %		0000	
• ⊕ 🗐 Checklist for Usage of Involved Objects		0 %		0000	
• ⊕ 🗐 Confirm Metadata in Target System with Data Transfer Prep		0 %		0000	

Figure 4.105 Define Object Mapping

Now click **Execute** to go to the screen shown in Figure 4.106 with the following details:

❶ Original TLOGO

❷ Original object name description

❸ Target TLOGO

❹ Target or new object name that will be in SAP BW/4HANA

❺ Transfer status

❻ Special options for transfer

❼ Metadata status

❽ Icons for grid management

❾ Buttons for showing and hiding objects

❿ Icons for package assignment and carveout scenarios

Figure 4.106 Object Mapping: Initial screen

Now you can decide if you need to have an inbound ADSO for the PSA table; if so, you need to provide a new ADSO name in the **Name** input field. The inbound ADSO will create TRFN and DTP additionally and then look like the screen in Figure 4.107.

Note that you need to make sure the S4E_SAP source system is created in ODP–SAP with the SAPI context in the target system and ensure that ZPRRTDS is replicated and exposed to ODP there.

199

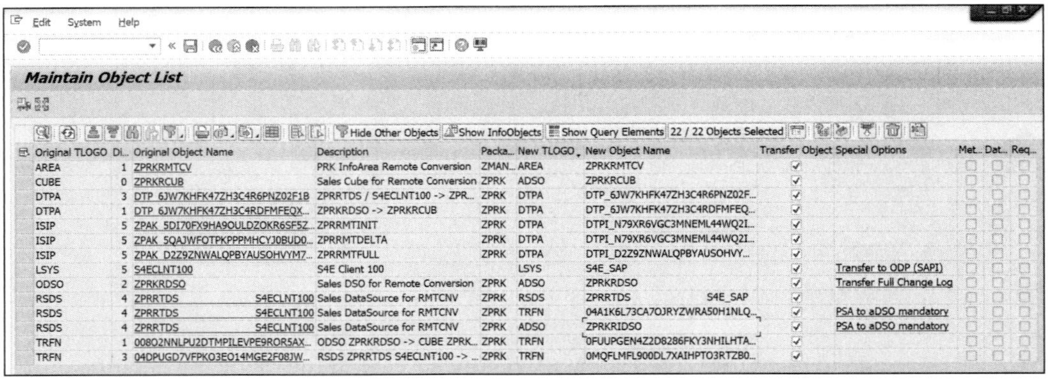

4   Remote Conversion

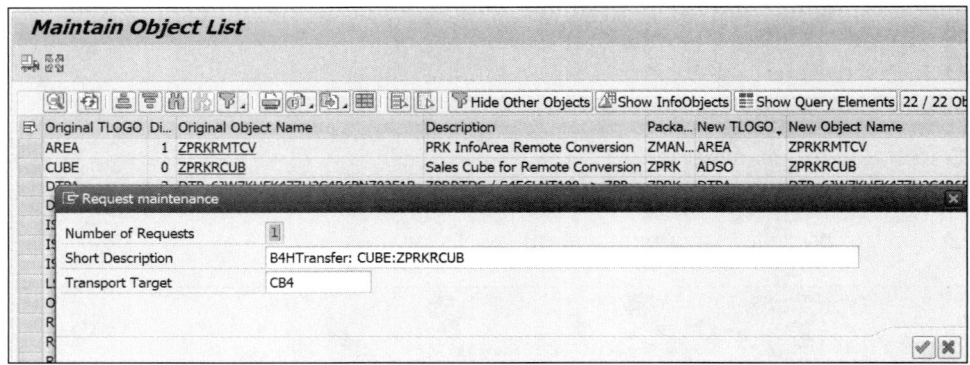

Figure 4.107 Inbound ADSO

Now you can see that TRFN and DTP are already created. Once everything is fine, you can choose ![] to see the popup shown in Figure 4.108.

Figure 4.108 Transport Request Popup

You can give **Number of Requests** from 1 to 9 based on your total object count, and it will help you to split the objects into different transport request numbers. For our demo, choose **1**. You'll see a ![Request created] message in the message bar. Then, if you choose the ![] icon again, you'll see the tool-generated transport request number, as shown in Figure 4.109. Click ![].

Figure 4.109 Generated Transport Request Number

If you have too many objects, you can group the transport request using icon, shown in Figure 4.110.

TLOGO Types on Separate Transports
Object Type
Multiple selection
Check

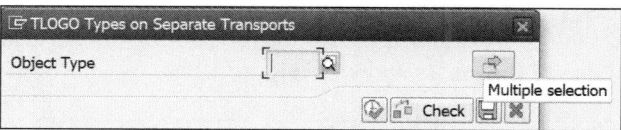

Figure 4.110 TLOGO Grouping

Now you can decide the order of your grouping. Note that the tool has a preference, so you can't expect the same grouping of TLOGO. Save and start the **Resume in Background** mode using the icon, and you'll see the status shown in Figure 4.111.

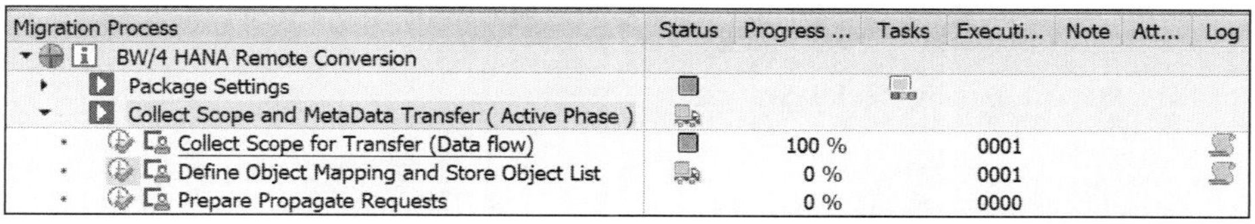

Migration Process
Status
Progress ...
Tasks
Executi...
Note
Att...
Log
BW/4 HANA Remote Conversion
Package Settings
Collect Scope and MetaData Transfer (Active Phase)
Collect Scope for Transfer (Data flow)
100 %
0001
Define Object Mapping and Store Object List
0 %
0001
Prepare Propagate Requests
0 %
0000

Figure 4.111 Execution of Object Mapping

Then, execute twice, and you can see the status as green, as shown in Figure 4.112.

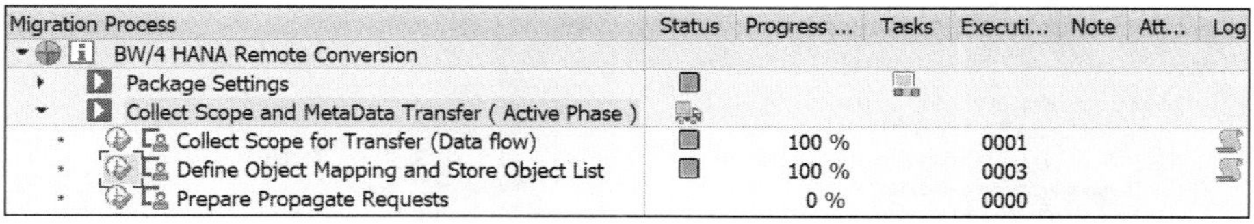

Migration Process
Status
Progress ...
Tasks
Executi...
Note
Att...
Log
BW/4 HANA Remote Conversion
Package Settings
Collect Scope and MetaData Transfer (Active Phase)
Collect Scope for Transfer (Data flow)
100 %
0001
Define Object Mapping and Store Object List
100 %
0003
Prepare Propagate Requests
0 %
0000

Figure 4.112 Object Mapping Status

As the step is green, you're ready to start the next task.

4.9.4 Prepare Propagate Request

Execution of the **Prepare Propagate Request** will open the screen shown in Figure 4.113.

Propagate
Edit
Goto
System
Help
Selection of DTPs to be propagated
Set 'Confirm Execution' for Selected DTP(s)
Set 'Insert into Scope' for Selected DTP(s)
Confirm Execution
Insert into Scope
Logically Delete DTP
Data Transfer Process
Scope List
Warnings
Object Status
Source Object
TLOGO
Target Object
TLOGO
DTP_6JW7KHFK47ZH3C4RDFMFEQXD8
In Scope
Active
ZPRKRDSO
ODSO
ZPRKRCUB
CUBE

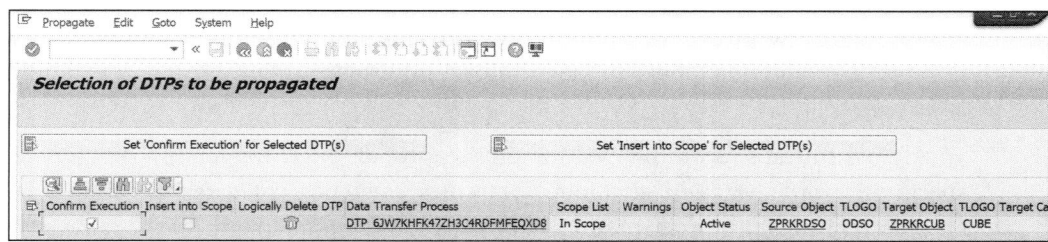

Figure 4.113 Prepare Propagate Request

Now make sure the **Confirm Execution** checkbox is checked (it's selected by default). Then, choose the **Set 'Confirm Execution' for Selected DTP(s)** button. If you choose this option, based on the objects selected, you might get a popup saying the DTP maintenance status is complete. Once you see that popup, choose **Enter and run as a background job**, and you'll see the status of the job, as shown in Figure 4.114.

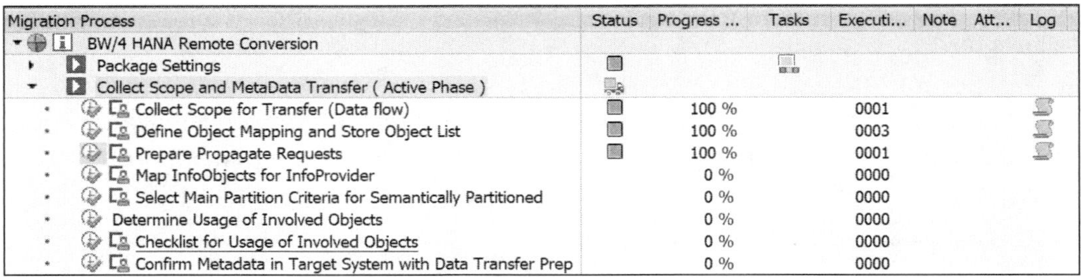

Figure 4.114 Prepare Propagate: Job Status

4.9.5 Map InfoObject for InfoProvider

The next step is to manually execute **Map InfoObjects for InfoProvider**. We have used the execute icon in this task list item and once it's started it will be in the running state and after few minutes the task list item is executed and has a green status, as shown in Figure 4.115.

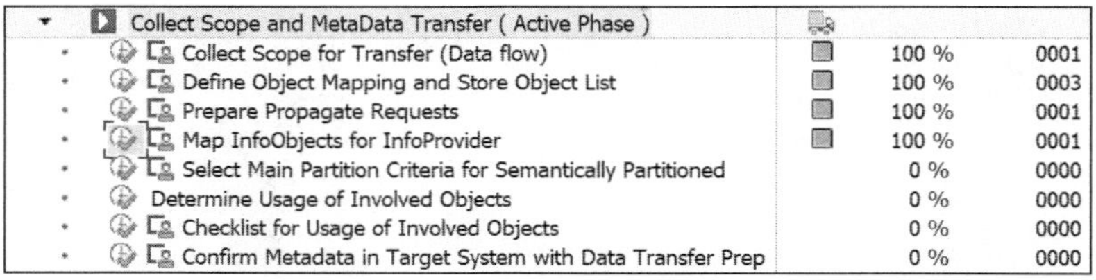

Figure 4.115 Map InfoObjects for InfoProvider

This task enables the essential InfoObject mapping for transferring a MultiProvider to a CompositeProvider. The mapping is crucial as the CompositeProvider demands a uniform mapping of navigation attributes. A consistent mapping is necessary for navigation attributes to align with the attribute-carrying characteristic. Additionally, the CompositeProvider no longer supports InfoSet InfoObjects, requiring a new mapping approach.

To address this, it's recommended for the creation of a new reference characteristic specifically for navigation attributes. This new reference characteristic should be employed in all MultiProvider transfers where a consistent mapping for the same navigation attribute is lacking. The task not only facilitates the creation of a new reference characteristic but also offers the option to generate mapping proposals from previous transfers.

When you have the edit parameter icon here, you'll see the window with InfoObject Compatibility Check for InfoProviders. If you see the list of InfoProvider columns and the New InfoObject column, then you must assign new InfoObjects or work with a field name (global unique name needs to be set here). You'll have these two icons for creating the InfoObjects.

4.9.6 Select Main Partition Criteria for Semantically Partitioned

The next step is select the **Select Main Partition Criteria for Semantically Partitioned**. In this task, you can choose the main partition InfoObject for the new ADSO in the semantic group you're creating. Because semantic groups can only have one partition InfoObject, it's important to make this selection. However, the filters from other partition InfoObjects are still included in the corresponding ADSO as filter criteria. Choose that row and execute using . When completed, it will have a green status, as shown in Figure 4.116.

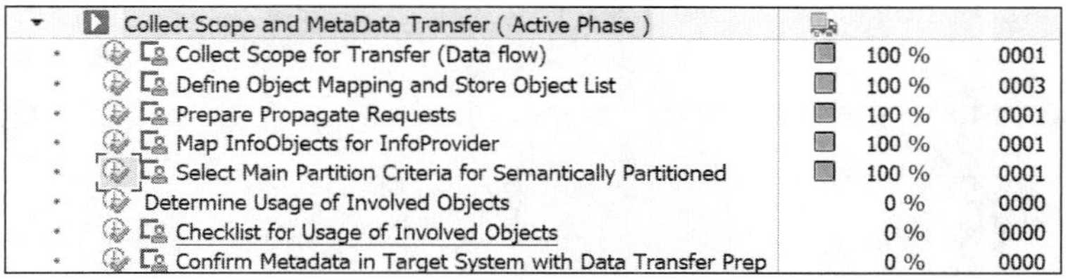

Figure 4.116 Select Partition Criteria

4.9.7 Determine Usage of Involved Objects

The next task, **Determine Usage of Involved Objects**, examines the objects generated for InfoProviders and conducts scans that involve routines, planning functions, and more. It aims to identify any problematic code that could potentially lead to errors during the transfer process. By scrutinizing these elements, it ensures a smoother transition and minimizes the risk of encountering issues post-transfer. Execute the task to see the status in Figure 4.117.

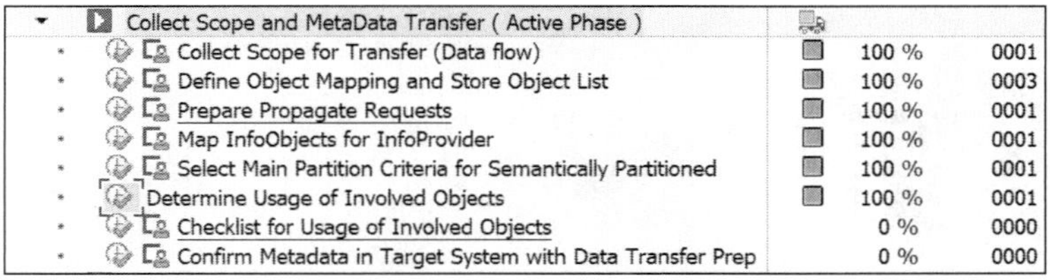

Figure 4.117 Usage of Involved Objects

4.9.8 Checklist for Usage of Involved Objects

In this task, a comprehensive list of objects or findings is presented for review and resolution. It encompasses various aspects such as the usage of generated objects and InfoProviders that may no longer be available after the transfer. Additionally, it involves scrutinizing customer code to identify any modifications resulting from the transfer process. Execute the **Checklist for Usage of Involved Objects**. In the task list **Checklist for Involved Objects**, you'll need to make sure that the coding for **BADI** and **AUTH** is already taken care and resolved. You'll see the screen in Figure 4.118 by default. If the coding is already done, choose the **Resolved** checkbox for both **BADI** and **AUTH**.

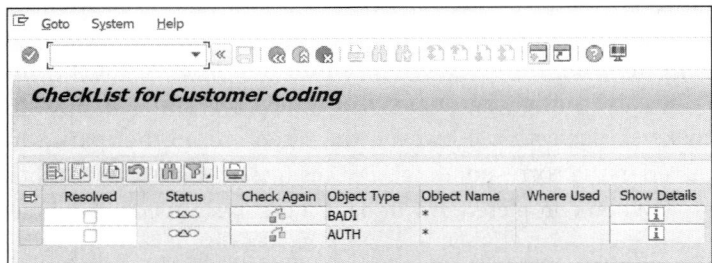

Figure 4.118 Checklist for Customer Coding Initial Screen

The status of the execution is green, as shown in Figure 4.119.

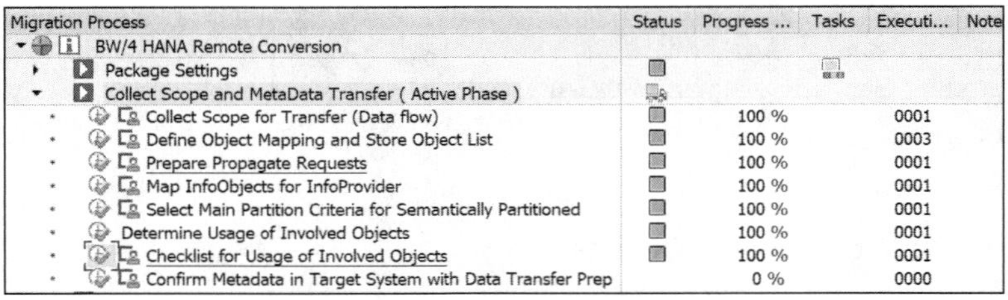

Figure 4.119 Checklist for Usage of Involved Objects Status

If you open the task list run using Transaction STCO2 and choose refresh, you can see the same status is synced up, as shown in Figure 4.120.

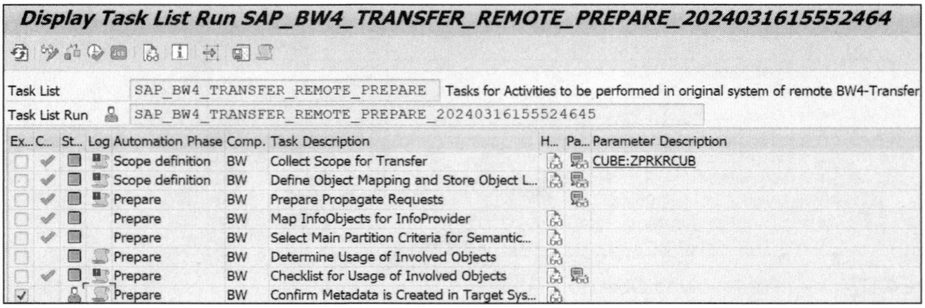

Figure 4.120 Transaction STC02 Status

4.9 Execution

4.9.9 Confirm Metadata in Target System with Data Transfer Prep

Before starting the next step, you need to release the transport request created in the **Define Object Mapping** step. Open Transaction SE01 and provide the transport request number (**BW4K900636**) that was generated by the SAP BW/4HANA conversion cockpit. Open the transport request, as shown in Figure 4.121. The transport request lists all the collected objects; for example, in the **InfoCube** folder, you can see the InfoCube that was collected (**ZPRKRCUB**).

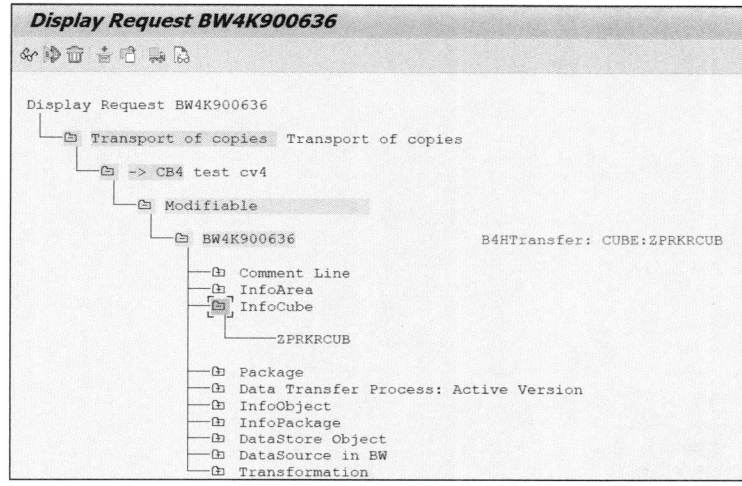

Figure 4.121 Transport Request Release in the SAP BW Sender System

Once you choose ![icon] to release the transport request after validating if it was sent to the target SAP BW/4HANA system with the Basis team, you'll see the status of the transport request release, as shown in Figure 4.122.

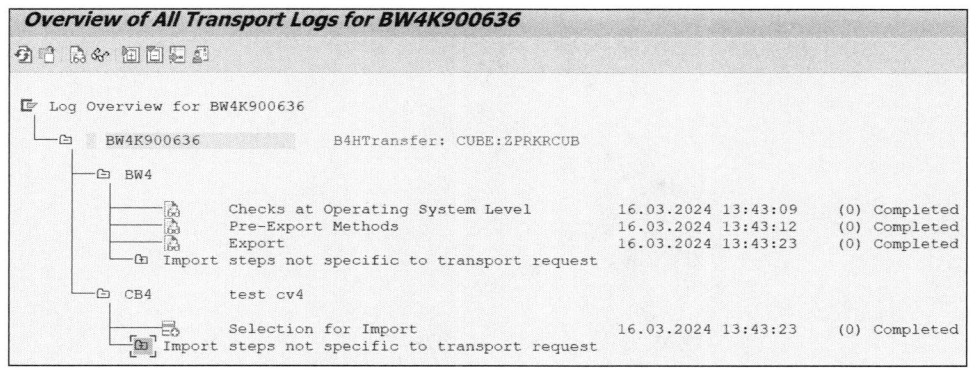

Figure 4.122 Post transport Request Release

Once the transport request is released, the next step is to log on to the target SAP BW/4HANA system, try to search for the released transport request number

205

4   Remote Conversion

(BW4K900636) by using Transaction STMS_IMPORT, and import the transport request using ![truck icon]. Then, the transport request import will start in the target SAP BW/4HANA system. See Figure 4.123 for the options to use during import. You'll use the **Options** tab to choose how to do the transport request import; in our case, we choose all checkboxes except **Overwrite Originals**. Finally, choose ![checkmark].

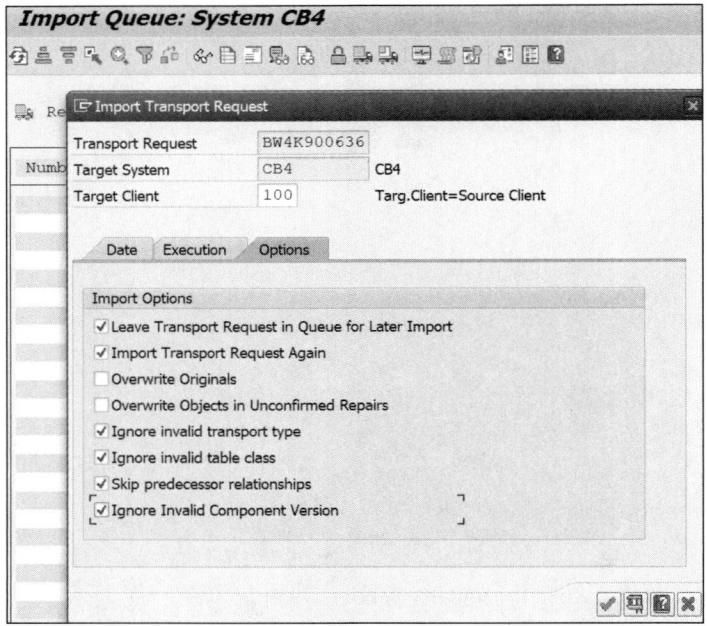

Figure 4.123 Transaction STMS_IMPORT

Once the transport request is imported in SAP BW/4HANA, you can view the logs using the ![icon] icon. You'll be taken to the overall import status shown in Figure 4.124.

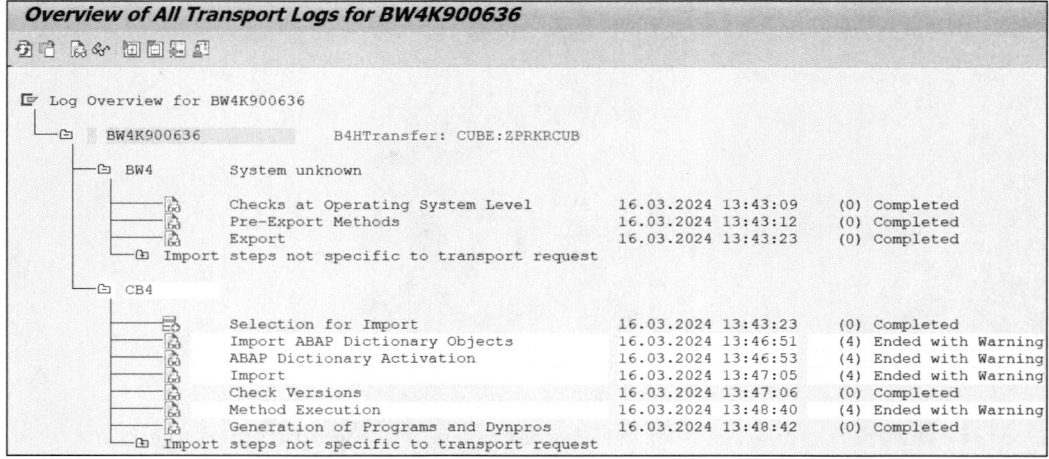

Figure 4.124 Import Status in SAP BW/4HANA

4.9 Execution

Open the SAP BW modeling tools in Eclipse, and you can see that the InfoCube got converted to an ADSO in the SAP BW modeling tools. The associated DSO is also converted to an ADSO, and you can see the flow in Figure 4.125.

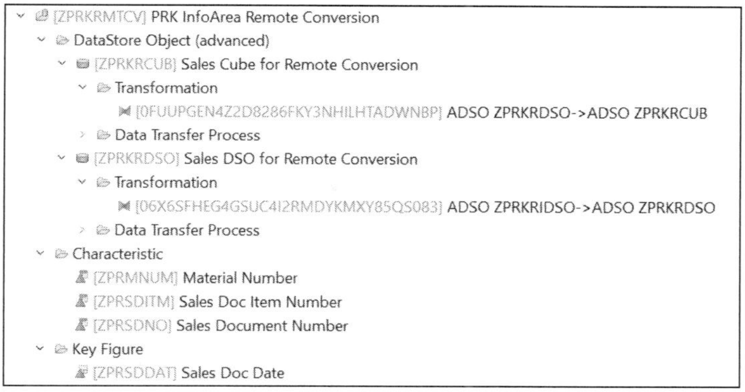

```
[ZPRKRMTCV] PRK InfoArea Remote Conversion
  DataStore Object (advanced)
    [ZPRKRCUB] Sales Cube for Remote Conversion
      Transformation
        [0FUUPGEN4Z2D8286FKY3NHILHTADWNBP] ADSO ZPRKRDSO->ADSO ZPRKRCUB
      Data Transfer Process
    [ZPRKRDSO] Sales DSO for Remote Conversion
      Transformation
        [06X6SFHEG4GSUC4I2RMDYKMXY85QS083] ADSO ZPRKRIDSO->ADSO ZPRKRDSO
      Data Transfer Process
  Characteristic
    [ZPRMNUM] Material Number
    [ZPRSDITM] Sales Doc Item Number
    [ZPRSDNO] Sales Document Number
  Key Figure
    [ZPRSDDAT] Sales Doc Date
```

Figure 4.125 SAP BW/4HANA: After Import on Objects

To validate the conversion of the data flow effectively, it's crucial to conduct a thorough comparison between the InfoCube flow from the legacy SAP BW 7.x sender system and the corresponding transformed ADSO flow within the SAP BW/4HANA environment. In the transition to SAP BW/4HANA, notable adjustments occur. First, all three InfoPackages are removed from the target SAP BW/4HANA system due to the lack of compatibility with InfoPackages in this updated version. This necessitates the migration of InfoPackage-based selections to the single DTP for continued functionality.

Second, a new inbound ADSO is implemented, specifically tailored to accommodate the data from the PSA table. This addition is complemented by the integration of new DTP and transformation processes, which significantly boost data processing efficiency within the SAP BW/4HANA system, particularly for handling data within the inbound ADSO. Formerly managed selections via InfoPackages are seamlessly transitioned to the newly introduced DTP, functioning in conjunction with the inbound ADSO.

Third, as part of the conversion and import process, an ADSO with the same name as the original InfoCube is generated based on the classical ODSO framework (this name was selected by the tool in the object mapping phase). This facilitates the retention of essential data structures and ensures continuity in data handling processes. The culmination of these transformations results in the creation of an ADSO, identified as **ZPRKRCUB**, derived from the initial InfoCube. This entire conversion process is shown in detail in Figure 4.126, which shows the sender system data flow ❶ and the SAP BW/4HANA system converted data flow ❷. Consequently, the revised data flow in the SAP BW/4HANA system consists of one inbound ADSO and two ADSOs, effectively replacing the functionalities of the ODSO and InfoCube within the post-conversion landscape.

207

4 Remote Conversion

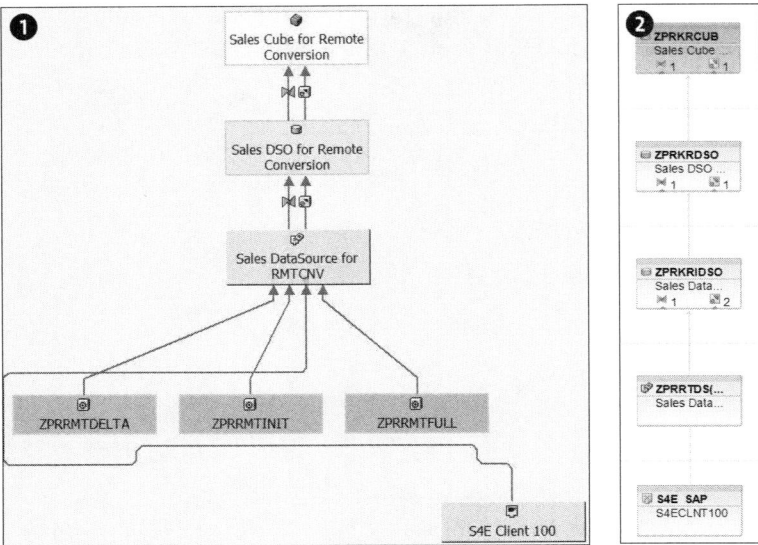

Figure 4.126 SAP BW Sender and SAP BW/4HANA Target

Do the metadata validation if all the transports are successful, and then finally execute this task. You'll get a popup that allows you to see that the InfoCube became an ADSO. Let's now review the ADSO properties of each converted object in the target SAP BW/4HANA system. The InfoCube has the **Data Mart DataStore Object** property, as shown in Figure 4.127.

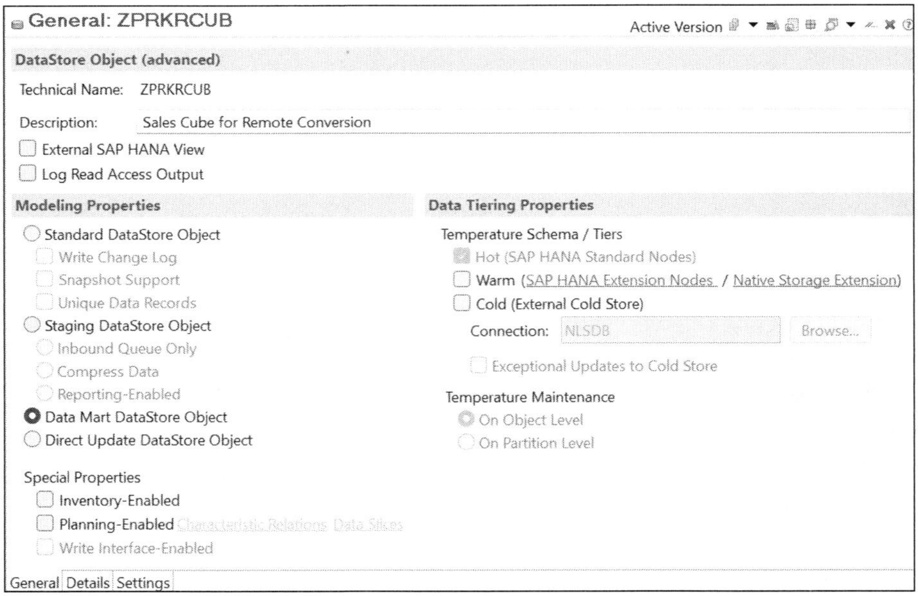

Figure 4.127 InfoCube to ADSO Conversion

You'll notice that the DSO has transitioned into a standard ADSO with a change log, as shown in Figure 4.128.

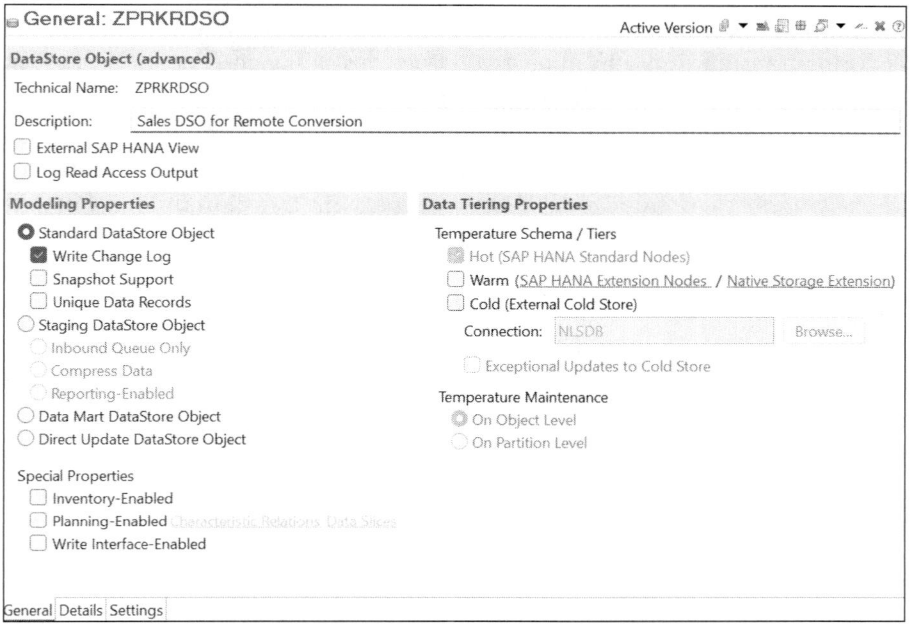

Figure 4.128 DSO to ADSO Conversion

You can observe that the PSA has transformed into a **Staging DataStore Object**, featuring an **Inbound Queue Only**, as shown in Figure 4.129.

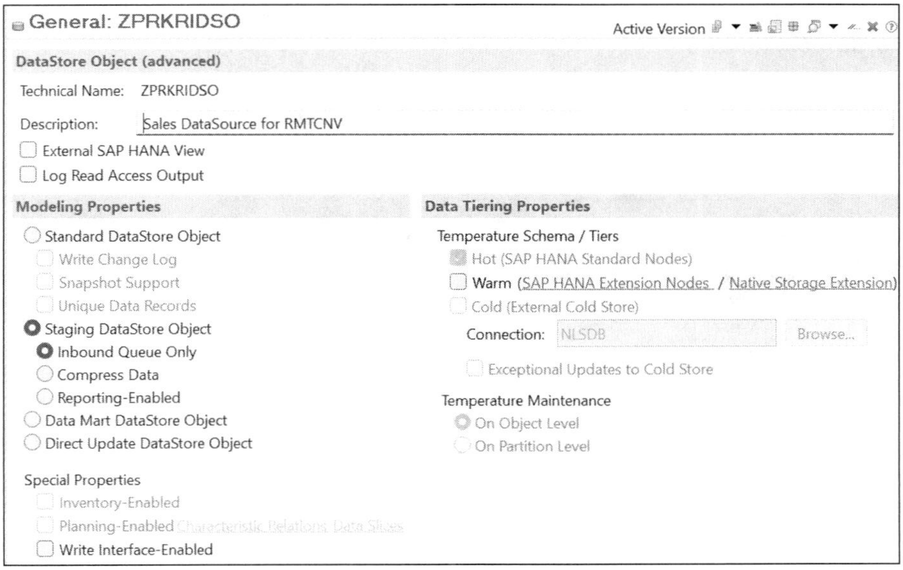

Figure 4.129 Staging DataStore Object for PSA and DataSource

Once you've ensured that all objects are active within your target SAP BW/4HANA system and haven't overlooked any DataSources, InfoObjects, ADSOs, InfoCubes, DTPs, or transformations, you can proceed to confirm the item. But remember, extensive validation of metadata transports in SAP BW/4HANA is imperative, and the target system checks during transport import can be quite time-consuming and demanding. Given the considerable time required for overall transport import, especially in large, complex projects where one may encounter 60,000, 90,000, or even beyond 100,000 objects grouped in TLOGOs across multiple transport requests, each import poses its unique challenges. It's essential to address any dependent transports and any corrective custom transports prior to tool-based transport imports. Ensure all custom tables, ABAP reports, and relevant objects necessary for activation are included. In the case of multiple transport requests, strive for successful status completion for each, as repeated reimportation may be necessary for activation of dependent objects. Ultimately, all transport requests should conclude with a green status, and all transported objects to SAP BW/4HANA must be available as active versions. Validate for any missing transformation rules and DTP filters to ensure completeness. Once all objects are verified as active within the target SAP BW/4HANA system, select the option to **Confirm** the metadata. Technically, if the tool-based generated transport request is successful, the SAP BW/4HANA system will be created without the data, as shown in Figure 4.130.

Figure 4.130 Confirmation

After confirmation, the **Confirm Metadata in Target System with Data Transfer Prep** step shown in Figure 4.131 has a green status.

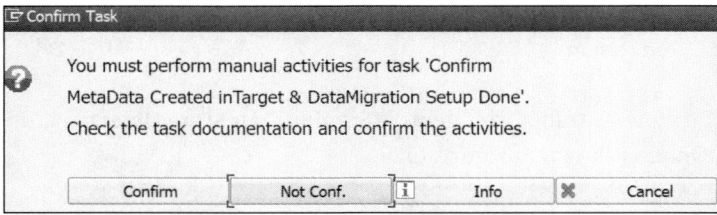

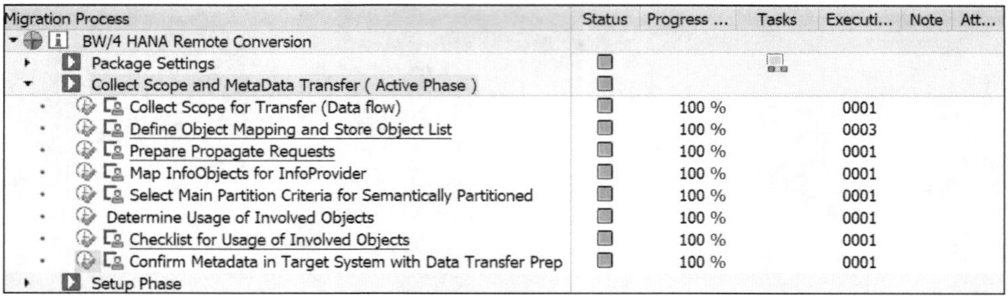

Figure 4.131 Confirm Metadata: Status

4.9.10 Setup Phase

Once all objects have been converted and validated in the target SAP BW/4HANA environment, the setup phase can commence. This phase entails reading object mappings established in the preceding phase and storing them in conversion control tables to facilitate mapping between the sender and target systems. Subsequently, the phase identifies table pools for migration based on these mappings and transfers package settings from the sender system to the target system.

The tool then executes a series of activities within the target SAP BW/4HANA system to ensure semantic consistency between sender system objects (e.g., InfoCubes) and newly generated ADSOs, while also identifying cubes for migration. Additionally, it endeavors to generate fact and dimension tables in the target system, serving as storage for data before transmission to ADSOs. This phase maps all objects between sender and target systems, scrutinizing any differences in Data Dictionary (DDIC) definitions.

Finally, the phase prepares the tool for conversion setup, scrutinizing conversion-related tables and fields and ensuring mapping consistency throughout the process. This comprehensive approach ensures a smooth transition and data integrity in the migration to SAP BW/4HANA.

> **Best Practice**
>
> If the transport imports are successful, and the objects are active in the target system, then the setup phase is likely to encounter minimal errors. There will be rigorous checking of DDIC consistency between sender and target systems during this phase. The activities in the setup phase itself will be conducted thorough DDIC checks, verifying that the data structures align correctly between the two systems. This will ensure that mapping is consistent.

Once the confirmation of metadata is completed, you'll see the **Setup Phase** step, as shown in Figure 4.132.

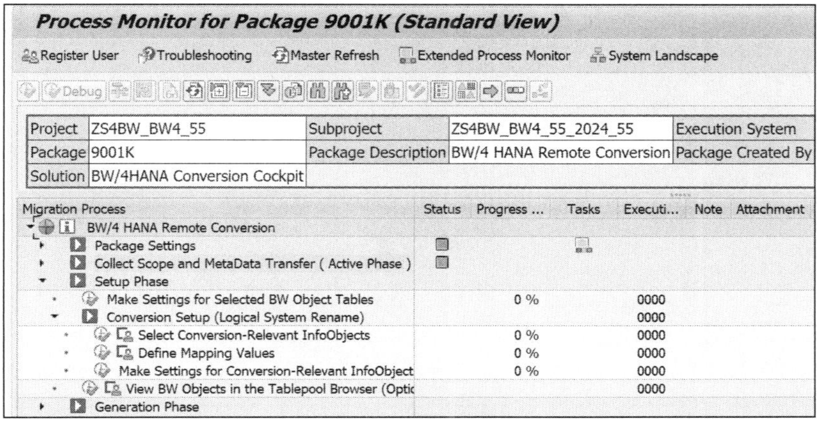

Figure 4.132 Initial Setup Phase

4 Remote Conversion

Though there seem to be fewer activities in the standard view, multiple tasks are running; to see those, you must be in extended mode. In that mode, you'll see the additional tasks that the tool will execute in the **Make Settings for Selected BW Object Tables** and other group activities in the process tree shown in Figure 4.132. If you click the icon, the popup shown in Figure 4.133 will appear where you can choose **Switch to Extended View** and click .

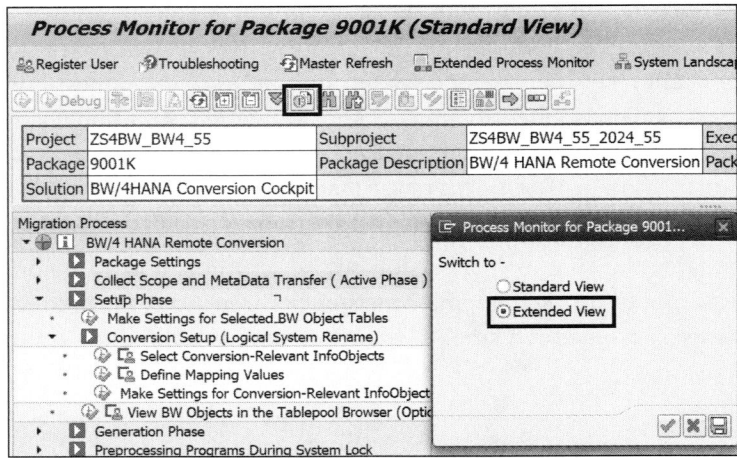

Figure 4.133 Extended View Selection

Once you change to the extended view, you'll be able to see additional tasks (under **Make Settings for Selected BW System**), as shown in Figure 4.134.

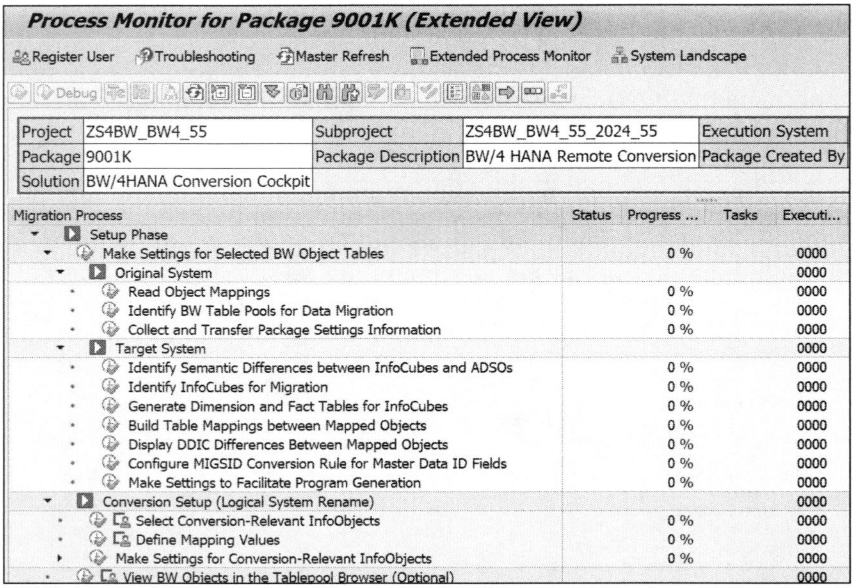

Figure 4.134 Setup: Extended View

Now you choose the top group activity, **Make Settings for Selected BW Object tables**, and it will start all other activities below it. If there is any error, the activity will stop; you can check the logs using ![], fix those errors, and restart using ![]. You can use the extended process monitor icon ![] anytime to see the status of the individual jobs. Now let's start the group activity by choosing ![] in front of **Make Settings for Selected BW Object Tables**, as shown in Figure 4.135.

Migration Process	Status	Progress …	Tasks	Executi…
▼ ▶ Setup Phase (Active Phase)				
▼ Make Settings for Selected BW Object Tables		0 %		0001
▼ ▶ Original System				0000
• Read Object Mappings		0 %		0000
• Identify BW Table Pools for Data Migration		0 %		0000
• Collect and Transfer Package Settings Information		0 %		0000
▼ ▶ Target System				0000
• Identify Semantic Differences between InfoCubes and ADSOs		0 %		0000
• Identify InfoCubes for Migration		0 %		0000
• Generate Dimension and Fact Tables for InfoCubes		0 %		0000
• Build Table Mappings between Mapped Objects		0 %		0000
• Display DDIC Differences Between Mapped Objects		0 %		0000
• Configure MIGSID Conversion Rule for Master Data ID Fields		0 %		0000
• Make Settings to Facilitate Program Generation		0 %		0000
▼ ▶ Conversion Setup (Logical System Rename)				0000
• Select Conversion-Relevant InfoObjects		0 %		0000
• Define Mapping Values		0 %		0000
▶ Make Settings for Conversion-Relevant InfoObjects		0 %		0000
• View BW Objects in the Tablepool Browser (Optional)				0000

Figure 4.135 Setup Phase: List of Activities

Let's review the basics of each activity in this phase before starting the execution:

- **Make Settings for Selected BW Object Tables**
  This is a group activity aimed at initiating subsequent activities in both the sender and target systems. In standard mode, only this activity is visible, but switching to EXTENDED mode reveals all associated subactivities. It entails conducting read mappings and DDIC checks between the sender and target systems, as well as configuring relevant settings accordingly.

- **Read Object Mappings**
  This task relies on the completion of previous object mappings. It retrieves and stores the object mappings in the conversion control tables, which are essential for preparing the migration phase.

- **Identify BW Table Pools for Data Migration**
  In the process of migrating data from the original (sender) system to the SAP BW/4HANA target (receiver) system, there's a critical step of identifying the relevant SAP BW table pools. This automated task is aimed at efficiently determining which table pools are essential for the migration process. It ensures that only necessary data is transferred, optimizing the migration process. This activity occurs within the original (sender) system, where the necessary information is gathered to facilitate the seamless transition of data to the target system.

- **Collect and Transfer Package Settings Information**
  The process of transferring data from the original (sender) system to the target (receiver) system requires the collection and transfer of package settings information.

This task ensures efficient gathering and transmission of package settings from the sender to the receiver system. It's crucial for maintaining consistency and coherence in data handling across systems. This activity is executed in both the original sender and target receiver systems, facilitating seamless communication and data transfer between the two. This activity automatically identifies DDIC information from table pools identified in the preceding **Identify BW Table Pools for Data Migration** task. It stores this information in the control table of the original (sender) system and transfers details about SAP BW objects to the control tables of the target (receiver) system for migration.

- **Identify Semantic Differences between InfoCubes and Mapped ADSOs**
  This task compares the semantic variances between InfoCubes in the sender system and ADSOs in the receiver system. It focuses on pinpointing structural and sequential disparities between InfoCubes and the corresponding mapped ADSOs. These differences are then stored in the control system tables. This activity is carried out within the receiver system.

- **Identify InfoCubes for Migration**
  This task identifies pertinent InfoCubes in the sender system for conversion into ADSOs in the SAP BW/4HANA target system. It operates in both the sender and receiver systems, automatically filtering the list of InfoCubes from all SAP BW objects scheduled for transfer. Subsequently, the following task generates fact and dimension tables for each identified InfoCube in the receiver system.

- **Generate Dimension and Fact Tables for InfoCubes**
  This task automatically creates dimension and fact tables in the receiver system (SAP BW/4HANA) for the selected list of InfoCubes identified in the previous activity.

- **Build Table Mappings between Mapped Objects**
  This task automatically establishes mappings between BW objects in SAP BW designated for transfer in the sender system and their corresponding SAP HANA–optimized counterparts in the SAP BW/4HANA target system. Examples of such mappings include ODSOs to ADSOs, InfoCubes to ADSOs, ADSOs to ADSOs, PSAs to ADSOs, and Open Hub Destinations to Open Hub Destinations. This activity takes place within the target receiver system. There will be multiple errors in this activity if the transports and metadata aren't correct between the sender and target system as it first confirms the presence of target object tables in the receiver system. Then, it scrutinizes any potential DDIC differences between the mapped tables in both the sender and receiver systems.

- **Configure MIGSID Conversion Rule for Master Data ID Fields**
  This task automatically confirms whether the S table and RS* tables, which store master data, already contain data in the receiver system. It operates within the receiver system.

- **Make Settings to Facilitate Program Generation**
  It's important to note that with the transition to SAP BW/4HANA and SAP BW 7.5 versions, there's a shift in the request ID generation concept. The traditional RSSM-

based requests are replaced by Transaction Sequence Numbers (TSNs). Therefore, when migrating data from classic objects to SAP HANA–optimized objects in the SAP BW/4HANA system, the old SID-based requests must be converted to TSN values. This task automatically arranges subactivities to address inconsistencies in the conversion control table within the receiver system. It also assigns conversion rules to manage the renaming of request IDs to TSN values. This activity takes place exclusively within the receiver system.

Once the group activity (**Make Settings for Selected BW Object Tables**) is triggered using 👁, all the associated subsequent tasks are automatically triggered upon completion of the predecessor task. If an error occurs, the status will be marked as red; however, if the status is green, the next successor activity will commence automatically. This automated progression halts only if there is a need for manual input parameters. You can see that tasks in the original system activity are all green now in Figure 4.136.

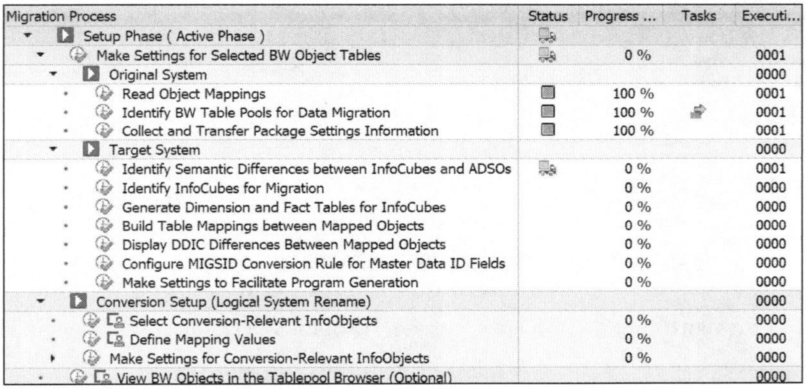

Figure 4.136 Setup Phase: Execution Status

You can choose the 📊 Extended Process Monitor to view the stats of the activities that are running and completed. Once you choose **Extended Process Monitor**, you'll see the node details, as shown in Figure 4.137.

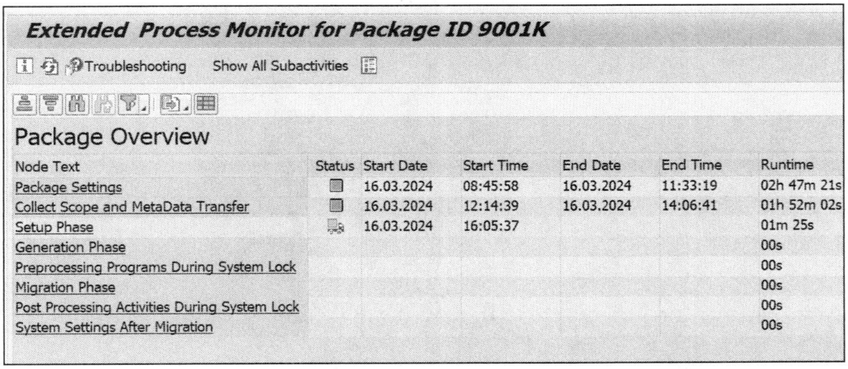

Figure 4.137 Extended Process Monitor

4   Remote Conversion

You choose the relevant phase in the node text. Right now we're in the setup phase, which is highlighted, so you need to click on that row to see the activities in that step, as shown in Figure 4.138.

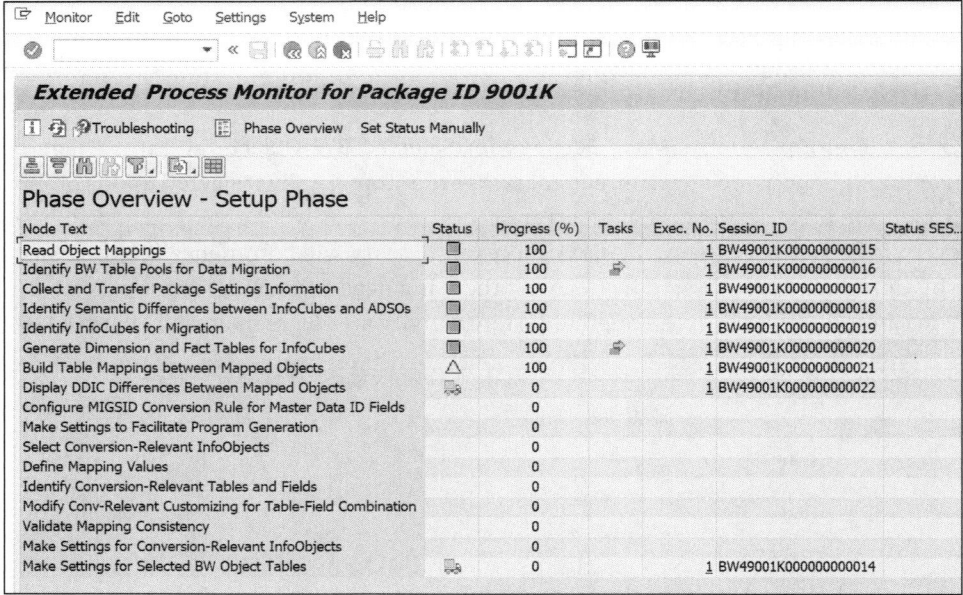

Figure 4.138 Status of the Setup Phase: Extended Process Monitor

If you choose any row with the icon, you can see the details of the node, as shown in Figure 4.139.

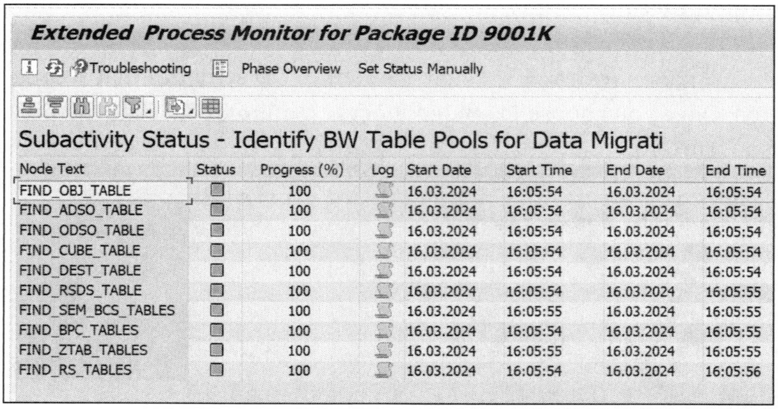

Figure 4.139 Tables Status

If you choose to refresh the process tree, the status can be seen in Figure 4.140.

Migration Process	Status	Progress ...	Tasks	Executi.
▼ ▶ Setup Phase (Active Phase)				
▼ 🕮 Make Settings for Selected BW Object Tables	⚠	100 %		0001
▼ ▶ Original System				0000
• 🕮 Read Object Mappings	☐	100 %		0001
• 🕮 Identify BW Table Pools for Data Migration	☐	100 %	📑	0001
• 🕮 Collect and Transfer Package Settings Information	☐	100 %		0001
▼ ▶ Target System				0000
• 🕮 Identify Semantic Differences between InfoCubes and ADSOs	☐	100 %		0001
• 🕮 Identify InfoCubes for Migration	☐	100 %		0001
• 🕮 Generate Dimension and Fact Tables for InfoCubes	☐	100 %	📑	0001
• 🕮 Build Table Mappings between Mapped Objects	⚠	100 %		0001
• 🕮 Display DDIC Differences Between Mapped Objects	☐	100 %		0001
• 🕮 Configure MIGSID Conversion Rule for Master Data ID Fields	☐	100 %		0001
• 🕮 Make Settings to Facilitate Program Generation	⚠	100 %		0001
▼ ▶ Conversion Setup (Logical System Rename)				0000
• 🕮 Select Conversion-Relevant InfoObjects		0 %		0000
• 🕮 Define Mapping Values		0 %		0000
▶ 🕮 Make Settings for Conversion-Relevant InfoObjects		0 %		0000
• 🕮 View BW Objects in the Tablepool Browser (Optional)				0000

Figure 4.140 Setup Status

Now you can choose the next step, which has input parameters where you need to enter the details manually and then choose to execute to proceed further. We'll move to the **Conversion Setup** group activity with the following tasks:

- **Select Conversion-Relevant InfoObjects**
  During this task, you can choose the InfoObjects you want to customize for conversion. This activity takes place in both the target (receiver) and execution systems. This task updates the list of InfoObjects in the conversion control table. When selecting an InfoObject for conversion, you assign a rule and rule version to it, stored in the conversion control table. The same rule can be used for multiple InfoObjects, each with a unique combination of rule and version for conversion mappings.

- **Define Mapping Values**
  During this task, you can specify new values for existing InfoObject values by inputting a unique combination of conversion rule and rule version. Furthermore, you have the option to verify the consistency of your mappings using the functionalities provided in the user interface. This activity takes place exclusively in the target (receiver) system.

- **Make Settings for Conversion-Relevant InfoObjects**
  This is a group activity that will start the sequence of activities in background. We'll walk the details of executing this activity in the next section.

4.9.11 Select Conversion-Relevant InfoObjects

You can choose **LOGSYS_NEW**, which is a conversion-relevant InfoObject, as shown in Figure 4.141.

4 Remote Conversion

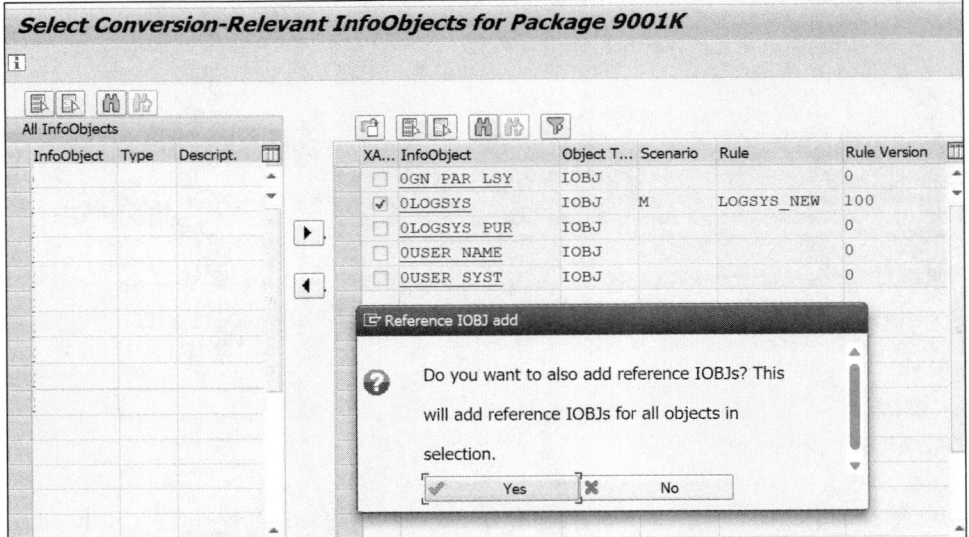

Figure 4.141 Conversion

Once you choose **Yes** in the popup, the option shown in Figure 4.142 is available.

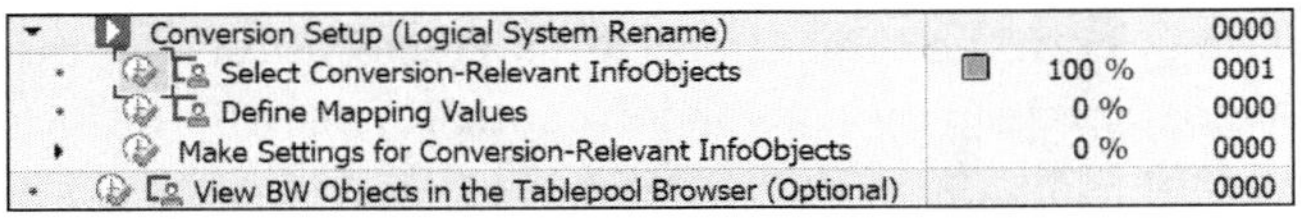

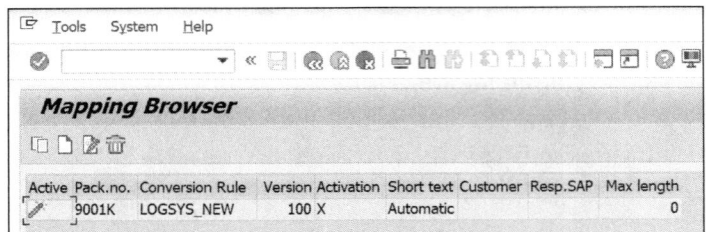

Figure 4.142 Conversion-Relevant InfoObject

4.9.12 Define Mapping Values

You'll see the conversion rule with the mapping status is active by default; ensure it's active , as shown in Figure 4.143, in the **Mapping Browser**.

Figure 4.143 Mapping Browser

Once you go back using the icon, you'll see the job's status, as shown in Figure 4.144.

The next step is to execute the group activity, which is **Make Settings for Conversion-Relevant InfoObjects**, as shown in Figure 4.145.

218

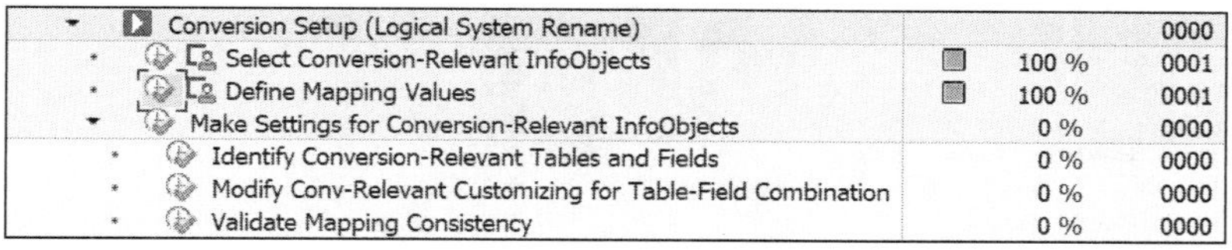

Figure 4.144 Define Mapping Values

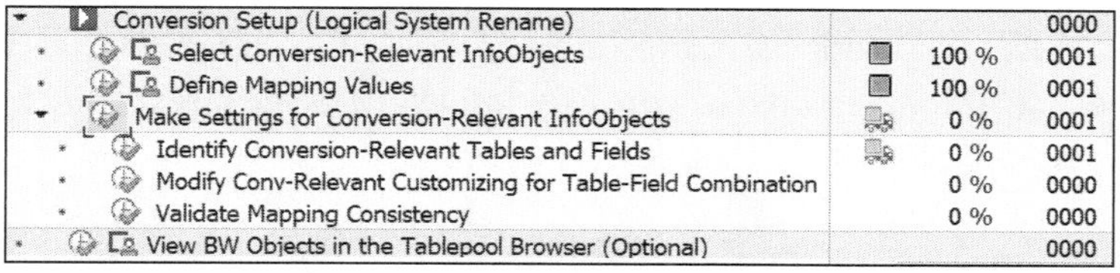

Figure 4.145 Group Activity Execution

You'll perform the steps described in the following subsections.

4.9.13 Make Settings for Conversion-Relevant InfoObjects.

Executing the **Make Settings for Conversion-Relevant InfoObjects** group activity initiates a sequence of background tasks. As each task automatically triggers its successor, no additional user intervention is needed. This group activity encompasses the following tasks, as shown in Figure 4.145:

- **Identify Conversion-Relevant Tables and Fields**
  This task automatically scans the where-used list for the InfoObjects chosen for conversion. It then compiles a list of tables and fields relevant for conversion based on the InfoProviders found in the where-used list. This activity occurs in both the target (receiver) and execution systems.

- **Modify Conv-Relevant Customizing for Table-Field Combination**
  This task uses the data gathered from the preceding step, **Identify Conversion-Relevant Tables and Fields**, and enhances it with table-level information from the original (sender) system. This activity takes place exclusively in the target (receiver) system.

- **Validate Mapping Consistency**
  This task automatically conducts multiple checks to verify the mappings defined in the preceding activity: **Define Mapping Values**. It operates within the target (receiver) system.

View BW Objects in the Tablepool Browser (Optional)
During this task, you can obtain a comprehensive overview of all SAP BW objects selected for conversion or migration.

The status of the completed execution is shown in Figure 4.146.

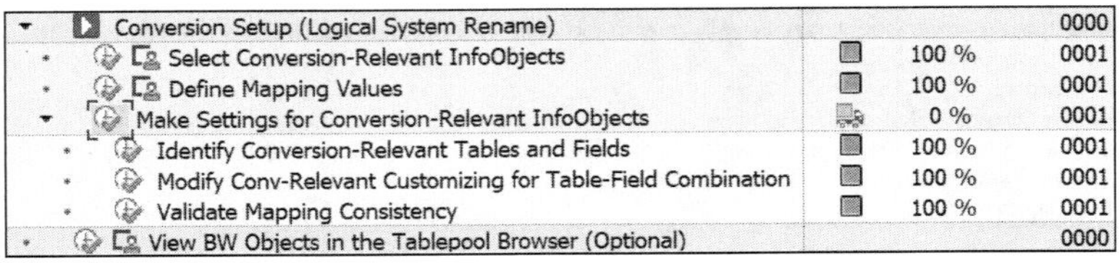

4 Remote Conversion

▼ ▶ Conversion Setup (Logical System Rename)			0000
· ⚙ 🗒 Select Conversion-Relevant InfoObjects	■	100 %	0001
· ⚙ 🗒 Define Mapping Values	■	100 %	0001
▼ ⚙ Make Settings for Conversion-Relevant InfoObjects	▣	0 %	0001
· ⚙ Identify Conversion-Relevant Tables and Fields	■	100 %	0001
· ⚙ Modify Conv-Relevant Customizing for Table-Field Combination	■	100 %	0001
· ⚙ Validate Mapping Consistency	■	100 %	0001
· ⚙ 🗒 View BW Objects in the Tablepool Browser (Optional)			0000

Figure 4.146 All Activity

If you need to view SAP BW objects and associated tables, execute the **View BW Objects in the Tablepool Browser (Optional)**, which will provide the details of the target table in a new window, as shown in Figure 4.147.

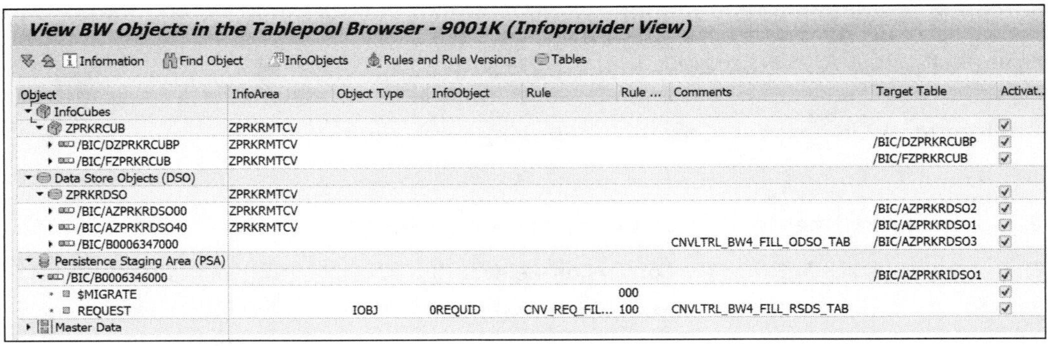

Figure 4.147 Table Browser

4.9.14 Generation Phase

Once all activities are done in the setup phase, you'll be in the generation phase, as shown in Figure 4.148. We'll cover each of these phases in the following sections.

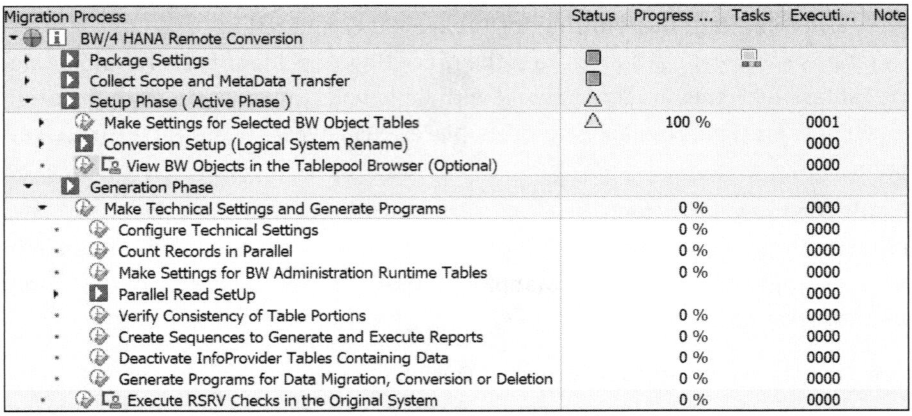

Figure 4.148 Generation Phase

220

Make Technical Settings and Generate Programs

Upon execution of this group activity, it will start other successor activities.

Configure Technical Settings

This activity automatically determines how the tool should manage tables listed in the conversion control table during conversion, migration, and deletion processes. It operates in both the original (sender) and target (receiver) systems during migration, and in the execution system during conversion. During this task, each conversion- or migration-relevant table is analyzed, and configurations are made such as counting rows, specifying key changes, setting megabytes per package for data, and defining the insert mode of tables. These configurations are then stored in the control table for reference.

Count Records in Parallel

This activity automatically tallies the record count for tables earmarked for conversion, migration, or deletion, and records these counts in the control tables. The count is performed concurrently using the default number of background jobs set in the standard shipment. You also have the option to adjust the default number of background jobs in the process monitor. This activity operates within either the original (sender) system or the execution system.

Make Settings for BW Administration Runtime Tables

This activity ensures consistency in release versions between the original (sender) and target (receiver) systems for SAP BW administration runtime tables. It operates in both the original (sender) and target (receiver) systems.

Now choose and execute the group activity **Make Technical Settings and Generate Programs**. The status of the activities in the generation phase are shown in Figure 4.149.

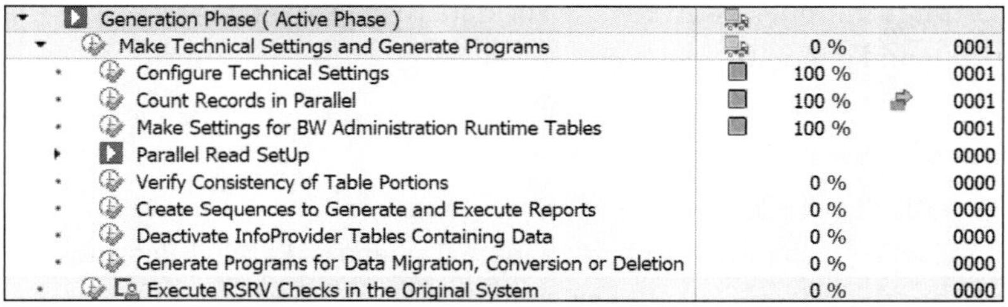

▼ ▶ Generation Phase (Active Phase)		
▼ ☼ Make Technical Settings and Generate Programs	0 %	0001
• ☼ Configure Technical Settings	100 %	0001
• ☼ Count Records in Parallel	100 %	0001
• ☼ Make Settings for BW Administration Runtime Tables	100 %	0001
▶ ▶ Parallel Read SetUp		0000
• ☼ Verify Consistency of Table Portions	0 %	0000
• ☼ Create Sequences to Generate and Execute Reports	0 %	0000
• ☼ Deactivate InfoProvider Tables Containing Data	0 %	0000
• ☼ Generate Programs for Data Migration, Conversion or Deletion	0 %	0000
• ☼ Execute RSRV Checks in the Original System	0 %	0000

Figure 4.149 Generation Phase: Activities

Parallel Read SetUp

This activity is an optional task that requires manual inputs in the analysis and preparation of tables for parallel reading. Implementing parallel reads is beneficial when the sender system contains a large volume of data to be transferred, as it helps optimize read performance by allocating additional parallel work processes during the reading phase.

Analyze and Prepare Tables for Parallel Read

This activity prepares the system for parallel read. You'll have input fields to specify a threshold value above which tables are chosen for parallel read. For instance, if you input 1,000,000, tables with more than a million records will be selected for parallel read. Additionally, you can provide the portion size, which determines the number of portions created for parallel read for each table. For example, if a table holds 5,000,000 records and the portion size is 500,000, this activity generates 10 portions for parallel read. During the read phase, 10 background jobs run in parallel to read the table contents and write them into a cluster table. Optimizing read operations during migration significantly reduces downtime in the sender SAP BW system.

Implementing Parallel Read

When you execute the **Analyze and Prepare Tables for Parallel Read** activity, you can give your threshold value and portion value, and then execute. The tool will take some time to scan the tables and then finally presents a list of tables suitable for parallel reading, based on the number of records surpassing the threshold value entered on the selection screen. If no settings are specified, the activity automatically selects the most relevant parallel read field and sets a default number of consistent portions, although you have the flexibility to choose different parallel read fields for each table and specify the portion count. Save the parallel read settings for the selected tables by clicking the **Save Split Settings** button. Proceed to generate and save the portions for the tables intended for parallel data reading. Monitor the progress of portion generation in the monitor. (The next activity, **Monitor Parallel Read Status**, will help you see this.) If inconsistent portions are generated, consider using a different parallel read field. Use the **Delete Parallel Read Settings** button to remove the parallel read settings and portions for selected tables. For tables with inconsistent portions, choose the **Delete Inconsistent Portion Settings** button. Define the consistency tolerance percentage using the **Define Default Settings** option. Finally, refresh the data to view the latest status of the parallel read settings and generated portions.

Verify Consistency of Table Portions

This activity automatically checks the consistency of table portions designated for parallel read, as established in the **Analyze and Prepare Tables for Parallel Read** task. It operates within either the original (sender) system or the execution system.

Create Sequences to Generate and Execute Reports

This report automatically generates sequences and task numbers for the creation and execution of conversion or migration reports. It operates in both the original (sender) and target (receiver) systems during migration, and in the execution system during conversion.

4.9 Execution

Deactivate InfoProvider Tables Containing Data

This activity configures settings to prevent redundant data writes to an InfoProvider, thereby ensuring consistency during data transfer. It operates exclusively in the target (receiver) system.

Generate Programs for Data Migration, Conversion, or Deletion

This activity automates the process of gathering settings from all preceding preparation tasks and generates programs to facilitate the migration, conversion, or deletion of data for relevant tables. These programs are housed within the /1LT/BW_GEN_OBJ development class in the /1LT/BW namespace. Depending on the scenario, the activity executes in different contexts: for migration scenarios, it operates in both the original (sender) and target (receiver) systems; for conversion scenarios, it runs in the execution system; and for deletion scenarios, it also functions within the execution system.

Once you start **Verify Consistency of Table Portions**, the successor activities will be executed by default, as shown in Figure 4.150.

Figure 4.150 Generate Programs for Data Migration

Execute RSRV Checks in the Original/Target/Execution System

After **Generate Programs for Data Migration, Conversion or Deletion** is completed, you'll be in the **Execute RSRV Checks in the Original System** step. It's essential for checking any issues in SID tables and other data management tables, and it's next to be executed, as shown in Figure 4.151. Because it will take a long time to execute the system-wide Transaction RSRV check from the tool, you have the option to save the package and execute it separately in nonpeak hours.

Figure 4.151 RSRV Tool-Based Execution

This report generates and executes (or saves) an analysis and repair of SAP BW objects (Transaction RSRV) package within the standard Transaction RSRV. Following this, you can load and execute the RSRV analysis package in the Transaction RSRV to validate the consistency of SAP BW objects included in the current run in the SAP BW sender system. Choose **Execute RSRV Checks in Original System** to go to the screen shown in Figure 4.152.

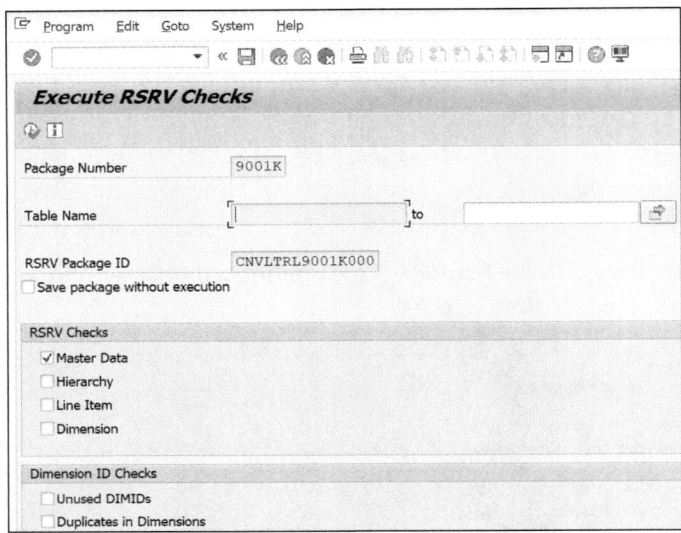

Figure 4.152 Execute RSRV Checks: Initial Screen

Choose **Save package without execution**, as shown in Figure 4.152.

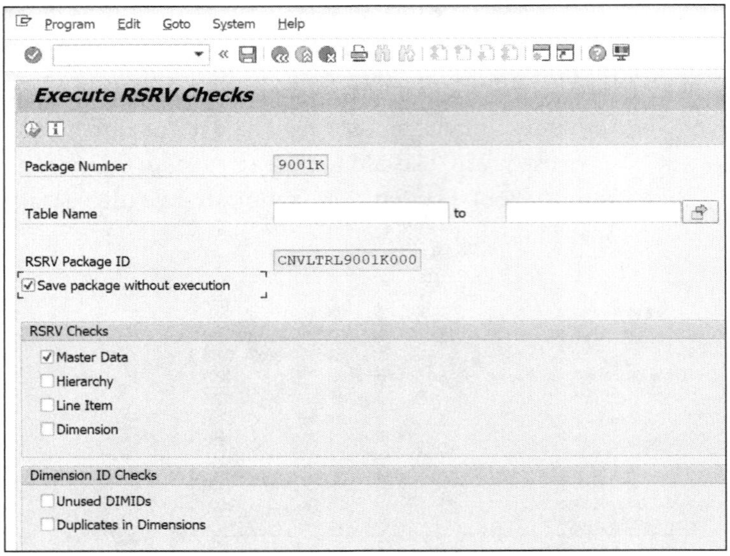

Figure 4.153 Save Package without Execution

4.9 Execution

You can note the package ID **CNVLTRL9001K000**, which can be used to test in Transaction RSRV offline. Now choose [icon], and go back to see the status, as shown in Figure 4.154.

Generation Phase (Active Phase)			
Make Technical Settings and Generate Programs	☐	100 %	0001
Configure Technical Settings	☐	100 %	0001
Count Records in Parallel	☐	100 %	0001
Make Settings for BW Administration Runtime Tables	☐	100 %	0001
Parallel Read SetUp			0000
Verify Consistency of Table Portions	☐	100 %	0001
Create Sequences to Generate and Execute Reports	☐	100 %	0001
Deactivate InfoProvider Tables Containing Data	☐	100 %	0001
Generate Programs for Data Migration, Conversion or Deletion	☐	100 %	0001
Execute RSRV Checks in the Original System	☐	100 %	0001
Preprocessing Programs During System Lock			
Migration Phase			
Post Processing Activities During System Lock			
System Settings After Migration			

Figure 4.154 Generation Phase

The generation phase activity is completed.

4.9.15 Preprocessing Programs During System Lock

This step will focus on creating a delta clone in all connected SAP ERP and SAP S/4HANA systems, propagating the request within the SAP BW sender, and finally locking the system before migration. You can see the activities list shown in Figure 4.155.

Project	ZS4BW_BW4_55	Subproject	ZS4BW_BW4_55_2024_55	Execution System
Package	9001K	Package Description	BW/4 HANA Remote Conversion	Package Created B
Solution	BW/4HANA Conversion Cockpit			

Migration Process	Status	Progress ...	Tasks	Executi..
Setup Phase	⚠			
Generation Phase (Active Phase)	☐			
Preprocessing Programs During System Lock				
Copy Delta Queues from SAPI to ODP Technology		0 %		0000
Confirm that delta was loaded for the cloned DataSources		0 %		0000
Check Whether Delta Queues Can Be Copied from SAPI to ODP		0 %		0000
Confirm the INIT SIMU Procedure		0 %		0000
Make Settings to Prepare to Lock all Data Changes		0 %		0000
Validate Migration Configuration		0 %		0000
Synchronize Delta Queues between SAPI and ODP		0 %		0000
Extract from PSAs of DataSources and Error Stacks		0 %		0000
Master Data Activation		0 %		0000
Propagate requests		0 %		0000
Lock All Data Changes		0 %		0000
Prepare Request Mapping for Transfer		0 %		0000
Activate Task Sequences		0 %		0000
Configure Execution Dependencies for Migration		0 %		0000
Delete Tasks Assigned to Empty Tables		0 %		0000
Migration Phase				

Figure 4.155 Preprocessing Programs During System Lock: Initial Tasks

4.9.16 Copy Delta Queues from SAPI to ODP Technology

This task involves cloning the delta queues within the connected SAP BW source systems to transition to ODP technology for data extraction. To understand the terminology, note the following: an ODP delta queue is called a *subscription*, with each subscription having a *subscriber* typically represented by a SAP BW system. Furthermore, a subscription includes a *subscriber process*, which can either be a SAP BW DataSource and source system (for InfoPackage loads) or a DTP. It's essential to check the SAP ERP system after this step to ensure everything is functioning correctly.

Before proceeding, ensure that the SAP BW source system has implemented the necessary notes provided by the Note Analyzer (refer to SAP Collection Note 2383530), focusing on the `Source_System_for_SAP_BW4HANA` scope. While efforts are made to minimize the impact of these changes, certain extensions may be required to facilitate queue cloning in the source systems.

One of the most frequently asked questions in the SAP BW/4HANA conversion project will be "How much downtime do I need in an SAP ERP system?" This step is one of the critical and debated topics in the project. When the SAP Notes in the source system are properly implemented, then the advantage of this approach is that queue cloning can occur without affecting operational processes in SAP ERP, eliminating the need to lock SAP ERP against data changes. However, failure to implement the required notes in SAP ERP may result in the exclusion of the corresponding source system from this task. In such cases, initialization simulations on the new DataSources from SAP BW/4HANA will be necessary to synchronize the delta. Keep in mind that depending on the specific extractor, this may require restricting new data entries against the corresponding queues until the delta simulation on the new DataSource is completed. If SAP Note 1855474 is implemented (included in the Note Analyzer list), the source system doesn't need to be changeable during the cloning procedure. It's also important to note that some extractors don't support extraction from more than one target system; refer to SAP Note 1932459 for a list of these. Additionally, ensure that delta loading functions properly for all DataSources within scope that have a delta initialization. Also check if the DataSource supports parallel data loading; if not, then you need to plan for the delayed extraction after go-live.

Upon completion of this task, it's necessary to request new deltas for all DataSources and synchronize the delta queues in the connected SAP BW source systems in the subsequent step, **Synchronize Delta Queues between SAPI and ODP**. This step is crucial for establishing a clear demarcation between the data that has already been loaded into SAP BW and the data that has yet to be loaded. Furthermore, the **Synchronize Delta Queues between SAPI and ODP** step ensures that deltas have been correctly loaded for all cloned DataSources.

Let's check the status of Transaction ODQMON (Monitor Delta Queues) in SAP ERP or SAP S/4HANA before execution of this activity, as shown in Figure 4.156.

4.9 Execution

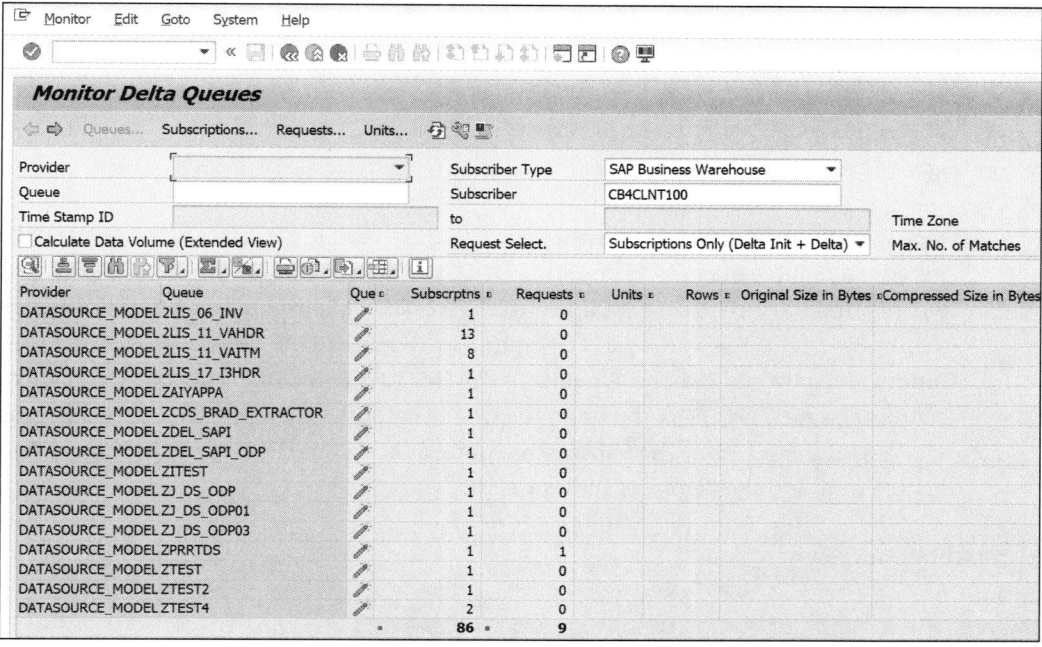

Figure 4.156 Transaction ODQMON Status: SAP S/4HANA

If you search for the involved DataSource (**ZPRRTDS**), it won't be available in the **Queues** view, as shown in Figure 4.157.

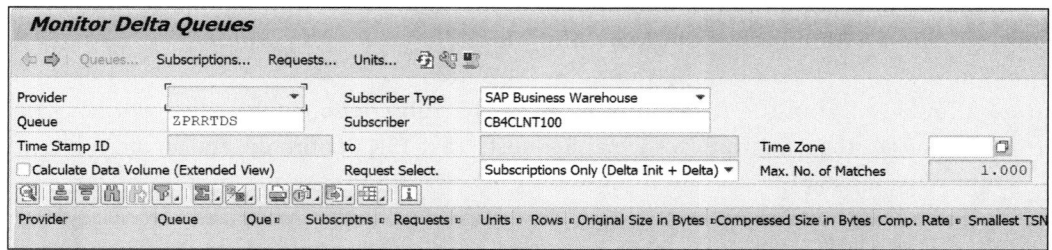

Figure 4.157 Transaction ODQMON: Before Clone

Now execute the **Copy DataSource from SAPI to ODP** activity in the process tree, and you'll see the activity has a completed status, as shown in Figure 4.158. If there are any errors, you need to fix the issues and repeat until the step is green. Note that not exposing to ODP is a common error.

> **Best Practice**
>
> It's best practice to make sure all the delta execution for the involved DataSource works before this step. You can execute the successful delta in the sender system at least one time before starting this step.

227

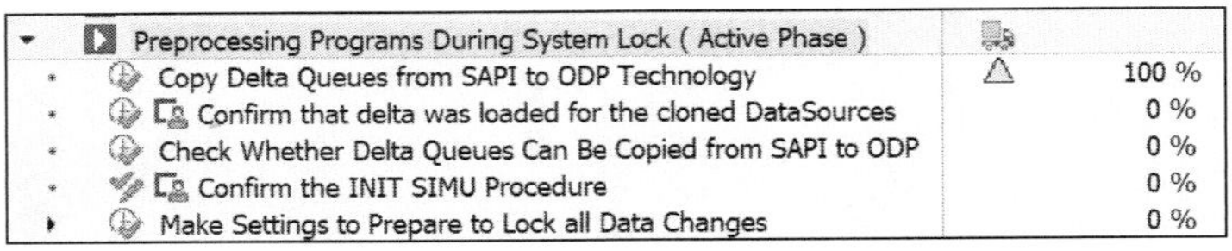

4 Remote Conversion

	Preprocessing Programs During System Lock (Active Phase)		
•	Copy Delta Queues from SAPI to ODP Technology	⚠	100 %
•	Confirm that delta was loaded for the cloned DataSources		0 %
•	Check Whether Delta Queues Can Be Copied from SAPI to ODP		0 %
•	Confirm the INIT SIMU Procedure		0 %
▸	Make Settings to Prepare to Lock all Data Changes		0 %

Figure 4.158 Status

Once the activity is completed with green or yellow status, you can see the queue with cloned DataSource name in Transaction ODQMON in the SAP ERP or SAP S/4HANA system. In our demo, we're seeing DataSource **ZPRRTDS** for **Subscriber CB4CLNT100**, which is our new SAP BW/4HANA target system (see Figure 4.159). You can see that it's in **Queues** perspective and the **Provider** is **DATASOURCE_MODEL**, and the DataSource technical name, **ZPRTDS**, is shown in the **Queue** column. The queue status is active, as shown with the 🖉 icon. The **Subscrptns** column shows **1** as the **Copy Delta Queue** step is successful, and the **Requests** shows **1** as well.

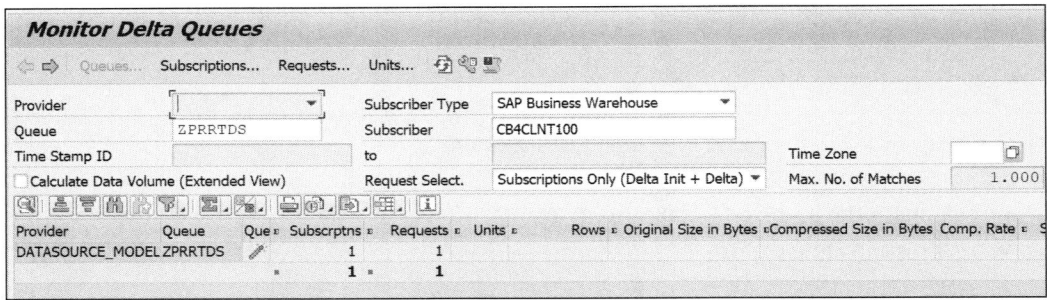

Figure 4.159 Transaction ODQMON: After Clone

Double-click on the DataSource to go to the **Subscriptions** tab, which will have the details of the subscription such as **Requests**, **Last TSN Confirmed**, and so on, as shown in Figure 4.160.

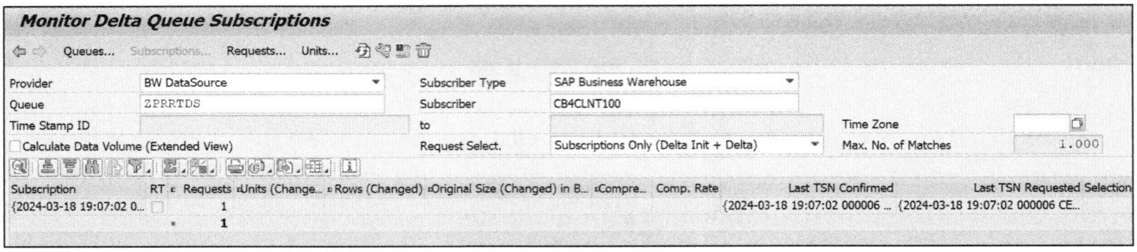

Figure 4.160 Subscription Status

Select the subscription row to get to the **Requests** screen, which will show the composite request number and the status. You can also see the extraction status and the **Extraction Mode**; in our case, this is an initialization request performed by the tool: **Delta Initialization without data transfer** (see Figure 4.161).

4.9 Execution

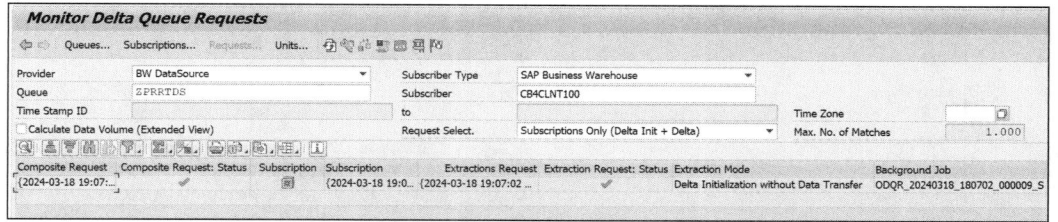

Figure 4.161 Request Status

If you need to see the **Composite Request** description, then you can choose the technical settings on/off icon . Now the composite request will have the name **DELTA_ CLONE**, as shown in Figure 4.162.

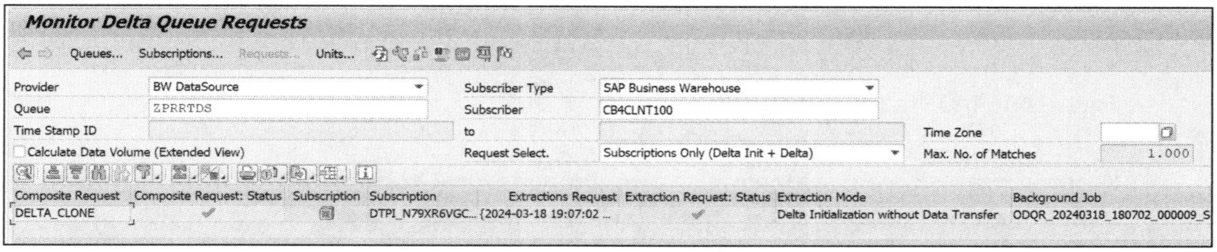

Figure 4.162 Clone Details

4.9.17 Confirm That Delta Was Loaded to Cloned DataSources

This step requires manual confirmation. Before proceeding, ensure that you execute the delta for all InfoPackages in the SAP BW sender system included in the scope. It's crucial that the execution status is green for each, even if the delta returns zero records. This execution of the delta for all DataSources post-clone is mandatory. You can see the delta InfoPackage with **Update Mode** set as **Delta Update** in Figure 4.163.

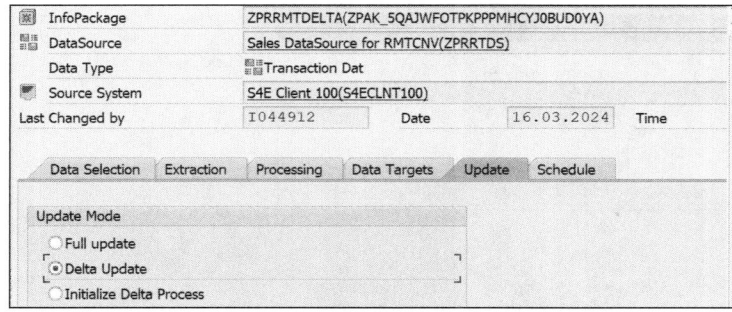

Figure 4.163 Delta InfoPackage in the Sender SAP BW System

Upon executing the delta load, the InfoPackage will check for any new and changed records in Transaction RSA7 and will try to load them to the SAP BW system. In our demo, there was no delta, and, as you can see in Figure 4.164, the delta load is successful with zero records.

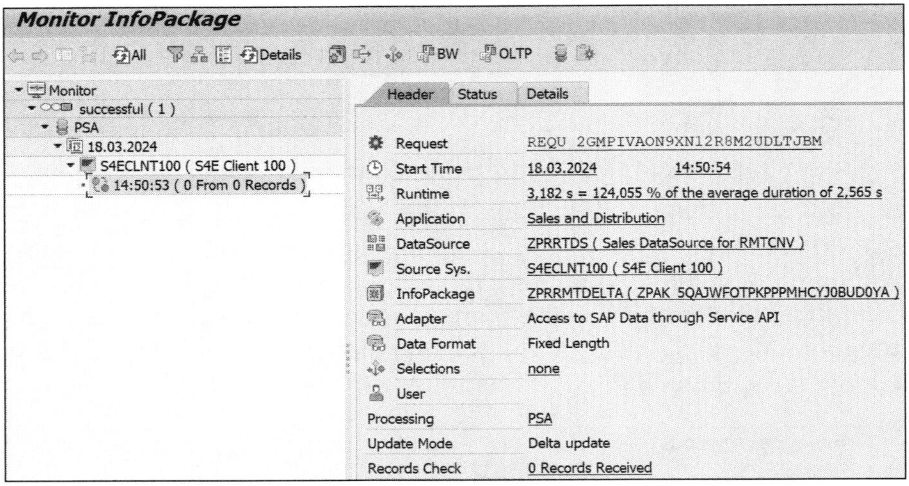

4 Remote Conversion

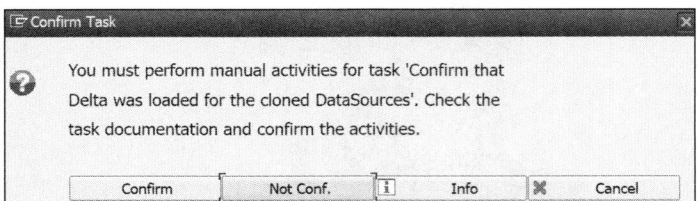

Figure 4.164 Delta Load Post Clone

> **Best Practice**
>
> When handling a large volume of DataSources, it's good practice to have a dedicated process chain to include all the delta load InfoPackages. You can trigger those process chain before the Clone step, and then the same process chain can be triggered after the Clone step. This will ensure that no delta is missed.

Once the delta execution is completed for all the cloned DataSources, you can execute the manual task **Confirm that Delta was loaded for the cloned DataSources**. In the popup that appears as shown in Figure 4.165, choose the **Confirm** option to complete this activity.

Figure 4.165 Confirmation

The activity status is green, as shown in Figure 4.166.

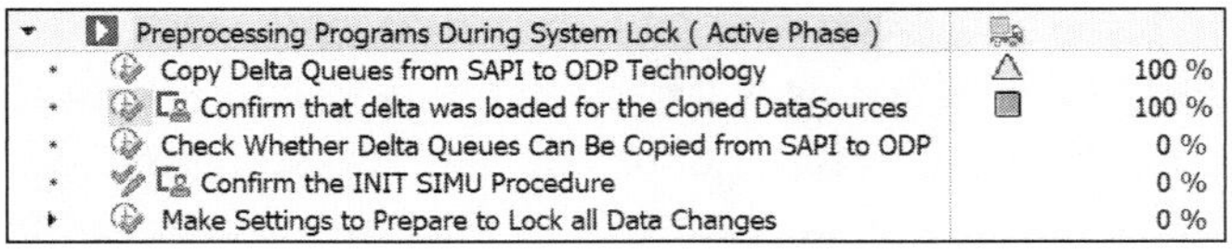

Figure 4.166 Activity Status

4.9.18 Check Whether Delta Queues Can Be Copied from SAPI to ODP

The next step is **Check Whether Delta Queues Can Be Copied from SAPI to ODP**. Execute this step which can be ran in background, as shown in Figure 4.167.

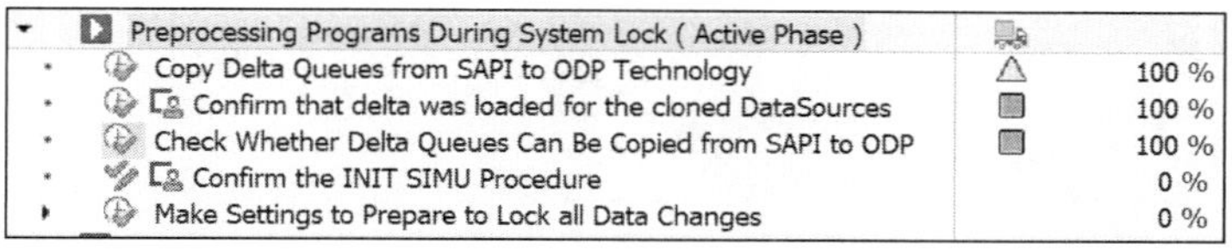

```
▼   ▶  Preprocessing Programs During System Lock ( Active Phase )
·   ⊕     Copy Delta Queues from SAPI to ODP Technology              △    100 %
·   ⊕ ᶜ₂  Confirm that delta was loaded for the cloned DataSources   ▢    100 %
·   ⊕     Check Whether Delta Queues Can Be Copied from SAPI to ODP  ▢    100 %
·   ✔ ᶜ₂  Confirm the INIT SIMU Procedure                                   0 %
▶   ⊕     Make Settings to Prepare to Lock all Data Changes                 0 %
```

Figure 4.167 Check Whether Delta Queues Can Be Copied

4.9.19 Confirm the INIT SIMU Procedure

The next step is **Confirm the INIT SIMU Procedure**. Use the ✔ icon to execute this step, and you'll see the popup shown in Figure 4.168.

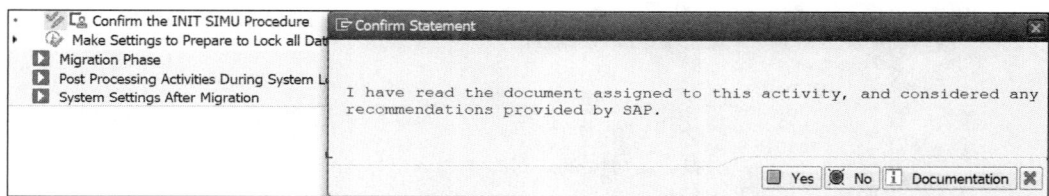

Figure 4.168 Init SIMU Popup

Once you choose **Yes**, the status will be updated, as shown in Figure 4.169.

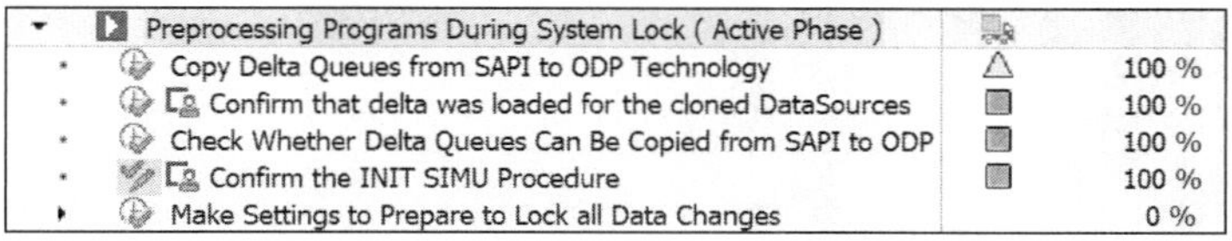

```
▼   ▶  Preprocessing Programs During System Lock ( Active Phase )
·   ⊕     Copy Delta Queues from SAPI to ODP Technology              △    100 %
·   ⊕ ᶜ₂  Confirm that delta was loaded for the cloned DataSources   ▢    100 %
·   ⊕     Check Whether Delta Queues Can Be Copied from SAPI to ODP  ▢    100 %
·   ✔ ᶜ₂  Confirm the INIT SIMU Procedure                            ▢    100 %
▶   ⊕     Make Settings to Prepare to Lock all Data Changes                 0 %
```

Figure 4.169 INIT SIMU Status

Next, we're ready to start the group activity that will synchronize the delta queue and then populate the data mart InfoProvider in the SAP BW sender system.

4.9.20 Make Settings to Prepare to Lock All Data Changes

Make Settings to Prepare to Lock all Data Changes is a group activity, as shown in Figure 4.170. We'll walk through its steps in this section. It will trigger all subsequent activity in background mode. If it's successful, the next activity will be started; otherwise, it will stop with a red status to fix and resume. The SAP BW sender system will have the lock on the scope object for the data load once this activity is completed.

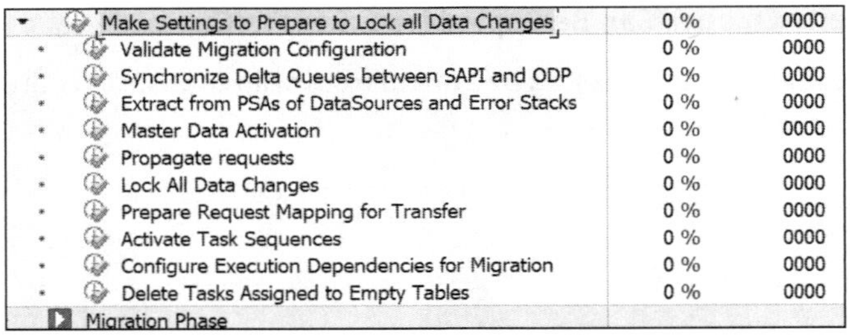

4   Remote Conversion

Figure 4.170 Make Settings to Prepare to Lock All Data Changes

Validate Migration Configuration

Click [icon] to start the group activity, as shown in Figure 4.171. This will first validate the migration configuration.

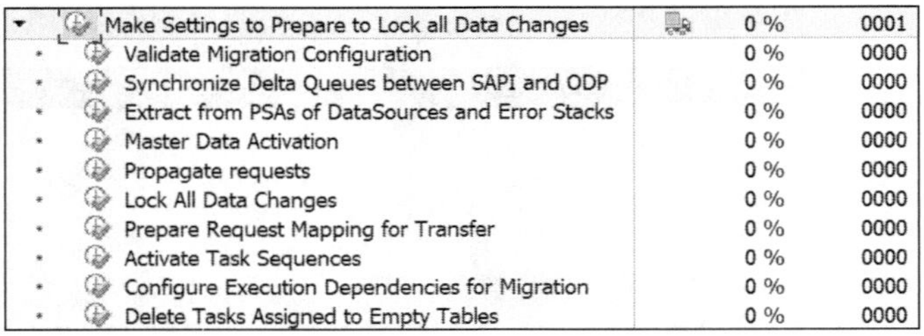

Figure 4.171 Start Activity

Synchronize Delta Queues between SAPI and ODP

This step will synchronize the cloned delta in the source SAP ERP or SAP S/4HANA system, as shown in Figure 4.172.

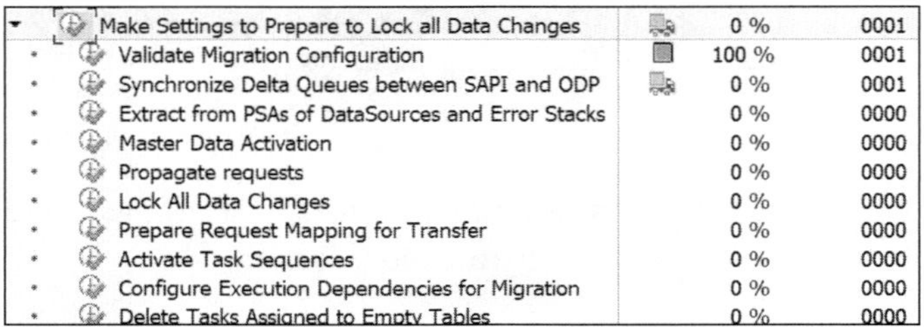

Figure 4.172 Synchronize Delta

You can see that **Synchronize Delta Queues between SAPI and ODP** is successful with warnings (Figure 4.173). During this step, you might get errors to execute delta loads; upon execution of those loads, the errors will be fixed in the next run. After this step is completed, you can't execute any manual delta loads in the system, and the delta will be locked after this step.

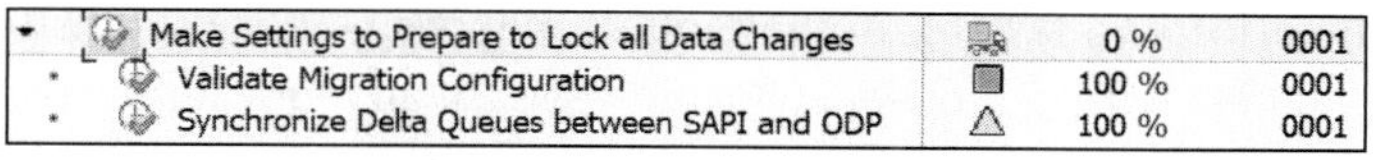

Figure 4.173 Delta Synchronization Completion

Log on to Transaction ODQMON in the SAP ERP system, and check the **Subscriptions** view to ensure the **Queue** has been cloned (**ZPRRTDS**). A **Subscription** with **DTPI*** will be created, as shown in Figure 4.174.

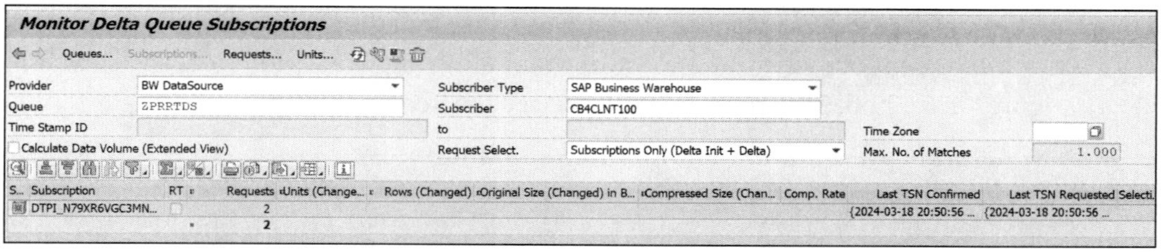

Figure 4.174 Transaction ODQMON: Subscriptions View

If you open the **Requests** view for the subscription, you can see that the **Composite Request** for **DELTA_SYNC** is now visible, as shown in Figure 4.175.

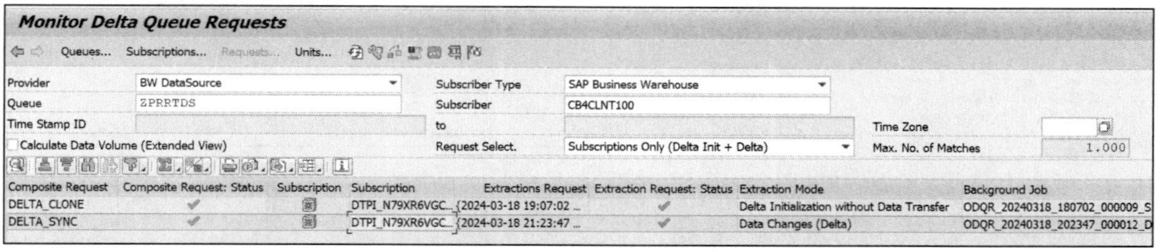

Figure 4.175 DELTA_SYNC in Transaction ODQMON

Next, we'll explain the successor steps that follow **Synchronize Delta Queues between SAPI and ODP**.

Extract from PSAs of DataSources and Error Stacks

This task manages preparations for old InfoProviders by executing all delta DTPs that extract from the PSA of a DataSource, particularly if the PSA won't be transferred as a field-based ADSO. In such instances, all requests from the PSA must be propagated into InfoProviders by delta DTPs. Additionally, it involves creating (if necessary) and executing error DTPs for all nonempty error stacks. Even if the error DTP requests fail, the error stack will be emptied, as they aren't transferred.

After the error DTP process is completed, the tool proceeds with the activation of master data. Following this, the requests from the DTP for the data mart are propagated to all data targets. Subsequently, the tool locks all InfoProviders within the scope against any data changes. With the data now secured, it becomes ready for copying into the new ADSOs. The tool then performs a task to fill the request mapping table, establishing connections between request GUIDs, request SIDs, and TSNs for all request types of an InfoProvider. Following this, the next activity initiates the activation of the task execution sequence. Subsequently, another activity configures table, task, and sequence dependencies essential for migration. Finally, the last activity automatically identifies and deletes tasks of empty tables in the original system, as well as tables set as inactive in the target (receiver) system. This ensures a streamlined and efficient data migration process, eliminating unnecessary clutter and inactive elements from both systems.

In the **Extended Process Monitor** screen, we can see that the group activity has triggered the subsequent activities, and most of them are completed with green, as shown in Figure 4.176. Choose the refresh icon to see the updated status.

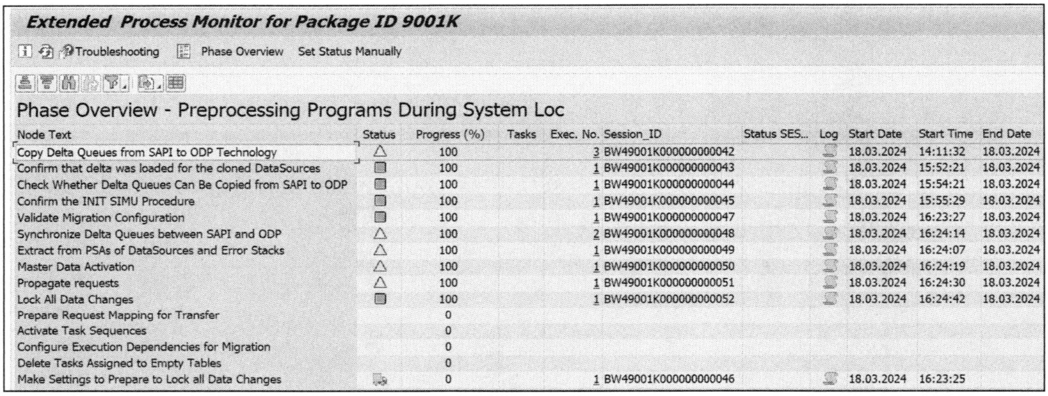

Extended Process Monitor for Package ID 9001K

Troubleshooting Phase Overview Set Status Manually

Phase Overview - Preprocessing Programs During System Loc

Node Text	Status	Progress (%)	Tasks	Exec. No.	Session_ID	Status SES...	Log	Start Date	Start Time	End Date
Copy Delta Queues from SAPI to ODP Technology	△	100		3	BW49001K000000000042			18.03.2024	14:11:32	18.03.2024
Confirm that delta was loaded for the cloned DataSources	■	100		1	BW49001K000000000043			18.03.2024	15:52:21	18.03.2024
Check Whether Delta Queues Can Be Copied from SAPI to ODP	■	100		1	BW49001K000000000044			18.03.2024	15:54:21	18.03.2024
Confirm the INIT SIMU Procedure	■	100		1	BW49001K000000000045			18.03.2024	15:55:29	18.03.2024
Validate Migration Configuration	■	100		1	BW49001K000000000047			18.03.2024	16:23:27	18.03.2024
Synchronize Delta Queues between SAPI and ODP	△	100		2	BW49001K000000000048			18.03.2024	16:24:14	18.03.2024
Extract from PSAs of DataSources and Error Stacks	△	100		1	BW49001K000000000049			18.03.2024	16:24:07	18.03.2024
Master Data Activation	△	100		1	BW49001K000000000050			18.03.2024	16:24:19	18.03.2024
Propagate requests	△	100		1	BW49001K000000000051			18.03.2024	16:24:30	18.03.2024
Lock All Data Changes	■	100		1	BW49001K000000000052			18.03.2024	16:24:42	18.03.2024
Prepare Request Mapping for Transfer		0								
Activate Task Sequences		0								
Configure Execution Dependencies for Migration		0								
Delete Tasks Assigned to Empty Tables		0								
Make Settings to Prepare to Lock all Data Changes		0		1	BW49001K000000000046			18.03.2024	16:23:25	

Figure 4.176 Extended Process Monitor: Status

After some time, you'll see that all the activities are in green status, as shown in Figure 4.177.

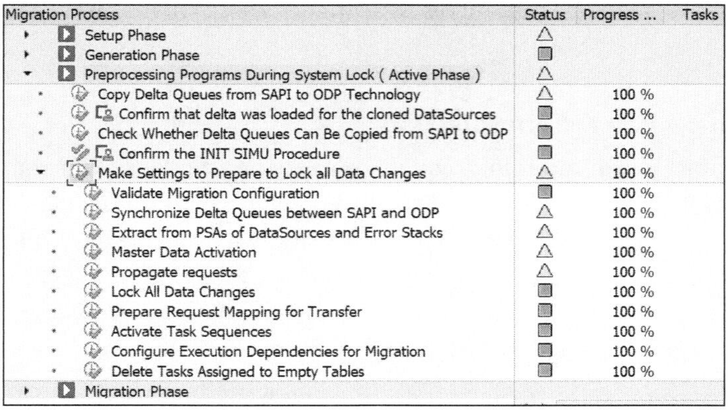

Migration Process	Status	Progress ...	Tasks
Setup Phase	△		
Generation Phase	■		
Preprocessing Programs During System Lock (Active Phase)	△		
Copy Delta Queues from SAPI to ODP Technology	△	100 %	
Confirm that delta was loaded for the cloned DataSources	■	100 %	
Check Whether Delta Queues Can Be Copied from SAPI to ODP	■	100 %	
Confirm the INIT SIMU Procedure	■	100 %	
Make Settings to Prepare to Lock all Data Changes	△	100 %	
Validate Migration Configuration	■	100 %	
Synchronize Delta Queues between SAPI and ODP	△	100 %	
Extract from PSAs of DataSources and Error Stacks	△	100 %	
Master Data Activation	△	100 %	
Propagate requests	△	100 %	
Lock All Data Changes	■	100 %	
Prepare Request Mapping for Transfer	■	100 %	
Activate Task Sequences	■	100 %	
Configure Execution Dependencies for Migration	■	100 %	
Delete Tasks Assigned to Empty Tables	■	100 %	
Migration Phase			

Figure 4.177 Preprocessing Program: Overall Status

We're now ready to start the migration phase.

4.9 Execution

4.9.21 Migration Phase

The migration phase is used to transfer the data from the sender SAP BW system to the target SAP BW/4HANA system. The activities in this phase are shown in Figure 4.178.

```
Migration Process                                            Status  Progress ...  Tasks  Executi...
▼ ⊕ ⓘ  BW/4 HANA Remote Conversion
  ▶   ▶ Package Settings                                       ☐                   🖥
  ▶   ▶ Collect Scope and MetaData Transfer                    ☐
  ▶   ▶ Setup Phase                                            △
  ▶   ▶ Generation Phase                                       ☐
  ▶   ▶ Preprocessing Programs During System Lock (Active Phase)  △
  ▼   ▶ Migration Phase
      ·  ⊕ 🗐 Make Runtime Settings for Migration                      0 %              0000
      ·  ⊕ 🗐 Reset Cluster Data Deletion Date                                           0000
      ·  ⊕    Start Migration                                          0 %              0000
      ·  ⊕ 🗐 Monitor: Transformation Status (Optional)                                  0000
      ·  ⊕ 🗐 Confirm Data Selection and Unlock Data Loading            0 %              0000
      ·  ⊕    Verify Task Completion                                   0 %              0000
  ▶   ▶ Post Processing Activities During System Lock
  ▶   ▶ System Settings After Migration
```

Figure 4.178 Migration Phase: Before

4.9.22 Make Runtime Settings for Migration

To start with the activity, choose **Make Runtime Settings for Migration** under the **Migration Phase** to see the screen shown in Figure 4.179. Based on your background job's availability, you can maintain the total jobs to be used for the migration in the sender SAP BW system (**No. of Jobs-Original**) for reading and writing into cluster table. You also need to maintain the jobs (**No. of Jobs-Target**) in the target SAP BW/4HANA system for converting the data from cluster tables to ADSO tables.

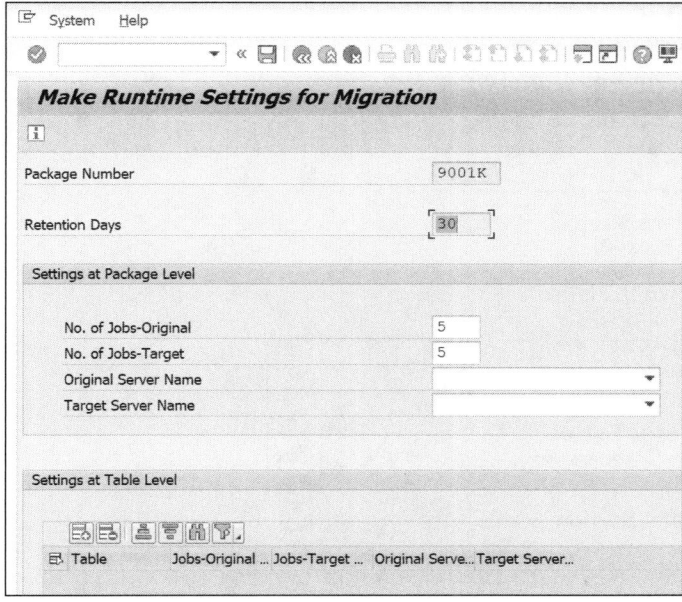

Figure 4.179 Migration Settings

235

Best Practice

If you have a huge data volume for the data migration, it's best practice to ensure you have hundreds of background work processes in the sender SAP BW system and target SAP BW/4HANA system. You'll provide more job counts in both systems in this step.

Once you make the settings and return, you'll see the green status of this activity in the process tree, as shown in Figure 4.180.

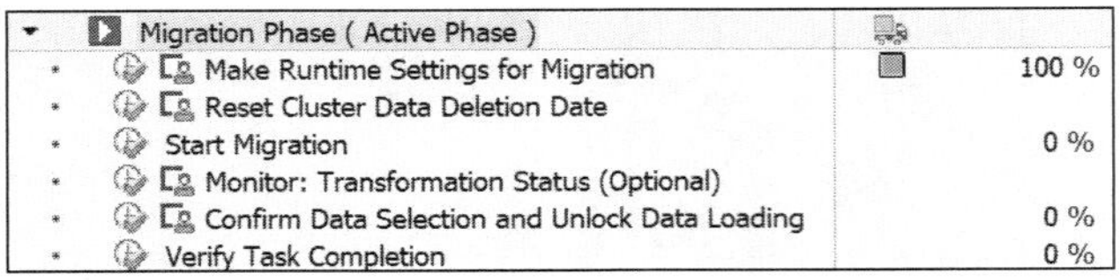

Migration Phase (Active Phase)
Make Runtime Settings for Migration
100 %
Reset Cluster Data Deletion Date
Start Migration
0 %
Monitor: Transformation Status (Optional)
Confirm Data Selection and Unlock Data Loading
0 %
Verify Task Completion
0 %

Figure 4.180 Migration Settings

4.9.23 Reset Cluster Data Deletion Date

When the data is read for the migration from the SAP BW sender system, its stored in the cluster tables of the sender and target systems and occupies certain space. After a certain time, you need to delete that table content. Using this activity, you can set that date; for our demo, the date is after three months, as shown in Figure 4.181.

Program
Edit
Goto
System
Help
Reset Deletion Date for Cluster Data
Package Number
9001K
Deletion Date for Cluster Data
29.06.2024

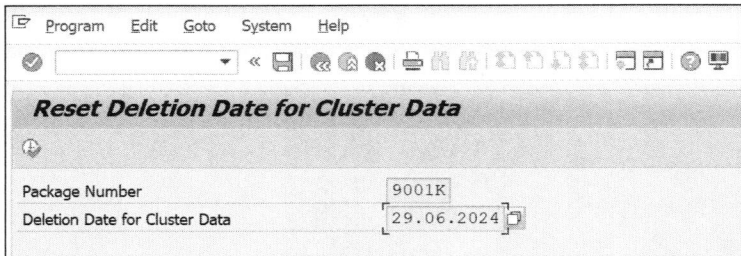

Figure 4.181 Reset Cluster Date

The status is green, as shown in Figure 4.182.

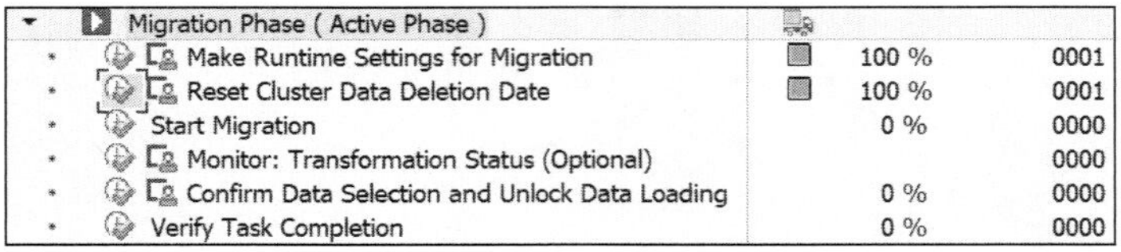

Migration Phase (Active Phase)
Make Runtime Settings for Migration
100 %
0001
Reset Cluster Data Deletion Date
100 %
0001
Start Migration
0 %
0000
Monitor: Transformation Status (Optional)
0000
Confirm Data Selection and Unlock Data Loading
0 %
0000
Verify Task Completion
0 %
0000

Figure 4.182 Rest Cluster Data Deletion: Status

4.9.24 Start Migration

Now to start the data migration, you need to execute the **Start Migration** activity, as shown in Figure 4.183.

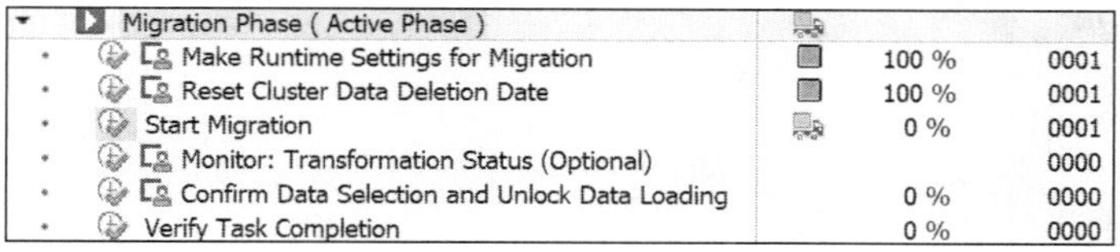

▶ Migration Phase (Active Phase)		
· Make Runtime Settings for Migration	100 %	0001
· Reset Cluster Data Deletion Date	100 %	0001
· Start Migration	0 %	0001
· Monitor: Transformation Status (Optional)		0000
· Confirm Data Selection and Unlock Data Loading	0 %	0000
· Verify Task Completion	0 %	0000

Figure 4.183 Start Migration

Once the migration is started, you'll see the 🚚 icon that indicates the data transfer is happening actively. To monitor migration, you can choose the next activity.

4.9.25 Monitor: Transformation Status (Optional)

Executing the **Monitor: Transformation Status (Optional)** activity will take you to the screen shown in Figure 4.184, which will show the current status of the migration. You'll see 🚚 on the screen to indicate the specific action that happens. Migration will have the following stages:

- **Read Status**
  Used to read the data from the SAP BW sender system and update the cluster table.
- **Move Cluster Status**
  Used to move to the cluster table in the target system.
- **Mig. (migration) Status**
  Used to migrate the data in the target SAP BW/4HANA system.

You need to make sure that all these have a green status for successful migration.

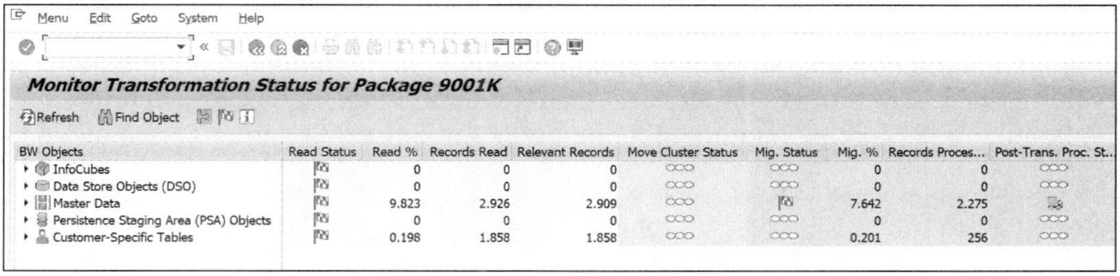

BW Objects	Read Status	Read %	Records Read	Relevant Records	Move Cluster Status	Mig. Status	Mig. %	Records Proces...	Post-Trans. Proc. St...
▶ InfoCubes		0	0	0	⚪⚪⚪	⚪⚪⚪	0	0	⚪⚪⚪
▶ Data Store Objects (DSO)		0	0	0	⚪⚪⚪	⚪⚪⚪	0	0	⚪⚪⚪
▶ Master Data		9.823	2.926	2.909	⚪⚪⚪		7.642	2.275	🚚
▶ Persistence Staging Area (PSA) Objects		0	0	0	⚪⚪⚪	⚪⚪⚪	0	0	⚪⚪⚪
▶ Customer-Specific Tables		0.198	1.858	1.858	⚪⚪⚪	⚪⚪⚪	0.201	256	⚪⚪⚪

Figure 4.184 Migration Status

You can also use Transaction CNVLTRL_MONI_C, which will prompt you for the package number for monitoring. In our case, we'll provide the current active package, as shown in Figure 4.185.

4 Remote Conversion

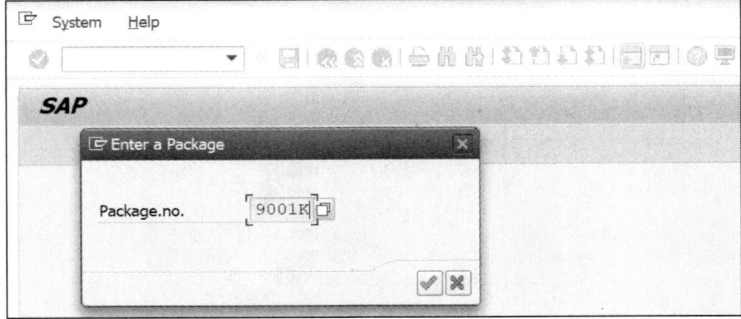

Figure 4.185 Open Package

The status of package 9001K is shown in Figure 4.186. This monitor will show the SAP BW object level details such as **InfoCubes**, **Data Store Objects (DSO)**, and **Master Data**, along with the details of **Persistent Staging Area (PSA) Objects** and **Customer-Specific Tables**.

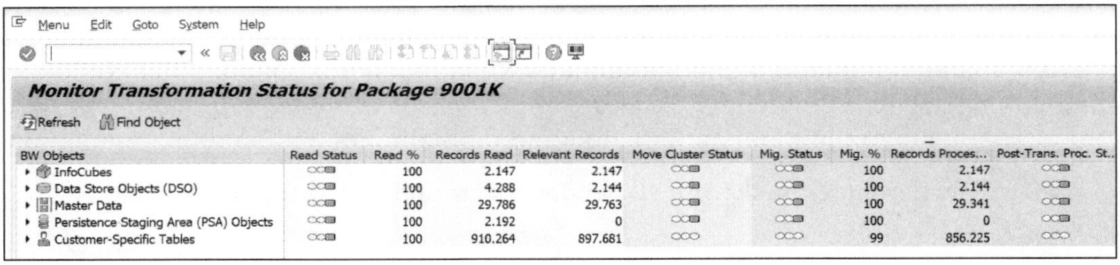

Figure 4.186 Migration Monitoring

You can refresh the status using the icon and expand the nodes below InfoCubes, DSOs, and other listed SAP BW objects to know the details of the tables and the migration status, as shown in Figure 4.187. You can download the data from **Menu • Download Transformation Data** and save it to your local file for the data conversion validation between sender and target system.

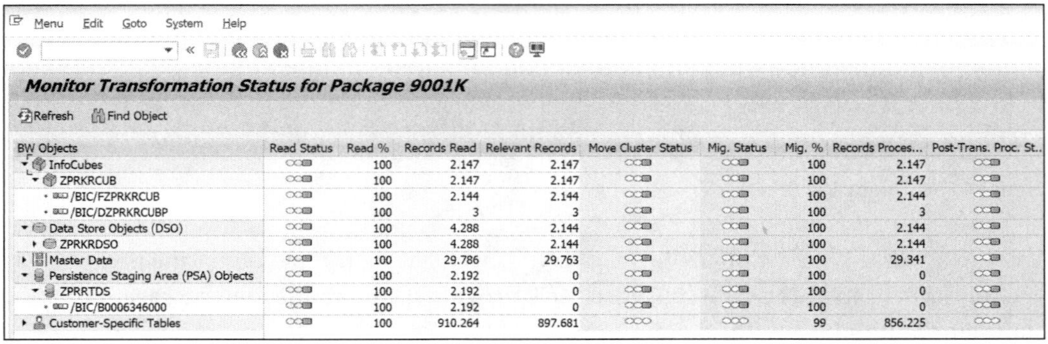

Figure 4.187 Migration Status: Table Level

4.9 Execution

Once all the data is migrated, you'll see the green status in Figure 4.188.

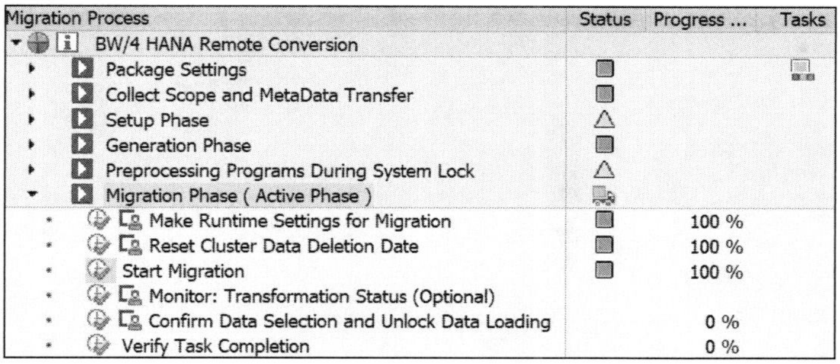

![Migration Process status table]

Figure 4.188 Overall Migration Status

4.9.26 Confirm Data Selection and Unlock Data Loading

You'll have the popup to confirm to **Unlock Loading and Data Target Changes**, as shown in Figure 4.189.

Figure 4.189 Unlock Data Loading

Status of the data unlock is shown in Figure 4.190.

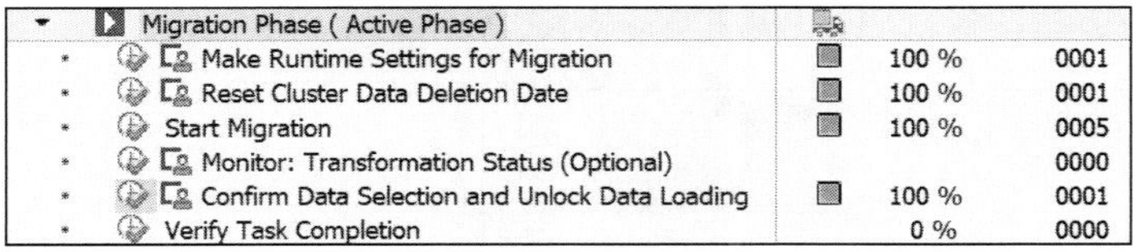

![Unlock Data Loading Status table]

Figure 4.190 Unlock Data Loading Status

4.9.27 Verify Task Completion

Execution of the **Verify Task Completion** activity will check if all the associated tasks have been completed in the sender system and target system. If so, this will become green, as shown in Figure 4.191.

239

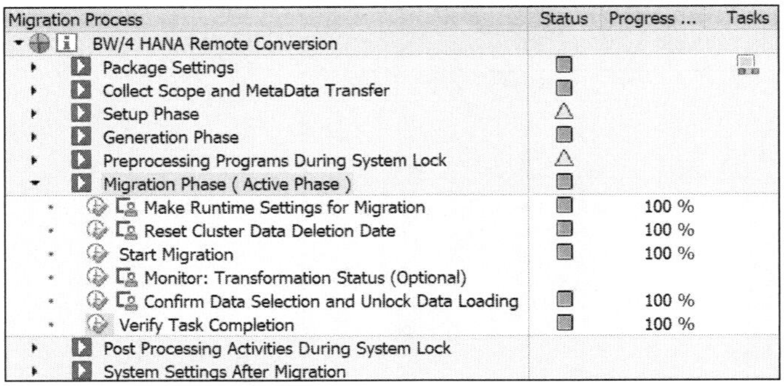

4 Remote Conversion

Figure 4.191 Verify Task Completion

4.9.28 Post Processing Activities During System Lock

This phase encompasses a series of activities aimed at rebuilding hierarchies, resetting number ranges, and configuring queries for generation. Additionally, it provides an option to create the Transaction RSRV package and execute checks in the target system, followed by viewing migration statistics, as shown in Figure 4.192.

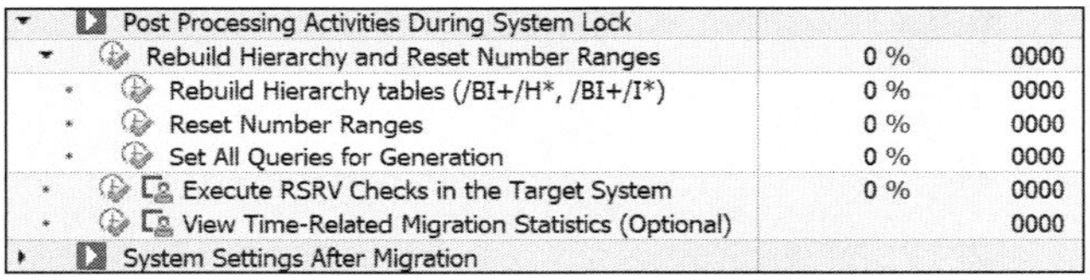

Figure 4.192 Post-Processing Activity: Before Execution

The next step is mostly on the post-processing activities. It's a group activity, so once it's started, you can see the status as running in Figure 4.193.

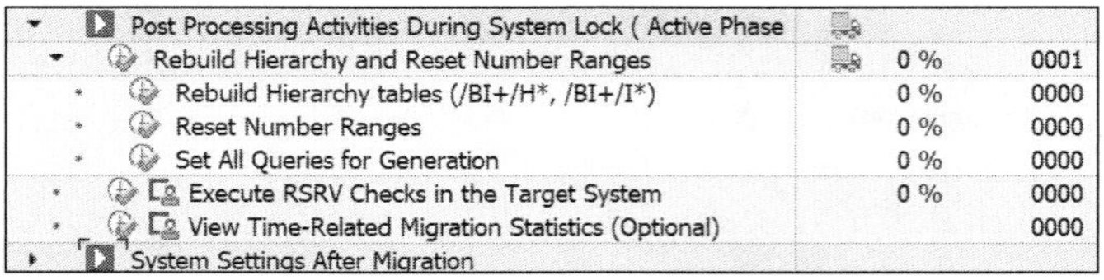

Figure 4.193 Post-Processing Activities: During Execution

After completion of all the activities, you can see the green status in Figure 4.194.

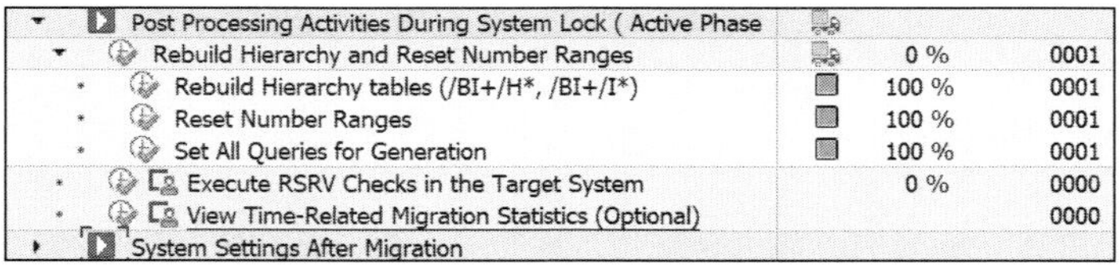

Figure 4.194 Rebuild Hierarchy and Reset Number Ranges: Status

4.9 Execution

4.9.29 Execute RSRV Checks in the Target System

This screen is similar the previous Transaction RSRV package creation, but the tool will create the package for the target SAP BW/4HANA system during this activity. You need to execute this activity with **Save Package without execution** and note the package ID for analysis in the target system later, as shown in Figure 4.195.

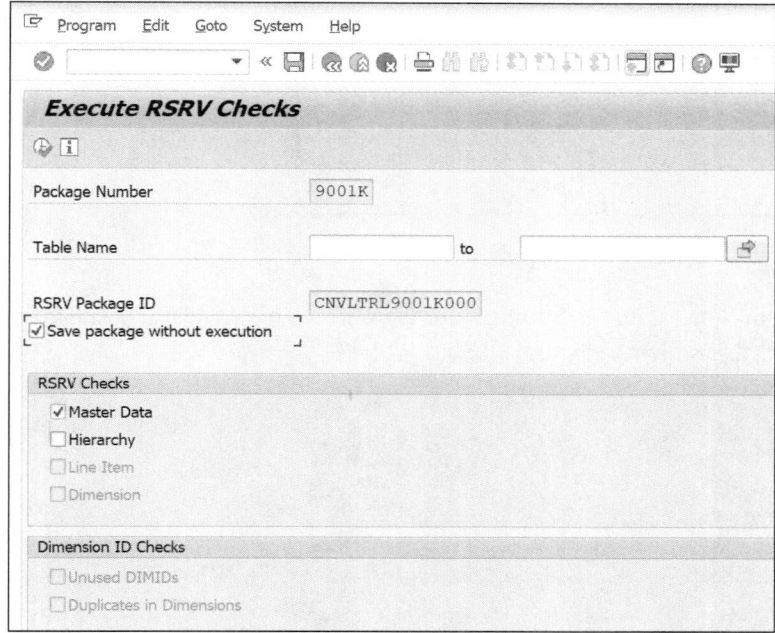

Figure 4.195 Transaction RSRV Checks: Parameter Selection

After confirming, you can see the status is green, as shown in Figure 4.196.

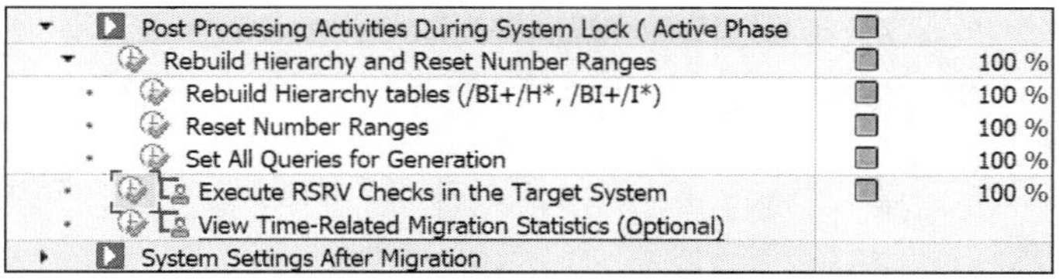

Figure 4.196 Transaction RSRV Check Status

4.9.30 View Time-Related Migration Statistics (Optional)

The next activity will allow you to see the time-related migration statistics. In the **Post Migration Settings** window, you have the option to review the duration required for migrating InfoProviders and SAP BW–relevant tables from the original (sender) system to the target (receiver) system, as shown in Figure 4.197. This information provides insights into the efficiency and progress of the migration process, aiding in the assessment of overall performance and resource allocation, This will help to prepare the system with optimal resource in further cycles.

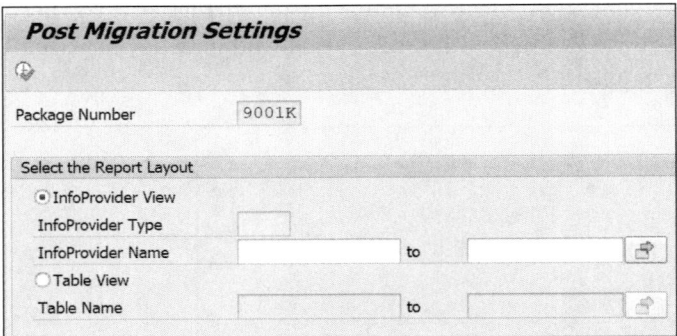

Figure 4.197 Time-Related Migration Statistics

You can see the details based on the InfoProvider View, which has the data for the InfoCube and InfoObject, as shown in Figure 4.198.

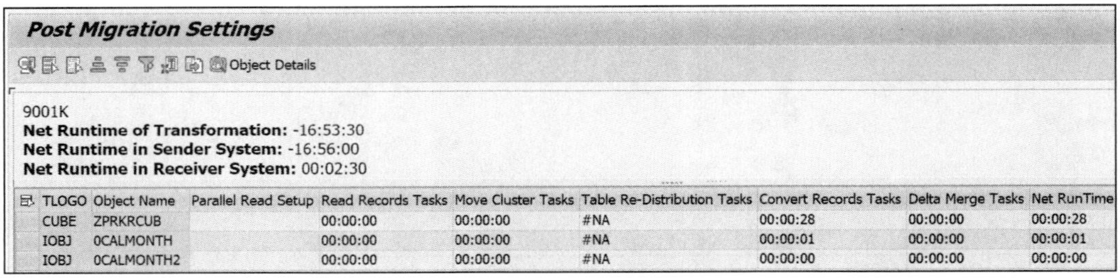

Figure 4.198 InfoProvider View Settings

When you go back, you can see the green status of the activity, as shown in Figure 4.199.

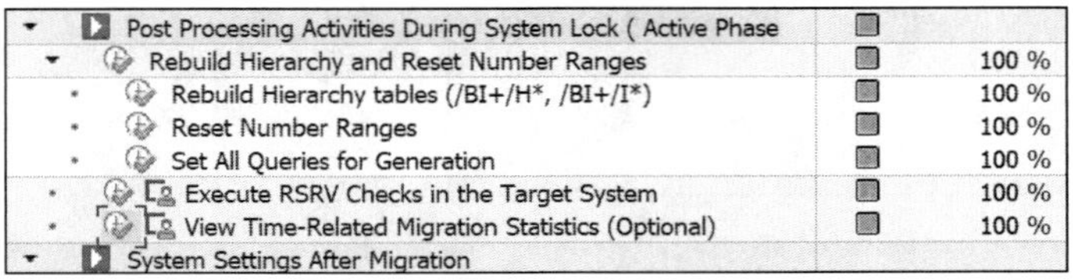

Figure 4.199 Post-Processing Activities: Final Status

4.9.31 System Settings After Migration (Active Phase)

This is a group activity, and, once it's triggered, you can see the successor activities, as shown in Figure 4.200.

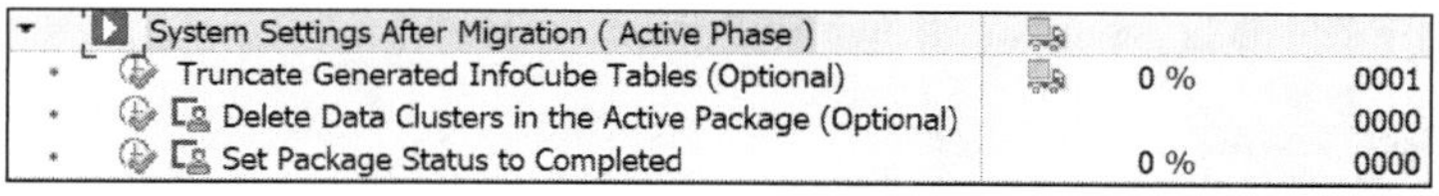

4.9 Execution

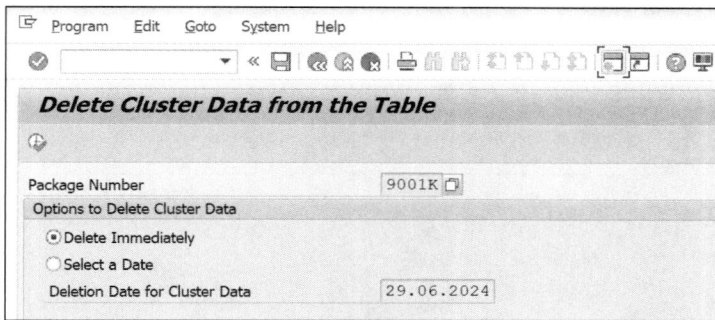

Figure 4.200 System Settings After Migration

The **Truncate Generated InfoCube Tables (Optional)** activity is an optional activity, as the name says, and it will clear the data from temporary tables, including dimension and fact tables for InfoCubes, as well as shadow tables used for SAP BW administration runtime tables. Truncating this data frees up memory in the target (receiver) system, optimizing performance and resource utilization. Selecting **Delete Immediately** will delete the cluster table at that time, as shown in Figure 4.201. For the **Deletion Date for Cluster Data** field, you're seeing the date as **29.06.2024** based on the previous selection. In projects, it's preferable to leave this cluster table for more than three months, so you can choose the **Select Date** option and choose the date you'd like to keep the cluster data until.

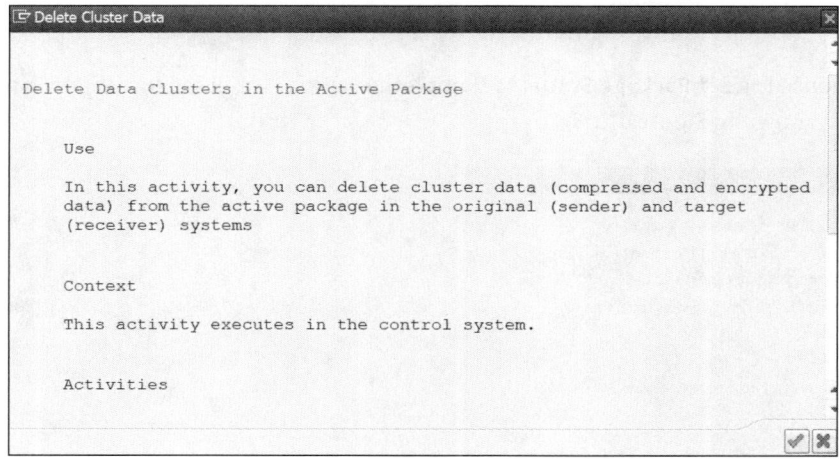

Figure 4.201 Delete Cluster Data: Selection

This triggers a popup, as shown in Figure 4.202.

Figure 4.202 Popup for Deletion Confirmation

243

Once you choose [✓], the status is green, as shown in Figure 4.203.

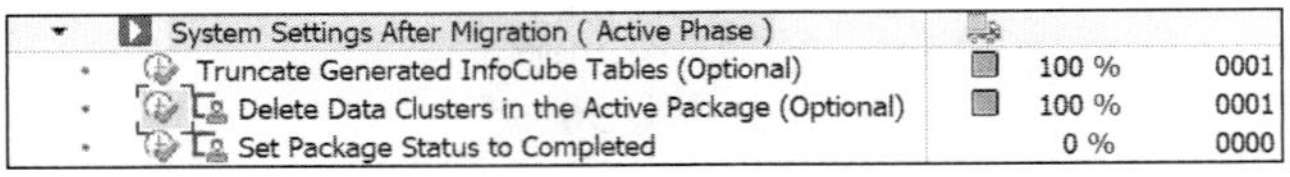

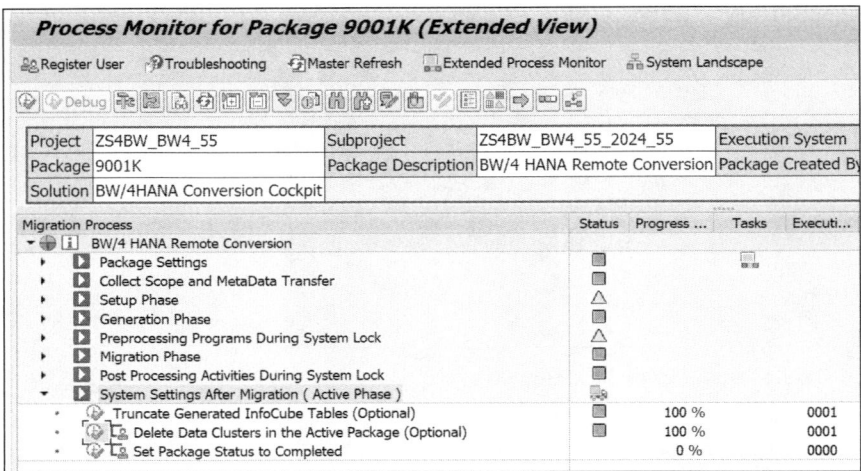

Figure 4.203 Status of Cluster Data Deletion

The next activity is to **Set Package Status to Completed**, as shown in Figure 4.204. It's a manual activity that doesn't need not be executed immediately, so keep this package open for the checks and analysis of data migration. You can execute this activity after all the testing and reconciliation is completed in the target SAP BW/4HANA system and you've confirmed that there is no missing data. But remember, this has to be done at some point because the system won't allow you to create another package if any package is empty.

Figure 4.204 Set Package Status to Be Completed

Once you choose the **Set Package Status to Completed** option, you'll see the popup for confirmation shown in Figure 4.205.

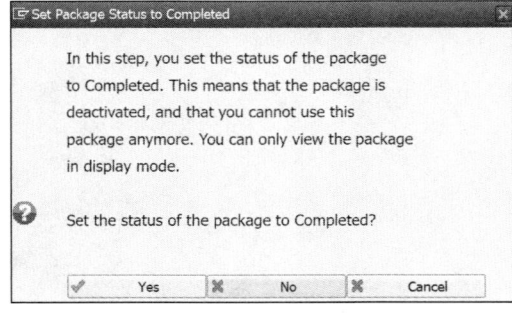

Figure 4.205 Popup for Package Confirmation

4.9 Execution

Once confirmed by selecting **Yes**, the status will be green, as shown in Figure 4.206.

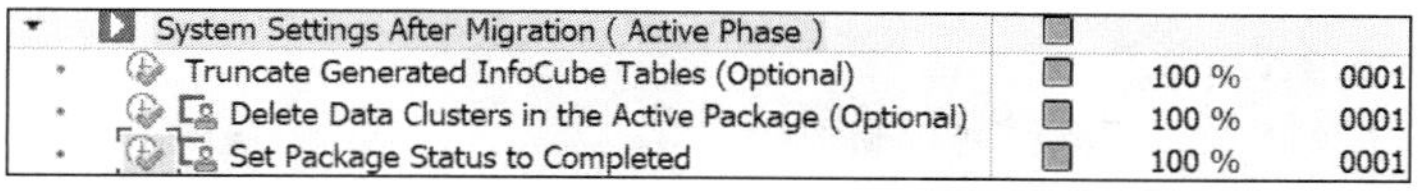

System Settings After Migration (Active Phase)			
Truncate Generated InfoCube Tables (Optional)	■	100 %	0001
Delete Data Clusters in the Active Package (Optional)	■	100 %	0001
Set Package Status to Completed	■	100 %	0001

Figure 4.206 Package Completion: Status

The overall status of package **9001K** after all the phases have been completed successfully is either green ■ or yellow △, as shown in Figure 4.207.

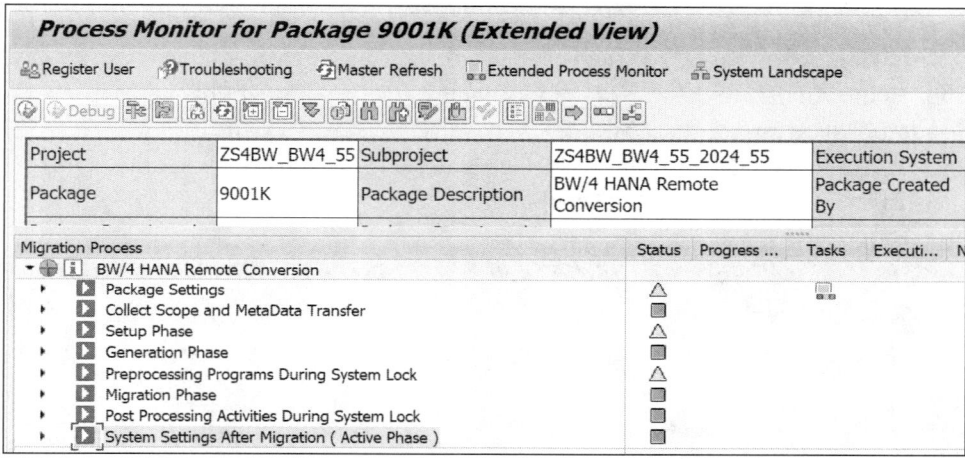

Figure 4.207 Overall Phase Status

After confirming all the activities and validating the data in the target system, you can deactivate the package by choosing the **Package ID (9001J)** row and clicking on the deactivate icon. The **Deactivate Package** popup will appear, as shown in Figure 4.208.

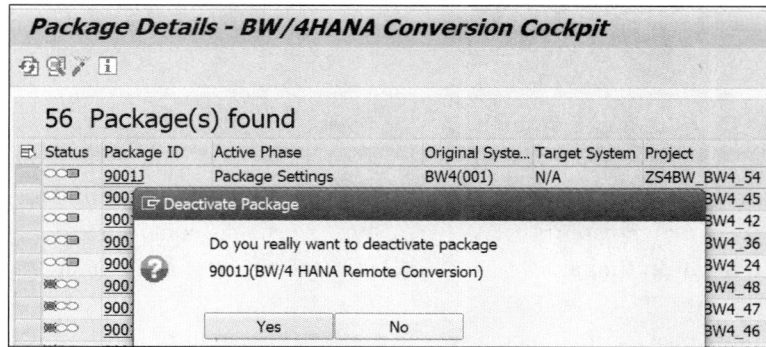

Figure 4.208 Deactivate Package

245

4   Remote Conversion

Once the package is deactivated, you'll see the ✱ icon under the **Status** column, as shown in Figure 4.209.

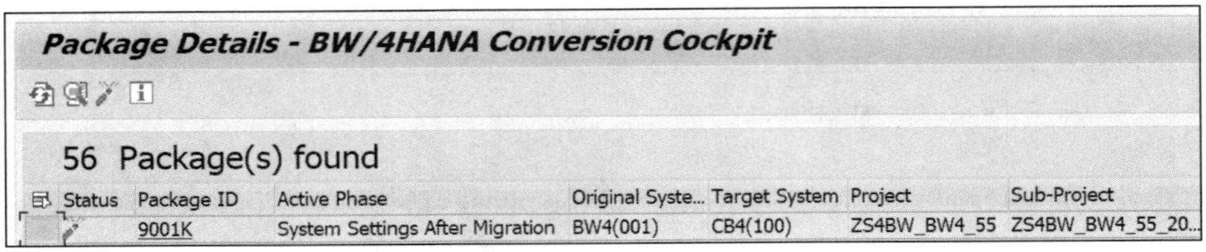

Figure 4.209 Package: After Deactivation

Access Transaction STC02 for the specific task list run, and you can see that the last step, **Finalization**, with the task of **Unlock Loading and Data Target Changes** is green ▪. In addition, you can see that **Task List Run** has a yellow status ⚠ and the locked icon 🔒, indicating the run is **Finalized**, as shown in Figure 4.210. A this point, you can't open this task list again and make any further changes.

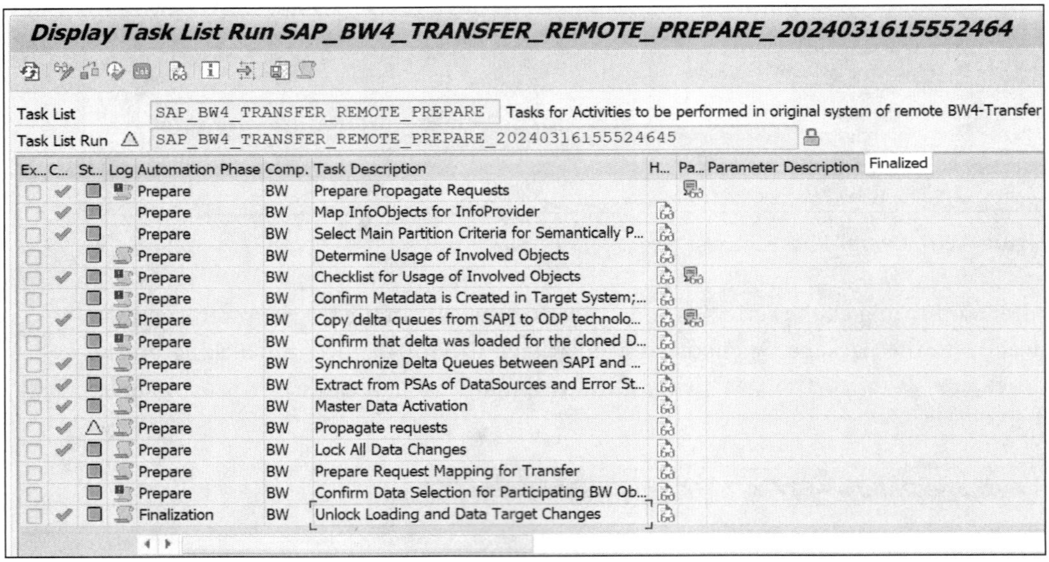

Figure 4.210 Transaction STC02 Status: After Package Completion

During the remote conversion phase, the transports will have specific after import handling. To ensure that normal transports in SAP BW/4HANA work fine, you need to deactivate the remote transfer in the target SAP BW/4HANA system after conversion via report RS_B4HANA_TRANSFER_REM_END. Provide the active task list run name in the sender SAP BW system and execute it. You'll see the message shown in Figure 4.211.

246

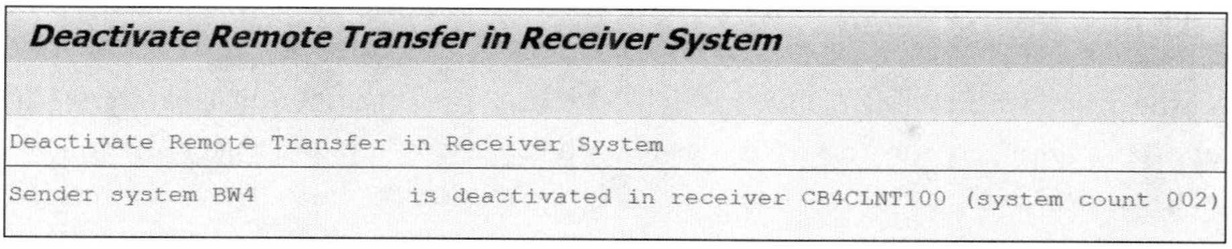

```
Deactivate Remote Transfer in Receiver System

Deactivate Remote Transfer in Receiver System
Sender system BW4            is deactivated in receiver CB4CLNT100 (system count 002)
```

Figure 4.211 Status of Remote Transfer End Report

4.10 Post-Conversion

After completing the conversion process, it's essential to initiate the shell conversion for SAP BW queries. A successful shell conversion ensures that all queries are properly provisioned in the SAP BW/4HANA system. Additionally, conclude any pending Analysis Process Designer (APD) conversions or standard authorization conversions. Post-conversion, it's crucial to ensure that transports aren't failing and to thoroughly check for InfoPackage filters, DTP filters, transformation routines, transformation fields, and all other essential objects between the sender and target systems to maintain data integrity. This might take some time, so you need to plan for this activity. Make sure you handle the parallel development using the proper transports.

Following the transport of custom development objects (e.g., DDIC or ABAP), it's necessary to make adjustments in the original system within the object directory (TADIR). Refer to SAP Notes 2458155 and 2690211 for guidance. Similarly, upon executing a scope transfer of SAP BW objects (e.g., InfoObjects and DSOs), it's crucial to adjust the original system in the object directory (TADIR). Relevant information can be found in SAP Notes 2458155 and 2690211. Use the Authorization Transfer Tool per SAP Note 2468657 to adjust user roles. In addition, update queries with unsupported technical InfoObjects (e.g., Request ID) using SAP Note 2479683.

Next, verify the success of the first delta load, and ensure that no delta records are missed. Subsequently, conduct detailed user acceptance testing (UAT) across relevant cycles to validate data consistency between objects and query results in both the sender and target SAP BW/4HANA systems. This testing phase may extend over several months to ensure thorough validation. The testing will focus on data loads and reporting from critical interfaces. Any issues identified during testing should be promptly addressed.

Upon completion of testing and parallel runs along with the existing SAP BW system, the decision to go live with the SAP BW/4HANA system can be made. Following this, the new SAP BW/4HANA system will be operational for all data loading and reporting purposes, providing an updated and optimized platform for data management and analysis.

4.11 Useful Reports and Tables

During the prepare and realize phases, you'll get many errors that require standard SAP BW reports to help activate and fix issues related to the conversion. Check the following list to see which report can be used in which conversion scenario. There are specific conditions to using these reports, so you need to ensure that you read them and follow those conditions carefully. It's generally not recommended to modify the status tables based on SAP BW/4HANA conversion, so we're not providing the backend tables related to the status management in this book. If these reports don't fix the issues, and you still have data consistency issues, then it's best practice to create an incident with SAP and follow up further You can refer to the following SAP reports during the SAP BW/4HANA conversion:

- **RS_B4HANA_TRANSFER_REM_RESET**
  If you like to reset or delete the SAP BW/4HANA remote conversion-based objects in the target system, this report needs to be executed in the sender SAP BW system, The execution will delete the system-generated objects in the target system and reset the control table flags. This will enable you to send the same objects one more time. This can be used in remote and shell conversion, and you need to execute this report in the original sender SAP BW system. You need to make sure that the SAP BW/4HANA conversion cockpit–based run is active, or the package must be active to execute this report, and you need to give the equivalent task list run name for the reset.

- **RS_B4HANA_RESET_DELTA_CLONE**
  This report is executed in the source SAP ERP or SAP S/4HANA system. If you have a situation where you see many errors after the delta sync step or if you need to rest the delta clone in the sender system, this report will be useful. This will reset the status of the delete queue clone, which will enable you to execute the clone step once again. You need to ensure that the task list run will be in change mode.

- **RS_B4HANA_REQUEST_RESET**
  Once the SAP BW/4HANA conversion cockpit–generated transport request is imported in the SAP BW/4HANA system, the control tables will be updated with the metadata transfer status, which won't allow you to repeat. This report will help you remove the entry from the control tables so that you can reimport the request again. This report is executed in the target SAP BW/4HANA system, and it can be used in remote and shell conversions. But if you've transferred the data to the object, you can't use this report.

- **RS_B4H_CLEANUP_METADATA**
  If the conversion is completed and data validation is successful, you can use this report to remove all the unnecessary InfoCube metadata from the target SAP BW/4HANA system. You shouldn't run this report if there are any transfers or migration isn't completed and validated. This report is used in remote and shell conversions, and it must be executed in the target SAP BW/4HANA system.

Some useful general activation reports in SAP BW include the following:

- For activation of InfoObjects, use RSDG_IOBJ_ACTIVATE_ALL.
- For activation of DataSources, use RSDS_DATASOURCE_ACTIVATE_ALL.
- For activation of transformations, use RSDG_TRFN_ACTIVATE.
- For activation of InfoCubes, use RSDG_CUBE_ACTIVATE.
- For activation of standard ODS, use RSDG_ODSO_ACTIVATE.
- For activation of ADSOs, use RSDG_ADSO_ACTIVATE.
- For activation of MultiProviders, use RSDG_MPRO_ACTIVATE.
- For activation of HCPR, use RSDG_HCPR_ACTIVATE.

To test a conversion, you might need to create the InfoProviders with data for testing purposes. In the SAP BW sender system, we have two reports that help you create the sample cube and DSO you can use for testing the conversion:

- To create the sample cube with data, use CUBE_SAMPLE_CREATE.
- To create the sample DSO with data (based on SFLIGHT), use RSDSO_SAMPLE_ADSO.

Following are some of the tables that can be used during the remote conversion:

- **SAP BW 7.x tables**
    - RSDS: Table for SAP BW 7.x DataSources
    - RSOLTPSOURCE: Table to check SAP BW 3.x DataSources
    - RSDIOBJ: Table with list of all InfoObjects in the system
    - RSDCUBE: Table with list of all InfoCubes in the system
    - RSDODSO: Table with list of all ODSs in the system
    - RSLDPIO: Table will link to the DataSource with the InfoPackage
    - RSTRAN: Table with list of all transformations
    - RSBKDTP: Table with list of all DTPs
    - RSPCCHAIN: Table with list of all process chains
    - RSBOHDEST: Table with list of all open hub destinations
    - RSRREPDIR: Table with list of all SAP BW reports
    - RSBKREQUEST: Table with all DTPs based on classical Transaction RSSM–based requests
    - RSDDSTAT: Table with statistics information
- **SAP BW/4HANA tables**
    - RSDS: Tables for SAP BW 7.x DataSources
    - RSOADSO: Table that stores the ADSOs
    - RSOHCPR: Table with list of all SAP HANA CompositeProviders
    - RSTRAN: Table with details of transformations

- RSBKDTP: Table with details of the DTPs
- RSPMREQUEST: Table with new requests based on TSNs

4.12 Summary

With the completion of this chapter, you're now ready to convert SAP BW 7.x to SAP BW/4HANA using remote conversion. You now have detailed insights on the system landscape preparation, and you know the activities to be done in the prepare phase and realization phase of the conversion. You're now aware of the task list and the SAP BW/4HANA conversion cockpit. You're ready to execute the process tree in the SAP BW/4HANA conversion cockpit and know how to check and fix any issues. You've learned about transport requests that are created from the SAP BW/4HANA conversion cockpit tool and how to import them in the target SAP BW/4HANA system and check for the conversion status. You've learned how the data migration happens in remote conversion and how to check the backend tables for data reconciliation. All of this knowledge will enable you to start the SAP BW/4HANA migration using remote conversion for your specific use case. In the next chapter, we'll focus on converting to SAP BW/4HANA using shell conversion.

Chapter 5
Shell Conversion

This chapter outlines the detailed procedure for implementing SAP BW/4HANA shell conversion. Be sure you've read about the system preparation and prerequisites discussed in Chapter 3. This conversion can be implemented separately, or it can be continued for certain objects after remote conversion, which was discussed in Chapter 4. Upon completing this chapter, you'll be proficient in executing SAP BW/4HANA shell conversion and have a comprehensive understanding of task lists and associated activities in the system landscape.

Shell conversion involves the conversion of SAP Business Warehouse (SAP BW) 7.x system to SAP BW/4HANA without data. You'll start learning about the overview of shell conversion in Section 5.1. Then, we'll go over the details of the system landscape for shell conversion in Section 5.2. SAP BW system compatibility is discussed in Section 5.3, and you'll learn about SAP BW system provisioning in Section 5.4. The precheck and simulation for the shell conversion are discussed in Section 5.5. The actual execution of the shell conversion is covered in detail in Section 5.6. You'll learn about the post-conversion activities after shell conversion in Section 5.7.

When an organization operating on the SAP BW 7.x version with any database contemplates the transition to SAP BW/4HANA, the choice of the shell conversion method becomes particularly appealing in specific scenarios. This approach gains prominence when the database size is relatively small, and the organization decides against migrating data to the new SAP BW/4HANA system. In such cases, the primary objective is the conversion of classical SAP BW 7.x objects to their counterparts in the advanced SAP BW/4HANA architecture.

The shell conversion method stands out for its streamlined and less intricate nature. One of its key advantages lies in the expedited implementation process, attributed to the fact that data migration isn't a part of the equation. By bypassing the need to transfer existing data to the new system, organizations can significantly reduce the complexity of the migration process and accelerate the overall transition timeline.

This approach offers a pragmatic solution for organizations that prioritize efficiency and want to swiftly adopt the enhanced capabilities of SAP BW/4HANA without the overhead of extensive data migration efforts. By concentrating on the conversion of objects, the shell conversion method provides a pragmatic and resource-efficient pathway to embrace the benefits of SAP BW/4HANA.

5 Shell Conversion

5.1 Overview of Shell Conversion

Before starting the details of SAP BW/4HANA migration let's discuss a few business scenarios where the migration to SAP BW/4HANA will be required.

5.1.1 Business Cases

There are two options for going with shell conversion:

- If the system's database size is minimal, and there's a need for a swift project cycle without loading data into the new SAP BW/4HANA immediately post-conversion, the Shell conversion is a suitable option. A well-structured project plan is crucial in this scenario, outlining the comprehensive list of objects earmarked for conversion. These objects typically include start objects such as process chains, MultiProvider, InfoCubes, DataStore objects (DSOs), InfoSets, and more.

- In alternative scenarios, a combination of shell conversion and remote conversion may be employed. For instance, after completing the remote conversion to transition SAP BW objects such as InfoCubes and DSOs to advanced DSOs (ADSOs) with data in the target SAP BW/4HANA system, the shell conversion approach can be used to gather SAP Business Explorer (SAP BEx) queries related to the converted objects in the target landscape. In such instances, system preparation and prechecks for remote conversion may have already been undertaken. Installation of XML-based notes for both remote and shell conversion might be necessary, followed by the direct execution of the task list. It's imperative to complete the remote conversion prior to initiating the shell conversion to prevent objects from being in a locked state.

5.1.2 Shell Conversion Flow

When the organization needs to convert the classical SAP BW objects to SAP BW/4HANA objects without the transfer of data, then the shell conversion needs to be used. This conversion is applicable from SAP BW release 7.0 to 7.5. The primary goal of the shell conversion is to update and adapt the structure and components of classical SAP BW objects to make them compatible with SAP BW/4HANA. This allows organizations to leverage the new capabilities of SAP BW/4HANA without the immediate need to migrate the data. It's particularly useful when a complete data migration isn't immediately feasible or necessary, but the organization still wants to benefit from the enhanced features of SAP BW/4HANA.

As mentioned earlier, to perform a shell conversion, a parallel landscape is required that includes both a SAP BW 7.x system (the sender system) and a SAP BW/4HANA system (the target system). The parallel landscape allows for the transformation and adaptation of objects in a controlled environment without affecting the production system. The process involves converting classical SAP BW objects, such as InfoProviders, transformations, and data transfer processes (DTPs), to their equivalent SAP BW/4HANA objects. This may include updating data models, transforming structures, and aligning configurations with SAP BW/4HANA standards. The shell conversion flow is shown in Figure 5.1.

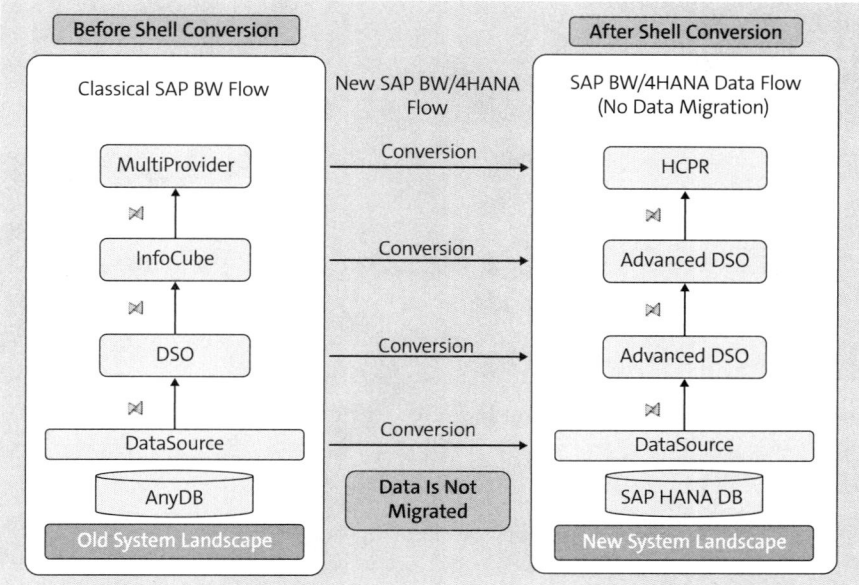

Figure 5.1 Shell Conversion Flow

5.2 System Landscape for Shell Conversion

Shell conversion requires the sender SAP BW and target SAP BW/4HANA system landscape. If the SAP BW system is connected to SAP ERP or SAP S/4HANA, then the landscape will need those connectivities too, as shown in Figure 5.2.

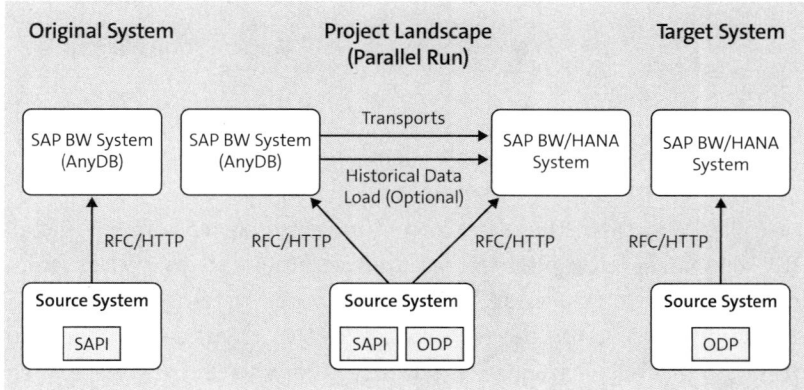

Figure 5.2 Shell Conversion System Landscape

5.3 SAP Business Warehouse System Compatibility

Before starting the shell conversion, you need to first check the sender and target system compatibility. You can find the details of the minimum supported release for shell conversion in Table 5.1.

Release	Minimum Support Release	Recommended Support Release
SAP BW 7.0	SP 28 and Higher	Latest
SAP BW 7.01	SP 12 and Higher	Latest
SAP BW 7.02	SP 10 and Higher	Latest
SAP BW 7.3	SP 10 and Higher	Latest
SAP BW 7.31	SP 10 and Higher	Latest
SAP BW 7.4	SP 12 and Higher	Latest
SAP BW 7.5	SP 12 and Higher	Latest

Table 5.1 Prerequisites: Sender System Support Release

Refer to Chapter 3 to understand the detailed system preparation activities. The preparation is the same for both remote and shell conversion.

Shell conversion involves two systems: the sender, which is the original SAP BW 7.x system, and the target SAP BW/4HANA system. Let's see the support packages for target SAP BW/4HANA systems. It's recommended to go for the latest SAP BW/4HANA release, as listed in Table 5.2.

Release	End of Maintenance	Next Release
SAP BW/4HANA 2.0	12/31/2024	SAP BW/4HANA 2021
SAP BW/4HANA 2021	12/31/2027	SAP BW/4HANA 2023
SAP BW/4HANA 2023	12/31/2030	SAP BW/4HANA 2026

Table 5.2 Target SAP BW/4HANA System Version Options: Shell Conversion

When you plan for SAP BW/4HANA, these are the minimum supported versions. Install the target SAP BW/4HANA, and complete the required activities such as initial system setup and connectivity to all SAP ERP and SAP S/4HANA source systems through operational data provisioning (ODP). Add the required SAP HANA smart data integration (SDI) system for the DB Connect–based system. If there is any flat file interface, then you'll need to add that too. Transfer the global settings, and make sure you have currency and units available in the target SAP BW/4HANA system.

5.4 SAP Business Warehouse System Provisioning

Before beginning the shell conversion process, it's important to ensure that you've completed the specific tasks described in the following sections. This helps in ensuring a smooth transition and minimizing any potential issues during the conversion process.

5.4.1 Target SAP BW/4HANA System Provisioning

Install the target SAP BW/4HANA system, and complete all the required post-installation activities (refer to Chapter 4, Section 4.2.3, for more details). Here is the high-level summary:

- Acquire all essential software packages for SAP BW/4HANA by using maintenance planner for a seamless download process.
- Use the Software Provisioning Manager (SWPM) to initiate the installation of a new SAP BW/4HANA system.
- Perform system setup and customization tasks to configure the environment according to specific requirements.
- Establish new source system connections within SAP BW/4HANA for all pertinent data flows, integrating ODP, SAP HANA smart data access (SDA), and SDI source systems.
- Install the necessary SAP Notes, and execute manual activities crucial for the shell conversion in the SAP BW/4HANA system by using Note Analyzer to trigger the required actions. The installation of the SAP Notes based on the transformation XML will avoid errors during the transport import.

5.4.2 Sender System Provisioning

In the sender SAP BW 7.x system, you'll carry out all the activities based on the discover and prepare phase. These steps are crucial to prepare the sender system for the SAP BW/4HANA conversion. Figure 5.3 shows the details of the discover and prepare phase.

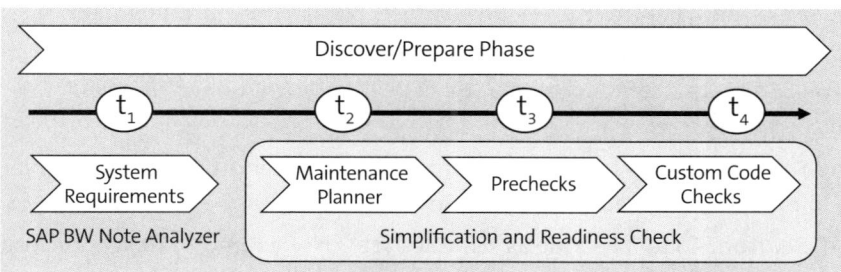

Figure 5.3 Prepare Phase

5.4.3 Sender and Target System Landscape Preparation

Begin by implementing the required SAP Notes, ensuring you use the most recent XML from SAP Note 2383530 "Conversion from SAP BW to SAP BW/4HANA." Extract the XML files from the attachment in the SAP Note. Next, execute the Note Analyzer with the listed XML files in the sender, target, and source systems. Verify that all SAP Notes listed in the result are in green status. See Figure 5.4 for more details. You can see the following XML files:

- **SAP_BW4HANA_Shell_Conversion_(Original_System)_Latest_Date**
- **SAP_BW4HANA_Shell_Conversion_(Target_System)_Latest_Date**
- **Source_System_for_SAP_BW4HANA_Latest_Date**

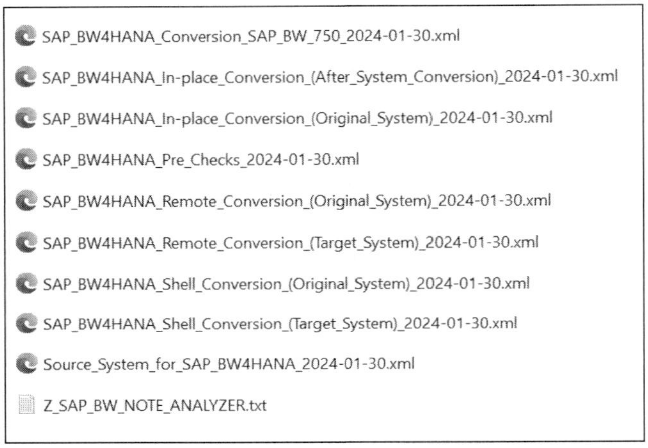

```
SAP_BW4HANA_Conversion_SAP_BW_750_2024-01-30.xml
SAP_BW4HANA_In-place_Conversion_(After_System_Conversion)_2024-01-30.xml
SAP_BW4HANA_In-place_Conversion_(Original_System)_2024-01-30.xml
SAP_BW4HANA_Pre_Checks_2024-01-30.xml
SAP_BW4HANA_Remote_Conversion_(Original_System)_2024-01-30.xml
SAP_BW4HANA_Remote_Conversion_(Target_System)_2024-01-30.xml
SAP_BW4HANA_Shell_Conversion_(Original_System)_2024-01-30.xml
SAP_BW4HANA_Shell_Conversion_(Target_System)_2024-01-30.xml
Source_System_for_SAP_BW4HANA_2024-01-30.xml
Z_SAP_BW_NOTE_ANALYZER.txt
```

Figure 5.4 Extracted XML files from SAP Note 2383530

For example, to prepare a landscape, you need to execute the Note Analyzer and install all the required notes in the landscape. Execute Transaction SE38, select **Z_SAP_BW_NOTE_ANALYZER**, and execute the report based on the XML listed in the relevant systems, as shown in Figure 5.5:

❶ Install the precheck files in the original SAP BW system:
  - For SAP BW from 7.3 to 7.4, use *SAP_BW4HANA_Pre_Checks_[last_update].xml*.
  - For SAP BW 7.5, the prechecks can't be installed isolated but are bundled with the SAP BW/4HANA transfer cockpit, so use *SAP_BW4HANA_Conversion_SAP_BW_750_[last_update].xml*.

❷ Install the SAP BW/4HANA transfer cockpit and Data Migration Integration Server (DMIS) add-on in the original SAP BW system:
  - For SAP BW from 7.3 to 7.4 acting as sending system for a remote conversion, use *SAP_BW4HANA_Shell_Conversion_(Original_System)_[last_update].xml*.
  - For SAP BW 7.5, use *SAP_BW4HANA_Conversion_SAP_BW_750_[last_update].xml*.

5.5 Prechecks and Simulation

❸ Install ODP updates in each system:
- For the source systems connected to SAP BW such as SAP ERP, SAP S/4HANA, SAP Customer Relationship Management (SAP CRM), and SAP Supplier Relationship Management (SAP SRM), use *Source_System_for_SAP_BW4HANA_[last_update].xml*.

❹ Install the SAP BW/4HANA transfer cockpit and DMIS add-on in the target SAP BW/4HANA system:
- For SAP BW/4HANA acting as receiving system for a remote conversion, use *SAP_BW4HANA_Shell_Conversion_(Target_System)_[last_update].xml*.

❺ Install all the SAP Notes for the transformation based on the XML from SAP Note 2603241. This needs to be done in the target SAP BW/4HANA system. Refer to SAP Note 2603241, "Overview and summary of the most important SAP Notes in the context of BW." Download the notes from *SAP_BW_Transformation_[Last_Update].xml* in *your target SAP BW/4HANA* system.

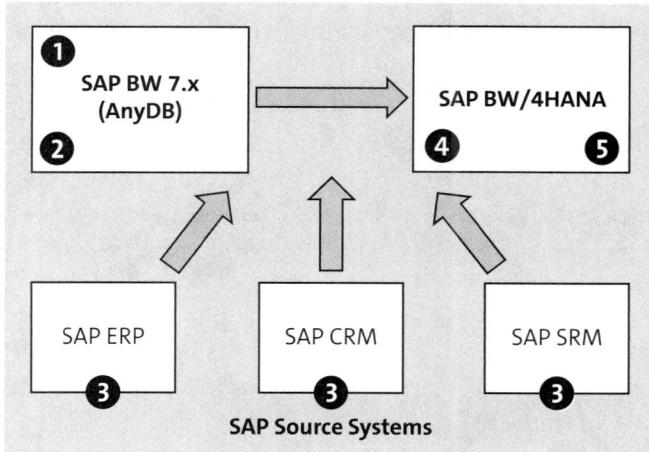

Figure 5.5 System Landscape Preparation

5.5 Prechecks and Simulation

Once all the SAP Notes are implemented, verify the installation status of the SAP BW/4HANA transfer cockpit, and proceed to perform the readiness check along with other precheck activities, as previously discussed. You won't be in the realization phase until you've completed all the prechecks.

Authorization is required for the following authorization objects: S_RS_B4H, S_TC, S_RFC, and S_RFC_ADM. Additionally, you must possess authorization to read any selected object intended for the transfer.

Perform the following activities:

- Based on the precheck result, fix all the issues related to the classical SAP BW 3.x objects and any issues found in the **Before Migration** folder, where you can see the details of the issue and the options.
- Check the Unicode and Basis configuration of the system.
- Ensure that the transport path is established with the target SAP BW/4HANA system.
- If your organization is only using shell conversion, then finalize the list of scope selection objects such as process chains, MultiProviders, InfoSets, InfoCubes, DSO, Open Hub , and so on.
- Read the simplification items to know the impact of the SAP BW/4HANA conversion.
- If your organization is using shell conversion after remote conversion, then have a list of SAP BEx query names that needs to be transferred.
- Check for the support of the reporting tool used in SAP BW/4HANA.

One important precheck is the simulation of scope objects. You can use the task list **SAP_BW4_TRANSFER_CHECK_SHELL** for this using Transaction STC01, as shown in Figure 5.6.

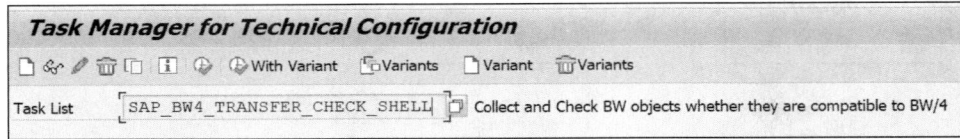

Figure 5.6 Scope Simulation

You'll see the activities that are part of the simulation check. You can provide the SAP BW objects and execute the job to start the analysis. Fix the errors if any before the actual scope selection, as shown in Figure 5.7.

Figure 5.7 Scope Simulation Shell

5.6 Realization Phase

After ensuring that all SAP Notes are implemented across the entire system landscape, check the installation status of the SAP BW/4HANA transfer cockpit in the SAP BW sender system. Proceed to conduct the Readiness Check along with other precheck activities, as discussed earlier. Once completed, you'll be ready to initiate the realization phase. Figure 5.8 shows more details.

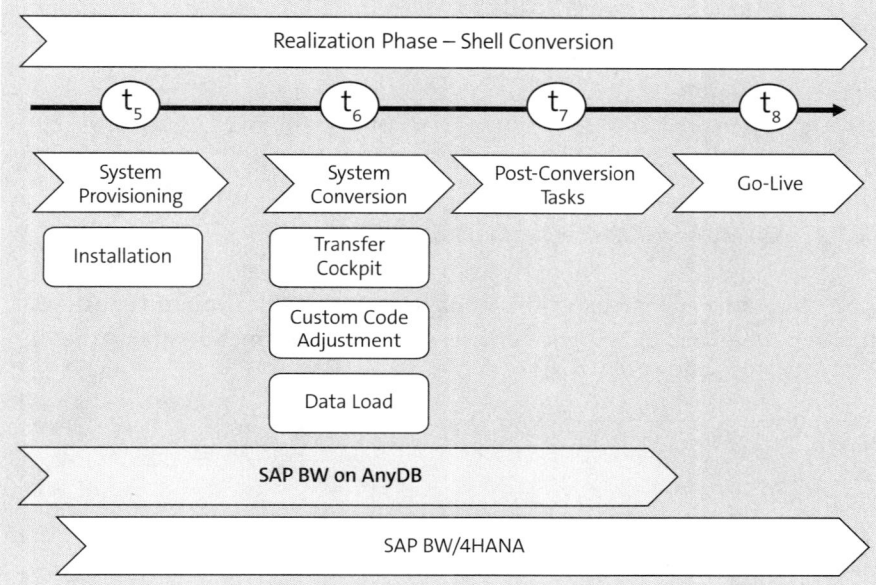

Figure 5.8 Realization Phase: Shell Conversion

If you've already provisioned the SAP BW/4HANA system, made sure all the checks are completed, and finished all the custom code checks, you can now start the shell conversion. The next step is to finalize the scope objects that you need to send to the target system.

5.6.1 Scope Object Identification

The initial stage of the shell conversion involves identifying the list of scope objects. In our scenario, we'll use the InfoCube (**ZPRC1_S**) as the test case. This InfoCube will gather all dependent objects in the tool during execution. Let's examine the data flow for this InfoCube first. In the SAP BW sender system, data exists, but it won't be migrated during the shell conversion process. Figure 5.9 shows the InfoCube (**ZPRC1_S**) data flow.

5 Shell Conversion

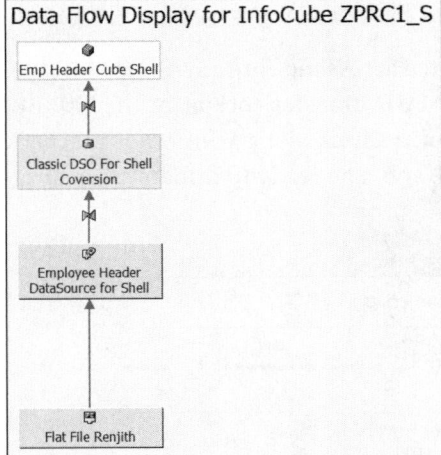

Figure 5.9 Scope Object Flow in the Sender System

Selecting the InfoCube and choosing the **Manage** option enables you to view the data within the InfoCube. This allows you to access detailed information about the InfoCube's data, as depicted in Figure 5.10.

ZPRENAMES	ZPRENUMS	ZPRESTATS	ZPREAGES	ZPRESALS
Emily	1	PA	26	100.000,00
Brandon	2	NY	28	120.000,00
Samantha	3	IL	25	90.000,00
Tyler	4	CA	29	110.000,00
Olivia	5	FL	30	130.000,00
Fiaz	6	OH	28	120.000,00
Alexis	7	GA	27	105.000,00
Austin	8	MI	35	150.000,00
Haley	9	NC	33	130.000,00
Dylan	10	NY	28	120.000,00
Ryan	11	FL	27	105.000,00
Dave	12	WA	26	100.000,00
Henry	13	NJ	30	130.000,00
Shankar	14	PA	28	120.000,00

Figure 5.10 InfoCube Data

The next step is to start the scope selection. You can create the shell conversion in two ways: by using the SAP BW/4HANA transfer cockpit or by using a task list. Let's start with the steps for using the SAP BW/4HANA transfer cockpit:

1. Create a new shell conversion task list from SAP BW/4HANA transfer cockpit using Transaction RSB4HCONV, as shown in Figure 5.11.

5.6 Realization Phase

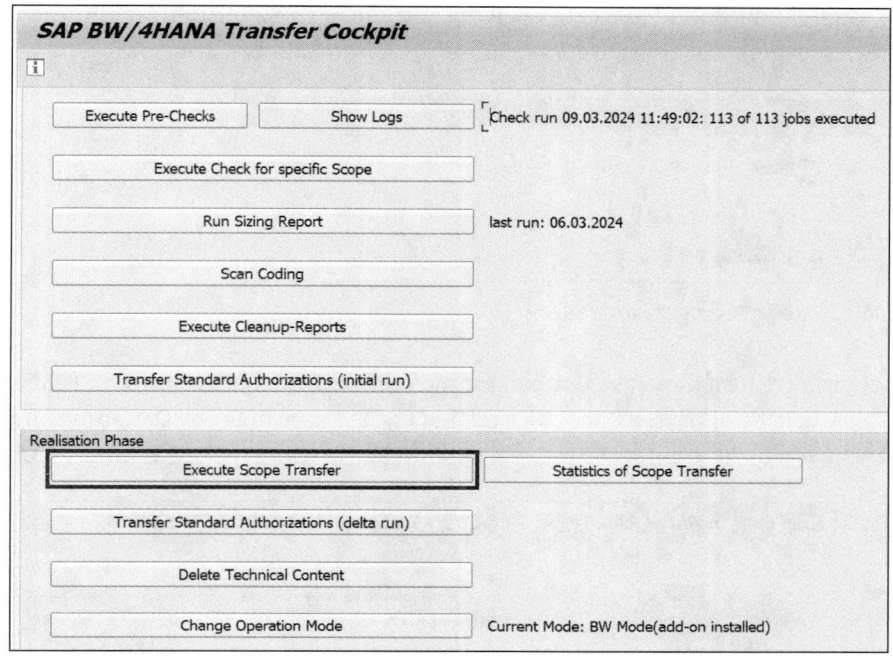

Figure 5.11 Execute Scope Transfer

2. Choose the **Execute Scope Transfer** option to see the popup shown in Figure 5.12.

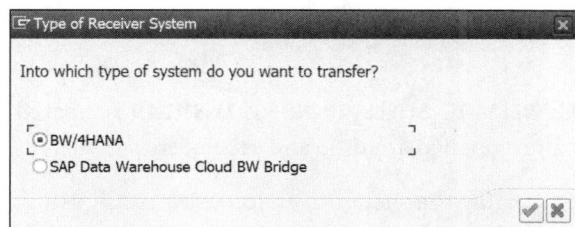

Figure 5.12 Receiver System Selection

3. Choose **SAP BW/4HANA** and confirm, and you'll see the selection for **Transfer Scenario**, as shown in Figure 5.13. Choose **Shell (Metadata only)** and continue.

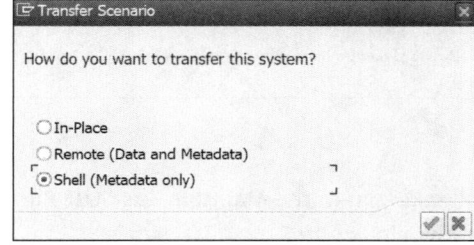

Figure 5.13 Transfer Scenario Selection

261

5   Shell Conversion

4. Next, you'll get the option to create a new task list run or show existing task list runs, as shown in Figure 5.14. Select **New** which will create the new task list.

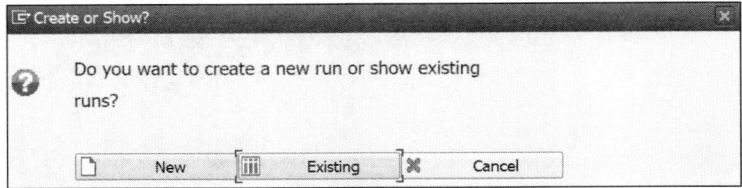

Figure 5.14 Create or Display Task List Runs

5. The tool will create a new **Task List Run** for the shell conversion, as shown in Figure 5.15. The task list will have six activities related to the phase of **Scope Definition** and **Prepare**. If you need to enter the manual parameter, you can see the icon.

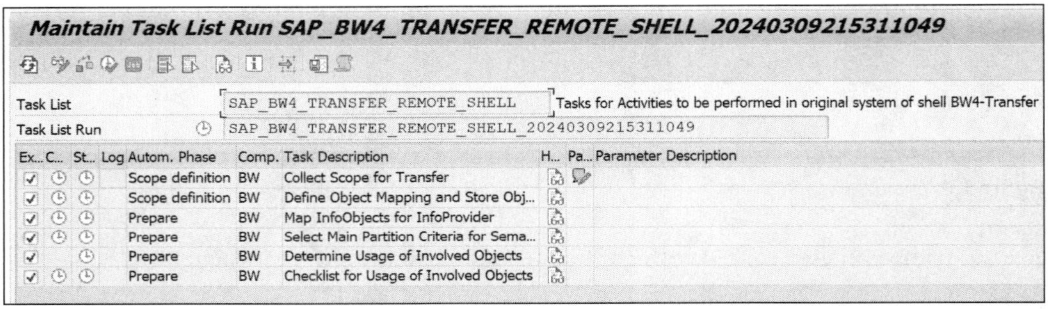

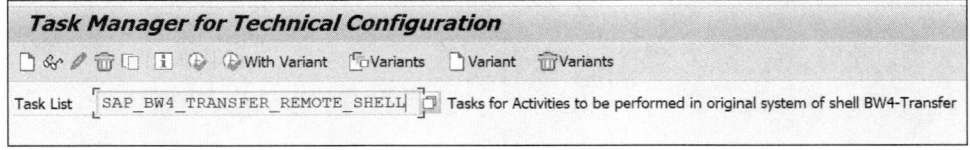

Figure 5.15 Shell Conversion Task List

6. Task List run **SAP_BW4_TRANSFER_REMOTE_SHELL_20240309215311049** is created. You can save this task list, and it can be opened anytime and processed.

Next, we'll walk through the steps for creating the shell conversion using a task list:

1. If you like to use the task list to start the shell conversion, execute Transaction STC01, and then execute **SAP_BW4_TRANSFER_REMOTE_SHELL**, as shown in Figure 5.16.

Figure 5.16 Task List for Shell Conversion

2. The execution of the preceding task list will take you to the **Maintain Task List Run** screen with the following information, as shown in Figure 5.17:

5.6 Realization Phase

❶ Shell conversion task list name
❷ Task list run name for the shell conversion
❸ Parameters for the task list item
❹ Icons for executing and resuming the job
❺ Icon for saving the task list run

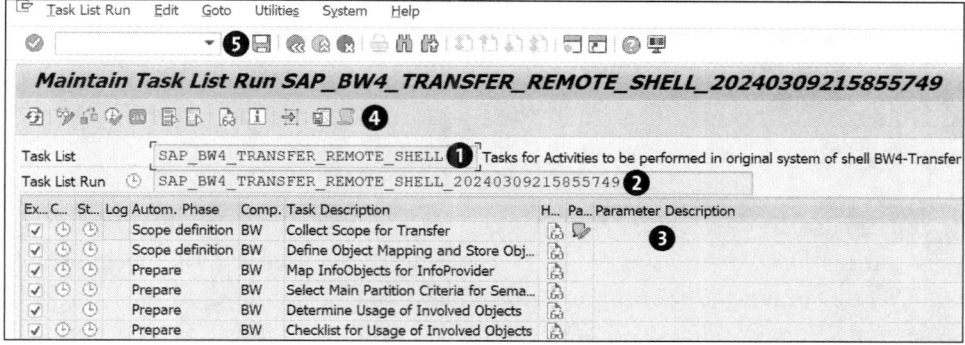

Figure 5.17 Task List Run

5.6.2 Scope Selection Execution

After saving the task list run we created in the previous section, follow these steps:

1. Click on the icon in the **Scope Definition** item, and you'll see **Select start objects** screen shown in Figure 5.18.

Figure 5.18 Select Start Objects Screen

2. Using , you can provide the Remote Function Call (RFC) destination for the SAP BW/4HANA system, as shown in Figure 5.19.

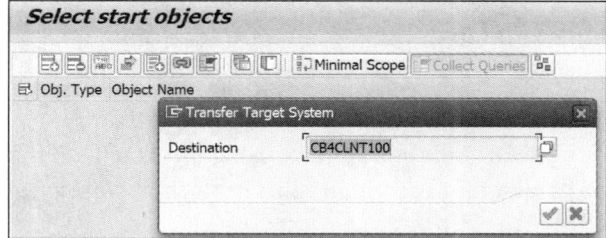

Figure 5.19 Target System RFC Destination

263

3. Once you continue using the ✓ icon, you'll see the option to select the scope object. Using the 📄 icon, you can start this process, as shown in Figure 5.20.

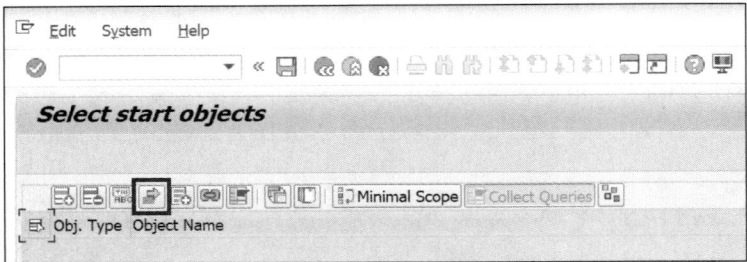

Figure 5.20 TLOGO and Scope Objects Selection

4. You can use the 📄 icon to insert the scope objects. Begin by specifying the TLOGO; in our instance, we're using the InfoCube, with its associated TLOGO being **CUBE**. Select that entry, and then input the technical name of the InfoCube requiring conversion. In our example, we'll use the InfoCube **ZPRC1_S**, as illustrated in Figure 5.21.

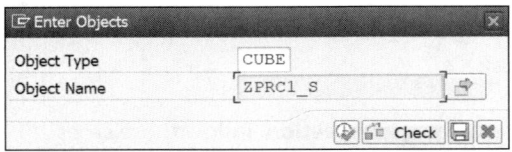

Figure 5.21 Scope Objects Input

5. Once you've selected the necessary scope objects, their details will populate the screen. You can add multiple scope objects and multiple TLOGOs by repeating the same steps. The selected objects will be displayed as rows, showing **Obj. Type** and **Object Name** (see Figure 5.22). In our demo, you can see the InfoCube name **ZPRCI_S**.

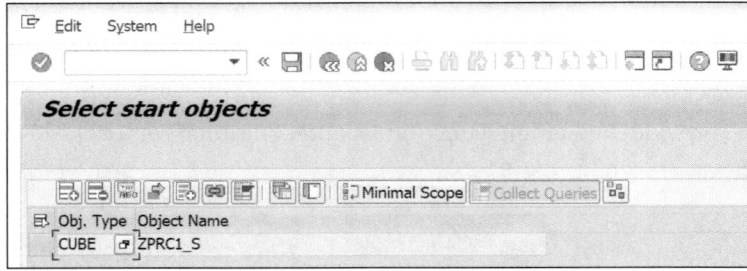

Figure 5.22 Final List of Start Objects

6. Save the start objects using the 💾 icon, and then go back using 🔙. You'll see the InfoCube name (**ZPRC1_S**) in the **Parameter Description** column, as shown in Figure 5.23.

5.6 Realization Phase

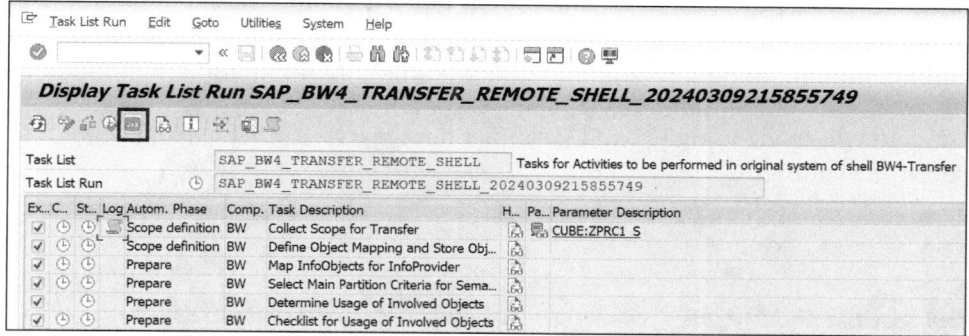

Figure 5.23 Executing Shell Conversion

7. Now click on [Job], and you'll see the job will be in the running state ⚙ (see Figure 5.24).

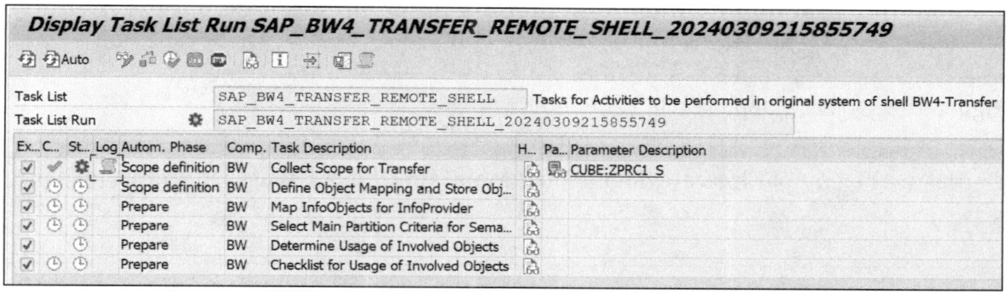

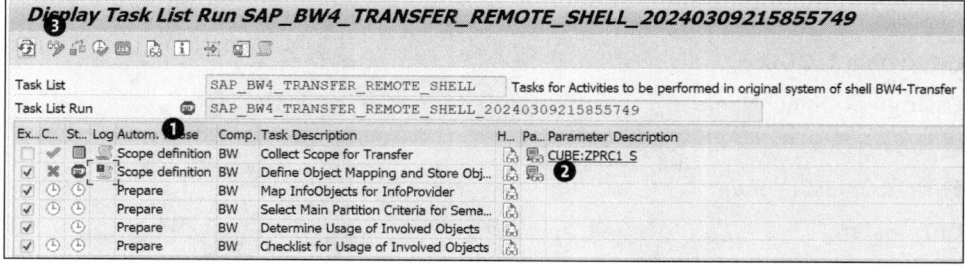

Figure 5.24 Executing Scope Definction in Shell Conversion

8. Once the job is completed you'll see the status, as shown in Figure 5.25:

   ❶ The status will be green, which means the activity is completed successfully.

   ❷ You'll see the **Stop** icon in the next row (**Define Object Mapping and Store Obj**), along with the 🗔 icon, which means you need use the 📝 icon to edit the parameter. Once you edit, you'll get into object mapping.

   ❸ You can see the display/change icon to update the parameter.

Figure 5.25 Scope Transfer: Job Status

265

5 Shell Conversion

5.6.3 Define Object Mapping and Store Object List

This activity will collect all the dependencies for the start object, attempt to associate the new TLOGO for the old objects, and assign the name according to the tool design. Get into edit mode using , and you'll see the icon for the object mapping, as shown in Figure 5.26.

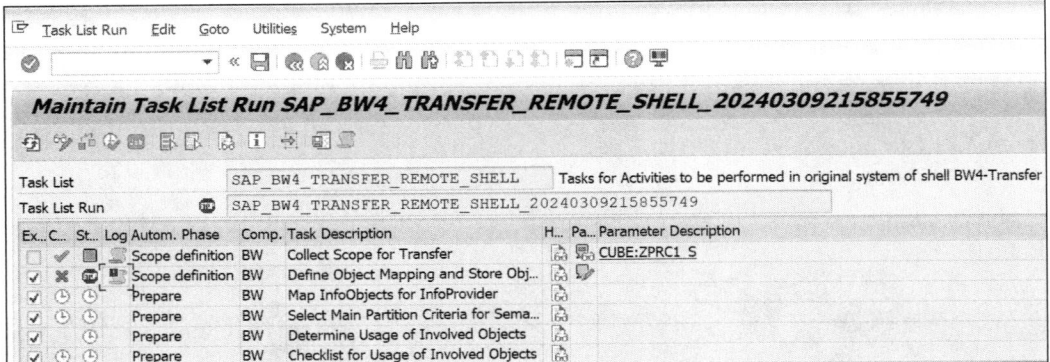

Figure 5.26 Object Mapping: Edit Parameter

Choosing the edit parameter will open the **Maintain Object List** screen shown in Figure 5.27.

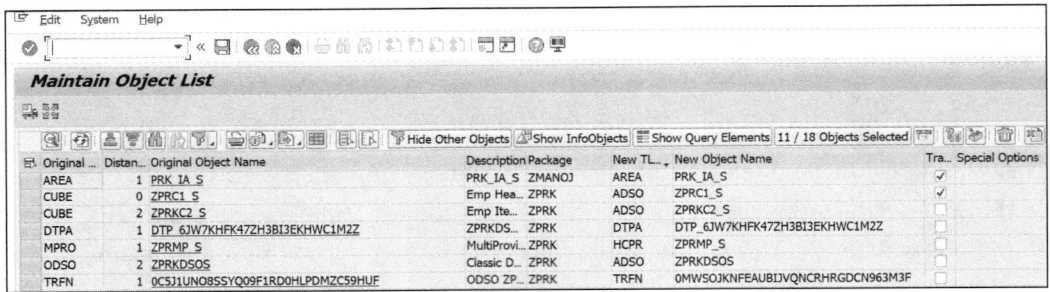

Figure 5.27 Object Mapping: Maintain Object List

Now click on the icon to see the other fields (Figure 5.28):

❶ Original TLOGO

❷ Original Object Name

❸ New TLOGO

❹ New Object Name

❺ Transfer

❻ Special Options

❼ Icons to assign package

266

5.6 Realization Phase

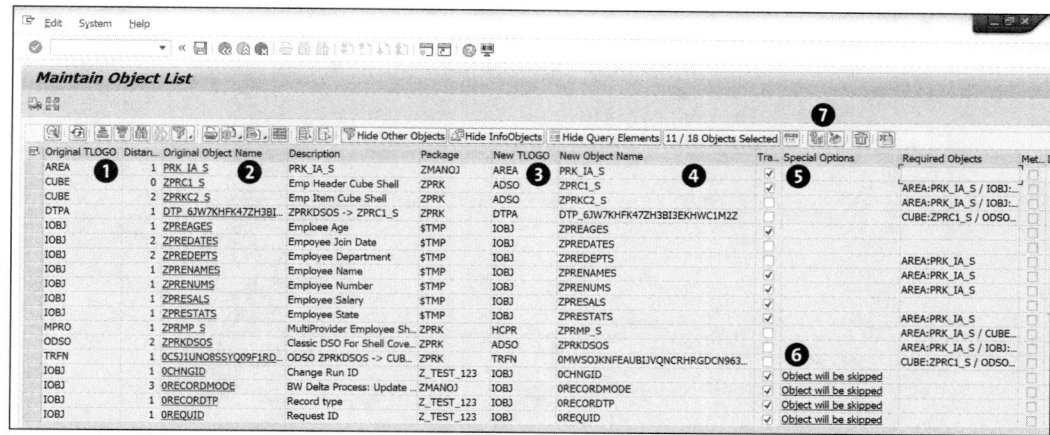

Figure 5.28 InfoObject and Missing Fields

You can see that the IOBJ and the InfoCube has been selected. You can modify the layout using ⊞ and push required objects using the ⇅ icons to the right. Choose continue ✓ when finished, as shown in Figure 5.29.

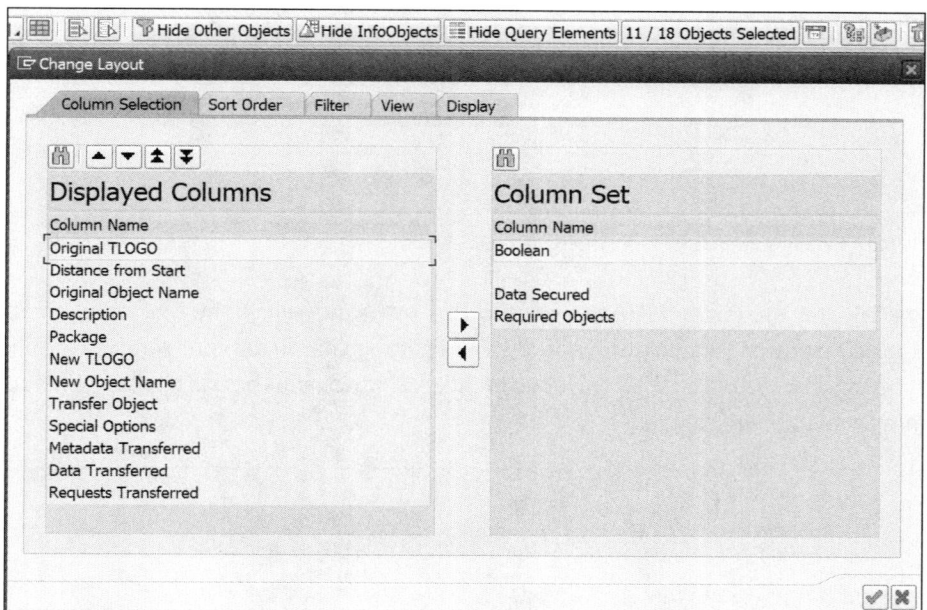

Figure 5.29 Change Layout Options

After the change of the layout, the result will be now shown as in Figure 5.30. You can choose **DTPA**, **ODSO**, and **TRFN** for the transfer, which were not selected previously.

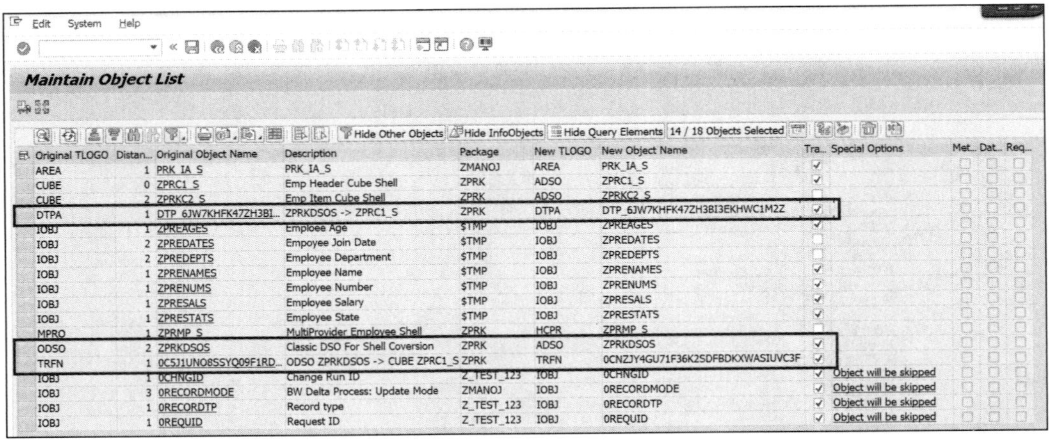

5　Shell Conversion

Figure 5.30 Choose ADSOs

See if there are any unassigned packages by using the 📦 icon. The InfoObjects (**IOBJ**) can be seen assigned to **$TMP** in the **Package** column in Figure 5.31.

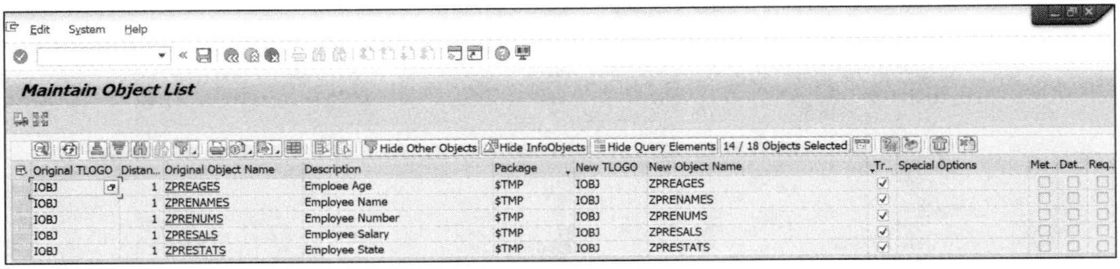

Figure 5.31 Unassigned Objects

Choose all the unassigned objects rows with **$TMP**, and select the 📦 icon. When prompted for the package name, choose a valid package name for your sender system, and you can choose or enter all the objects that will now have the package assigned to them, as shown in Figure 5.32.

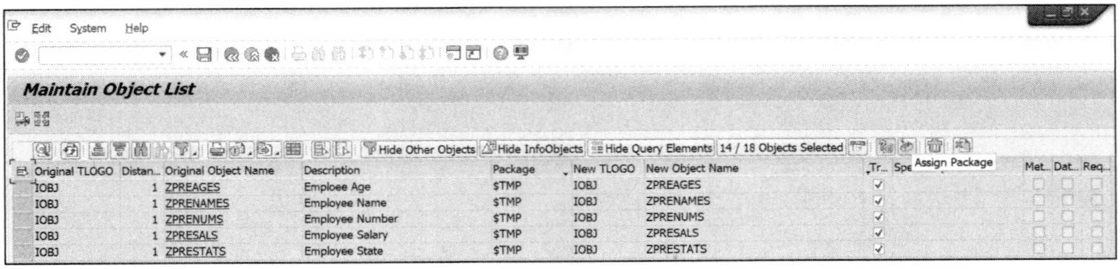

Figure 5.32 Package Assignment Selection

Once the package has been assigned, you can remove the filter on the unassigned objects. This action will display all objects, along with their package assignments. For

268

5.6 Realization Phase

example, you can observe that **IOBJ** is assigned to the **ZPRK** package. There's no requirement to assign a package for objects that aren't enabled for transfer, such as **ZPREDATES** and **ZPREDEPTS**. Additionally, **LSYS** won't have a package assignment by default, though it's not shown in this demo (see Figure 5.33).

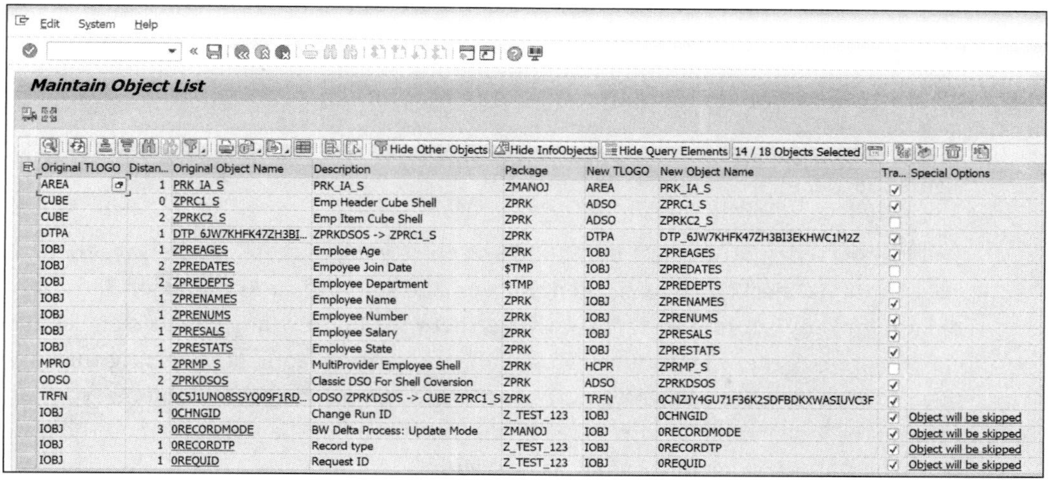

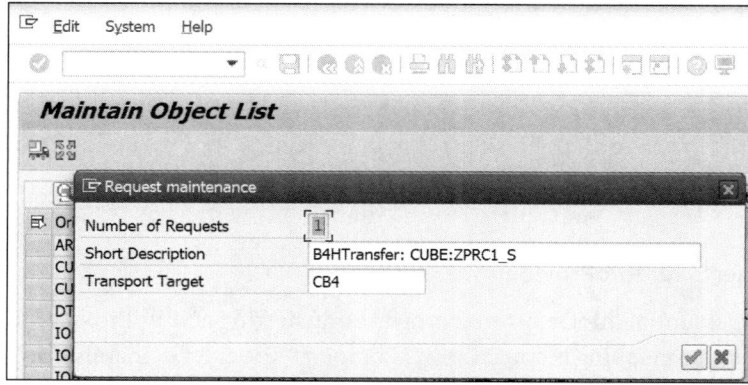

Figure 5.33 Package Assignment

The subsequent action involves creating a transport request to gather these objects in the SAP BW sender system. Click on the icon, and it will prompt you to enter the request number up to 9 based on the object count. Select the target system or transport group in the **Transport Target** input field where you intend to transport the objects, as depicted in Figure 5.34.

Figure 5.34 Transport Request Maintenance

One you choose click continue, you'll see ☑ Request created at the bottom of the screen, and the tools-based generated transport request number will be shown in the **Request** field, as shown in Figure 5.35.

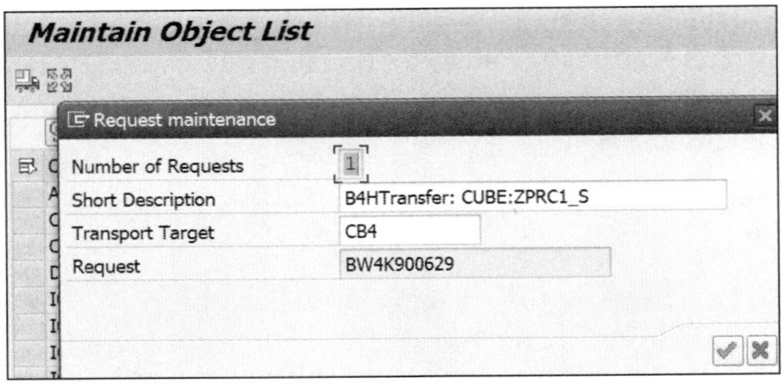

5 Shell Conversion

Figure 5.35 Generated Transport Request

You can see transport request **BW4K900629**. Once you see these details, save them in the **Maintain Object List** screen. Then, go back, and start the background job. If the tool did not find any errors apart from warnings that will not cause any inconsistencies and can be ignored, then the tool will stop with the warning icon [!], implying that you need to look into the warning and fix any issues. After fixing the problems, you can resume the job again to confirm that you've taken care of the warnings, and the next time, the status will be green, as shown in Figure 5.36.

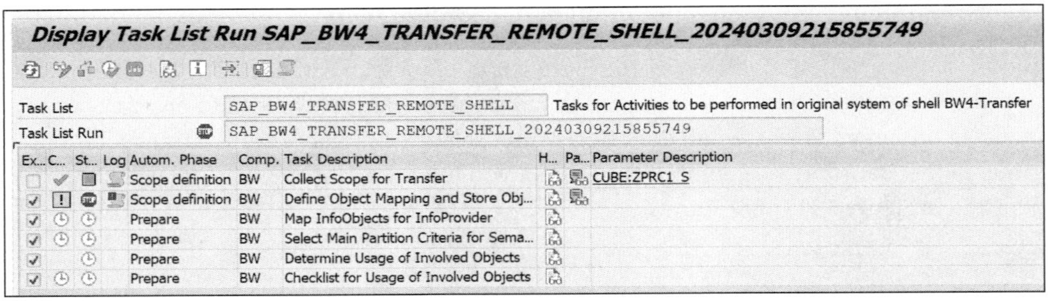

Figure 5.36 Task List Status

Now execute again to complete all the task list line items, which are explained in the following detailed subsections.

5.6.4 Map InfoObjects for InfoProvider

This task enables the essential InfoObject mapping for transferring a MultiProvider to a CompositeProvider. The mapping is crucial as the CompositeProvider demands a uniform mapping of navigation attributes. A consistent mapping is necessary for navigation attributes to align with the attribute-carrying characteristic. Additionally, the CompositeProvider no longer supports InfoSet InfoObjects, requiring a new mapping approach.

To address this, we recommend the creation of a new reference characteristic specifically for navigation attributes. This new reference characteristic should be employed in all MultiProvider transfers where a consistent mapping for the same navigation attribute is lacking. The task not only facilitates the creation of a new reference characteristic but also offers the option to generate mapping proposals from previous transfers.

You'll see the **InfoObject Compatibility Check for InfoProviders** window. If you see the list of **InfoProvider** columns and the **New InfoObject** column, then you must assign new InfoObjects or work with a field name(a globally unique name needs to be set here). You'll have these two icons for creating the InfoObjects.

5.6.5 Select Main Partition Criteria for Semantically Partitioned Objects

This task allows you to choose the primary partition InfoObject for the newly created ADSO in the generated semantic group. Because semantic groups accommodate only a single partition InfoObject, this selection is essential. However, it's worth noting that the remaining partition InfoObject filters are still incorporated as filter criteria for the corresponding ADSO.

5.6.6 Determine Usage of Involved Objects

In this task, a thorough examination is conducted on the generated objects within InfoProviders. The scrutiny extends to various components such as routines, planning functions, and similar elements. The primary objective is to scrutinize the existing code for issues that could potentially trigger errors during the transfer process. By conducting a comprehensive analysis, the task aims to proactively identify and address any problematic code to ensure a smooth and error-free transition. This diligent review helps preemptively mitigate any issues related to routines, planning functions, and other elements. Figure 5.37 shows the status of the execution. In our demo, the status is running . If there is no error, it will complete with the green status.

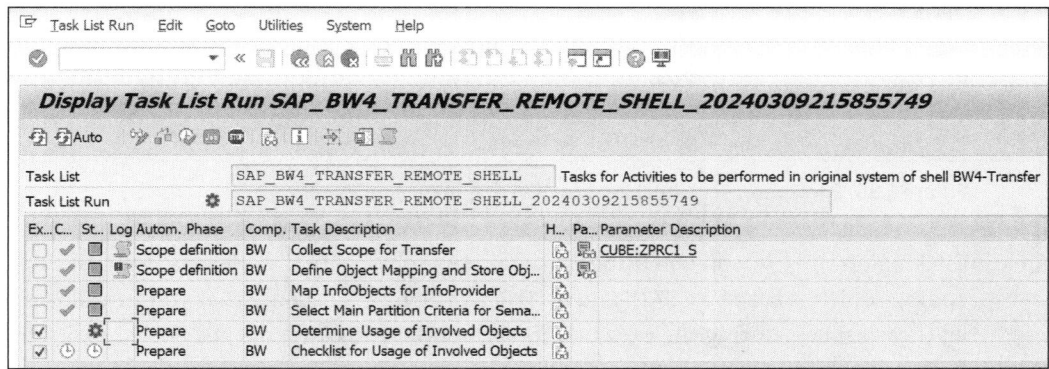

Figure 5.37 Determine Usage of Involved Objects

5.6.7 Checklist for Usage of Involved Objects

This task presents a compilation of objects or discoveries that necessitate verification or resolution. It encompasses the examination of both the usage of generated objects and references to InfoProviders that may no longer exist post-transfer. Additionally, it emphasizes the need to scrutinize customer code for any alterations resulting from the transfer process. This examination is essential, particularly in instances where there are changes such as the exchange of InfoObjects or a shift in queries from running on a CompositeProvider after the transfer, as opposed to being based on a DSO or InfoCube. You can see the details in Figure 5.38.

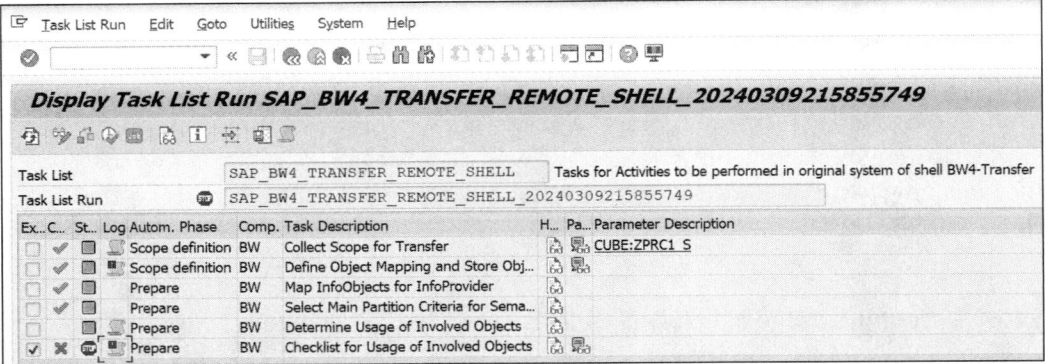

Figure 5.38 Checklist for Usage of Involved Objects

When the tool stops in the manual activity that requires a parameter, then you need to choose to see the parameter for the activity, as shown in Figure 5.39.

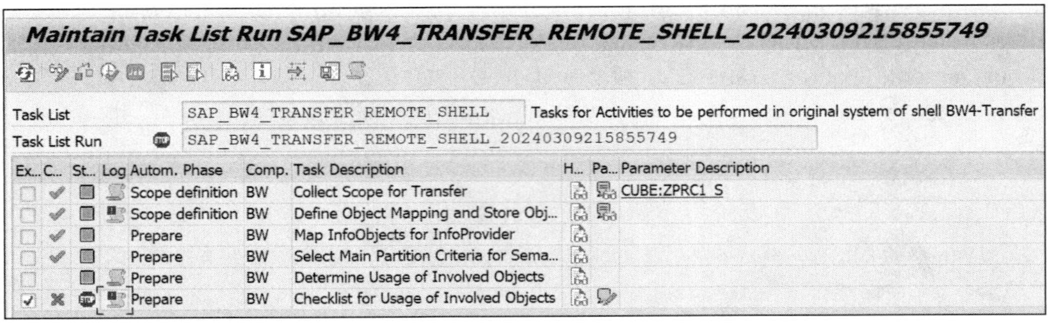

Figure 5.39 Parameter for Checklist for Usage of Involved Objects

Clicking on the icon will bring up the screen to validate the checklist for customer coding, which is based on **BADI** and **AUTH**. It's essential to ensure that you've validated all these aspects during the prework. In such cases, you can mark the checkbox in the **Resolved** column. The default screen will appear as shown in Figure 5.40.

5.6 Realization Phase

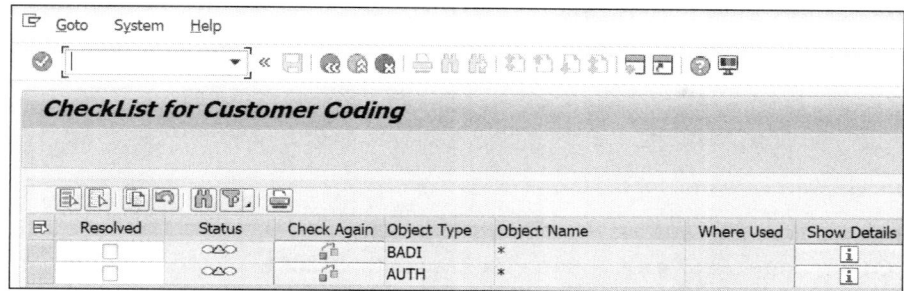

Figure 5.40 Object Type and Resolution Checks

The **Resolved** column is shown checked for both object types in Figure 5.41.

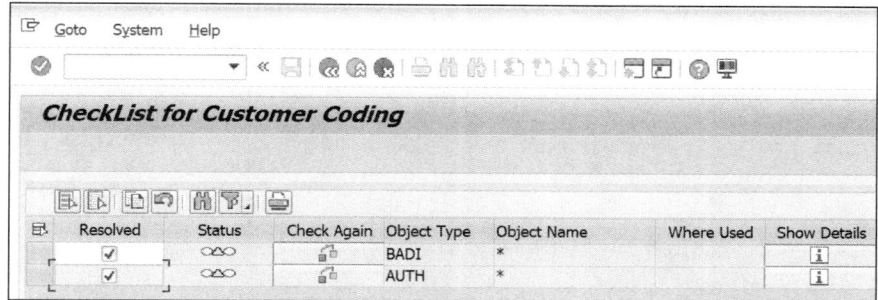

Figure 5.41 Status of Resolution for Customer Coding Checks

Once you click the icon, the job will be executed, and the final status can be seen, as shown in Figure 5.42.

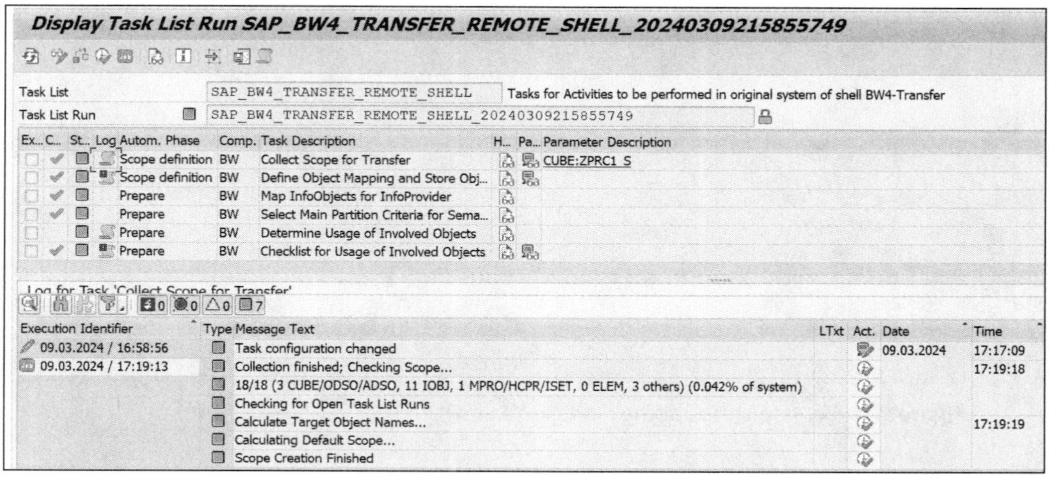

Figure 5.42 Shell Conversion: Overall Status

5 Shell Conversion

5.7 Post-Conversion Activities

The first post-conversion activity takes place in the target system. Check if all the required source systems, such as ODP–SAP, are created in the target SAP BW/4HANA system. Make sure you have other source systems as well, such as flat files. You need to have the SDI-based system for DB Connect.

Next, we'll focus on the sender system. Export the transport request that was generated by the Conversion Cockpit tool. You can use Transaction SE01 to display the transport request and view its contents. The demo transport request is shown in Figure 5.43.

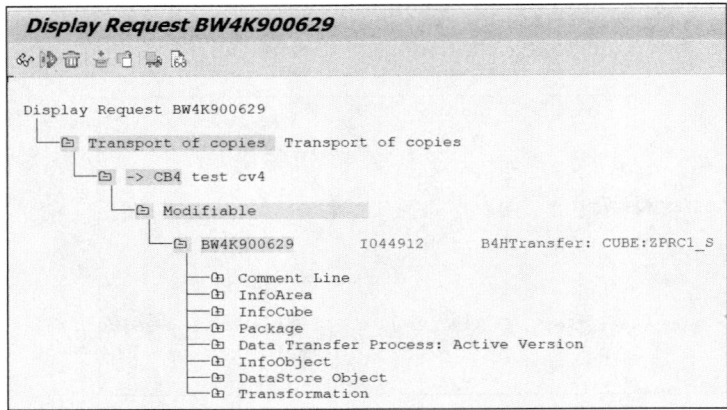

Figure 5.43 Tool-Generated Transport Request

Once you release the transport request using 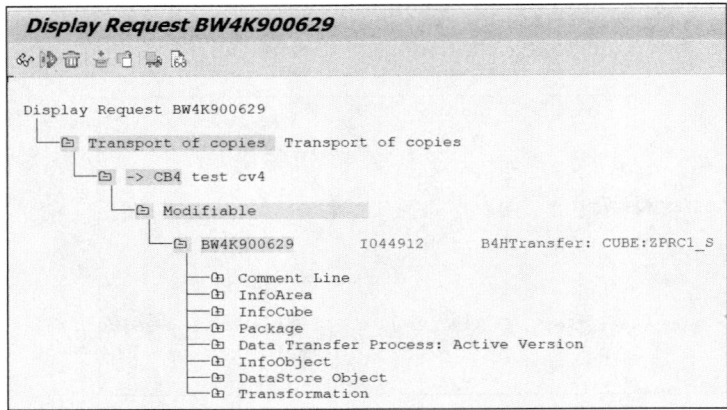, the logs can be seen. The target SAP BW/4HANA system is **CB4**, as shown in Figure 5.44.

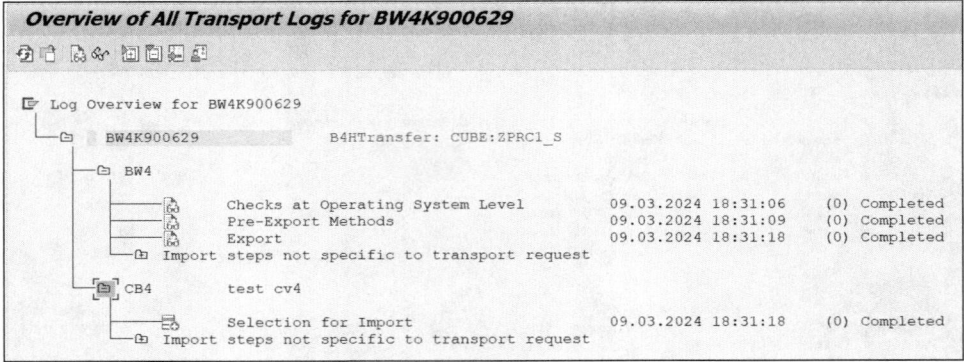

Figure 5.44 Release Status of the Transport Request

In the target system, access Transaction STMS_IMPORT, and proceed to import the transport request generated by the Conversion Cockpit tool. Ensure that you select the correct client and verify that the options shown are appropriately chosen (see Figure 5.45).

5.7 Post-Conversion Activities

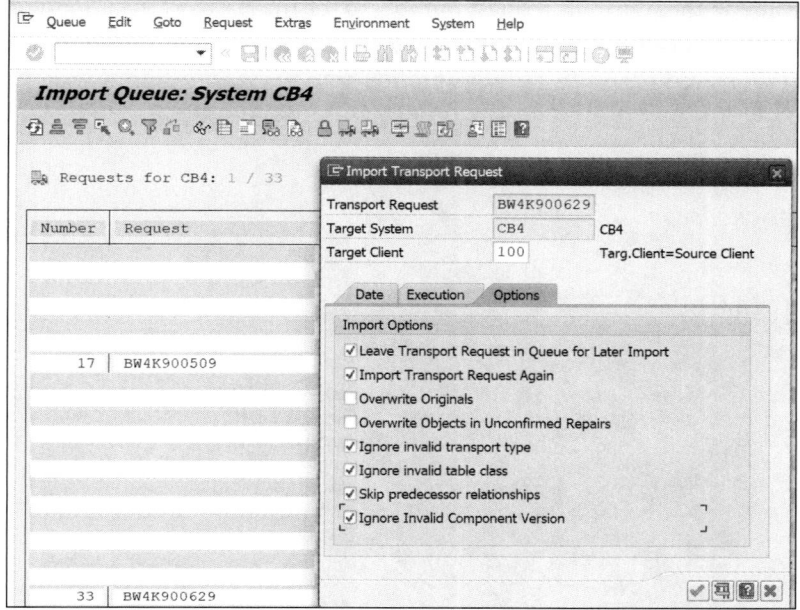

Figure 5.45 Transport Request Import Process in Target SAP BW/4HANA

Upon successful import of the transport request, the logs will be visible as depicted in Figure 5.46. In the event of a transport request failure, examine the logs, perform the necessary remediation, and attempt to reimport the transport request until it succeeds.

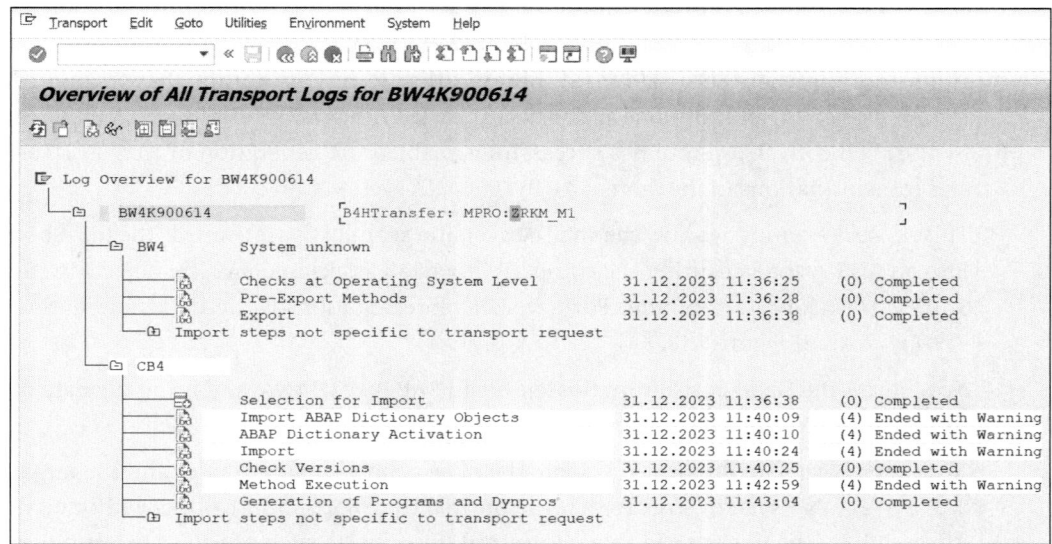

Figure 5.46 Transport Request Import Logs

You can double click on the transport request number to see the status of the import of each object within the transport request. You can see if the object is imported and generated, as shown in Figure 5.47.

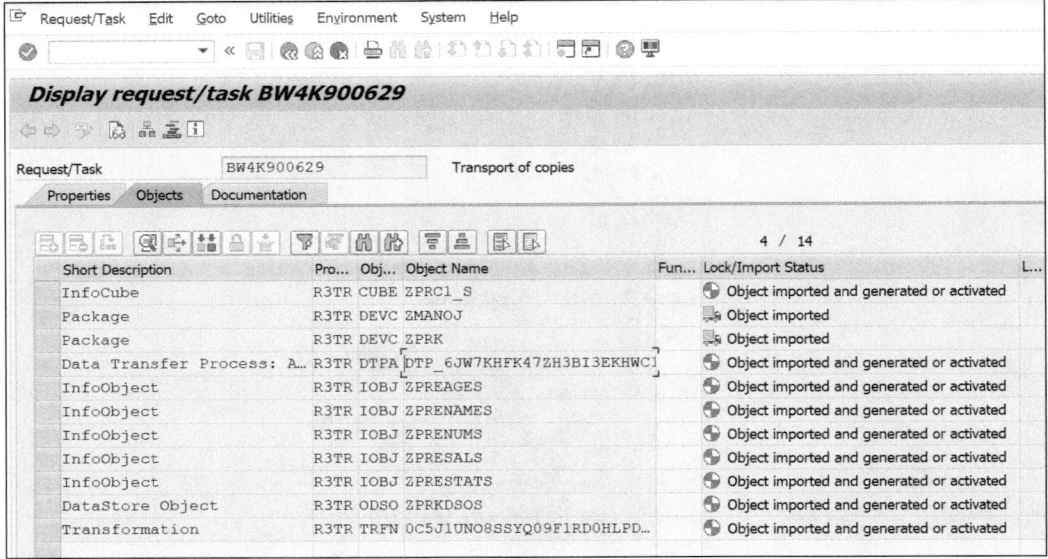

Figure 5.47 Transport Request Import Status in SAP BW/4HANA

Once the import of the tool-generated transport request is successful, log in to the SAP BW modeling tools for the SAP BW/4HANA system. Search for the ADSO named **ZPRC1_S**, which was created based on InfoCube ZPRC1_S. Note that the ADSO name is the same as the InfoCube name, as suggested by the tool in the object mapping step.

Upon locating the ADSO (**ZPRC1_S**), observe that it resides within the designated InfoArea (**PRK_IA**) and contains all the necessary InfoObjects. The TR import facilitated by the shell conversion tool has successfully enabled the generation of ADSOs, DTPs, and transformations in the target SAP BW/4HANA system.

This process exemplifies how the shell conversion seamlessly transforms the InfoCube into an ADSO while ensuring the compatibility of all associated SAP BW objects in the SAP BW/4HANA environment. For a visual representation of the data flow in SAP BW/4HANA, see Figure 5.48.

Now create the DataSource from the flat file in SAP BW/4HANA, and bring the data to the ADSO in the SAP BW/4HANA target system.

After the objects are imported, validate the objects between the sender and the target SAP BW/4HANA system. Ensure that the normal transport import works without any issues, adjust the queries that have technical InfoObjects such as 0REQUID, and perform the conversion of roles to the SAP BW/4HANA–based roles. Check if you have any Analysis Process Designer (APD) objects and convert accordingly.

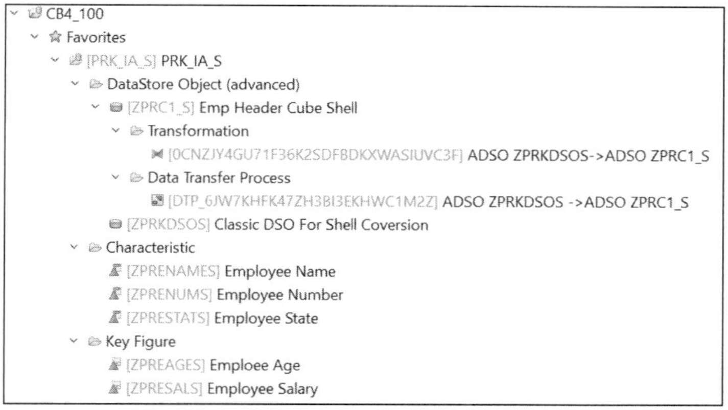

```
∨ 🗎 CB4_100
  ∨ ☆ Favorites
    ∨ 🗎 [PRK_IA_S] PRK_IA_S
      ∨ 🗀 DataStore Object (advanced)
        ∨ 🗇 [ZPRC1_S] Emp Header Cube Shell
          ∨ 🗀 Transformation
              ⋈ [0CNZJY4GU71F36K2SDFBDKXWASIUVC3F] ADSO ZPRKDSOS->ADSO ZPRC1_S
          ∨ 🗀 Data Transfer Process
              🗟 [DTP_6JW7KHFK47ZH3BI3EKHWC1M2Z] ADSO ZPRKDSOS ->ADSO ZPRC1_S
        🗇 [ZPRKDSOS] Classic DSO For Shell Coversion
      ∨ 🗀 Characteristic
          🗝 [ZPRENAMES] Employee Name
          🗝 [ZPRENUMS] Employee Number
          🗝 [ZPRESTATS] Employee State
      ∨ 🗀 Key Figure
          🗝 [ZPREAGES] Emploee Age
          🗝 [ZPRESALS] Employee Salary
```

Figure 5.48 Converted Objects in SAP BW/4HANA: After Shell Conversion

> **Useful Reports**
> Refer to Chapter 4, Section 4.11, for the list of reports that can be used in shell conversion and general SAP BW objects management.

5.8 Summary

With the completion of this chapter, you're now ready to execute the shell conversion for SAP BW/4HANA. You know the activities to be done in the prepare phase and realization phase of the conversion. You're now aware of the steps that are involved in the shell conversion. You've learned about the transport requests that are created from the tool and how to import them into the target SAP BW/4HANA system. With this knowledge, you can start the SAP BW/4HANA migration using shell conversion for your specific use case. In the next chapter, we'll focus on the in-place conversion to SAP BW/4HANA.

Chapter 6
In-Place Conversion

This chapter outlines the detailed procedure for implementing an SAP BW/4HANA in-place conversion. You need to have read about the system preparation and prerequisites discussed in Chapter 3. Upon completing this chapter, you'll be proficient in executing an SAP BW/4HANA in-place conversion and understand task lists and associated activities in the system landscape.

In this chapter, you'll learn how to convert the SAP Business Warehouse (SAP BW) 7.x system to SAP BW/4HANA using in-place conversion. We'll start with the in-place conversion overview in Section 6.1. Then, you'll learn about the system landscape preparation in Section 6.2. We'll explore the SAP BW system provision in Section 6.3. The sender system provisioning will be discussed in Section 6.4. Details regarding the realization phase are explained in Section 6.5. We'll explore the SAP BW/4HANA starter add-on in Section 6.6. You'll learn about the operating modes in Section 6.7. We'll walk through the steps to transfer standard authorization in Section 6.8. We'll delve into the details of in-place conversion in Section 6.9, followed by the process to start the in-place conversion in Section 6.10. You'll understand the steps involved in performing a scope transfer in Section 6.11, and then get a detailed step-by-step explanation of executing scope selection in Section 6.12. You'll learn about the transfer of standard authorization delta run in Section 6.13 and obsolete objects deletion in Section 6.14. Once all the activity is done, we'll cover the final system conversion steps in Section 6.15.

6.1 Overview of In-Place Conversion

In this section, we delve into the fundamental aspects of in-place conversion, building upon the groundwork laid in the previous chapter's preparation activities. Let's review some common business scenarios:

- **SAP BW 7.5 on SAP HANA**
  If your organization possesses a substantial volume of data, prefers transitioning to SAP BW/4HANA for a more modern SAP BW landscape, wants to maintain the same system ID (SID) for the SAP BW system even after the SAP BW/4HANA conversion, and wants full system conversion without much disruption to the existing business

processes, then the in-place conversion method might be the ideal choice. This approach becomes particularly advantageous if you're currently operating in SAP BW on an SAP HANA database. Opting for in-place conversion allows you to systematically transfer classical objects to SAP HANA-optimized objects, facilitating a step-by-step migration process. This sequential transfer approach helps avoid the complexities associated with remote conversion and provides a smoother transition to SAP BW/4HANA.

- **Non-SAP BW 7.5 on SAP HANA**
  However, if your system isn't yet on SAP BW 7.5 or SAP HANA database, an initial step involves first updating it to SAP BW 7.5 on SAP HANA. You can use the data migration option tool (DMO) tool to accomplish the SAP BW 7.5 upgrade and SAP HANA database migration in one step. DMO is a key option in Software Update Manager (SUM). Subsequently, you can proceed with the process of step-by-step in-place transfer of classical objects before transitioning to SAP BW/4HANA. The flexibility of this approach ensures a tailored and phased migration, aligning with your organization's specific reporting requirements and system preferences (see Figure 6.1).

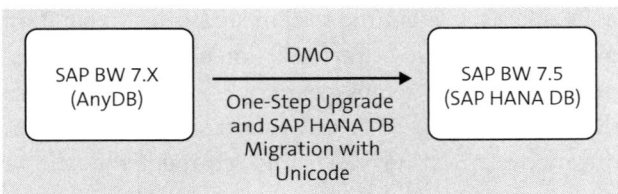

Figure 6.1 Path to SAP BW 7.5 on SAP HANA

- **In-Place conversion flow**
  When migrating to SAP BW/4HANA through the in-place conversion process, your existing system undergoes transformation into SAP BW/4HANA. During this conversion, the system's objects are gathered in a scope transfer and then transformed into SAP BW/4HANA–optimized objects. This scope transfer process involves converting DataStore Objects (DSO) and InfoCubes into Advanced DSOs (ADSOs), while MultiProviders and InfoSets are transformed into SAP HANA CompositeProviders (HCPR). InfoObjects and DataSources will remain the same. Additionally, InfoPackages are eliminated, with their selections being transferred to the data transfer process (DTP) layer. The SID remains intact throughout the conversion process. Data migration from old tables to new object-based tables occurs after the objects have been converted. Figure 6.2 shows a simplified representation of the conversion flow.

6.2 System Landscape for In-Place Conversion

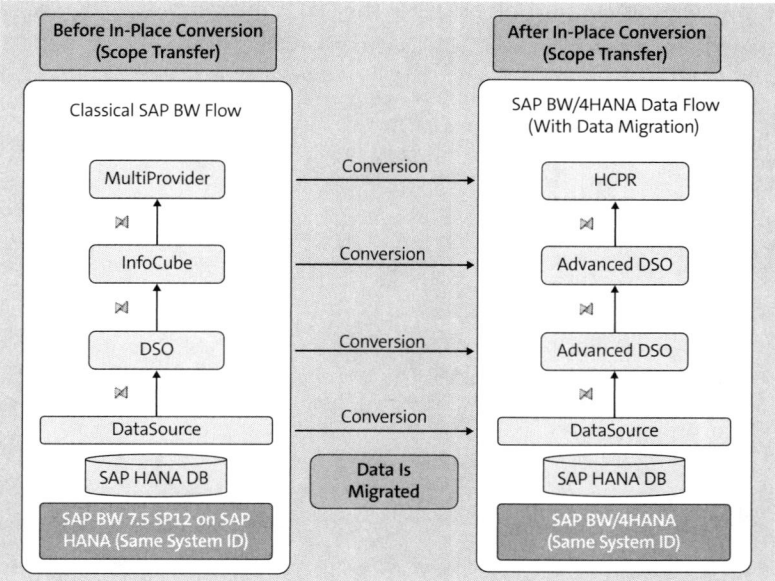

Figure 6.2 In-Place Conversion Flow

6.2 System Landscape for In-Place Conversion

Ensuring an optimal system landscape is crucial for the success of the in-place conversion. Typically, organizations opt for an N + 1 landscape configuration, where a separate project landscape runs in parallel. Once all scope transfers are finished, the subsequent post-conversion tasks involve transitioning from SAP BW on SAP HANA to the target SAP BW/4HANA system. It's worth noting that SAP BW/4HANA exclusively supports data loading through operational data provisioning (ODP). For reference, you can check the sample system landscape shown in Figure 6.3.

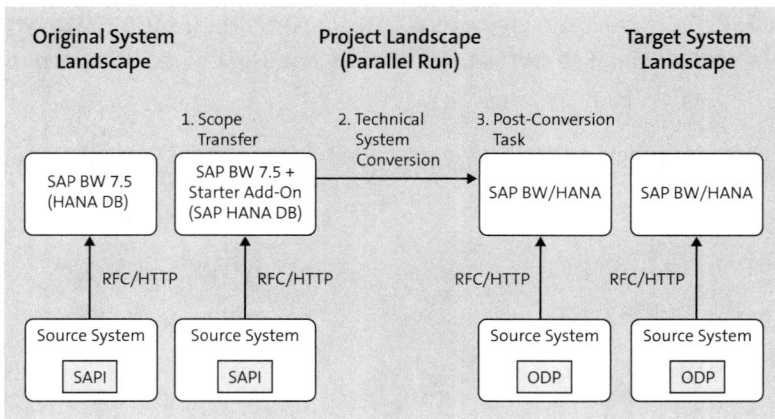

Figure 6.3 In-Place Conversion System Landscape

281

6   In-Place Conversion

You'll notice that in-place conversion will preserve the SID; for example, we consider SID as BWP post-conversion to SAP BW/4HANA, as shown in Figure 6.4.

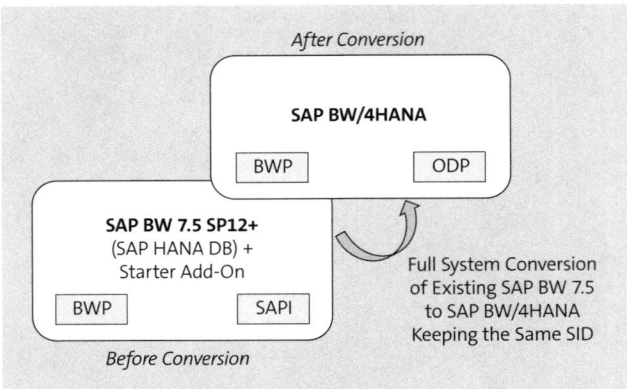

Figure 6.4 Same-System Conversion Logic

6.3   SAP Business Warehouse System Provisioning

Ensure that you determine the type of landscape you intend to use, whether it's an N or N + 1 landscape. Assuming you already have an SAP BW 7.5 on SAP HANA system, certain conditions must be verified for an in-place conversion. In the case of an in-place conversion, only one system is required, and the conversion is performed directly within the same system. If you're considering an in-place conversion, it's crucial to confirm that the system meets the required Support Package (SP) level outlined here:

- SAP BW 7.5 powered by SAP HANA SP 05 or higher. It's generally best practice to go for the latest support package. The minimum recommended is SP 12.
- Direct in-place conversion from SAP BW 7.50 to SAP BW/4HANA 2023 SP 00 or higher (DW4CORE 4.00 SP 0 or higher) is supported.
- As of SAP BW 7.50 SP 16, the in-place conversion is only supported with SAP BW/4HANA 2.0, SAP BW/4HANA 2021, and SAP BW/4HANA 2023 as a destination, as described in Table 6.1.

Release	End of Maintenance	Next Release
SAP BW/4HANA 2.0	12/31/2024	SAP BW/4HANA 2021
SAP BW/4HANA 2021	12/31/2027	SAP BW/4HANA 2023
SAP BW/4HANA 2023	12/31/2030	SAP BW/4HANA 2026

Table 6.1 Target SAP BW/4HANA Version

If the SAP BW 7.5 on SAP HANA system is up and running, then you'll have the required SAP HANA database. Check for your target SAP BW/4HANA version, and plan for the database upgrade and SAP BW 7.5 upgrade accordingly.

The minimum SAP HANA release required for compatibility with SAP BW/4HANA 2023 is specified as SAP HANA 2.0 SP 07 Rev. 71. This indicates that for a successful integration and optimal performance of SAP BW/4HANA 2023, the SAP HANA database must be at least at version 2.0 with Support Package 07, and it should have the revision level 71. This detailed specification ensures that the underlying SAP HANA infrastructure is aligned with the version and revision requirements, promoting a smooth and reliable operation of SAP BW/4HANA 2023. It's crucial to adhere to these minimum requirements to leverage the enhanced features and functionalities offered by the latest release of SAP BW/4HANA.

If you plan to go for SAP BW/4HANA 2021, then the minimum SAP HANA database is SAP HANA 2.0 SP 05. Refer to Chapter 3, Section 3.3, for more details.

Make sure that the current add-ons in your existing SAP BW 7.5 system are compatible with your SAP BW/4HANA target version. Check whether there are any third-party add-ons you need to verify with the vendor regarding compatibility.

Before transitioning to SAP BW/4HANA, understanding the supported release details of SAP BW is crucial. The conversion method choice introduces a dependency, requiring verification of the supported release in both the SAP BW sender and SAP BW target systems. The in-place conversion method streamlines the process, requiring only one system for the actual conversion. Ensure compatibility across sender and target systems to guarantee a smooth transition to SAP BW/4HANA, considering the unique requirements of each conversion approach. Let's check that for each of the conversion approaches.

To get the precheck tools, you need to have certain SAP Notes installed in your SAP BW system:

- Install SAP Note 2777672, "Revision 11: New Check Tool (7.3–7.5)." This will have essential corrections for the tools, which helps prevent many known errors.

- Ensure that you've successfully installed the SAP Readiness Check notes (refer to Chapter 3, Section 3.9). The XML file can be obtained from SAP Note 2575059 "SAP BW Note Analyzer Files for SAP Readiness Check for SAP BW/4HANA and SAP Datasphere, SAP BW Bridge." Make certain to have the *SAP_BW4HANA_Readiness_Check_[Latest].zip* XML file. Do the manual steps mentioned in the SAP Note.

- The next step is to check SAP Note 2383530 "Conversion from SAP BW to SAP BW/4HANA." Download *SAP_BW4_Transfer_Note_Analyzer_2024-01-30.zip* from the SAP Note, and check the XML files, as shown in Figure 6.5.

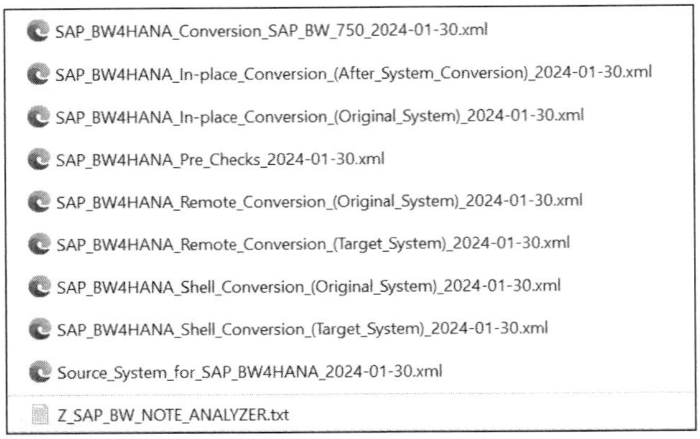

6 In-Place Conversion

- SAP_BW4HANA_Conversion_SAP_BW_750_2024-01-30.xml
- SAP_BW4HANA_In-place_Conversion_(After_System_Conversion)_2024-01-30.xml
- SAP_BW4HANA_In-place_Conversion_(Original_System)_2024-01-30.xml
- SAP_BW4HANA_Pre_Checks_2024-01-30.xml
- SAP_BW4HANA_Remote_Conversion_(Original_System)_2024-01-30.xml
- SAP_BW4HANA_Remote_Conversion_(Target_System)_2024-01-30.xml
- SAP_BW4HANA_Shell_Conversion_(Original_System)_2024-01-30.xml
- SAP_BW4HANA_Shell_Conversion_(Target_System)_2024-01-30.xml
- Source_System_for_SAP_BW4HANA_2024-01-30.xml
- Z_SAP_BW_NOTE_ANALYZER.txt

Figure 6.5 XML for Prechecks

It's essential to fully implement the SAP Notes derived from the XML files in Figure 6.5. These notes play a critical role in ensuring the smooth functioning of the system throughout the conversion process. See Figure 6.6 to identify the specific XML files that need to be implemented in your system:

❶ Install the precheck tool in this phase.

❷ Install the SAP BW/4HANA transfer cockpit in the SAP BW 7.5 system.
- For SAP BW 7.5 and in-place conversion, if you start now with the installation of the tool, use *SAP_BW4HANA_ Conversion_SAP_BW_750_[last_update].xml*.
- If you've already installed most of the notes before and only want the latest updates, continue to use *SAP_BW4HANA_In-place_Conversion_(Original_System)_[last_update].xml*.

❸ Install the ODP-based notes in all connected ERP systems.
- Use the *Source_System_for_SAP_BW4HANA_[last_update].xml*.

❹ Install the transformation notes in SAP BW/4HANA post-conversion.
- Use *SAP_BW4HANA_In-place_Conversion_(After_System_Conversion)_[last_update].xml*.

❺ Install the SAP BW/4HANA updates post-conversion.

> **Prerequisite Reading**
>
> Before proceeding into the next steps, we strongly encourage you to thoroughly explore Chapter 3. It's essential to read Chapter 3, Section 3.9 and Chapter 3, Section 3.10 to achieve significant milestones in the SAP BW/4HANA conversion process. Be sure to also read Chapter 3, Section 3.11, which provides insights into the simplification list.

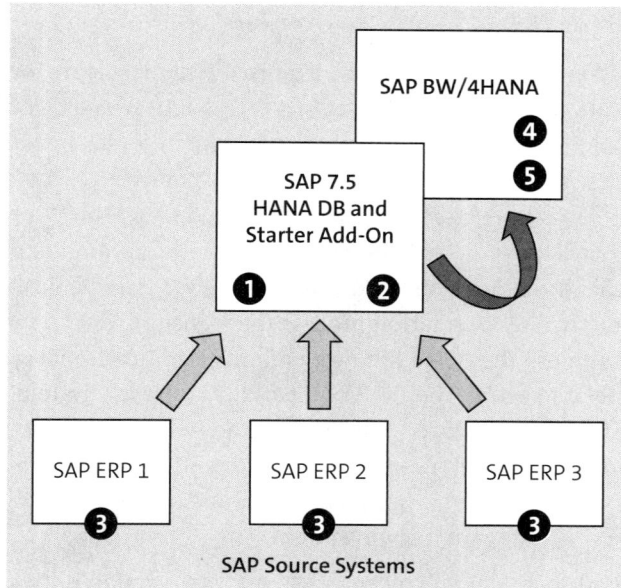

Figure 6.6 System Landscape and XML Files

Once all the SAP Notes are implemented properly, then you'll able to see the transfer cockpit using Transaction RSB4HCONV, as shown Figure 6.7.

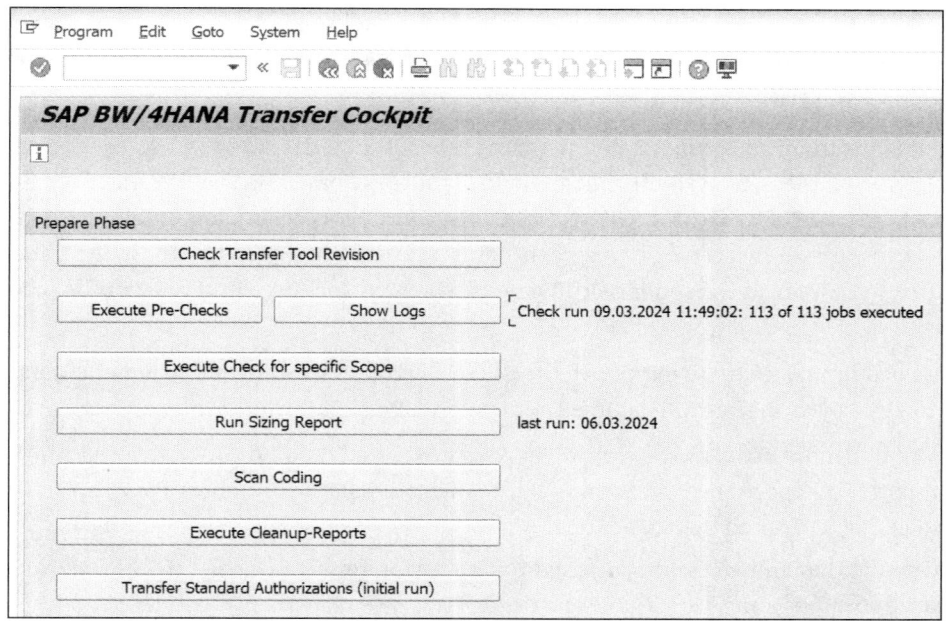

Figure 6.7 Transfer Cockpit in the SAP BW on SAP HANA System

6.4 Sender System Provisioning

In the realm of SAP BW/4HANA, ODP takes center stage as the primary infrastructure for extracting and replicating data from SAP (ABAP) applications to the SAP BW/4HANA data warehouse. This pivotal role positions ODP as the key mechanism for seamless data transfer within the SAP environment. Therefore, the incorporation of SAP BW/4HANA with SAP NetWeaver systems, such as SAP Business Suite, SAP S/4HANA, or SAP BW, relies on ODP. Ensuring ODP readiness is imperative for both inbound and outbound connections across all interconnected SAP and SAP BW systems. ODP ensures a seamless, unified, and efficient integration process for managing data flow between these systems, contributing to the overall interoperability and functionality of the SAP environment. You need to verify the ODP compatibility of each system, including SAP Business Suite, SAP S/4HANA, and SAP BW, to facilitate smooth data provision. The mandatory ODP readiness guarantees optimal functionality and data flow between SAP BW/4HANA and associated systems. Regularly assess and update the ODP status to maintain the integrity of the integrated environment.

ODP Data Replication API 2.0 is a functional enhancement to the first version of this interface released and the recommended application programming interface (API) for source systems. ODP Data Replication API 2.0 facilitates internal connections between SAP BW/4HANA, SAP BW release 7.3 or higher, SAP Data Services 4.2, and SAP HANA smart data integration (SDI; ABAP adapter). It supports various data provider types, including DataSources/extractors (ODP-SAPI context), ABAP Core Data Services (CDS) views (ODP-CDS Context), and InfoProviders of SAP BW/4HANA or SAP BW release 7.4 or higher (ODP-BW context). This API serves as a versatile tool for seamless integration across diverse DataSources within the SAP ecosystem.

As shown earlier in Figure 6.5, you need to install all notes in all your connected SAP ERP, SAP S/4HANA, or systems that are in the landscape based on the XML file *Source_system_for_SAP_BW/4HANA_Latest_Date.XML*. Once you've implemented the SAP Notes, make sure everything has a green status, and then you need to expose the DataSource to ODP in all the sender ERP systems. The next section will show you the process.

You need to make sure to expose all the SAP-Delivered DataSources and generic DataSources in all connected SAP source systems to ODP using the following reports:

- SAP-delivered DataSources can be exposed using a standard SAP delivered ABAP report. To expose the DataSource, call Transaction SE38, run report BS_ANLY_DS_RELEASE_ODP, and then execute it.
- Generic DataSource can be exposed to ODP using report RODPS_OS_EXPOSE. You need to expose one DataSource at a time.
- Once you've exposed the DataSource, you'll need to see that table ROOSATTR has the entry for the DataSource with the **EXPOSE_EXTERNAL** field value as **X**, as shown in Figure 6.8.

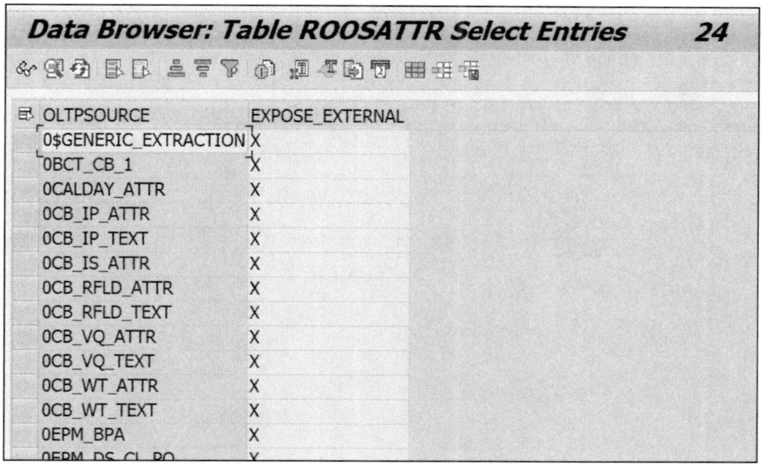

Data Browser: Table ROOSATTR Select Entries 24

OLTPSOURCE	EXPOSE_EXTERNAL
0$GENERIC_EXTRACTION	X
0BCT_CB_1	X
0CALDAY_ATTR	X
0CB_IP_ATTR	X
0CB_IP_TEXT	X
0CB_IS_ATTR	X
0CB_RFLD_ATTR	X
0CB_RFLD_TEXT	X
0CB_VQ_ATTR	X
0CB_VQ_TEXT	X
0CB_WT_ATTR	X
0CB_WT_TEXT	X
0EPM_BPA	X

Figure 6.8 ODP Expose Table

6.5 Realization Phase

After wrapping up all the preparatory tasks outlined in the preparation phase via the transfer cockpit, the subsequent phase is the realization phase. Remember to install the SAP BW/4HANA starter add-on in your SAP BW/4HANA system. This add-on is essential for the conversion to SAP BW/4HANA. Before diving into this phase, ensure all preparation steps for the conversion are completed thoroughly. It's crucial to ensure that you've obtained the necessary license for SAP BW/4HANA before delving into this phase.

The journey through the realization phase kicks off with the installation of the SAP BW/4HANA starter add-on. Before proceeding with the actual conversion of the SAP BW on SAP HANA system to the SAP BW/4HANA system, several preliminary activities need to be completed once all the scope transfer is completed. There will be multiple cleanups of unsupported objects before the system conversion using SUM.

Post-conversion to SAP BW/4HANA, there are post-conversion tasks awaiting completion to prepare and fine-tune the system. These tasks are pivotal in ensuring the seamless integration and optimal performance of the SAP BW/4HANA system. Finally, with these tasks accomplished, the SAP BW/4HANA system is ready to go live, marking the culmination of the conversion process. The activities in this phase can be seen in Figure 6.9.

The important thing in in-place conversion is to install the SAP BW/4HANA starter add-on in the SAP BW/4HANA system because you can only convert to SAP BW/4HANA if the add-on is installed. Before coming into this phase, make sure that you've done all the preparation activities.

6 In-Place Conversion

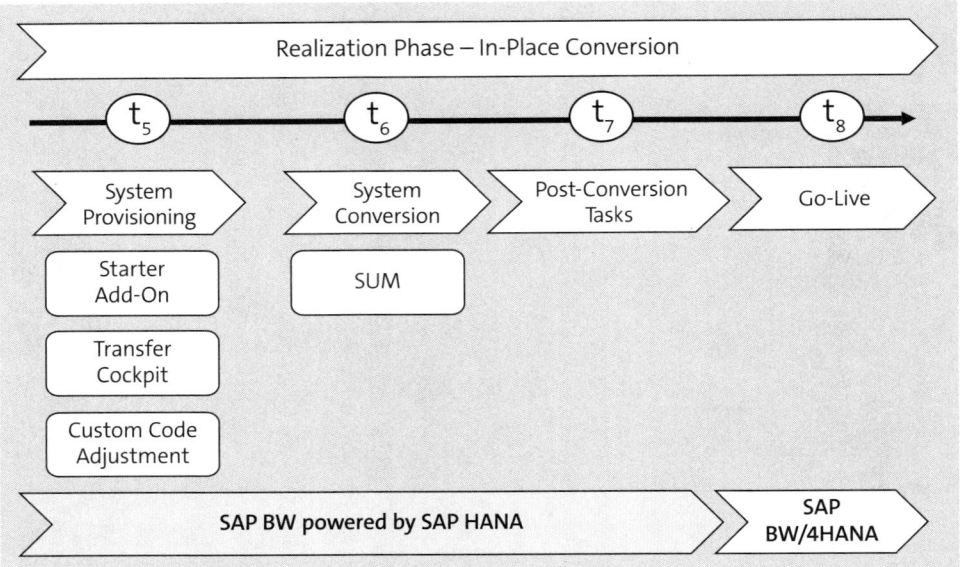

Figure 6.9 In-Place Realization Phase

6.6 Introduction to the SAP BW/4HANA Starter Add-On

In this section, you'll learn about the processes that occur during in-place conversion. We'll discuss installing the SAP BW/4HANA starter add-on and using the maintenance planner.

The SAP BW/4HANA starter add-on facilitates a smooth transition from existing object types, processes, and interfaces to the optimized ones in SAP BW/4HANA. The add-on is an extension within the SAP NetWeaver framework, and it plays a pivotal role in aiding the conversion process for SAP BW 7.5 systems operating on an SAP HANA database. Specifically designed to support this transition, the add-on becomes deployable from SAP BW 7.5 SPS 5 onward. Its primary objective is to facilitate the seamless preparation of the existing system for conversion into an advanced SAP BW/4HANA system.

6.6.1 Process during In-Place Conversion

To embark on this transition to SAP BW/4HANA, follow these steps within the SAP BW 7.5 system after the installation of the add-on:

- **Precheck for compatibility**
  - Conduct a thorough precheck to assess the compatibility of existing objects and processes with SAP BW/4HANA.
  - Identify any elements that may pose challenges during the transition.

- **Scope transfer with the SAP BW/4HANA transfer cockpit**
  - Use the transfer cockpit to transfer incompatible objects and processes to compatible ones.
  - Execute a scope transfer, encompassing data models and flows, and streamlining the migration process.
- **Switch to B4H mode**
  - Transition the system to an operating mode that exclusively allows objects and processes compatible with SAP BW/4HANA.
  - The switch to B4H mode prepares the SAP BW system for subsequent conversion steps. This switch can be executed using the transfer cockpit, with the task list initiated post the successful execution of the report.
- **System-wide cleanup**
  - Perform a detailed check to ensure that all system-wide cleanup activities have been successfully completed.
  - Address any residual issues or artifacts to ensure a clean environment for the subsequent phases.
- **Switch to Ready for Conversion mode**
  - Switch the system to an operating mode that facilitates the technical upgrade and prepares the SAP BW system for the final conversion to SAP BW/4HANA.
  - This transition to Ready for Conversion mode can be executed through the transfer cockpit.

6.6.2 SAP BW/4HANA Starter Add-On

In this section, you'll learn about the maintenance planner and SAP BW/4HANA starter add-on.

Maintenance Planner

To install the SAP BW/4HANA starter add-on, you can use the maintenance planner; you'll need support from the Basis team for this activity. You need to have the required prerequisites for using maintenance planner such as system landscape directory (SLD) and landscape management database (LMDB). We discussed this in detail in Chapter 3, Section 3.5, so check that out before proceeding here. Let's see how to install the SAP BW/4HANA starter add-on using maintenance planner now. Though you can download the maintenance planner from SAP Support Portal, it's recommended to use maintenance planner to get the SAP BW/4HANA starter add-on with generated *Stack.XML*. You also need to have the latest versions of the Support Package Manager (SPAM) and SAP Add-on Installation Tool (Transaction SAINT).

6   In-Place Conversion

Follow these steps:

1. Start the maintenance planner, and log on to the **SAP Maintenance Planner** portal, as shown in Figure 6.10.

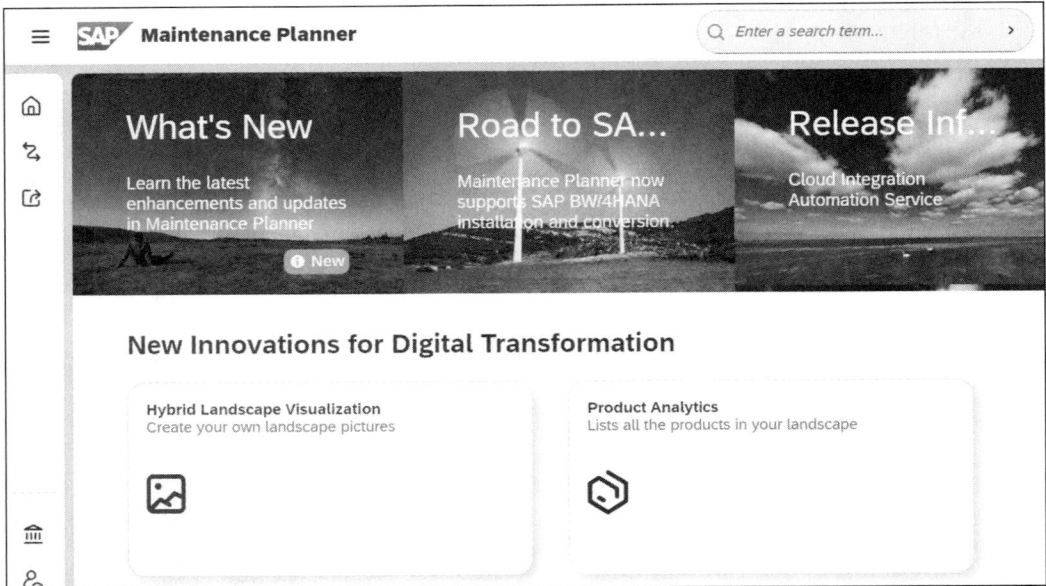

Figure 6.10 SAP Maintenance Planner Portal

2. You can enter your SAP BW 7.5 on SAP HANA SID or search for your SID, as shown in Figure 6.11.

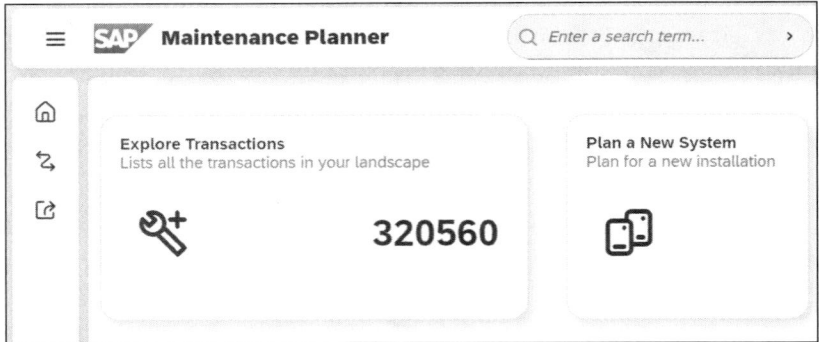

Figure 6.11 Plan a New System

6.6 Introduction to the SAP BW/4HANA Starter Add-On

3. Choose **Plan a New System**, and you can see the maintenance planner in Figure 6.12.

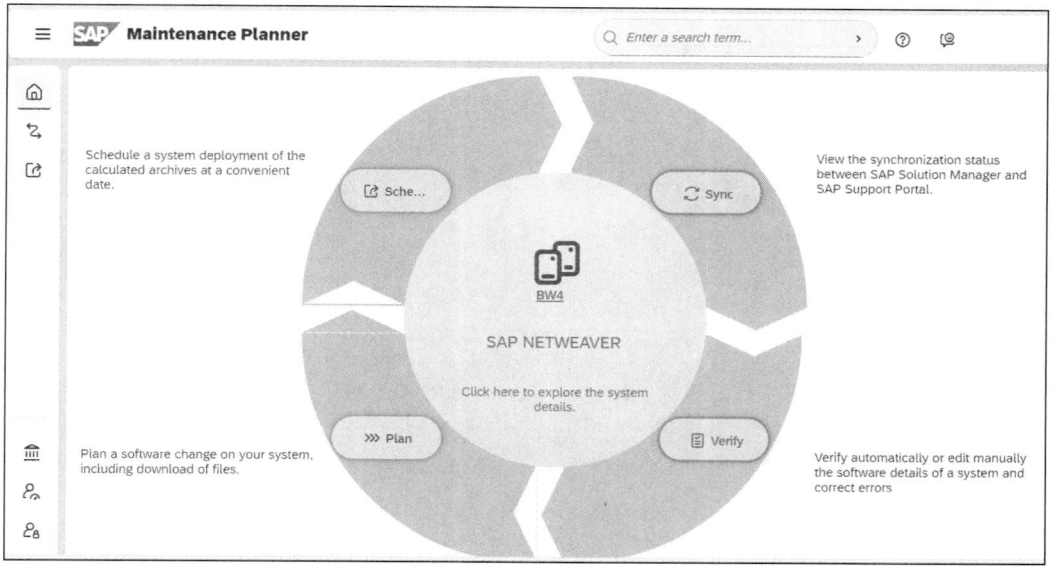

Figure 6.12 Maintenance Planner

4. Choose **Plan** to go to the screen shown in Figure 6.13. **Define Change** is the first option. You'll see the options for **SAP NetWeaver 7.5** as well.

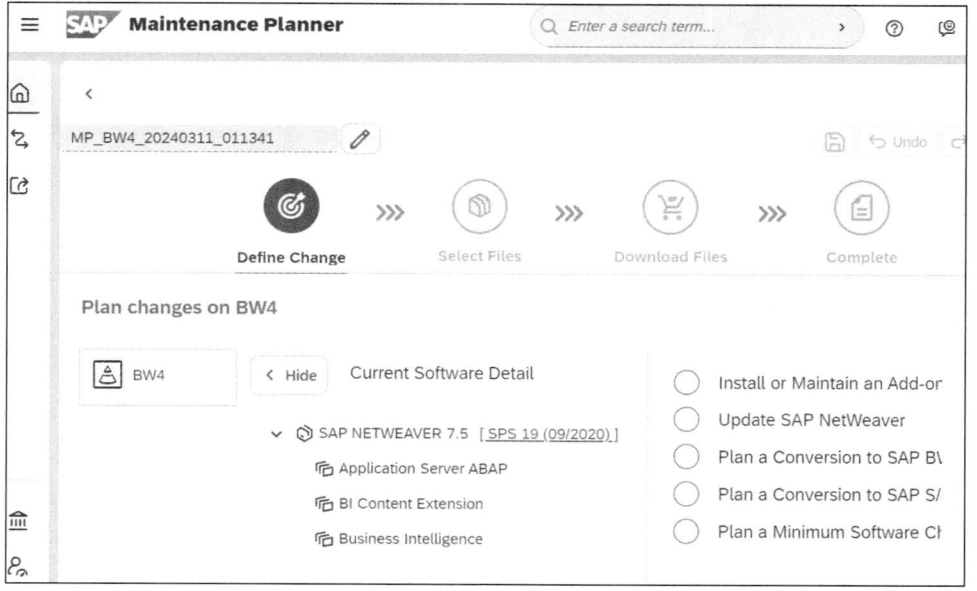

Figure 6.13 Define Change Option: Update SAP NetWeaver

291

6   In-Place Conversion

5. Choose **Update SAP NetWeaver** on the right side of the screen, and then select **SAP NetWeaver 7.5**, as shown in Figure 6.14 ❶.

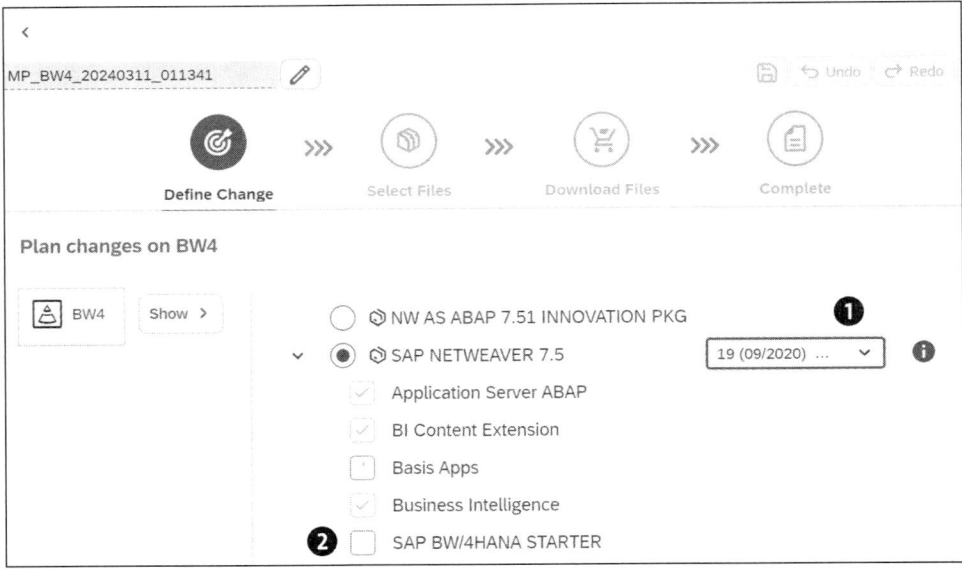

Figure 6.14 Plan Changes

6. Choose **SAP BW/4HANA STARTER** as shown in Figure 6.14 ❷. Click on **Confirm Selection**, and you'll see the screen shown in Figure 6.15.

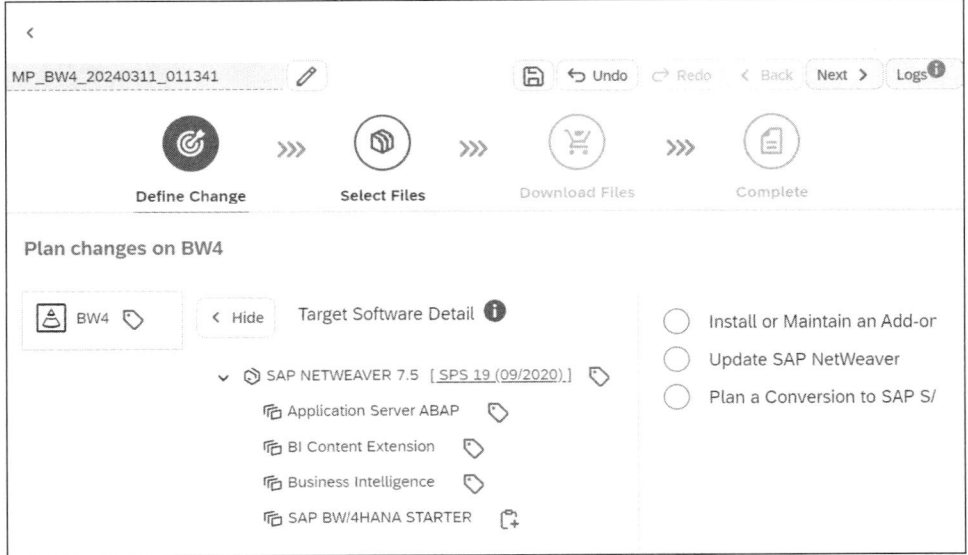

Figure 6.15 Starter Confirmation

7. Click **Next**, select the **OS/DB Dependent files** based on your installation, and choose **Confirm Selection** to go to the screen shown in Figure 6.16.

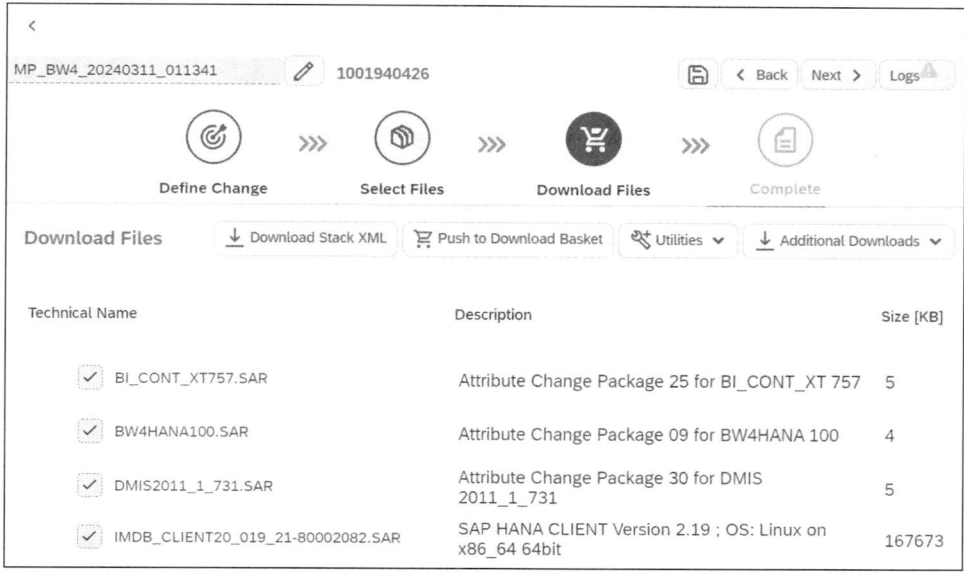

Figure 6.16 Stack XML Files

8. Choose **Download Stack XML**, and click **Next** to complete.

Once this activity is completed, you'll have the SAP BW/4HANA starter add-on in your SAP BW on SAP HANA system. In short, to install the SAP BW/4HANA starter add-on, first generate a *stack.xml* file using the maintenance planner tool and select the add-on. Then, use Transaction SAINT for installation. Make sure your SAP BW system doesn't have any add-ons that might cause issues during installation—check SAP Note 2189708 for the list. If you find unsupported add-ons, remove them before installing the SAP BW/4HANA starter add-on. Follow SAP Note 2246699 for more details. Once everything is ready, go ahead and install the SAP BW/4HANA starter add-on.

Installing the SAP BW/4HANA Starter Add-On

The installation process for the SAP BW/4HANA starter add-on takes place within the SAP BW system using Transaction SAINT. It involves the upload of the *stack.xml* file, previously generated in the maintenance planner, to facilitate a smooth installation. To initiate the installation, log in to client 000 of your SAP BW system, and access the add-on installation tool through Transaction SAINT. Then, select **Installation Package**, and choose **Load Package** to upload the necessary software packages. Ensure the *stack.xml* file is loaded, confirm the installation queue, and proceed with the installation. Upon completion, the SAP BW/4HANA software component (**BW4HANA**) should be successfully installed and ready for use. To see the details of our system after installation on the **Installed Software** screen, go to Transaction RSA1, and choose **System • Status • SAP Systems Data • Display** as shown in Figure 6.17:

❶ **SAP BW/4HANA Starter Add-On**

❷ **DMIS**

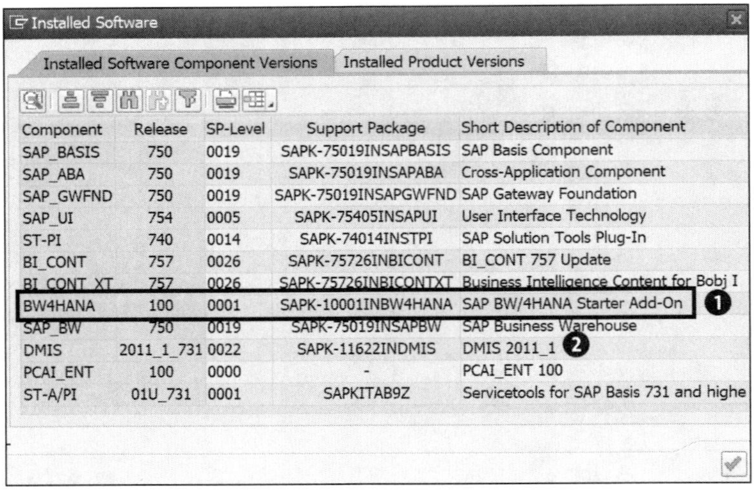
Installed Software
Installed Software Component Versions
Installed Product Versions
Component Release SP-Level Support Package Short Description of Component
SAP_BASIS 750 0019 SAPK-75019INSAPBASIS SAP Basis Component
SAP_ABA 750 0019 SAPK-75019INSAPABA Cross-Application Component
SAP_GWFND 750 0019 SAPK-75019INSAPGWFND SAP Gateway Foundation
SAP_UI 754 0005 SAPK-75405INSAPUI User Interface Technology
ST-PI 740 0014 SAPK-74014INSTPI SAP Solution Tools Plug-In
BI_CONT 757 0026 SAPK-75726INBICONT BI_CONT 757 Update
BI_CONT_XT 757 0026 SAPK-75726INBICONTXT Business Intelligence Content for Bobj I
BW4HANA 100 0001 SAPK-10001INBW4HANA SAP BW/4HANA Starter Add-On
1
SAP_BW 750 0019 SAPK-75019INSAPBW SAP Business Warehouse
DMIS 2011_1_731 0022 SAPK-11622INDMIS DMIS 2011_1
2
PCAI_ENT 100 0000 - PCAI_ENT 100
ST-A/PI 01U_731 0001 SAPKITAB9Z Servicetools for SAP Basis 731 and highe

Figure 6.17 SAP BW/4HANA Starter Add-On

The SAP BW/4HANA starter add-on becomes applicable starting from SAP BW 7.50 SP 5 and is exclusively designed for systems operating on SAP HANA. As discussed already, you need to know that installation of the SAP BW/4HANA starter add-on is subject to the prerequisite that supported add-ons are already installed in the system. Upon successful installation, your system transitions into compatibility mode, a state easily reversible to BW mode using Transaction RSB4HCONV. In this mode, objects that aren't compliant with SAP BW/4HANA remain accessible and all processes continue to function, but the transport/export of these objects is restricted, except for those included in the whitelist.

Simultaneously, while in compatibility mode, objects that are compliant with SAP BW/4HANA can be effortlessly created and edited. The system's functionality is maintained, offering flexibility for the ongoing development of SAP BW/4HANA–compliant components. To further enhance the transition, the SAP BW/4HANA Transfer Tool can be used to seamlessly transform noncompliant objects into their SAP BW/4HANA–compliant counterparts, streamlining the migration process and ensuring compatibility with the advanced features of SAP BW/4HANA.

Once the SAP BW/4HANA starter add-on is configured, it becomes essential to keep the system landscape up to date by upgrading to the latest version of the transfer cockpit. This process involves using the Note Analyzer along with the corresponding XML file, ensuring the implementation of all necessary SAP Notes. Given that SAP consistently enhances the tools required for conversion, it's strongly advisable to perform regular updates to the transfer cockpit. This can be achieved by running the Note Analyzer with the most recent XML files provided in SAP Note 2383530. Regular updates contribute to the optimization and effectiveness of the conversion process, aligning the system with the latest advancements in SAP BW/4HANA tools. We've seen the detailed steps to implement the XML in previous sections; for in-place conversions, refer to Section 6.3.

6.7 Operating Modes

Once the system has the SAP BW/4HANA starter add-on, you can see the following modes:

- **BW mode**
  In this mode, the system operates like a standard SAP BW system, running without any limitations. It functions similarly to SAP BW powered by SAP HANA, providing the same capabilities even without the SAP BW/4HANA starter add-on.

- **Compatibility mode**
  Once in this mode, it's generally recommended to adopt a forward approach rather than reverting. This state serves as the initial mode post-installation of the SAP BW/4HANA starter add-on. In this mode, the creation of unsupported object types isn't permissible, although modifications can still be made. Existing scenarios can seamlessly continue their operations, but for unsupported objects, transportation is only possible when they are added to a predefined whitelist using a specific task list.

- **B4H mode**
  In SAP BW 7.5 on SAP HANA, this mode is employed to mimic the functionality of SAP BW/4HANA. During this phase, there are no imports of unsupported object types, rendering the whitelist irrelevant. The system exclusively comprises SAP BW/4HANA–compatible objects, and while it operates with these new elements, the final preparations for the conversion haven't been completed.

- **Ready for Conversion mode**
  This mode requires downtime before system conversion. Once the final preparations are completed, and the system is deemed ready for conversion, certain components within the system cease to function (e.g., Service API [SAPI] source systems, virtual InfoProviders, SAP Business Planning and Consolidation [SAP BPC], etc.). It's crucial to plan the transition into this mode just before initiating the system conversion.

Let's continue our discussion of operating modes by learning how to switch them.

6 In-Place Conversion

6.7.1 Switching Modes

With the SAP BW/4HANA starter add-on, use Transaction RSB4HCONV to change the operational mode of your SAP BW system. This involves starting a task list via the ABAP Task Manager, which checks system status and lists required activities before the switch. If all prerequisites are met, the system smoothly transitions to the new mode. Simulation mode allows periodic checks for readiness before making the switch. Figure 6.18 shows the details of switching modes in SAP BW 7.5.

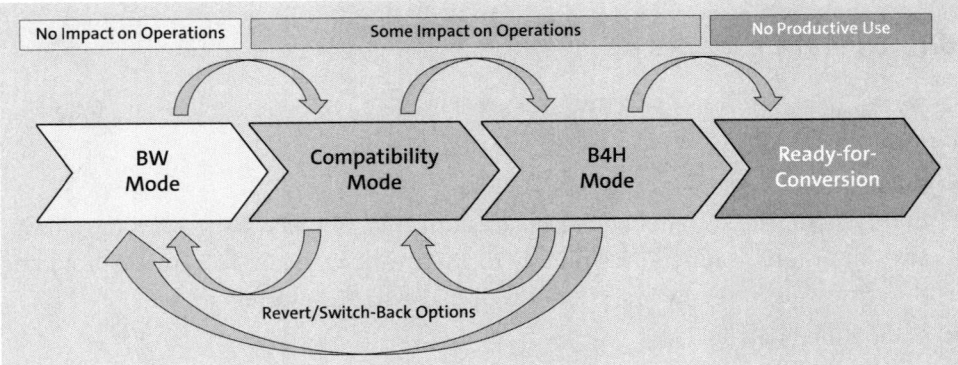

Figure 6.18 Switching Modes Demo

Follow these steps to switch modes:

1. Open the transfer cockpit using Transaction RSB4HCONV, as shown in Figure 6.19.

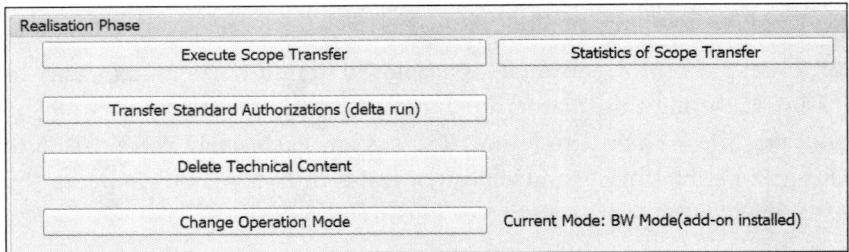

Figure 6.19 Transfer Cockpit: Change Operation Mode

2. Choose the **Change Operation Mode** option, and you'll see the screen shown in Figure 6.20. You can change between modes such as **BW Mode**, **Compatibility mode**, **B4H Mode**, and **Ready For Conversion mode**. Leave the other settings as is.

3. In the **Set the system to the new mode** area, you can either simulate using **Run Pre-Checks with dedicated mode** option or switch to a new mode using **Execute new mode**.

6.7 Operating Modes

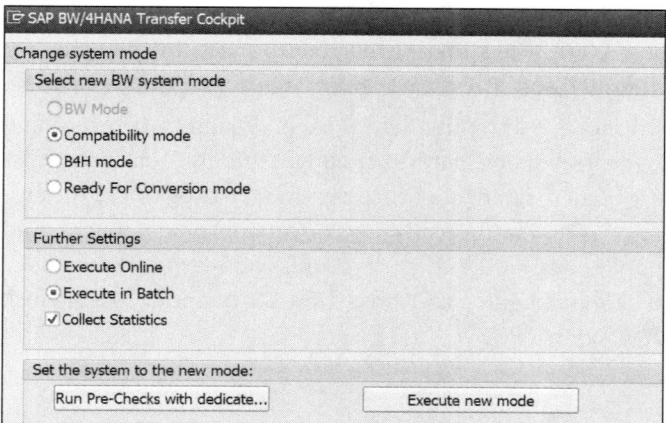

Figure 6.20 Operation Mode Change

6.7.2 Mode Switch and Prerequisites

The entire in-place conversion project will depend on how you use the task list and switch the mode; you'll be in one mode for a long time and one mode for less time, as shown in Figure 6.21:

- Transitioning from compatibility mode or B4H mode back to BW mode is flexible and can be done at any point. Upon completion of the switch, the system operates similarly to a standard SAP BW system, without the influence of the SAP BW/4HANA starter add-on. Notably, any modifications made during compatibility mode or B4H mode persist and aren't undone.

- Switching from compatibility mode to B4H mode is possible only when the system exclusively contains supported object types and the specified task list activities have been executed. If there are unsupported objects, the switch isn't executed until compliance is achieved. Upon meeting the switch conditions, a task list is initiated to either delete or migrate unsupported objects, establishing the system in B4H mode. Deletion includes SAP Business Explorer Analyzer (SAP BEx Analyzer) objects, data mining models, Analysis Process Designer (APD) analysis processes, InfoSpokes, Xcelsius Dashboards, SAP Crystal Reports, configuration for real-time data acquisition, and process chains related to obsolete objects. If the system is unfit for B4H mode, the report allows a return to compatibility mode, with irreversible changes initiated by the task list. Detailed task descriptions are available by going to Transaction STC01 and choosing the RS_COMPATIBILITY_TO_B4H task list.

- Switching from B4H mode to ready for conversion mode is contingent on the thorough completion of all-encompassing system-wide cleanup operations. This cleanup includes the deletion of BI content, specifically the D versions of all SAP BW objects; VirtualProviders designed for planning purposes; unsupported source systems; SAPI connections linking to other SAP BW systems; persistent staging area (PSA) tables; and error stacks. Furthermore, the switch initiates a cessation of all ongoing process chain schedules and real-time data acquisition daemons (with planned implementation in 2019). The system is eligible to transition to ready for conversion mode only when all these critical tasks have been successfully executed.

- Once ready for conversion mode is entered, there is no provision to revert to any other mode. This is due to the irreversible nature of changes initiated via the task list and the concurrent locking of all master data processes. In scenarios where the system faces constraints in accommodating ready for conversion mode, you should seek help from SAP support. For an in-depth understanding of each task involved, task descriptions are available in by going to Transaction STC01 and accessing task list RS_B4H_TO_READY4CONVERSION.

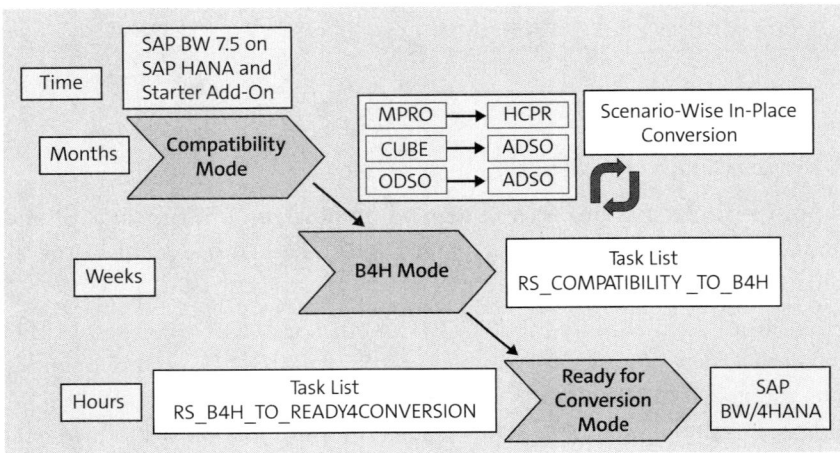

Figure 6.21 Timeline for Modes

6.7.3 Handling Unsupported Objects

During Compatibility mode in SAP BW/4HANA, the migration process is actively underway. This phase signifies a critical stage where the transition from legacy objects, such as InfoCubes and MultiProviders, to the new optimized data models takes place. Importantly, during Compatibility mode, any new development that necessitates the creation or modification of old-type objects is restricted. The focus is on completing the migration process smoothly, ensuring compatibility with the advanced features of SAP BW/4HANA.

While the creation or modification of old-type objects is generally prohibited, there are specific scenarios where exceptions can be made. Activating BI content or importing objects from the development system is allowed during Compatibility mode, enabling organizations to incorporate essential components seamlessly. In instances of emergency or critical requirements necessitating changes to old objects, an exception can be granted by adding the object to the whitelist. The whitelist, managed through the RS_B4HANA_WHITELIST_MAINTAIN report, serves as a mechanism to accommodate exceptional changes during this transitional phase, allowing for a more flexible approach to

6.7 Operating Modes

address urgent needs without compromising the migration process. Follow these steps:

1. Call Transaction SE38, and execute report RS_B4HANA_WHITELIST_MAINTAIN. You'll see the screen shown in Figure 6.22.

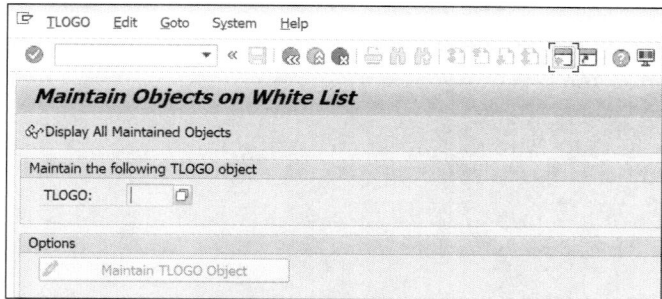

Figure 6.22 Maintain TLOGO

2. Choose to enter your **TLOGO** for maintenance, as shown in Figure 6.23. Here, enter "RSDS" which is for **DataSource.**

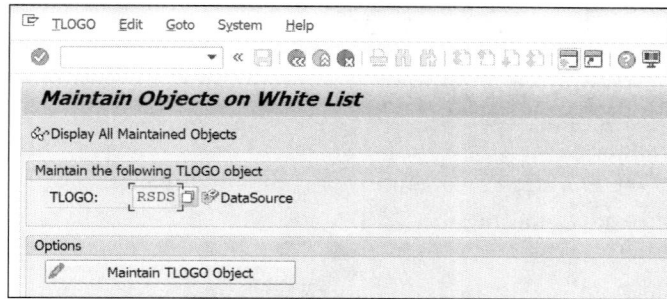

Figure 6.23 Maintain the Object List Entry: RSDS

3. Choose the **Maintain TLOGO Object** button, as shown in Figure 6.24.

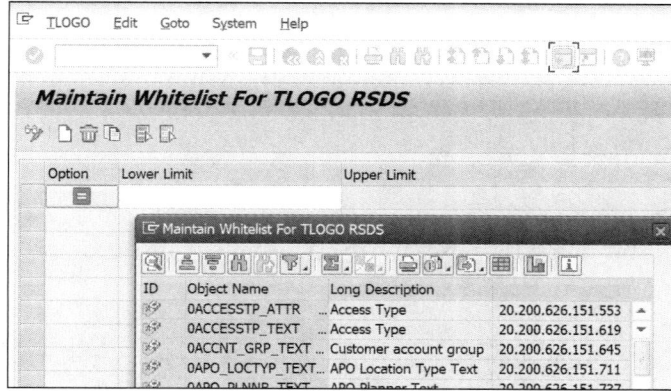

Figure 6.24 TLOGO Object Selection

299

4. Choose the TLOGO object and save it. You'll see the status shown in Figure 6.25. The **Option** column (with the **Equal to** icon, for our case) will be available. You can now provide the details of the selected objects, such as a specific DataSource, as shown in the **Lower Limit Column**.

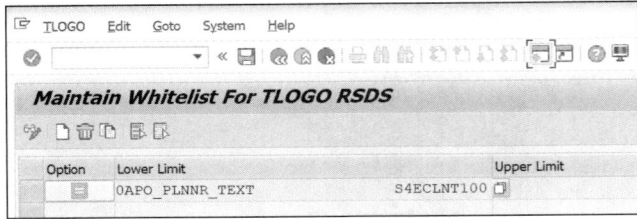

Figure 6.25 Final TLOGO

If you've maintained the object in this list, you'll be able to export it as a transport request.

6.7.4 Landscape Considerations

In the existing system landscape, we want to preserve the functionality of existing scenarios while aligning them for optimization on SAP HANA and for eventual conversion to SAP BW/4HANA.

Consider the following points for development systems:

- **Compatibility mode foundation**
  The development system sets the stage by operating in compatibility mode, laying the foundation for the subsequent journey.

- **Operational continuity**
  Existing scenarios seamlessly continue their operations without any disruptions, ensuring business processes remain unaffected.

- **Allowlist for modifications**
  When modifications are deemed necessary for existing scenarios, a critical aspect involves enlisting the pertinent object, distinguished by a supported object type, into the allowlist.

- **Staged conversion**
  Existing scenarios undergo a phased conversion, providing flexibility to delete unsupported objects post-conversion, as long as successor objects are present.

- **New scenario standards**
  A proactive approach dictates that all new scenarios are exclusively crafted with objects adhering to supported object types, setting the stage for enhanced compatibility.

- **Transition to B4H mode**
  As milestones are achieved, the development system seamlessly transitions to B4H mode, marking a pivotal step in the optimization process.

- **Ready for Conversion mode preparation**
  With preparatory activities culminating, the development system shifts to ready for conversion mode, paving the way for the subsequent system conversion to SAP BW/4HANA.

Consider the following points for test and production systems:

- **B4H mode transition conditions**
  The switch to B4H mode is contingent upon the successful transfer of all scenarios or objects featuring unsupported object types. Simultaneously, the system must absorb the transferred deletions related to unsupported objects seamlessly.
- **Strategic timing for system conversion**
  The transition to ready for conversion mode aligns with the system go-live in both the test and production environments. This synchronization ensures a methodical and controlled progression toward SAP BW/4HANA, safeguarding the integrity of data and operational continuity.

6.8 Transferring Standard Authorization

The streamlining of object types within SAP BW/4HANA introduces changes to authorization objects. During the conversion of an SAP BW system to SAP BW/4HANA, authorizations associated with object types not present in SAP BW/4HANA (e.g., InfoCubes) need to be substituted with authorizations tailored for equivalent object types (e.g., ADSO). The transfer cockpit incorporates an authorization transfer tool designed to facilitate the automated transfer of roles and standard authorizations.

In this section, you'll see how to transfer the standard authorization during in-place conversion. We'll cover the conversion rules, the scope transfer, and authorization conversion. You'll learn about the initial and delta run for the authorization transfer. We'll delve into the Simulation mode and finally explore how to execute the authorization transfer initial run

6.8.1 Conversion Rules for Authorization Objects

The transition from SAP BW to SAP BW/4HANA requires the conversion of authorization objects involving four distinct actions tailored to specific authorization objects:

Consider the following scenarios during the SAP BW/4HANA conversion process:

- **Assume**
  No action is required. These authorization objects will seamlessly transition 1:1 after the SAP BW/4HANA conversion.
- **Adjust**
  Review and modify values as needed. In instances where InfoCube/DSO names are in use, it's recommended to update them to reflect the new ADSO names.

- **Replace**
  Manual intervention is essential. Integrate a new authorization object (indicated in the **Replacement** column) into the roles, replacing the outdated one specified in the **Technical Name** column.

- **Obsolete**
  The authorization object becomes obsolete and is no longer necessary. Notably, objects marked as obsolete persist in the system (visible in Transaction SU21, under the *RS* folder). This retention occurs because old objects are imported during remote transfer and subsequently converted during activation. Specific objects, such as S_RS_BITM and S_RS_BTMP, which participate in a pilot program for the web application designer in SAP BW/4HANA, can't be deleted for regression reasons.

The copy principle of the authorization transfer tool is illustrated in Figure 6.26. The tool optimally uses the roles already established in your system. It functions by generating duplicates of these existing roles while ensuring the preservation of the original ones. Subsequently, conversion rules specific to authorization objects are implemented on these duplicated roles. Upon completion of the object conversion through the scope transfer tool, both the original roles and their created counterparts are assigned to users. Following the confirmation of successful authorization object conversion and the system's transition to SAP BW/4HANA, you have the flexibility to eliminate the original roles, streamlining the authorization structure in alignment with the updated framework.

Figure 6.26 Copy Principle in Authorization Transfer

6.8.2 Scope Transfer and Authorization Conversion

The authorization object conversion rules play a pivotal role in ensuring a seamless transition from SAP BW to SAP BW/4HANA. Let's delve deeper into the two distinct categories of these conversion rules:

- **Rules not dependent on other conversion actions:**
  - Assume: This rule signifies that no additional actions are required for the authorization objects. They seamlessly function 1:1 after the SAP BW/4HANA conversion.
  - Obsolete: Authorization objects marked as obsolete imply that they are no longer necessary post-conversion. Despite their presence in the system, these objects become redundant and are essentially phased out for regression reasons.

- **Rules dependent on other conversion actions:**
  - Adjust: This action involves a meticulous check and adaptation of values. If existing object names, such as InfoCube, DSO, and so on, are in use, they need to be modified to correspond with the new ADSO names in SAP BW/4HANA.
  - Replace: When an authorization object requires replacement, a manual task is initiated to introduce a new authorization object in place of the old one. The replacement column specifies the new object to be included in the roles, ensuring compatibility with SAP BW/4HANA.

For the second group of rules, actions are interlinked with the successful transfer of corresponding SAP BW objects using the scope transfer tool. This tool provides critical information for adjusting or substitutions in the authorization objects within designated roles. Key insights include the mapping of new names and types of converted InfoProviders and the identification of additional InfoProviders generated during the conversion, such as CompositeProviders for ADSO with navigational attributes. These steps are integral to maintaining a robust and secure authorization structure in alignment with the advanced framework of SAP BW/4HANA.

6.8.3 Initial and Delta Run

You'll see this option when doing in-place conversion. The execution of the authorization transfer tool is structured into two distinct phases, each catering to specific rule categories:

- **Initial run (preparation phase):**
  - Not dependent on scope transfer execution: This phase operates independently of the scope transfer execution, addressing rules falling under the first group.
  - Handles rules of the first group: The initial run focuses on executing rules that are self-contained and don't rely on the completion of scope transfer actions.
- **Delta run (realization phase):**
  - Dependent on scope transfer execution: Unlike the initial run, the delta run is contingent upon the successful execution of scope transfer activities.
  - Handles rules of the second group: This phase is specifically designed to manage rules that are interlinked with other conversion actions, such as adjustments and replacements. It aligns with the realization phase of the overall conversion process, ensuring a cohesive and systematic transfer of authorization objects in accordance with the second group of rules.

6.8.4 Simulation Mode

In the context of the authorization transfer tool, the simulation mode is a valuable feature that ensures a heightened level of user engagement and confirmation. When operating in simulation mode, the system refrains from immediately creating role copies.

6 In-Place Conversion

Instead, it follows a two-step process involving an initial run and a subsequent delta run for the transfer of authorization objects. During these runs, all role copies are prepared and stored in temporary tables, awaiting confirmation by the user.

This deliberate separation of role creation into distinct confirmation phases allows users to thoroughly review and validate the proposed changes before they are implemented in the live system. It serves as a preventive measure, ensuring that any unintended consequences or errors can be identified and addressed before the actual creation of new roles within the system.

Figure 6.27 illustrates the SAP authorization system objects that play a crucial role in facilitating a seamless and controlled transition during the execution of the authorization transfer tool in simulation mode. This cautious approach enhances the overall reliability and accuracy of the authorization object transfer process, contributing to a smoother transition to SAP BW/4HANA.

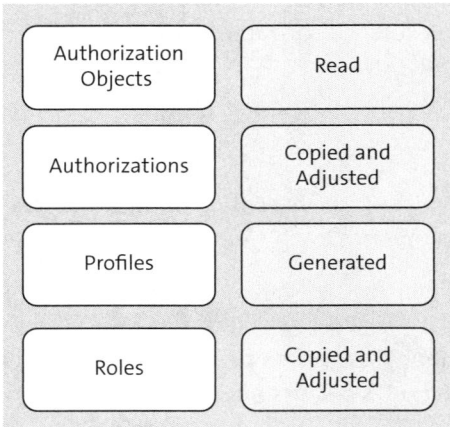

Figure 6.27 Authorizations

6.8.5 Authorization Transfer Process (Initial Run)

You need to make sure that you have the role for the conversion, **Z_BW_AUTH_CONV**, in your SAP BW 7.5 system, as shown in Figure 6.28.

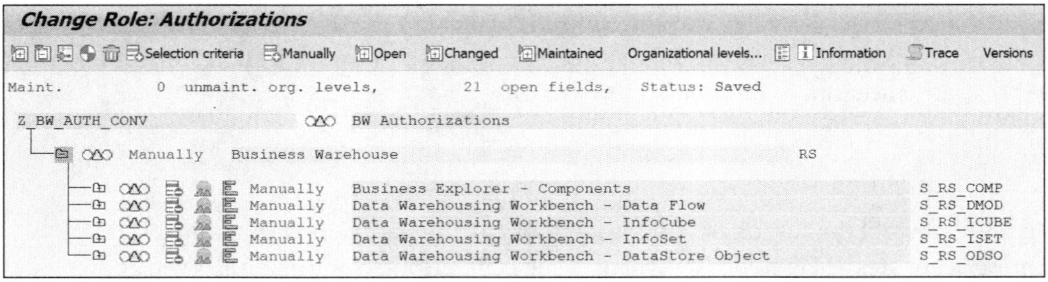

Figure 6.28 Role in SAP BW 7.5

6.8 Transferring Standard Authorization

Follow these steps to add this role:

1. Start the transfer cockpit using Transaction RSB4HCONV, as shown in Figure 6.29.

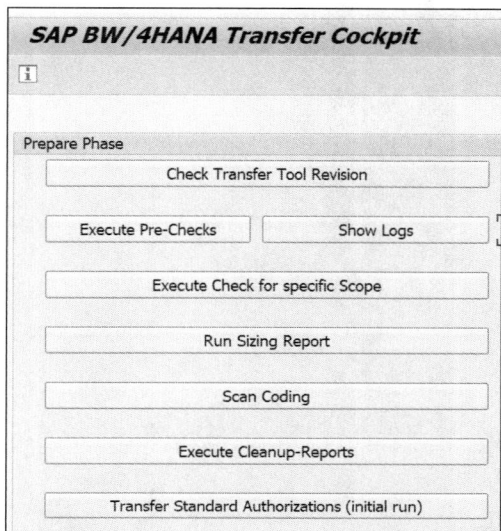

Figure 6.29 Transfer Cockpit: Initial Screen

2. Choose the **Transfer Standard Authorizations (initial run)** option, and you'll be prompted to provide the **Run Id**, as shown in Figure 6.30. Enter "Z_BW_AUTH_TR".

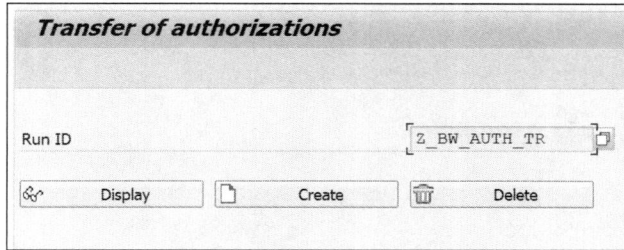

Figure 6.30 Run ID

3. Choose **Create**, provide the **Description**, and confirm. You'll see the **Transfer of authorizations Z_BW_AUTH_TR** screen, as shown in Figure 6.31. It will have multiple options such as **Initial Run**, **Delta Run**, **Generate**, and **Settings**. The grid will be blank initially.

4. Choose the **Add roles** button, add your role name ("Z_BW_AUTH_CONV"), as shown in Figure 6.32. Press [Enter].

305

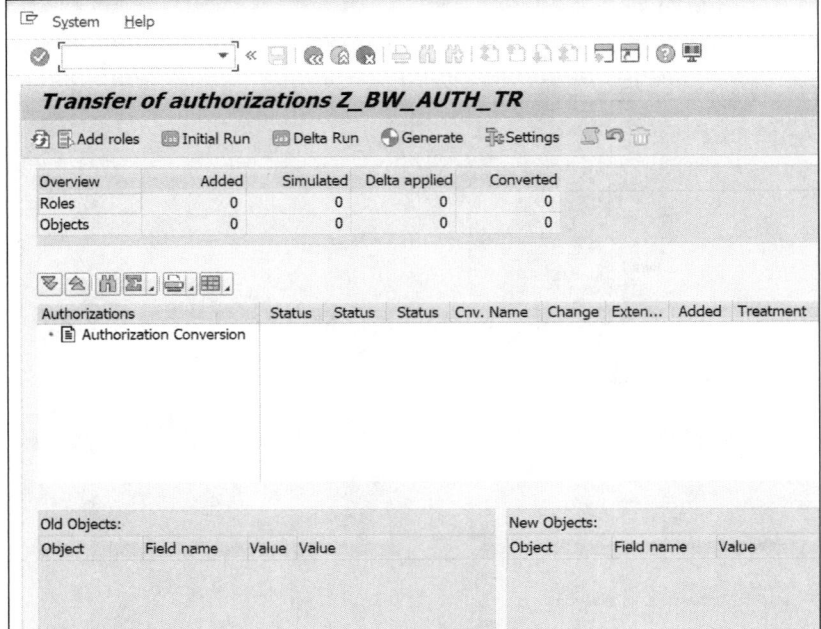

Figure 6.31 Authorization Transfer

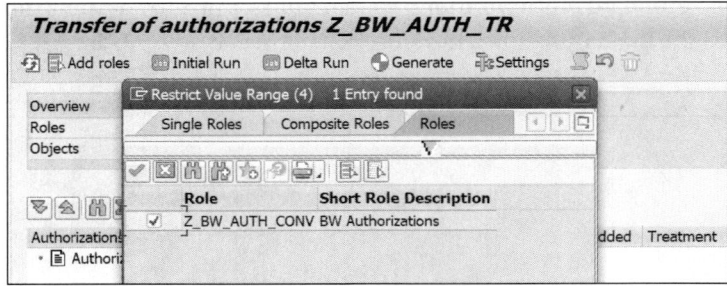

Figure 6.32 Add Role: Transfer of Authorization

5. Provide the value for **Role Suffix**, and you'll see a message that the authorization role is added to the conversion run, as shown in Figure 6.33.

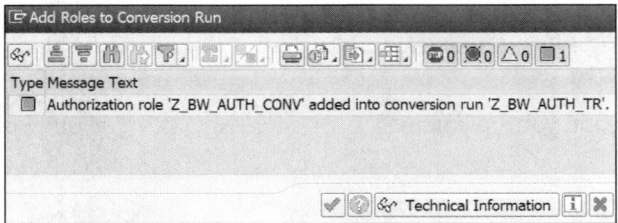

Figure 6.33 Role Addition

6.8 Transferring Standard Authorization

After the role is added, the **Transfer of Authorization** screen will show the following details (see Figure 6.34):

❶ Name of the run

❷ **Add roles** button to add the role

❸ Role name

❹ Converted name

❺ List of authorization objects in the role

❻ **Treatment** of authorization objects

❼ Authorization objects to choose

❽ **Old Objects** details (new isn't yet filled)

❾ **Initial Run** option

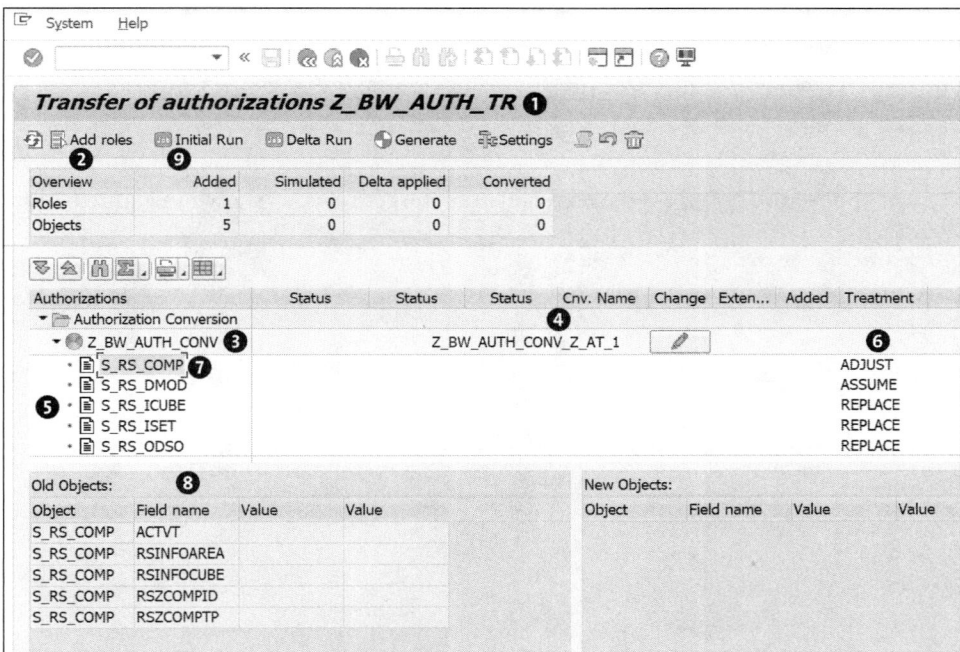

Figure 6.34 Transfer Options

6. Now choose the **Initial Run** icon, and you'll see the screen shown in Figure 6.35:

❶ Green simulation status

❷ **New Objects** information

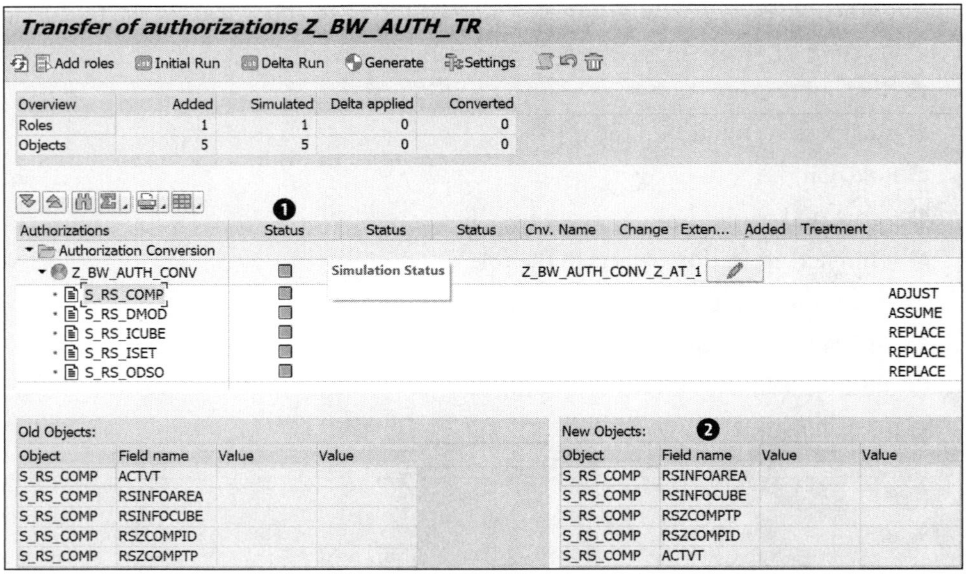

6 In-Place Conversion

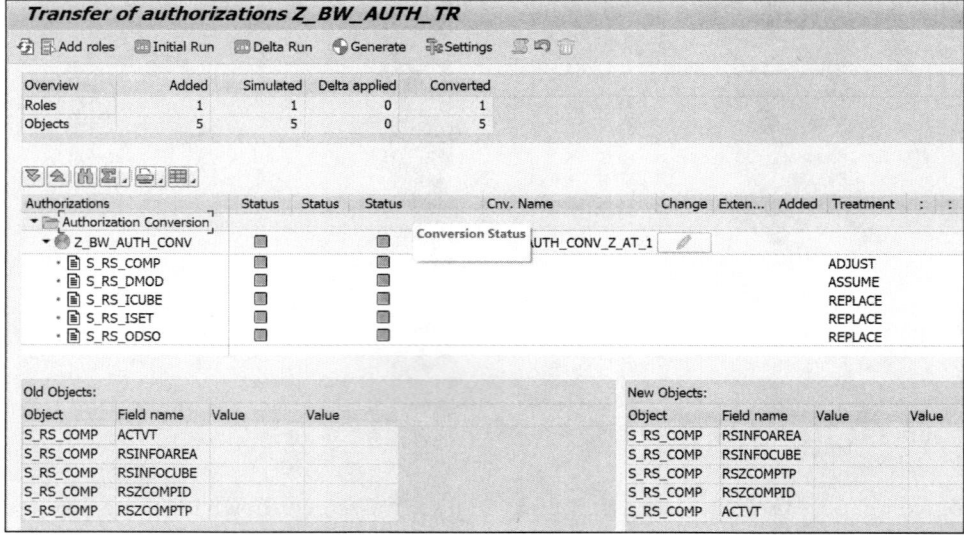

Figure 6.35 Initial Run

7. Now choose the Generate icon to generate target roles, as shown in Figure 6.36.

Figure 6.36 Conversion Status

You can see that **Conversion Status** is now green, and the **Status** in the middle is for delta. Once this is completed, continue with the object transfer using the scope transfer tool for the core in-place conversion activity, and complete the transfer of authorizations afterward.

6.9 Execution Overview

Once you've completed all the steps in the previous section, the next step is to execute the in-place conversion scope by scope, as we'll discuss in the following sections. For this demo, we'll be using the data model in SAP BW 7.5 on an SAP HANA system, as shown in the next section in Figure 6.37.

6.9.1 General Features

In transitioning to an environment optimized for SAP HANA, several object types undergo simplifications or replacements. Classic DSOs are succeeded by ADSOs, with the transition detailed in the document. Likewise, the move from old request management to new request management for ADSOs is documented. InfoCubes give way to ADSOs, while MultiProviders are replaced by SAP HANA CompositeProviders (HCPR). InfoSets are similarly replaced by SAP HANA CompositeProviders (HCPR). Semantic objects such as InfoCubes or classic DSOs are streamlined into groups of ADSOs. The transition from old CompositeProviders (COPR) to SAP HANA CompositeProviders (HCPR) is detailed. Source systems undergo a shift to ODP source systems, while SAP BW and SAP source systems are also replaced by ODP source systems. For database source systems, the shift is toward SAP HANA source systems. Processes involving InfoPackages and PSAs are replaced with DTPs and ADSOs. You can see the sample flow-based change in Figure 6.37. There are many other simplifications that are part of SAP BW/4HANA, and if the tool isn't converting them, then you need to do some adjustments and replacements.

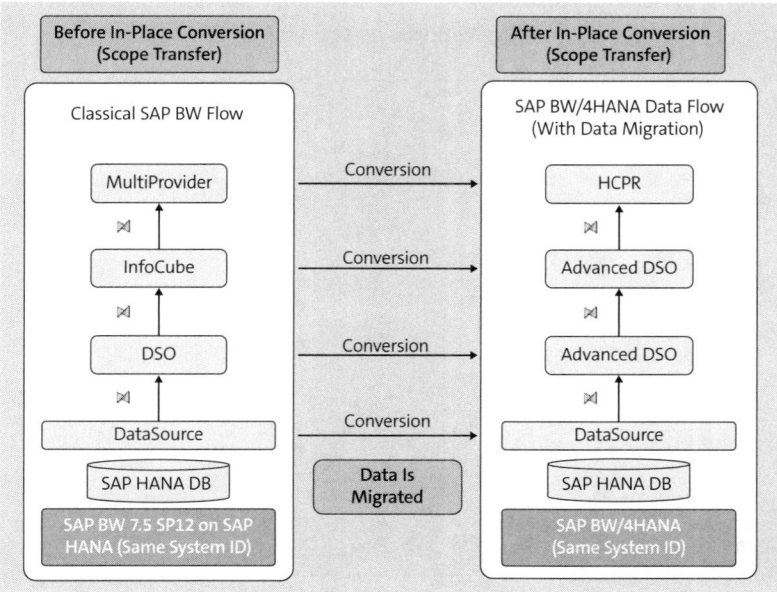

Figure 6.37 In-Place Conversion: High-Level Simplification Flow

Perform the specified adjustments for each object type as follows:

- Replace objects with corresponding successors for analysis authorization, considering transfer of authorization discussed earlier.
- Adjust or replace the InfoProvider for SAP BW queries/query views.
- Replace objects with corresponding successors in SAP BW workspaces.
- Adjust DataSources according to changes in source systems, and replace export DataSources from the Myself source system with ODP-BW source systems.
- For the DTP, adjust the source and target, replacing error DTP with Data Transfer Intermediate Storage (DTIS). During scope transfer, set the DTP to inactive, and create DTIS upon reactivation post-conversion to SAP BW/4HANA.
- Replace objects with corresponding successors in the data flow and SAP HANA analysis processes.
- In Integrated Planning, modify or replace the InfoProviders, and adjust planning functions.
- For near-line storage (NLS)/data archiving processes, adapt NLS request management and re-create processes for ADSOs.
- Switch to new request management for Open Hub Destinations.
- In process chains/variants, replace InfoPackages with DTPs, activate ADSOs, and replace APD processes with "dummy" process chains.
- Adjust source and target transformations, and modify DSO lookups.

6.9.2 Cleanup Execution

Before starting scope transfer, you need to run the cleanup. Enter Transaction RSB4HCONV, and choose **Execute Cleanup-Reports**, as shown in Figure 6.38.

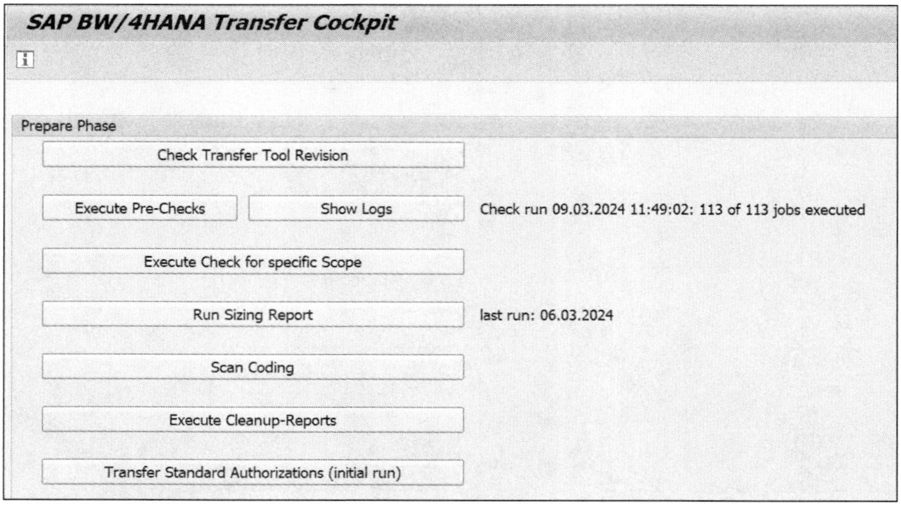

Figure 6.38 Cleanup

This will open task list **RS_B4H_CHK_CLEANUP**, which you can also open using Transaction STC01, as shown in Figure 6.39:

- RSTRAN_ROUT_RSFO_CHECK checks and fixes routine/formula inconsistencies in transformations.
- RSTRAN_TRFN_DELETE_ORPHAN_PROG removes obsolete transformation programs.
- Cleanup of table RSOOBJXREF ensures accurate object relationships for efficient processes such as SAP BW/4HANA transfer, preventing errors caused by missing data.

Additionally, make sure to perform the consistency checks with Transaction RSRV for smooth data transfer. This will make sure that the systems don't have any issues with data consistency.

Figure 6.39 Cleanup Reports

6.9.3 Data Transfer Logic

The data transfer during scope transfer for the new objects will be based on the following pattern:

- When transitioning from an InfoCube to an ADSO, compressed data from one partition is relocated to the active data table, while uncompressed data from another partition is transferred to the inbound table.
- In the case of converting from a standard DSO (classic) to an ADSO, the inbound data isn't transferred but undergoes activation into the active data before being transferred. Subsequently, the active data is moved to the corresponding active table within the ADSO. The change log also migrates accordingly to the new change log table, offering flexibility to retain either the complete log or only partial entries. Additionally, two compatibility views with the same DSO names are established to ensure continued interface compatibility during the transition. This will ensure that your reports based on the DSO work without any issues.

6 In-Place Conversion

6.9.4 SAP Business Warehouse Data Model

In this section, we'll explore the SAP BW data model used for the in-place conversion demo. You'll learn the basic modeling features, understand the delta queue status, and get a high-level overview of the data flow in the SAP BW 7.5 system. Knowing the SAP BW 7.5 data flow will help you correlate with the new SAP BW/4HANA objects that are converted after the in-place conversion is completed.

Consider that you have the following data model that extracts data from an SAP S/4HANA system: DataSource **ZPRIPDS**. This data model is based on an SAP S/4HANA system, is exposed to ODP, and is replicated into SAP BW 7.5. It has four InfoObjects and the data feed data into the PSA, DSO, and InfoCube.

Transaction RSA7 Status

Before the scope transfer, you can log on to the sender SAP ERP or SAP S/4HANA system and check the status of the SAPI-based delta queue in Transaction RSA7, as shown in Figure 6.40.

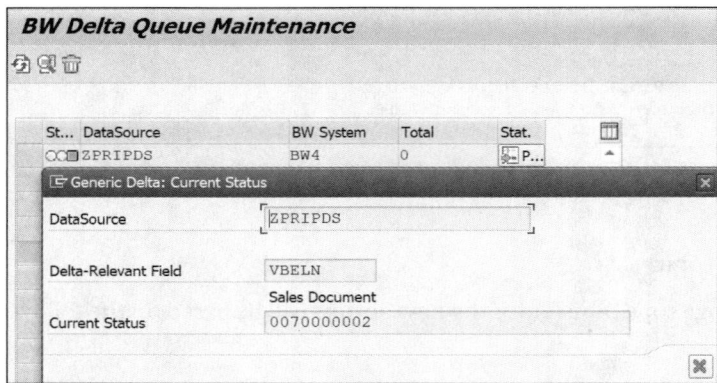

Figure 6.40 Delta Queue Status in the Sender System

SAP BW 7.5 System Objects

Let's walk through the SAP BW system data flow. Figure 6.41 shows the InfoCube (**ZPRKIPCV**) and DSO (**ZPRKIDSO**). This will be used in the scope transfer.

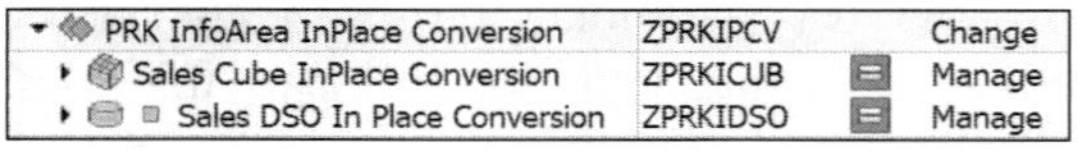

Figure 6.41 InfoCube and DSO

You can see the data flow of the InfoCube (**ZPRKICUB**) in Figure 6.42. It gets data from the DSO, and the DSO (**ZPRKIDSO**) fetches data from the DataSource (**ZPRIPDS**), which has two InfoPackages—one for Init and the other for delta.

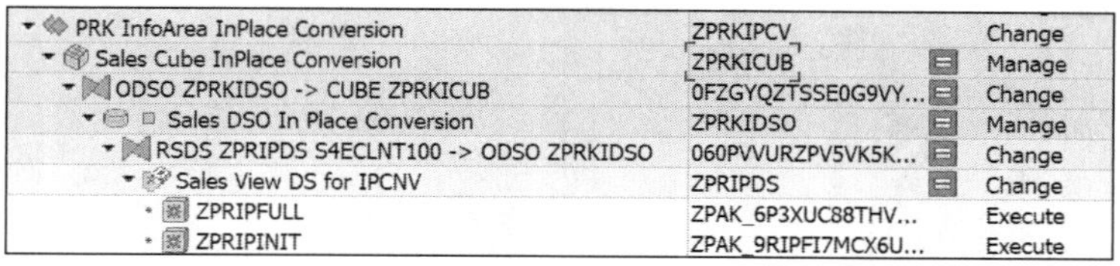

Figure 6.42 InfoCube Flow

You can see the details of the loaded request in the standard DSO (**ZPRKIDSO**) from the DataSource in Figure 6.43. It has **2192** records.

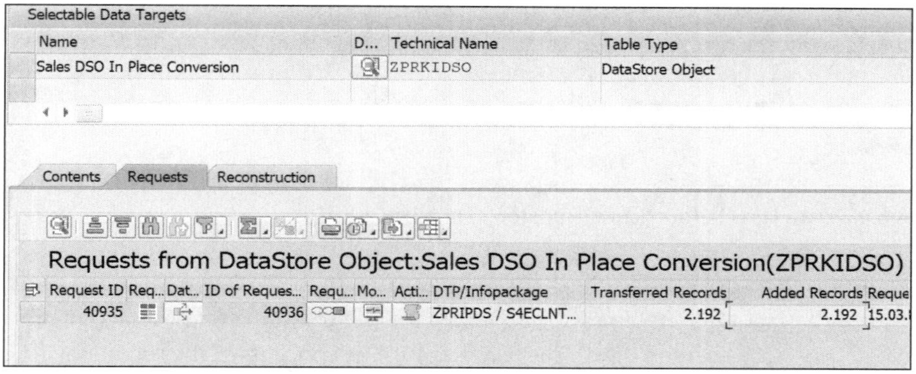

Figure 6.43 DSO Request

In the InfoCube management screen, you can see the **Request**, which was loaded from the DSO. There are **2148** records loaded into the InfoCube (**ZPRKICUB**), as shown in Figure 6.44.

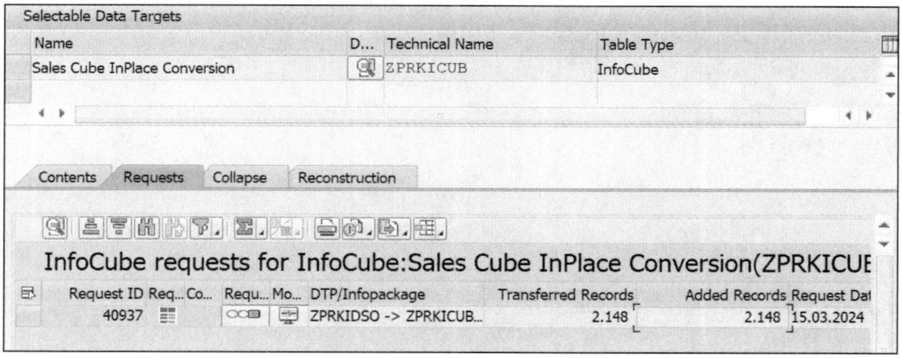

Figure 6.44 InfoCube Request

Overall Data Flow in SAP BW 7.5

Execute Transaction RSA1, and select **Choose Cube • Display Dataflow** to see the data flow of the InfoCube, as shown in Figure 6.45.

6 In-Place Conversion

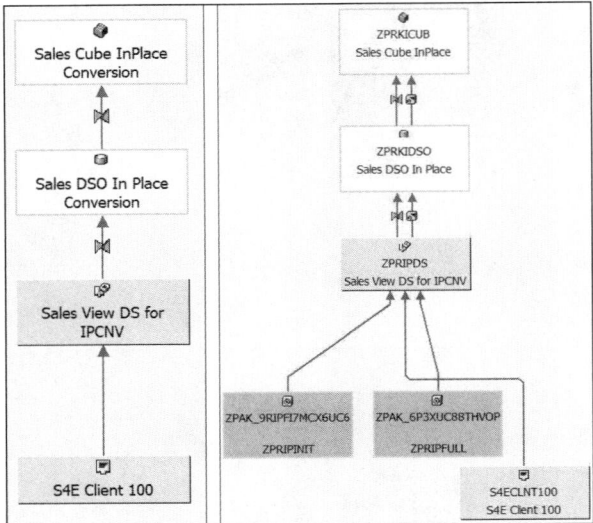

Figure 6.45 Data Flow before Scope Transfer

Before the scope transfer, try to check the tables related to the DSO and InfoCube. Here, we can see the fact table (table /BIC/FZPRKICUB) for the InfoCube and the active table (table /BIC/AZPRKIDS00) for the DSO (see Figure 6.46). The contents of these tables will be migrated during the data transfer.

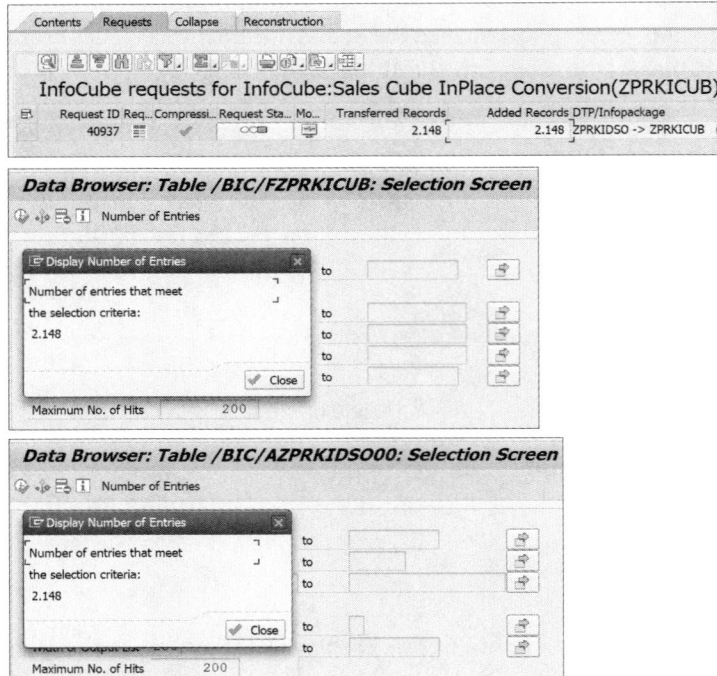

Figure 6.46 Before Scope Transfer InfoCube

6.10 Starting In-Place Conversion

You have multiple options to start the in-place scope transfer:

- **Using the transfer cockpit**
  Open Transaction RSB4HCONV. You'll see the options shown in Figure 6.47:

  ❶ Choose **Execute Scope Transfer**.

  ❷ Select the receiver system; in our case, it's **BW/4HANA**.

  ❸ Select the transfer scenario as **In-Place**.

  ❹ Choose **New** to create a new task list run.

  ❺ The new task list run for the in-place conversion is created.

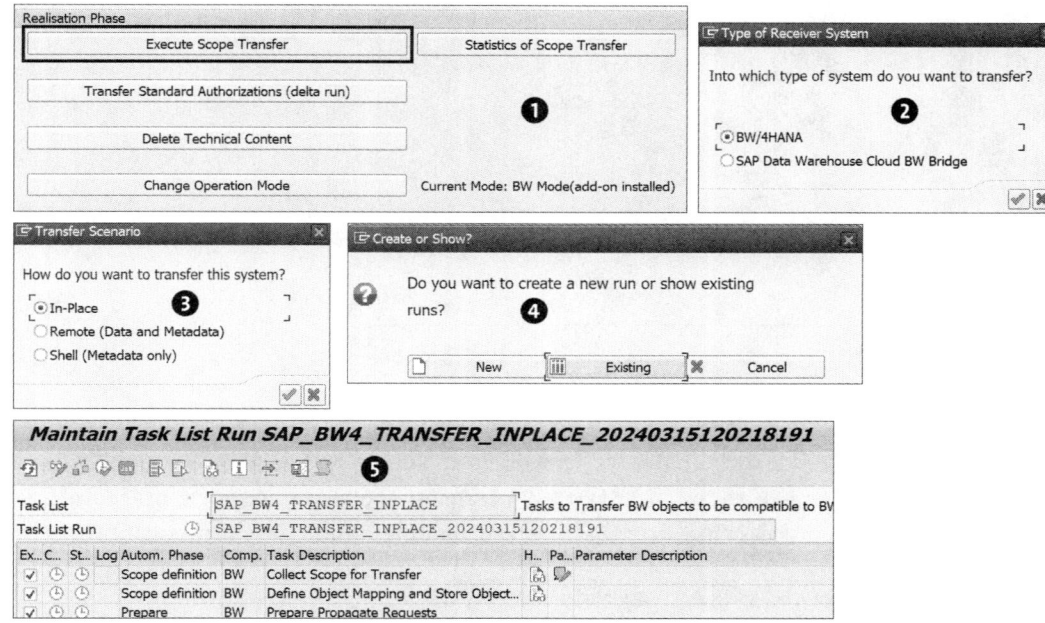

Figure 6.47 In-Place Conversion: Using the Transfer Cockpit

- **Using the standard task list name**
  Open Transaction STC01, enter the task list name "SAP_BW4_TRANSFER_INPLACE", and then execute it, as shown in Figure 6.48.

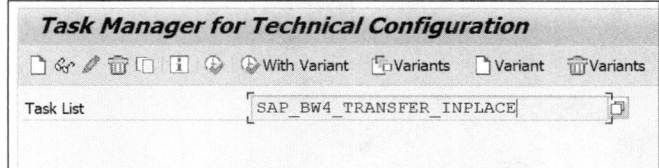

Figure 6.48 Task List Name for In-Place Conversion

315

6 In-Place Conversion

- **Using the data flow tool**
  In Transaction RSA1, choose the InfoProvider, right-click, and select **Display Dataflow**. You'll see the data flow for the InfoProvider, as shown in Figure 6.49. Once the data flow is confirmed, click the icon (**Start BW4 Transfer Tool**) to include the data flow as a scope transfer object in the task list for in-place conversion.

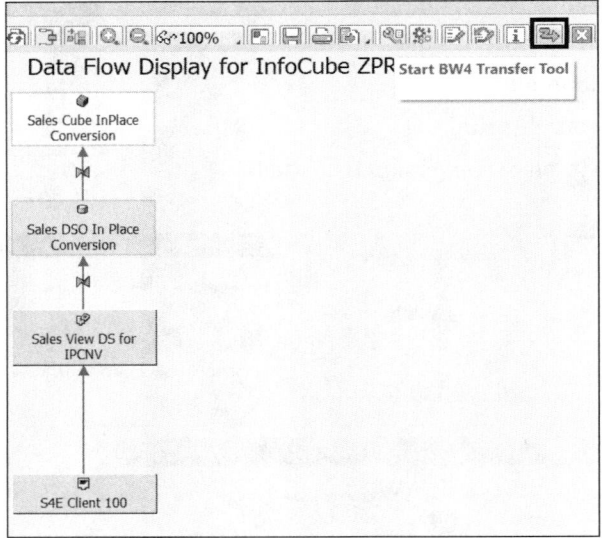

Figure 6.49 In-Place Conversion Using the Data Flow Tool

- **Process chain maintenance**
  Open any process chain that you need to transfer to SAP BW/4HANA using Transaction RSPC1. Click on the (**Start BW4 Transfer Tool**) icon, as shown in Figure 6.50, to add the process chain to the scope transfer object and start the task list run.

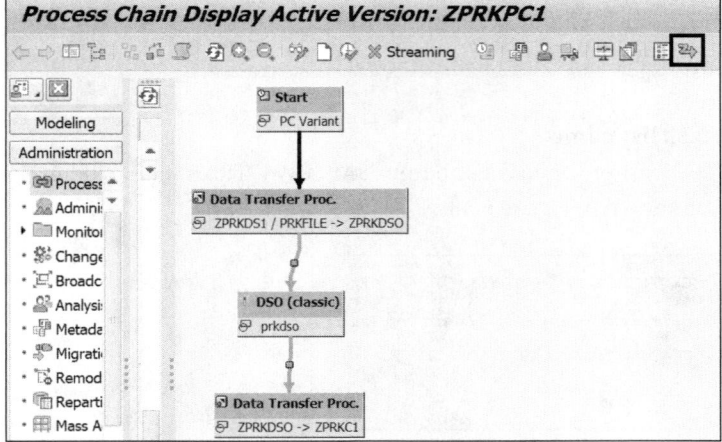

Figure 6.50 In-Place Conversion Using Process Chain Maintenance

6.11 Scope Transfer

You'll now see the steps to perform the scope transfer using in-place conversion. Use the transfer tool option for the data flow in Transaction RSA1, open the InfoProvider, and right-click on any of the InfoProviders. From the context menu that appears, select **Display Data Flow**. Choose the icon, as shown in Figure 6.51 ❶. Choose **BW/4HANA** as the receiver system ❷.

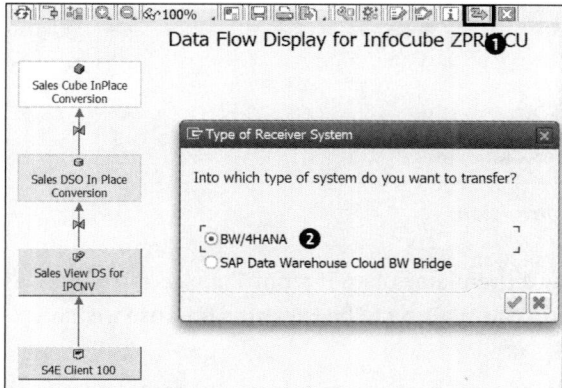

Figure 6.51 Scope Transfer

Once SAP BW/4HANA is selected, you'll see the **Transfer Scenario** popup. Choose the **In-Place** option, as shown in Figure 6.52.

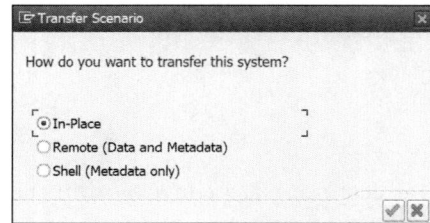

Figure 6.52 Selection of the Transfer Scenario

Confirming the transfer scenario as in-place will provide the **Check** or **Execute** radio buttons, as shown in Figure 6.53. Choose **Execution**, as its going to the conversion execution.

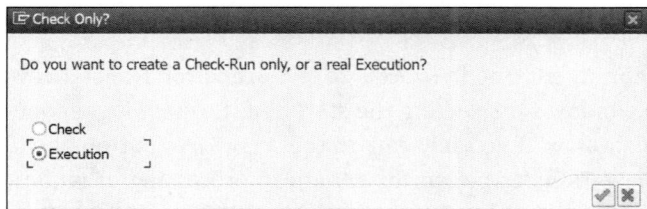

Figure 6.53 Options to Check or Execute

This will create a new **Task List Run** for the in-place conversion, as shown in Figure 6.54. This is the initial screen for the task list run named **SAP_BW4_TRANSFER_INPLACE_20240315120721719**.

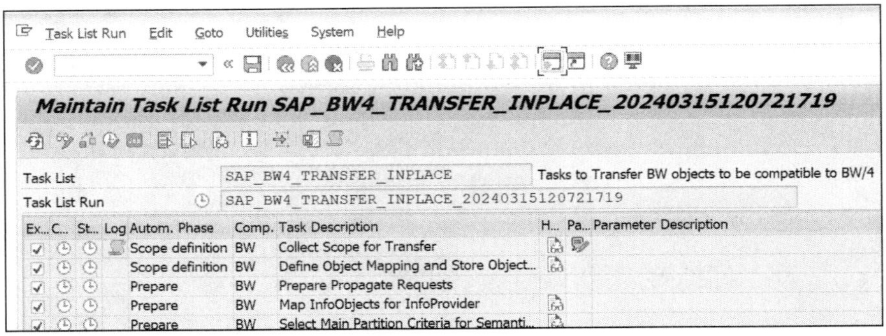

Figure 6.54 Task List Run for In-Place Conversion

You can first save this task list run and then later open it from Transaction STC02, as shown in Figure 6.55. The **Status** is shown as **Waiting to be Executed** because we haven't started the task yet.

Figure 6.55 Checking the Task List Run in Transaction STC02

6.12 Executing Scope Selection

In this section, we'll walk through the steps to execute in-place conversion in the sender system.

6.12.1 Collect Scope for Transfer

This task list run was created based the data flow, so it will have the data flow objects by default in the parameters. You need to first get into the edit mode of the task list run using , as shown in Figure 6.56.

When you click on the icon in edit mode for the **Collect Scope for Transfer** row, you'll be presented with the window for selecting the start object, which is prepopulated based on the selected data flow. If you navigate to the **Requests** button, it will activate the **Optimized Request Transfer** popup for enhanced processing of request transfers. You'll get a popup containing the SAP Note related to this feature. Click **OK** to enable the requests, as shown in Figure 6.57.

6.12 Executing Scope Selection

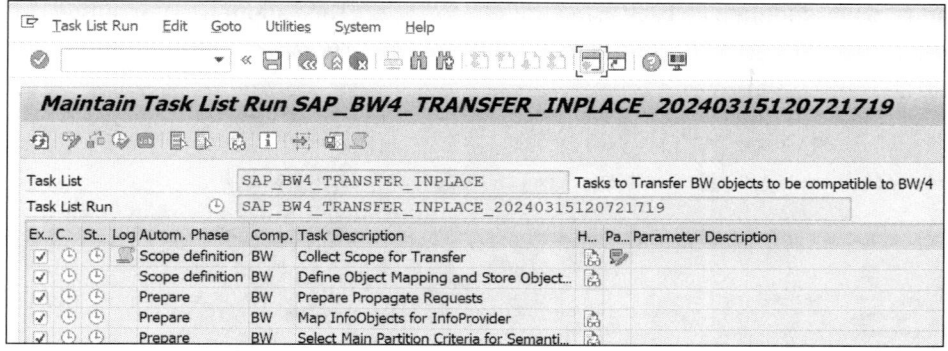

Figure 6.56 Scope Transfer: In-Place Conversion

Figure 6.57 Start Objects and Optimized Request Selection

Once you choose **OK**, the status of the selected start object can be seen in the rows with their corresponding object type, as shown in Figure 6.58.

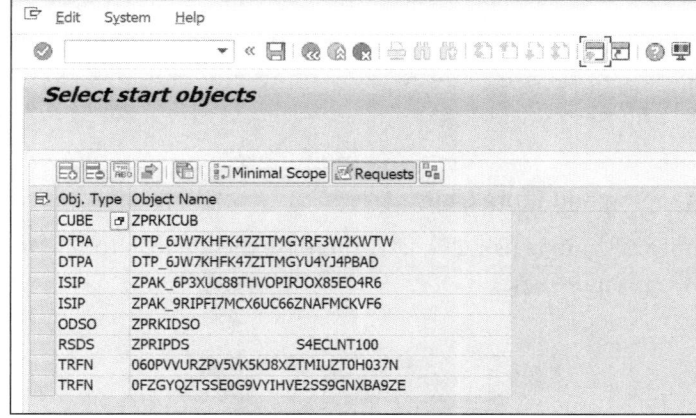

Figure 6.58 Selected Start Objects List

Choose **Save**, and then go back using the ⟲ icon to see the **Parameter Description** column with details of the start objects, as shown in Figure 6.59.

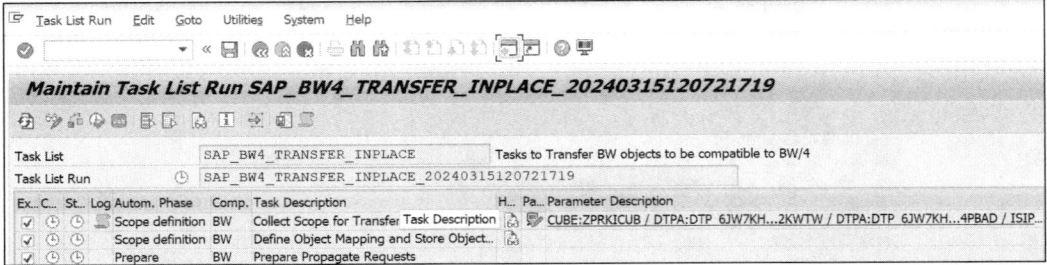

Figure 6.59 Scope Transfer: Before Execution

Save the task list and execute using 🔧, and the **Collect Scope for Transfer** will start running, indicated by the ⚙ icon in Figure 6.60.

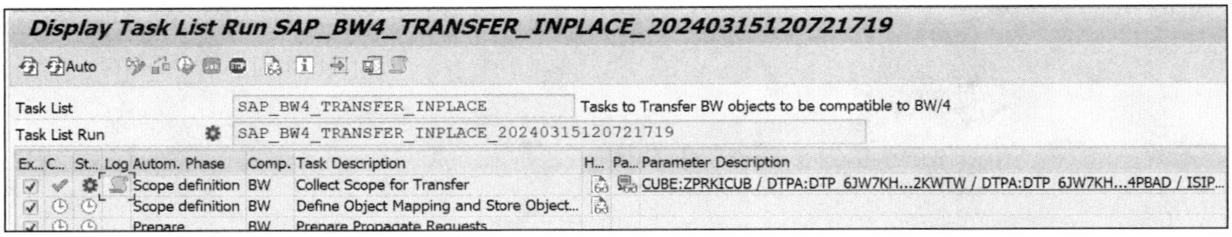

Figure 6.60 Scope Transfer: Running Status

Once the scope transfer is completed, it will have the green status, and the next activity, **Define Object Mapping and Store Object**, will be in red stop status, as shown in Figure 6.61.

Figure 6.61 Scope Transfer: Final Status

6.12.2 Define Object Mapping and Store Object List

To start this activity, get into edit mode using ✎, and then use the 🔍 icon to see the following options, as shown in Figure 6.62:

❶ Original TLOGO

❷ Original Object Name

❸ New TLOGO

❹ New Object name

6.12 Executing Scope Selection

❺ **Transfer Object**
❻ **Special Options**
❼ **MetaData Information**
❽ **Change Layout** icon
❾ **Save** icon

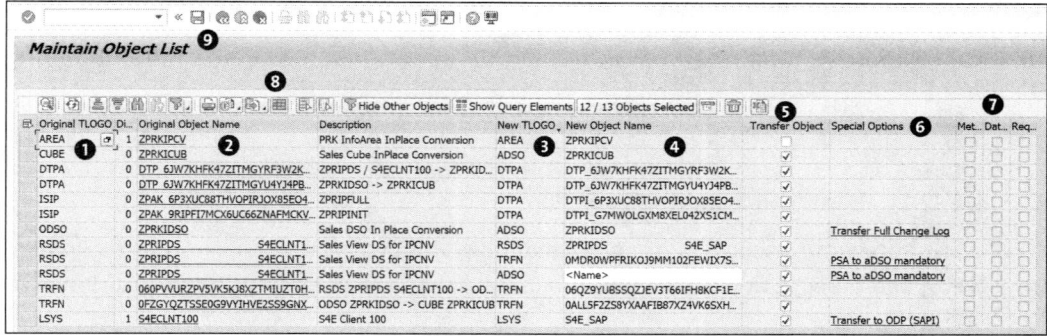

Figure 6.62 Object Mapping: Maintain Object List

Now, give the new ADSO name in the highlighted row where it's blank in Figure 6.62; for this demo, enter the name as "ZPRIBADSO", as shown in Figure 6.63.

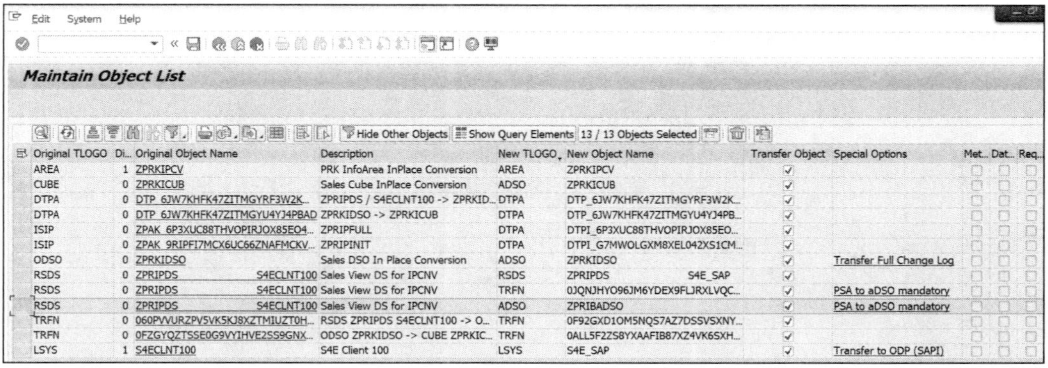

Figure 6.63 New Inbound ADSO Name

Once the object mapping is executed, it will have the ⓘ status, as shown in Figure 6.64.

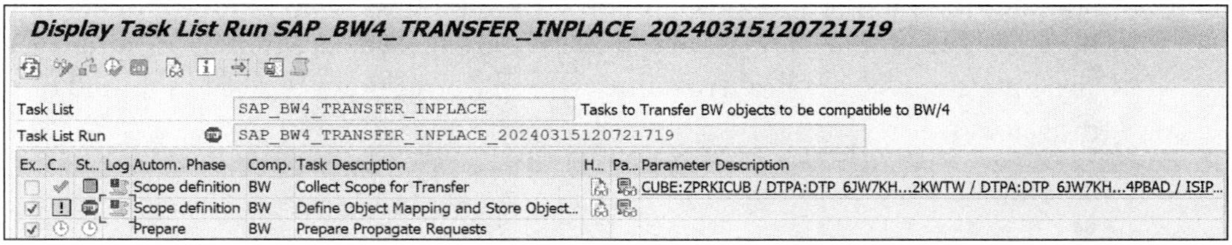

Figure 6.64 Object Mapping: Initial Status of Execution

6 In-Place Conversion

Now execute the object mapping again, and the overall status will turn green, as shown in Figure 6.65.

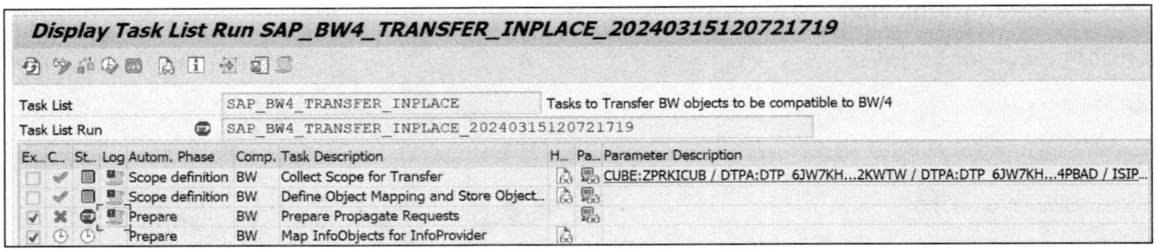

Figure 6.65 Object Mapping: Final Status

6.12.3 Prepare Propagate Requests

The successful execution of object mapping will take you to the activity to **Prepare Propagate Request**. Get into edit mode, and open the parameter screen, as shown in Figure 6.66.

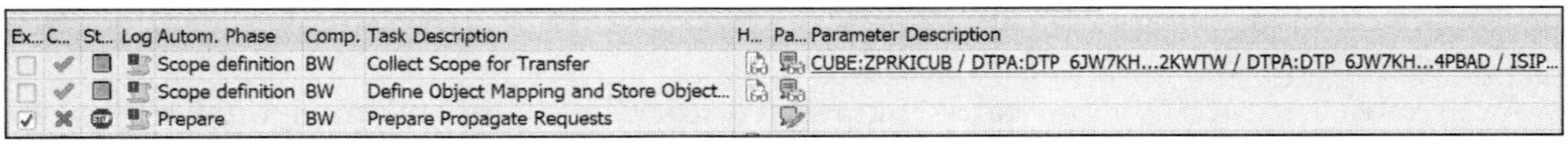

Figure 6.66 Prepare Propagate Request

In this step, the tool will list the DTPs that will send the data to the target InfoProviders. Validate the list of DTPs, and then choose **Set 'Confirm Execution' for Selected DTP(s)**, as shown in Figure 6.67. This will add the DTP to the list of propagation.

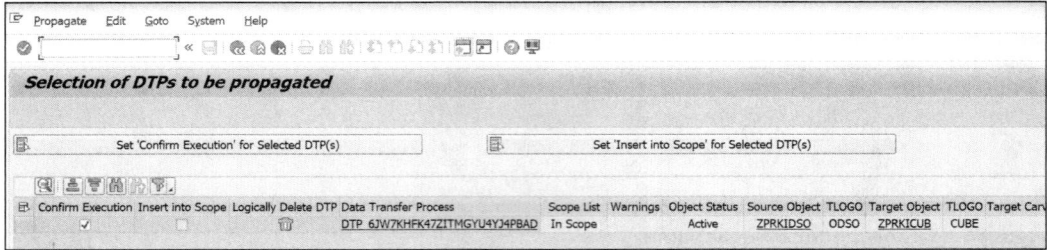

Figure 6.67 Confirm the DTPs to be Propagated

Once you choose **Confirm**, the tool does some checks and displays a popup saying **DTP maintenance completed**. Choose **Yes** (see Figure 6.68). If you see the **DTP maintenance is incomplete** message instead, then there is some error that you need to fix and confirm again.

6.12 Executing Scope Selection

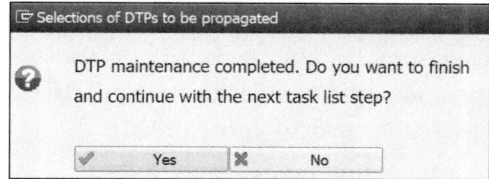

Figure 6.68 DTP Maintenance Confirmation

Choosing **Yes** will take you to the task list screen shown in Figure 6.69. The activity is ready to be executed.

Figure 6.69 Prepare Propagate Request: Execution

When you execute the **Prepare Propagate Request** using the [Job] icon, the tool will initiate the job. If no errors occur, the subsequent activity will proceed automatically. In our scenario, multiple activities have started and completed, as shown in Figure 6.70.

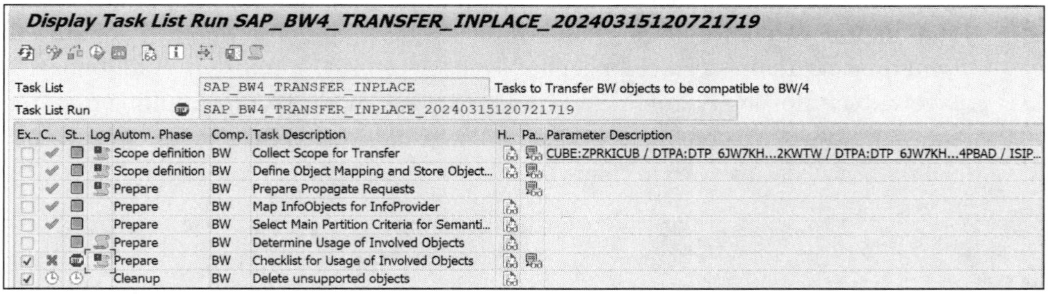

Figure 6.70 Execution Status of Multiple Activities

The details of the completed activity such as **Map InfoObjects for InfoProvider**, **Select Main Partition Criteria for Semantically Partitioned Objects**, **Determine Usage of Involved Objects**, and **Checklist for Usage of Involved Objects** are explained in the following subsections.

323

6 In-Place Conversion

6.12.4 Map InfoObjects for InfoProviders

This task lets you map InfoObjects for moving from a MultiProvider to a CompositeProvider. This mapping is important for the CompositeProvider to have consistent navigation attributes. Each attribute needs a matching map to work correctly.

6.12.5 Select Main Partition Criteria for Semantically Partitioned Objects

In this task, you can choose the main partition InfoObject for the new ADSO in the semantic group you're creating. Because semantic groups can only have one partition InfoObject, it's important to make this selection. However, the filters from other partition InfoObjects are still included in the corresponding ADSO as filter criteria.

6.12.6 Determine Usage of Involved Objects

This task examines the objects generated for InfoProviders and conducts scans that involve routines, planning functions, and more. It aims to identify any problematic code that could potentially lead to errors during the transfer process. Scrutinizing these elements helps to ensure a smoother transition and minimizes the risk of encountering issues post-transfer.

6.12.7 Checklist for Usage of Involved Objects

In this task, a comprehensive list of objects or findings is presented for review and resolution. It encompasses various aspects such as the usage of generated objects and InfoProviders that may no longer be available after the transfer. Additionally, it involves scrutinizing customer code to identify any modifications resulting from the transfer process. Execute the **Checklist for Usage of Involved Objects** step, and the initial status will be as shown in Figure 6.71.

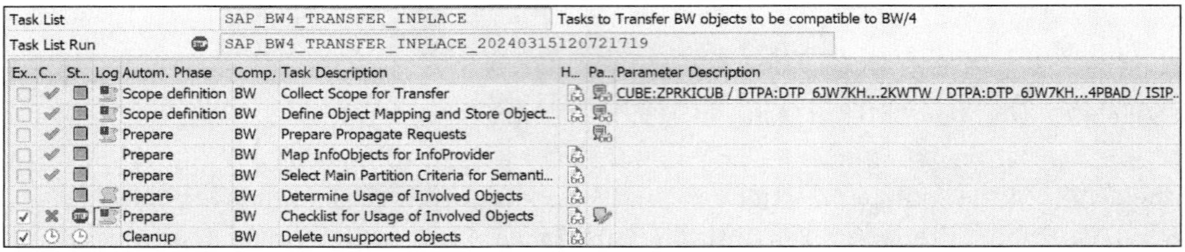

Figure 6.71 Checklist for Usage of Involved Objects

This lists the object type of BADI and AUTH and expects us to mention if the checks are resolved. You need to ensure that the **BADI** and **AUTH** and customer coding is already resolved during the prework; if so, the initial screen will be seen as shown in Figure 6.72.

6.12 Executing Scope Selection

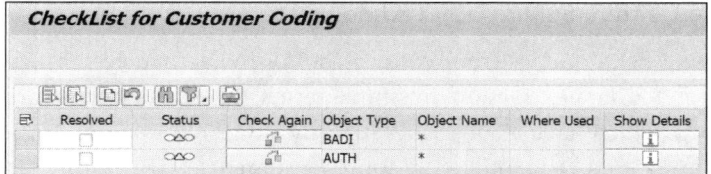

Figure 6.72 Customer Coding Checklist for BADI and AUTH

Choose the **Resolved** checkbox for **BADI** and **AUTH,** as shown in Figure 6.73.

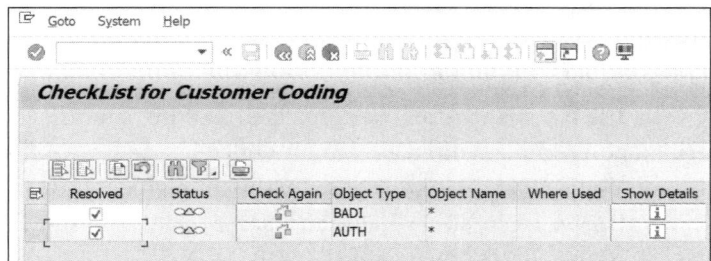

Figure 6.73 Confirmation of Customer Coding

Once the task list is seen, it's now ready to be executed, as shown in Figure 6.74.

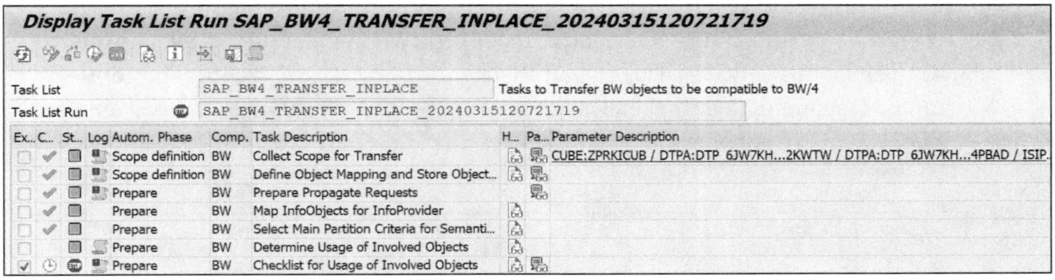

Figure 6.74 Checklist for Usage of Involved Objects: Pre-execution

The status of the execution is shown in Figure 6.75.

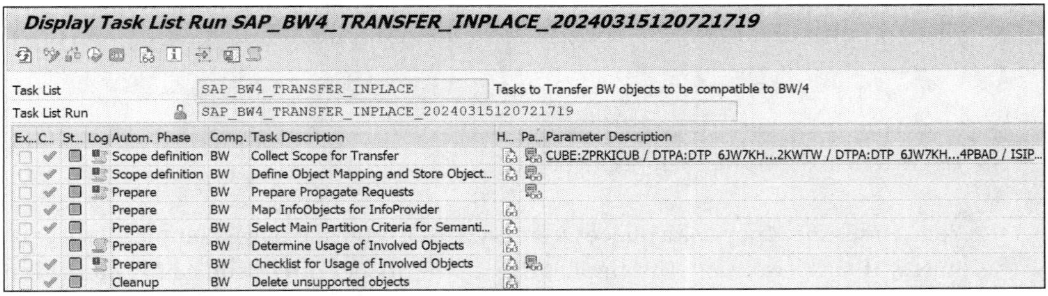

Figure 6.75 Final Status of Checklist of Involved Objects

325

6.12.8 Copy Delta Queues from SAPI to ODP Technology

This task involves copying delta queues from connected SAP BW source systems to switch to ODP technology for data extraction. In simple terms, ODP delta queues are renamed as subscriptions, which have a subscriber (usually a SAP BW system) and a subscriber process (e.g., an SAP BW DataSource and source system or a DTP). During the transfer process, if the original object is an InfoPackage for a SAPI source system, its SAPI delta queue is duplicated as an ODP subscription. If it's an InfoPackage for an ODP source system, the ODP subscription for the DataSource/Logsys target process is copied into a similar ODP subscription for the DTP target process.

Before executing the next step, check Transaction ODQMON (Monitor Delta Queues) in the sender system. DataSource **ZPRIPDS** is now available as shown in Figure 6.76.

To ensure compatibility with SAP BW/4HANA, all the connected SAP BW source systems must have the specified notes implemented as recommended by the Note Analyzer (refer to SAP Collection Note 2383530). The scope for this implementation is Source_System_for_SAP_BW4HANA, so in all SAP ERP and SAP S/4HANA systems, make sure that all the involved DataSources are exposed to ODP.

> **Downtime Requirement**
>
> Given the involvement of the SAP ERP system and the cloning of SAPI delta queues to ODP, questions may arise regarding the necessity of downtime in SAP ERP. Generally, there isn't a requirement for absolute SAP ERP downtime during this step as the tool can handle this very well. However, if there are specific business criticalities or uncertainties regarding this step, individual scope-based functional modules may require a brief downtime or nonposting period.

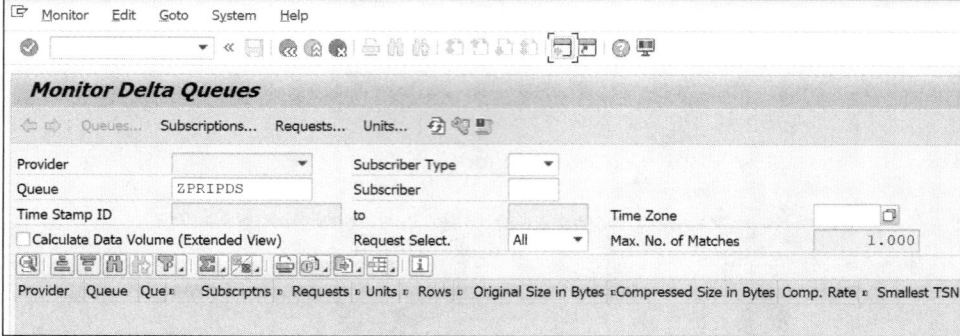

Figure 6.76 Delta Queue: Before Execution

You can see that **Copy delta queues from SAPI to ODP technology** is completed successfully with just a warning. This implies that all the queues can be copied to Transaction ODQMON (see Figure 6.77).

6.12   Executing Scope Selection

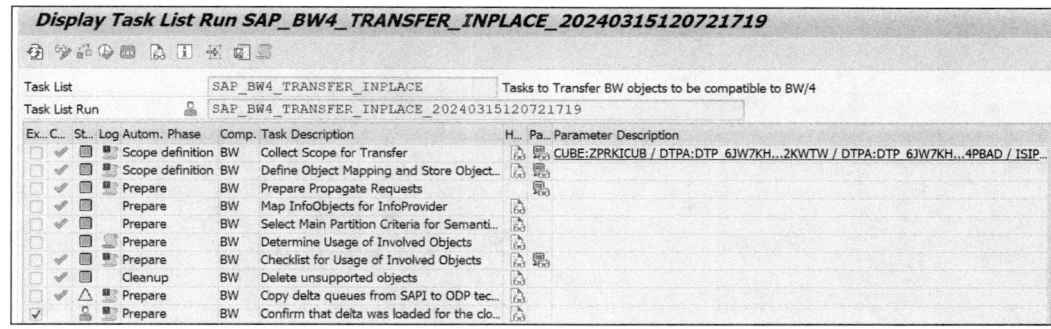

Figure 6.77 Copy Delta Queue Status: After Execution

In Transaction ODQMON, if you refresh the queues now, DataSource **ZPRIPDS** can be seen with the provider's name as **DATASOURCE_MODEL** and queue name as **ZPRIPDS**. The status is active with **1** active subscription, as shown in Figure 6.78.

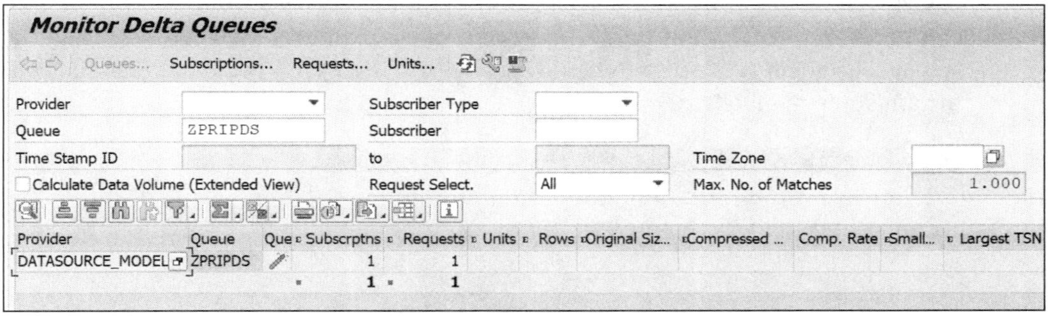

Figure 6.78 Transaction ODQMON: Copied Delta Queue Status

If you double-click on the queue row, it will take you to the **Subscriptions** view where you can see the inbound subscription with **DTPI** and the transaction sequence number details, as shown in Figure 6.79.

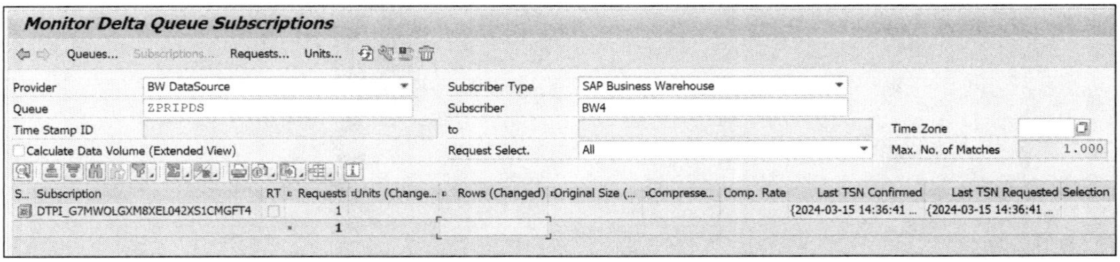

Figure 6.79 Transaction ODQMON: Subscription for Queue

Selecting the subscription **DTPI*** will navigate you to the **Requests** view where you can see the **Composite Request** labeled as **DELTA_CLONE**, indicating its creation during this process. The **Extraction Mode** is displayed as **Delta Initialization**, sourced from SAP, and

327

you can also view the generated background job based on the **ODQR*** listed in the same row, as shown in Figure 6.80.

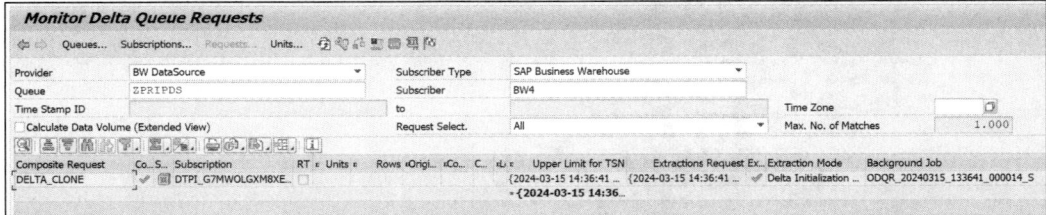

Figure 6.80 Transaction ODQMON Requests: Composite Request for DELTA_CLONE

6.12.9 Confirm Delta Was Loaded for the Cloned DataSources

Before starting this step, you need to make sure to execute the delta for the involved DataSource and ensure that the delta load is successful. Then, execute this manual step, and you'll see the popup shown in Figure 6.81. If delta was successful, then you need to choose **Confirm** based on the icon in edit mode (refer back to Figure 6.77), and the step will be green. If the delta is throwing an error, then you need to fix that and confirm this step manually.

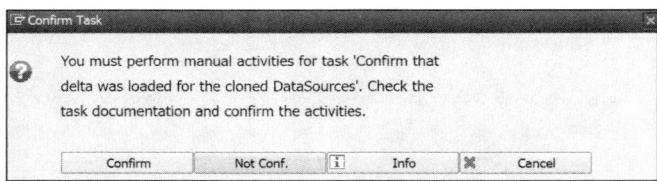

Figure 6.81 Confirmation of Manual Steps

6.12.10 Synchronize Delta Queues between SAPI and ODP

This task will allow the tool to synchronize the cloned delta queue in all the connected source systems. You need to ensure that you've executed the delta load successfully before this step. The status before delta execution is shown in Figure 6.82.

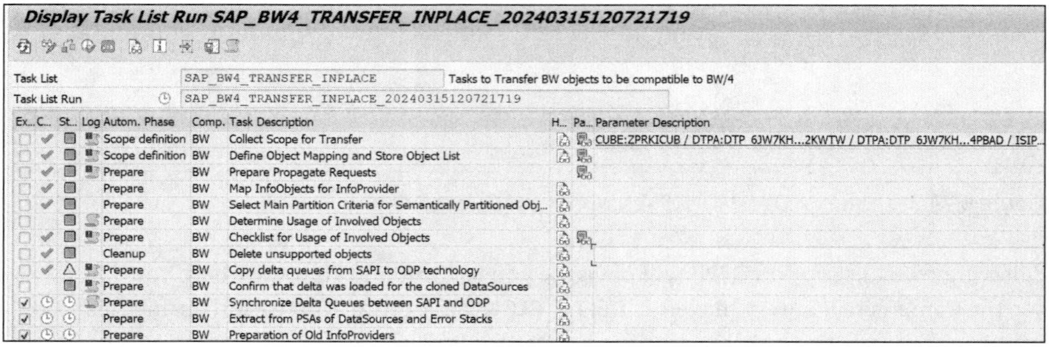

Figure 6.82 Synchronize Delta Queue: Before Execution

Now, execute the delta load for the InfoPackage that is involved. If the delta is completed, execute this step using ![Job]; if the status is green, log on to Transaction ODQMON in the SAP ERP system, and you'll see the composite request with the name **DELTA_SYNC**, as shown in Figure 6.83.

Figure 6.83 DELTA_SYNC in Transaction ODQMON

> **Delta Lock**
> Once this step is completed, you won't be able to execute the delta for the InfoPackage anymore; the delta execution will be locked from the completion of this step onward.

6.12.11 Extract from PSAs of DataSources and Error Stacks

This task involves preparing old InfoProviders in two important ways. First, it executes all delta DTPs to extract data from the PSA of a DataSource. This step is necessary when the PSA won't be transferred as a field-based ADSO. In such cases, all requests from the PSA must be moved into InfoProviders by delta DTPs. Second, it creates error DTPs if needed and then executes them for all error stacks that aren't empty. Even if the requests from error DTPs fail, the error stacks will be cleared. It's crucial to have empty error stacks because they aren't transferred. Now execute the **Extract from PSAs of DataSources and Error Stacks** task, and you can see the status as ⚙ (running), which means the activity is still in progress, in Figure 6.84.

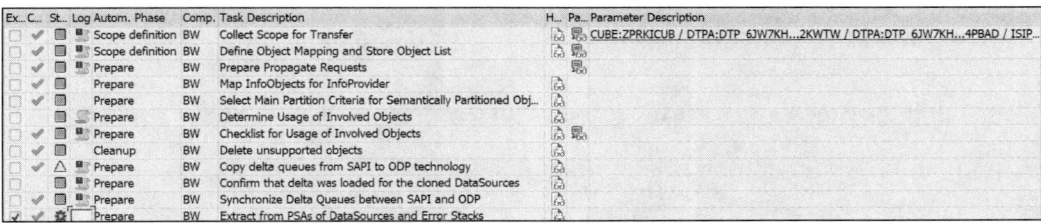

Figure 6.84 Extraction from PSA and Error Stack

Note that if there is no error, the next set of activities will be started by default, and the system will stop in lock step. Based on the message, you can release the transport request number or add the required objects, and then that step will be in green status. Let's see the details of those steps next.

6.12.12 Master Data Activation

Once the tool reaches this step, it will activate all the required master data. If there is any error in this step, then you need to fix the issues in specific InfoProviders and resume the activity using [Job].

6.12.13 Propagate Requests

This task handles automatic request propagation by collecting relevant DTPs from the scope list. It activates nonactivated requests of standard optimized DSOs (ODSOs) before initiating request propagation. It then automatically propagates requests by executing the relevant DTPs until no new data is available. Additionally, it conducts roll-up and compression of InfoCubes and carries out cleanup activities such as deleting temporarily created error DTPs.

6.12.14 Lock All Data Changes

This task involves locking all InfoProviders within the specified scope to prevent any further data alterations. Once locked, the data within these InfoProviders becomes static and is prepared for copying into the new ADSOs. This ensures that the information is stable and consistent before being transferred to the new data storage format. You can see that lock step isn't yet executed in Figure 6.85.

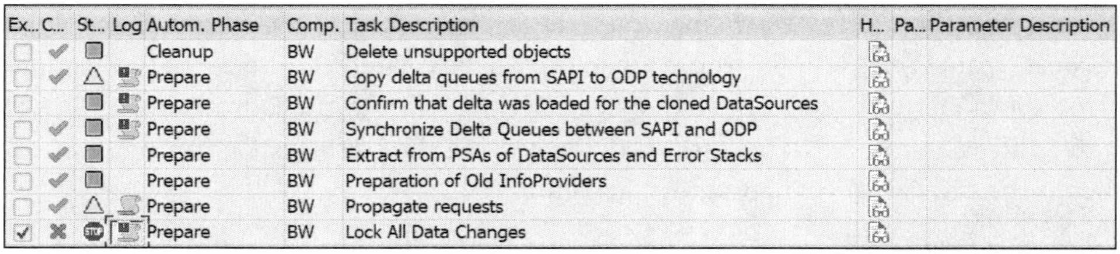

Ex...	C...	St...	Log	Autom. Phase	Comp.	Task Description	H...	Pa...	Parameter Description
☐	✓	☐		Cleanup	BW	Delete unsupported objects			
☐	✓	△	▪	Prepare	BW	Copy delta queues from SAPI to ODP technology			
☐		☐	▪	Prepare	BW	Confirm that delta was loaded for the cloned DataSources			
☐	✓	☐	▪	Prepare	BW	Synchronize Delta Queues between SAPI and ODP			
☐	✓	☐		Prepare	BW	Extract from PSAs of DataSources and Error Stacks			
☐	✓	☐		Prepare	BW	Preparation of Old InfoProviders			
☐	✓	△		Prepare	BW	Propagate requests			
✓	✗		▪	Prepare	BW	Lock All Data Changes			

Figure 6.85 Lock All Data Changes

After the execution, you can see the status, as shown in Figure 6.86.

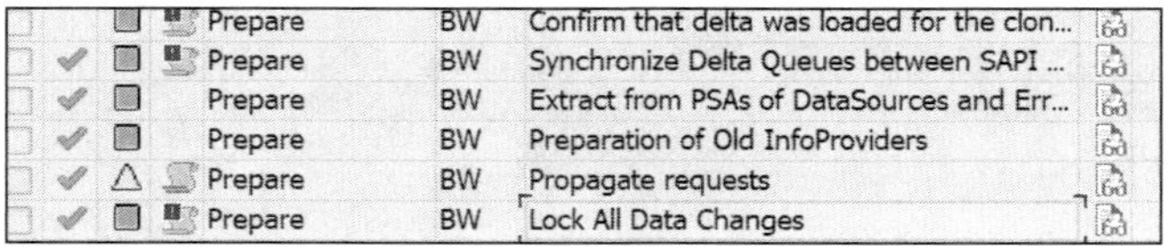

				Prepare	BW	Confirm that delta was loaded for the clon...			
	✓	☐	▪	Prepare	BW	Synchronize Delta Queues between SAPI ...			
	✓	☐		Prepare	BW	Extract from PSAs of DataSources and Err...			
	✓	☐		Prepare	BW	Preparation of Old InfoProviders			
	✓	△		Prepare	BW	Propagate requests			
	✓	☐	▪	Prepare	BW	Lock All Data Changes			

Figure 6.86 Lock All Data Changes: Status

6.12.15 Prepare Request Mapping for Transfer

In this task, the mapping control table for Request GUID, Request SID, and Transaction Sequence Numbers (TSNs) across all request types in an InfoProvider are populated, facilitating effective request tracking and management. For each InfoProvider and DataSource within scope, a parallel process initiates, identifying all requests associated with these objects and generating a dummy TSN for each. This mapping between old and new requests is then stored in the mapping control table. Typically, the TSN for most requests begins with 19000101000000 + (request SID as seconds), ensuring clear differentiation from requests created natively in the new request status and process management (RSPM). This distinction can be observed in the ADSO request management interface post-execution. In the scope transfer, enabling optimized request transfer will be optimal in execution.

6.12.16 Save Data

This task ensures the preservation of data from old InfoProviders by renaming the corresponding database tables. This precautionary measure prevents data loss prior to the subsequent deletion of the InfoProviders.

6.12.17 Transfer Metadata

In this task, the metadata of SAP BW/4HANA incompatible InfoProviders is transferred to compatible InfoProviders. Additionally, it involves adapting dependent objects such as transformations and DTPs to ensure compatibility with the new InfoProviders. This process ensures a smooth transition from outdated InfoProviders to compatible ones, preserving data integrity and facilitating seamless operations within the SAP BW/4HANA environment. Upon starting task using [Job], you can see the status as ⚙ (running) in Figure 6.87.

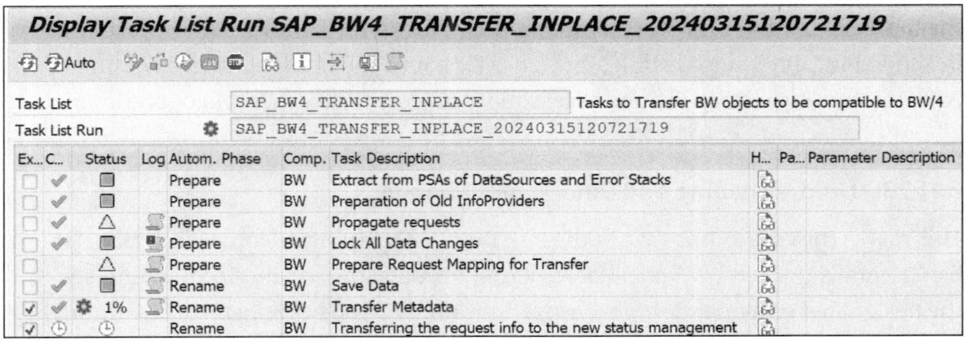

Ex...	C...	Status	Log	Autom. Phase	Comp.	Task Description	H...	Pa... Parameter Description
☐	✓	■		Prepare	BW	Extract from PSAs of DataSources and Error Stacks	📖	
☐	✓	■		Prepare	BW	Preparation of Old InfoProviders	📖	
☐	✓	△		Prepare	BW	Propagate requests	📖	
☐	✓	■		Prepare	BW	Lock All Data Changes	📖	
☐		△		Prepare	BW	Prepare Request Mapping for Transfer	📖	
☐	✓	■		Rename	BW	Save Data	📖	
✓	✓	⚙ 1%		Rename	BW	Transfer Metadata	📖	
✓	⏱	⏱		Rename	BW	Transferring the request info to the new status management	📖	

Figure 6.87 Transfer Metadata Execution

6.12.18 Transferring the Request Info to the New Status Management

The **Transferring the Request Info to the New Status Management** task transfers information from the old status management system to the new one. This information is crucial for all following steps and allows for the smooth administration of InfoProviders and InfoObjects, as shown in Figure 6.88. You can see the status as ✱ (running), or still in progress. Once its completed, then the tool can transfer the request information from the SAP BW 7.5–based request management tables into SAP BW/4HANA request management tables

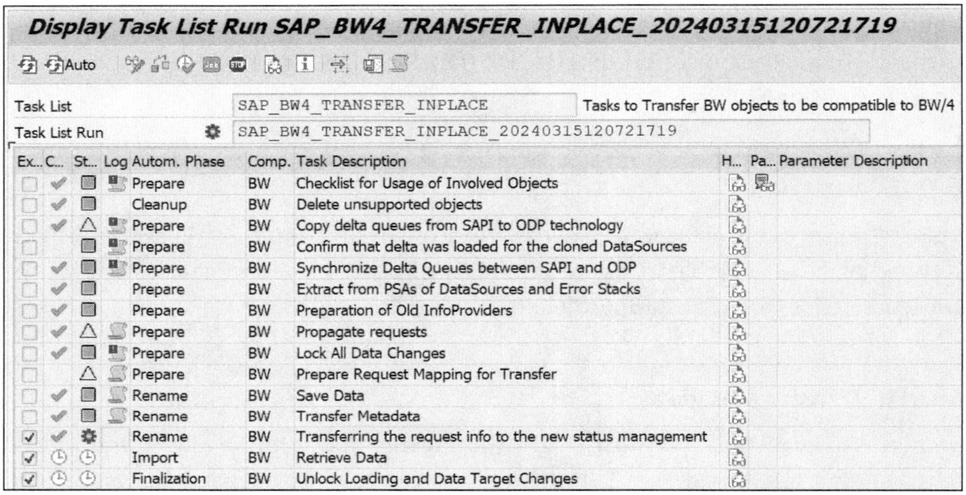

Figure 6.88 Transferring the Request Information to the New Status Management

6.12.19 Retrieve Data

In this task, the data previously saved from old InfoProviders, safeguarded by renaming their corresponding database tables, is retrieved. This data is then transferred into the newly created ADSOs. By completing this process, the historical information stored in the old InfoProviders seamlessly transitions into the updated data structure of the ADSOs, ensuring continuity and accessibility for future analyses and operations.

6.12.20 Unlock Loading and Data Target Changes

This marks the concluding task in the SAP BW/4HANA in-place conversion transfer task list. It entails the removal of all locks placed on DataSources and InfoProviders within the designated scope. By doing so, operations on these objects can resume without hindrance. Whether it involves the original objects or their corresponding new counterparts in the case of an in-place transfer, this step ensures the seamless continuation of activities on these essential data structures within the SAP BW/4HANA environment.

6.12 Executing Scope Selection

Figure 6.89 shows the phase finalization, and you can see the **Task Description** as **Unlock Loading and Data Target Changes**.

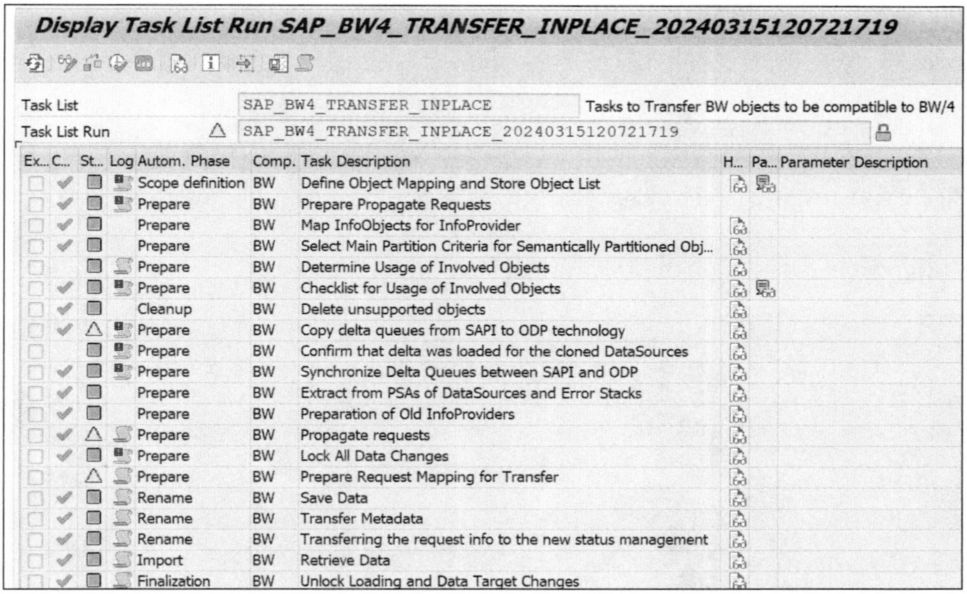

Figure 6.89 Overall In-Place Conversion Status

6.12.21 Post Transfer Checks

You have the option to access the converted ADSO **ZPRKIPV** within the SAP BW system and examine its data flow. The traditional objects, InfoCubes and ODSOs, have been replaced by the ADSO. This transition represents a significant evolution in data management within the SAP BW system, offering enhanced functionalities and capabilities for data storage and processing based on SAP BW/4HANA. You can see that new inbound ADSO (**ZPRIBADSO**) that is created for the DataSource-based PSA table in SAP BW 7.5, as shown in Figure 6.90.

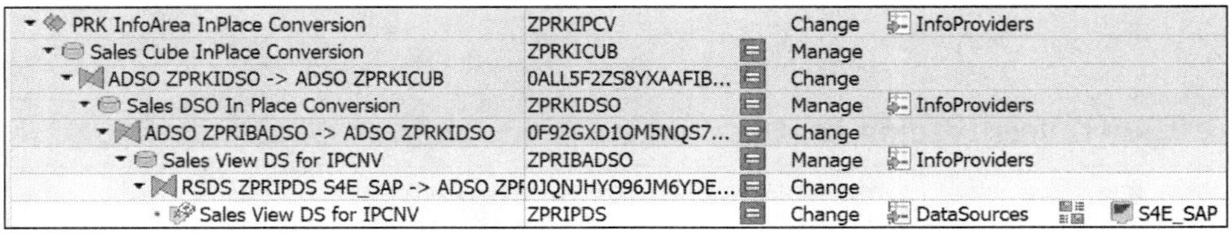

Figure 6.90 After In-Place Scope Transfer

You can open the SAP BW modeling tools to check the status of the ADSO, as shown in Figure 6.91.

6 In-Place Conversion

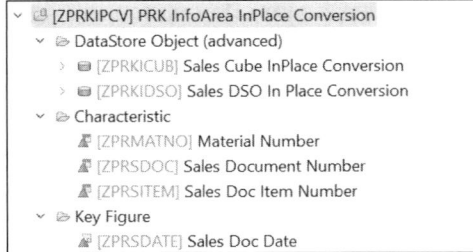

Figure 6.91 Updated InfoProviders

The updated data flow in the SAP BW GUI is shown in Figure 6.92.

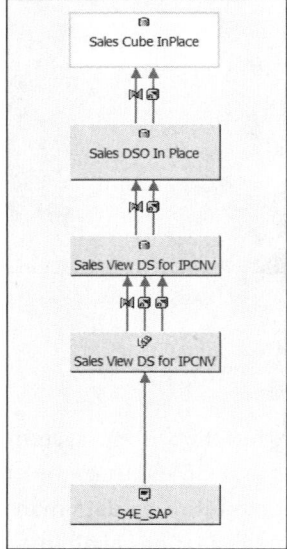

Figure 6.92 New Data Flow Based on ADSOs

In your SAP BW system, you can see new tables are created for InfoCube to ADSO. As shown in Figure 6.93, per the SAP BW/4HANA framework, the five standard tables are generated for the ADSO.

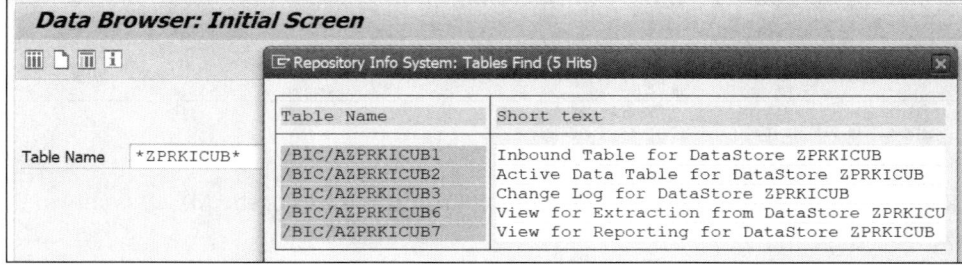

Figure 6.93 Generated ADSO Tables

6.12 Executing Scope Selection

Select the **Active Data Table for DataStore ZPRKICUB** (the converted ADSO) called **/BIC/ AZPRKICUB2** to view its contents, and you'll see the data stored in the active table, as shown in Figure 6.94.

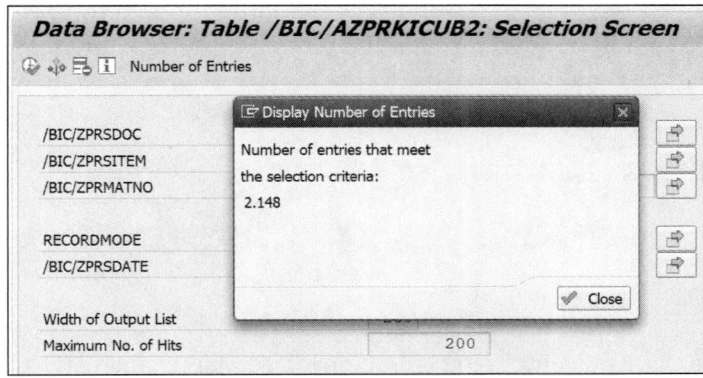

Figure 6.94 Check the Records in the ADSO Active Table

If you have the old request management details as a screenshot, you can conduct a thorough comparison of request management before and after conversion. For instance, to showcase our demo, you can examine how the conversion from an InfoCube (**ZPRKICUB**) to an ADSO (**ZPRKICUB**) impacts request management. This analysis provides valuable insights into the changes and improvements brought about by the conversion process; notice that the number of records (**2148**) didn't change, as shown in Figure 6.95.

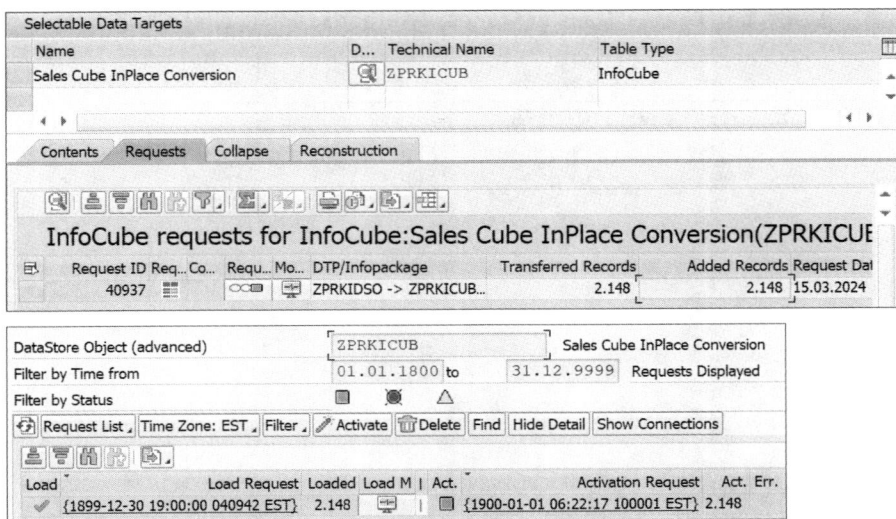

Figure 6.95 Request Management: InfoCube to ADSO before and after Scope Transfer

Similarly, check for the DSO (**ZPRKIDSO**) to ADSO comparison, as shown in Figure 6.96.

335

6 In-Place Conversion

Figure 6.96 Request Management DSO to ADSO

In large projects, ensure that you collect scope for all involved objects and perform scope transfer accordingly. It's crucial to maintain a properly ordered list of all scope transfers for execution in subsequent cycles. Additionally, make sure to gather all correction transports and forward them to the next system to maintain consistency and accuracy across cycles.

6.13 Transfer Standard Authorization (Delta)

Once the scope transfer is completed, you can complete the transfer of standard authorization in delta mode. To do so, open the tool in delta mode via Transaction RSB4HCONV. Navigate to **Realization Phase • Transfer Standard Authorization (Delta)**. Choose the run ID, and you'll see the option to execute the delta run, as shown in Figure 6.97.

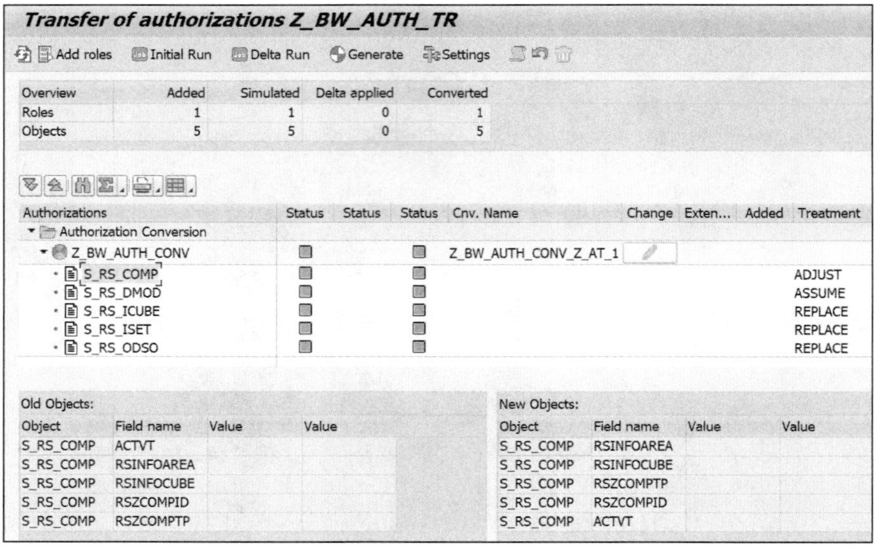

Figure 6.97 Transfer Run: Delta Status

336

6.14 Objects Deletion

After the execution, you can see the **Delta Status** is green, as shown in Figure 6.98. Choose **Generate**.

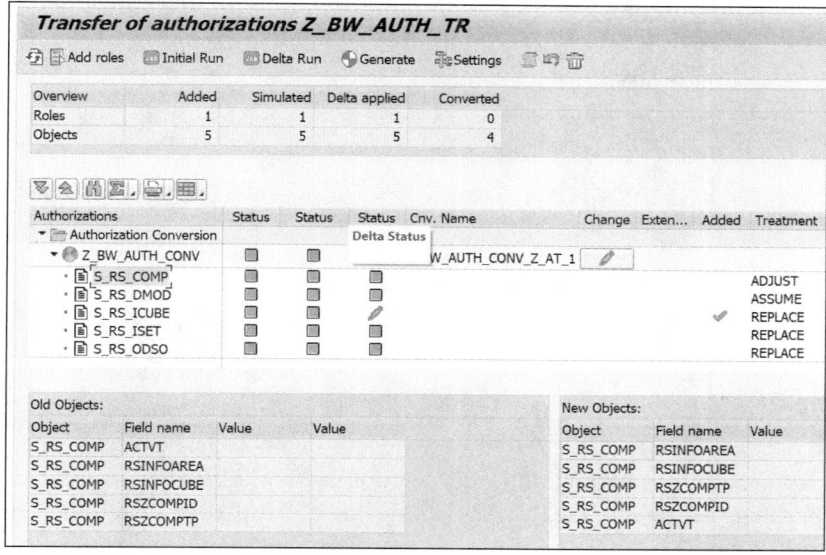

Figure 6.98 Delta Run Status

To maintain the BW4 Transfer Tool statistics, you can use Transaction SE38 and execute report RSB4H_STAT_MAINTIAN, The data is stored in table RSB4H_STAT. Using this report, you can move the statistics or delete the statistics data, as shown in Figure 6.99.

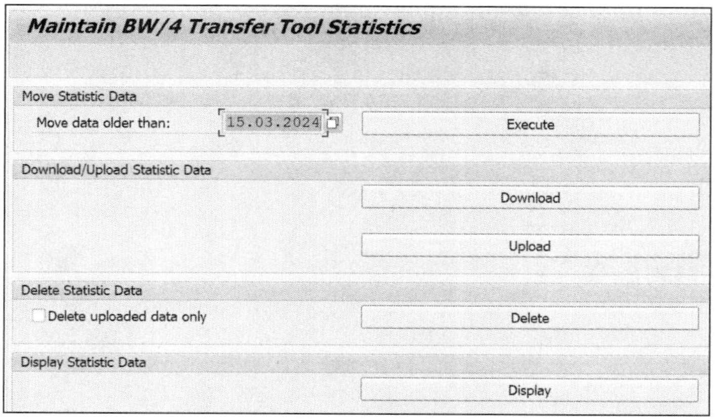

Figure 6.99 Maintain Statistics

6.14 Objects Deletion

Prior to the system conversion, it's essential to remove any unnecessary objects. We'll now explore the process of deleting these objects.

6 In-Place Conversion

6.14.1 Technical Content

To delete technical content, navigate to Transaction RSB4HCONV, and select **Realization Phase • Delete Technical Content** to arrive at the screen shown in Figure 6.100. If you need to mark InfoCubes, ODSOs, and MultiProviders for deletion, then click the **Include CUBE/ODSO/MPRO** checkbox.

Figure 6.100 Objects Deletion

6.14.2 Delete Other SAP Business Warehouse Objects

You can use report RS_DELETE_TLOGO to delete the required objects. Be careful when using this program because you can't undo any deletions, so make sure you really want to delete something before deleting. It's a good idea to back up your system before deleting a lot of things. Don't delete the Myself source system manually because the SAP BW system won't work without it. It will be deleted automatically when you switch to Ready for Conversion mode. When deleting objects in the test or production systems, you need to collect them into a transport request first. Some object collectors rely on the source system assignment for the user deleting them. To fix this, go to **Data Warehousing Workbench • Transport Connection • Edit • Source System Assignment**. Make sure to implement SAP Note 2705147 to get the latest version of program RS_DELETE_TLOGO and support deletion of all object types, as shown in Figure 6.101.

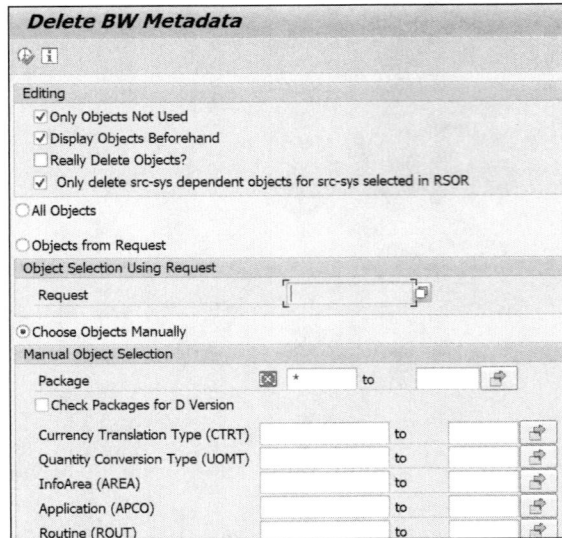

Figure 6.101 SAP BW Object Deletion Report and Selection Screen

6.14.3 Delete SAP Business Warehouse Queries

Now you can delete queries using SAP standard report RSZDELETE, as shown in Figure 6.102. You can do a search based on the **InfoProvider** or **Technical Name**.

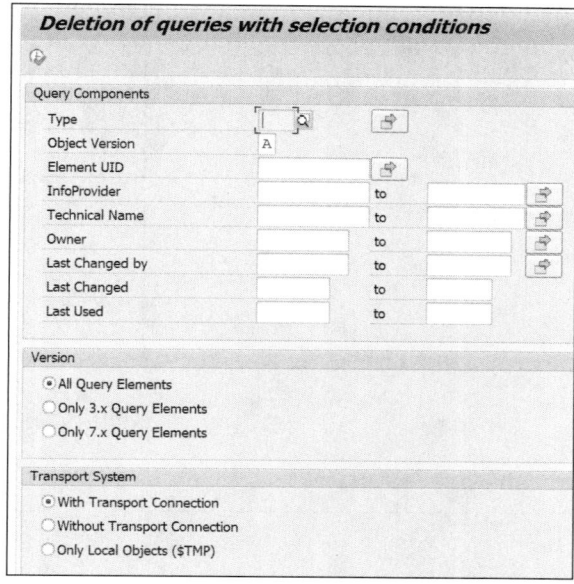

Figure 6.102 Query Deletion Report and Selection Screen

6.15 Final System Conversion

Before making the conversion, ensure that all prerequisites are met and any unsupported add-ons for SAP BW/4HANA are removed. Check for add-ons such as BPC, SEM-BCS, and so on as the system needs to be upgraded to SAP Business Planning and Consolidation for SAP S/4HANA. Check for other third-party add-ons too. You can use Transaction RSB4HCONV, which has the system change mode options shown in Figure 6.103.

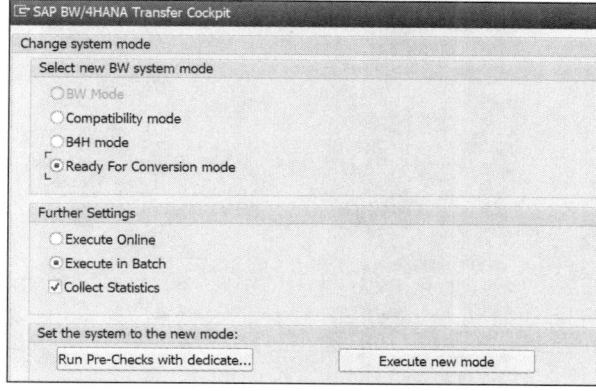

Figure 6.103 SAP BW System Mode Options

6　In-Place Conversion

Select **Ready For Conversion mode**, and the follow these steps:

1. Download SAP BW/4HANA via maintenance planner, as shown in Figure 6.104.

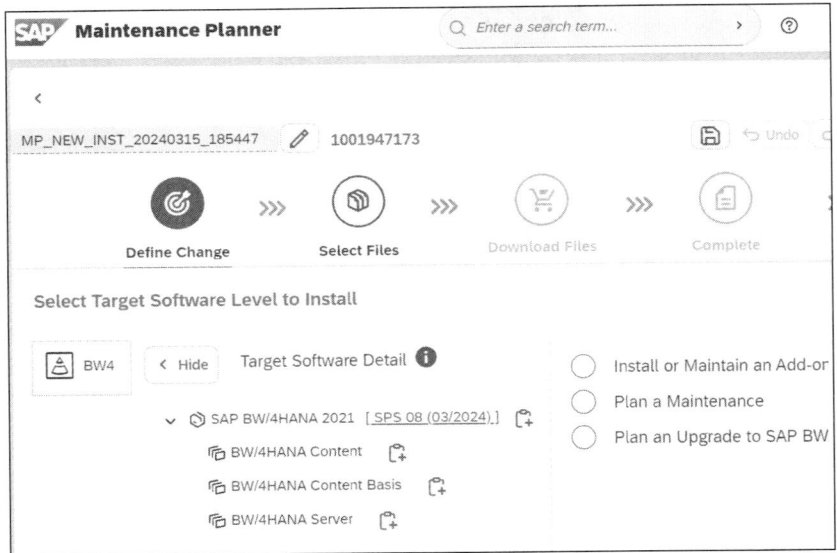

Figure 6.104　Maintenance Planner

2. Follow the wizard to choose your SAP BW/4HANA version and associated content, and generate the stack XML, as shown in Figure 6.105.

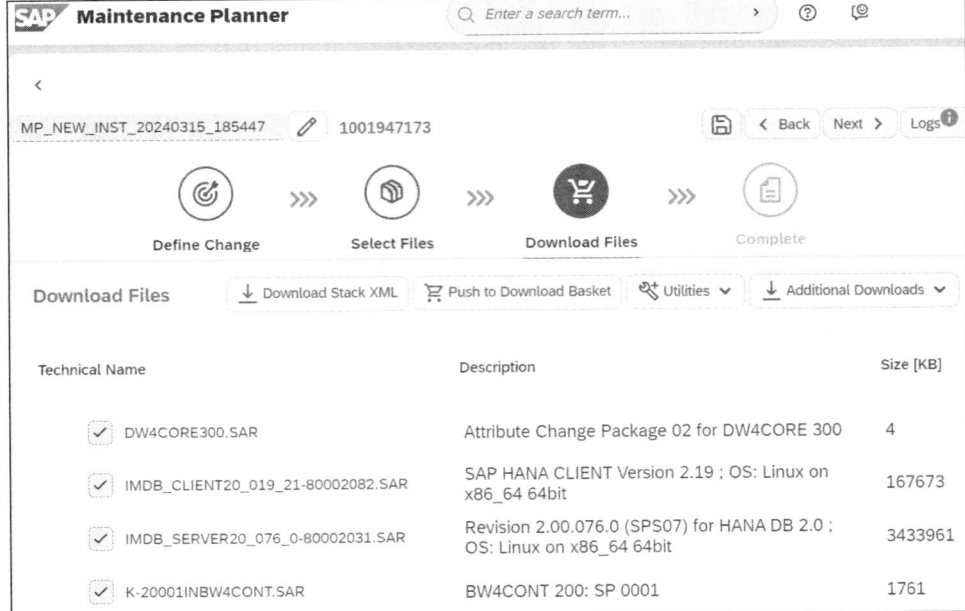

Figure 6.105　Stack XML

6.15 Final System Conversion

3. Perform the technical system conversion using Transaction SAINT.
4. Complete the following post-conversion activities:
   - After completing the system conversion to SAP BW/4HANA, several follow-up activities are necessary before the system can be fully used. These activities are consolidated into a task list designed for straightforward configuration and automated execution. It's essential to ensure that the system conversion process has been successfully completed before proceeding with these tasks. Before this step, update the transfer tool with the latest XML, use Transaction STC01 and call SAP_BW4_AFTER_CONVERSION, and complete the task list as shown in Figure 6.106. You can start the activity by choosing ![Job], and if there are any errors, you can resume using the same. This task list will have the different phases such as configuration, initialization, and cleanup. The activity will start with the installation of essential content post-conversion to SAP BW/4HANA. Then, you'll have the execution of the Simplified Migration report (RSDG_IOBJ_IOBC_MIGRATE_AS_TASK). Then, the tool will execute report RSADMIN_TO_IMG_XPRA to transfer the **RSADMIN** parameter to Customizing. You can see the option to transfer SAP BW Workspace **RSDDTREXAMIN** to **SPRO** with standard report RSDDTREXADMIN_TO_RSWSPCUST. You'll have some configuration activities such as adjustment of DTPs and transfer of the source system of SAP HANA smart data access (SDA) to SAP HANA or big data. Finally, the tool will provide the **Migrate the Table Groups for Table Placement** task list time. If this task list is completed, you can ensure that the after-conversion activity is completed.

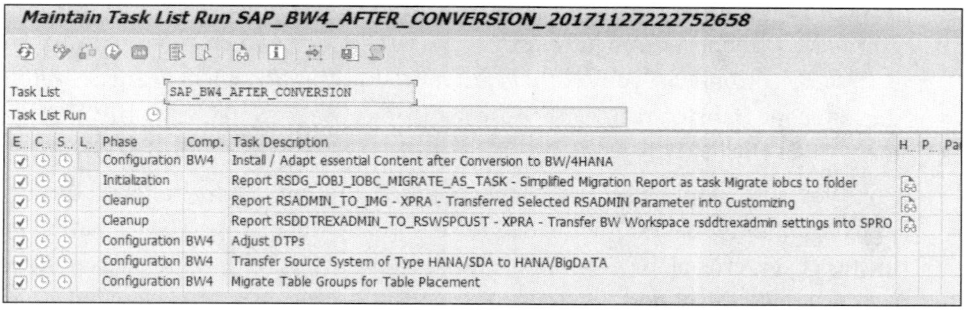

Figure 6.106 Post-Conversion Task List

   - Configure the launchpad with the task list SAP_FIORI_LAUNCHPAD_INIT_SETUP in Transaction STC01, and finish the configuration of the SAP BW/4HANA cockpit, as shown in Figure 6.107. You can see that the task list will have **Prepare**, **Configuration**, and **Postprocessing** phases. There will be a few manual parameters you can set using the ![icon] icon. This task list will enable you to create the Customizing request and local folder, activate the gateway OData and HTTP services, and set the profile and Internet Communication Manager (ICM) parameters. Finally, in the configuration, the task will add the launchpad URL to the favorites.

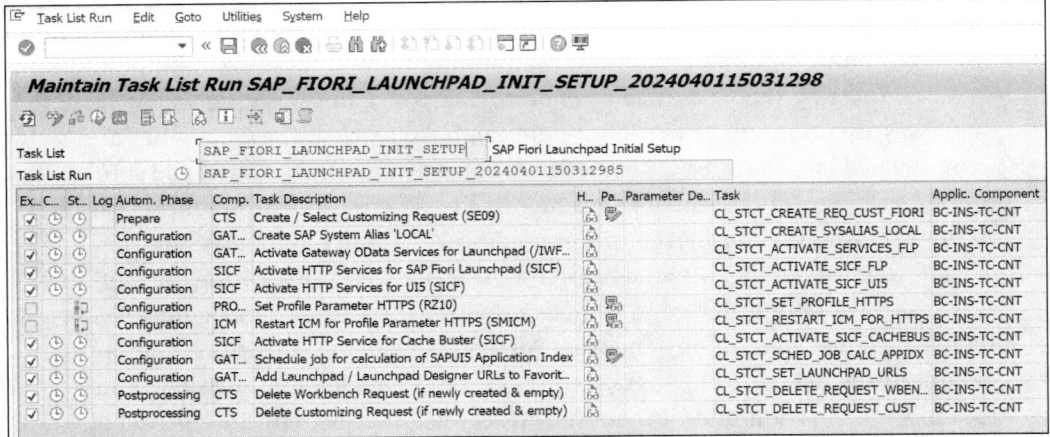

Figure 6.107 SAP Fiori Launchpad Initialization Setup

– Finally, adjust the profiles and roles to have any standard authorization in SAP BW/4HANA. Note that with SAP BW/4HANA, certain types of SAP BW queries will no longer be available. When transitioning to B4H mode, the system will automatically convert queries from SAP BW 3.x to SAP BW 7.x format. Inconsistent queries may need manual adjustment or could be deleted automatically. Queries referencing a Request ID (OREQUID) must be adjusted manually after conversion. Queries using audit characteristics and navigational attributes on DSOs or InfoCubes require manual adjustment to use attributes of the new request TSNs. Queries without a MultiProvider will be converted to corresponding queries on a CompositeProvider. Configuration or tools referencing such queries must be updated accordingly. If you see that certain DTPs with error stacks are deactivated during scope transfer, reactivate those using the post-conversion task list that we mentioned already in Figure 6.106.

– Once you complete the system conversion in the initial system, follow the same steps in subsequent systems such as development and quality. The execution of the task list is crucial in every system due to potential variations in local configurations and the likelihood of direct object creation within each system, whether it's in a testing or production environment. After the completion of a scope transfer, the newly acquired objects become accessible in SAP BW and can be managed like any other objects. This grants users the flexibility to modify and transport them across different systems. However, objects included in the scope selection and already associated with an ongoing transport request can't be transferred directly. In such instances, it's necessary to release these transports and then import them into the relevant test and production systems.

Object transfer is conducted based on the scope selection, necessitating a phased approach, especially in the rollout of SAP HANA–optimized scenarios within the SAP BW production environment. As each scenario becomes optimized and compatible with SAP BW/4HANA, and once the SAP BW system operates fully in B4H mode, the final step involves system conversion to complete the transition to SAP BW/4HANA and officially initiate the go-live process with the updated system configuration.

> **Useful Report**
>
> Report RS_B4HANA_INPLACE_CLEANUP can be used when the new objects have been deleted manually. The execution of this report will reset the control tables so that the transfer can be repeated.
>
> For the other standard reports that can be used in an SAP BW 7.5 system, refer to Chapter 4, Section 4.11.

6.16 Summary

With the completion of this chapter, you're now ready to execute the in-place conversion for SAP BW/4HANA. You know the activities to be done in both the prepare phase and realization phase of the conversion. You're now aware of the task lists and modes that are involved in the in-place conversion. Having learned all of these details will enable you to start the in-place conversion for your specific use case. By competing this chapter, you've learned all the options that are available to convert to SAP BW/4HANA. In the next chapter, we'll focus on the conversion of SAP BW or SAP BW/4HANA via SAP BW bridge.

Chapter 7
SAP BW Bridge for SAP Datasphere

This chapter outlines the detailed procedure for implementing SAP BW bridge shell conversion and SAP BW bridge remote conversion. You need to have read about the system preparation and prerequisites discussed in Chapter 3. Upon completing this chapter, you'll be proficient in executing SAP BW bridge–based shell and remote conversions and understand task lists and associated activities in the system landscape. You'll also have learned about SAP Datasphere and how SAP BW bridge is used for data provisioning.

SAP BW bridge, a feature enhancement of SAP Datasphere, offers SAP Business Warehouse (SAP BW) and SAP BW/4HANA users access to the public cloud. We'll start with the introduction of SAP Datasphere in Section 7.1. The introduction to SAP BW bridge is discussed in Section 7.2. We'll cover the SAP BW bridge shell conversion overview in Section 7.3. The prepare phase activities are covered in Section 7.3.1. Then, you'll learn about the system preparation in Section 7.3.2. The realization phase of SAP BW bridge for shell conversion execution is explained in Section 7.3.3. The step to connect the sender system to SAP BW bridge is explained in Section 7.3.4. Next, you'll learn about the creation of an SAP BW bridge project in SAP BW modeling tools in Section 7.3.5. You'll learn about the steps to create the ABAP cloud project in ABAP development tools in Section 7.3.6. We'll discuss the object simplification in Section 7.3.7, and then we'll focus on the prework for the SAP BW bridge–based conversion in Section 7.3.8. The shell conversion execution for SAP BW bridge is explained in Section 7.3.9. Then, you'll explore the remote conversion in Section 7.3.1 with the actual execution of the SAP BW bridge remote conversion explained in Section 7.4.2. Finally, you'll learn about the post-migration options in Section 7.4.3.

7.1 Introduction to SAP Datasphere

SAP Datasphere is a modern cloud-based data warehouse service provided as software as a service (SaaS) on the public cloud. At its core, SAP Datasphere offers a comprehensive suite of tools and functionalities designed to streamline and optimize data management processes within organizations. It operates on the principle of harmonization, enabling seamless integration and synchronization of data across the entire enterprise ecosystem. This includes a sophisticated array of services tailored to meet the diverse needs of modern businesses, spanning from robust data modeling capabilities to efficient data integration solutions. Moreover, SAP Datasphere extends its support beyond SAP-specific data, accommodating a wide range of DataSources, both SAP and non-SAP alike. This versatility ensures that organizations can leverage their existing data infrastructure while harnessing the power of SAP Datasphere to drive actionable insights and informed decision-making. Ultimately, SAP Datasphere serves as a cornerstone for businesses seeking to establish a unified and agile data warehouse architecture, empowering them to unlock the full potential of their data assets in today's digital landscape. With SAP Datasphere, businesses can effortlessly handle data integration, cataloging, semantic modeling, data warehousing, and workload virtualization across both SAP and non-SAP datasets. As a cornerstone of unified data warehouse architecture, SAP Datasphere empowers organizations to unlock their data assets' full potential in today's digital landscape. You can refer to the SAP Datasphere architecture in Figure 7.1.

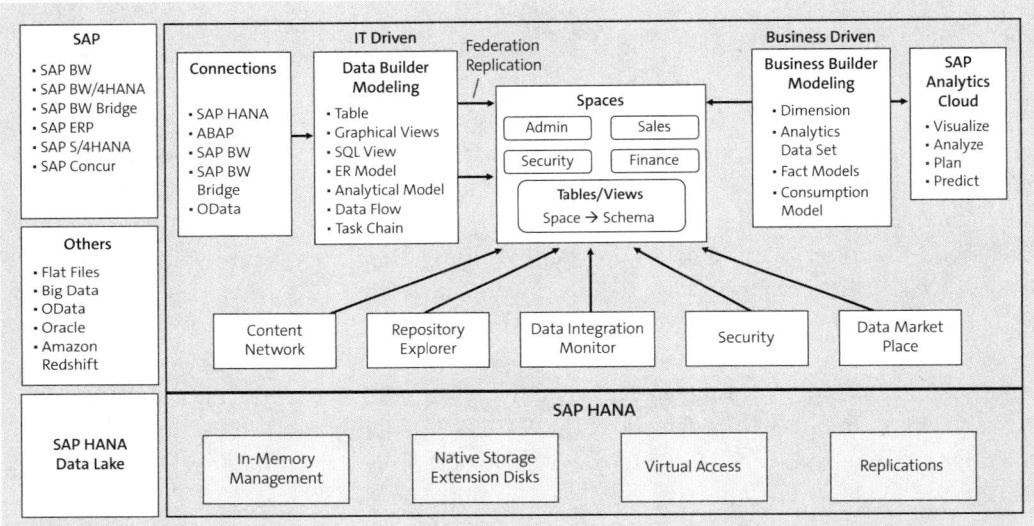

Figure 7.1 SAP Datasphere Architecture

Based on Figure 7.1, you can see that SAP Datasphere comprises key components such as the following:

- **Space management**
  Customize spaces for data organization based on the business needs.
- **Data integration and flow**
  Access and prepare data from various sources (SAP, non-SAP, on-premise, and cloud).
- **Data Builder**
  Define and update data models efficiently using the technical drivel tools.
- **Business Builder**
  Model scenarios for easy analysis by business users.

- **Administration and security**
  Manage connectivity settings and access control.
- **Semantic onboarding**
  Import semantically rich data objects.
- **Consumption**
  Connect and consume data models easily.

The detailed process of modeling in SAP Datasphere isn't the focus of this book; instead, we'll concentrate on how to perform a conversion to SAP BW bridge from SAP BW 7.x or SAP BW/4HANA.

7.2 Introduction to SAP BW Bridge

SAP's introduction of SAP Datasphere represents a strategic move toward offering cloud-based data solutions, catering to the evolving needs of businesses in the digital age. This initiative is particularly significant for SAP BW and SAP BW/4HANA users, providing them with a clear pathway to transition from traditional on-premises systems to a more adaptable and scalable cloud environment.

One of the primary objectives of SAP Datasphere is to enable organizations to retain the value of their investments in SAP BW while facilitating a seamless migration process to the cloud. By transitioning to SAP Datasphere, businesses can take advantage of cloud-native features and capabilities, ensuring they remain competitive and responsive to changing market dynamics.

A pivotal development in this journey was the launch of SAP BW bridge in November 2021. Positioned as an extension of SAP Datasphere, SAP BW bridge serves as a bridge between legacy SAP BW environments and the cloud-native capabilities of SAP Datasphere. It offers organizations a streamlined approach to migrate critical components such as data models, sources, transformations, and data transfer processes (DTPs) from their existing SAP BW landscapes to the SAP BW bridge environment.

One of the key benefits of SAP BW bridge is its capability to integrate these migrated components as remote tables within SAP Datasphere. This integration provides organizations with access to a robust self-service data and analytics platform hosted in the cloud. By leveraging remote tables, business users can seamlessly access and analyze data without the need for extensive IT intervention, empowering them to make informed decisions and derive valuable insights from their data.

Furthermore, SAP BW bridge extends the capabilities of SAP BW directly into the public cloud, offering connectivity and prebuilt business content for SAP BW–based data integration. This includes extractors from SAP ERP and SAP S/4HANA (the latest generation of SAP's ERP suite), enabling organizations to integrate data from their SAP systems seamlessly. Additionally, SAP BW bridge provides SAP BW functionalities for loading data into the cloud environment. This includes features such as partitioning, monitoring, and error handling, all tailored to meet the specific needs of organizations migrating to the cloud.

In essence, the integration of SAP BW bridge with SAP Datasphere represents a significant step forward for businesses seeking to unlock the full potential of their data assets. It not only accelerates the transition process from on-premise to cloud-based data solutions but also ensures that organizations can leverage their existing investments in SAP BW while embracing the agility and scalability offered by the cloud. By empowering business users with intuitive tools and functionalities, SAP BW bridge and SAP Datasphere pave the way for a data-driven future, where insights can be derived swiftly and decisively to drive business success. You can see the connectivity between SAP BW bridge and SAP Datasphere in Figure 7.2.

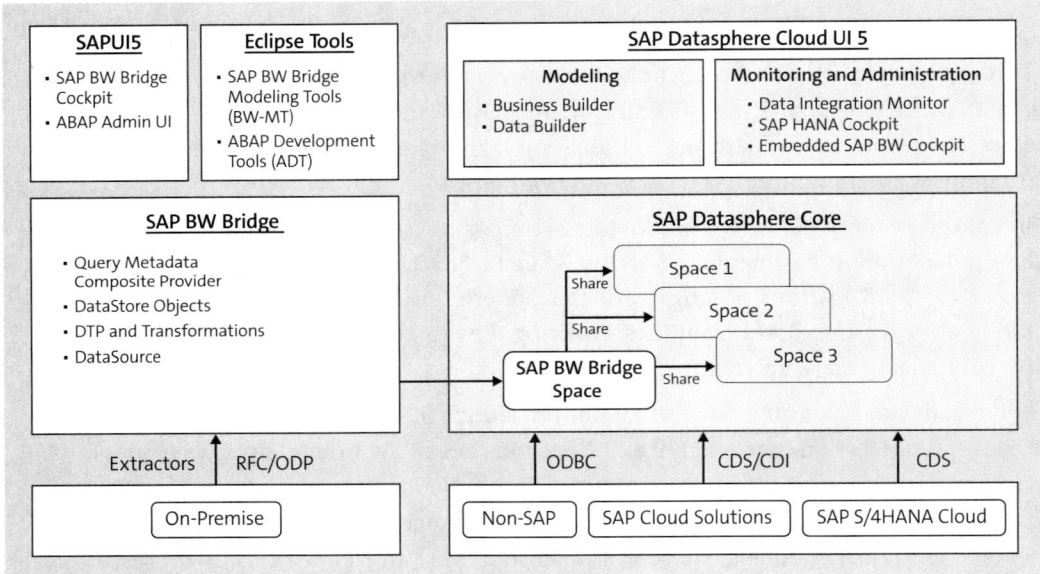

Figure 7.2 SAP BW Bridge and SAP Datasphere

SAP BW bridge introduces a streamlined approach to accessing its functionalities through two key tools: Eclipse for development and SAP BW bridge cockpit for monitoring and administration. Within Eclipse, developers can now leverage a new project type known as SAP BW bridge, specifically tailored to facilitate seamless connectivity with the SAP BW bridge environment. This integration is facilitated using service keys generated within the SAP Datasphere application, ensuring a smooth and efficient development experience.

The SAP BW bridge cockpit serves as a central hub for monitoring and managing various administrative tasks associated with SAP BW bridge. From this intuitive interface, users can oversee the execution of critical processes such as process chains and other essential activities. By providing real-time insights into system performance and status updates, the cockpit empowers administrators to maintain optimal efficiency and address any potential issues promptly. This comprehensive monitoring and administration capability further enhances the overall functionality and reliability of the SAP BW bridge ecosystem, ensuring seamless operation and maximum productivity for businesses embracing this innovative solution.

The functionality supported by SAP BW bridge offers a comprehensive suite of features essential for efficient data management. Through the SAP BW modeling tools, users can establish connections with SAP on-premise source systems using operational data provisioning (ODP) in various contexts, including ODP for SAP BW, SAP extractors, ABAP CDS views, and SAP Landscape Transformation. This connectivity enables seamless data extraction and integration from diverse sources, empowering users to access and use data from within the SAP ecosystem.

Data flow components such as DataStore objects (DSOs), transformations, and DTPs can be created and configured using the SAP BW modeling tools. These tools provide a user-friendly interface for designing and implementing data flows within the environment. Process chains, vital for orchestrating data loading and transformation tasks, can be developed and scheduled within the SAP BW bridge environment. The SAP BW bridge cockpit, directly accessible from the Data Integration Monitor in SAP Datasphere, facilitates the modeling and management of these process chains. Effective data management is facilitated through the ability to monitor and oversee data operations within InfoObjects and DSOs, ensuring data quality and integrity throughout the process. SAP BW bridge enables the seamless integration of its objects into the broader SAP Datasphere environment. DSOs and CompositeProviders can be imported into the SAP BW bridge space within SAP Datasphere, allowing for the creation of new data models that integrate diverse DataSources or associate master data views across different spaces. Furthermore, users can leverage a tailored version of the SAP BW/4HANA content add-on, optimized for use with SAP BW bridge. This business content streamlines the creation of data models, enhancing productivity and efficiency within the environment.

Before starting the SAP BW bridge conversion, you need to be aware of its limitations, SAP BW bridge within SAP Datasphere has limitations in application development, connectivity, and functionality. It lacks support for app building within SAP Datasphere and doesn't allow direct connection via SAP GUI. Connectivity options are limited to ODP source systems and push scenarios only, with queries not supported. Functionalities relying on the online analytical processing (OLAP) engine, such as analysis authorizations, are unavailable. Additionally, SAP BW bridge doesn't support planning capabilities, certain add-ons, and specific process types. It doesn't support the generation of external SAP HANA views nor allow SAP HANA calculation views as PartProviders of the CompositeProvider, other limitations include no support for cold store and DTO handling for DSOs, temporal joins for CompositeProviders, and variables as selection criteria for DTPs. Lastly, process chains in streaming mode aren't currently supported. Keep in mind that the listed limitations are current as of March 2024, and with the constant development in SAP Datasphere, this limitation may be drastically reduced in the near future. It's best practice to recheck the limitations before the start of your conversion project, Standard SAP Help will constantly be updated with the latest information to which you can refer.

7.3 Shell Conversion with SAP BW Bridge

Instead of undergoing a fresh installation within SAP Datasphere, SAP offers shell conversion for transitioning from SAP BW or SAP BW/4HANA to SAP Datasphere via the SAP BW bridge. This process involves using the SAP BW bridge cockpit to seamlessly migrate chosen data models into SAP Datasphere, leveraging the functionalities provided by the SAP BW bridge.

Shell conversion is compatible with SAP BW systems operating on SAP NetWeaver versions 7.30 to 7.51 (running on SAP HANA or AnyDB), as well as SAP BW/4HANA 2021. As you might be aware by now, the shell conversion doesn't facilitate the transfer and synchronization of existing datasets. Instead, users have the option to either load data from the original sources or load data from the SAP BW or SAP BW/4HANA system used for metadata transfer. Let's see the sequence of the shell conversion, as shown in Figure 7.3.

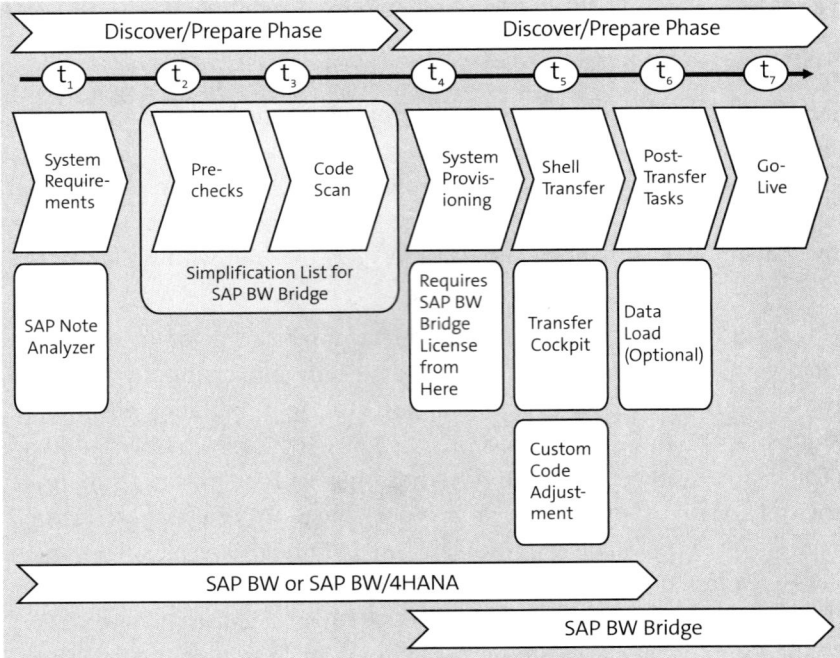

Figure 7.3 SAP BW Bridge Shell Conversion Sequence Flow

7.3.1 Prepare Phase Activities

In the prepare phase, you'll perform all the activities to set up the system for the conversion, the activities have been discussed in detail in Chapter 3. You need to ensure that all the preparation activities are completed before stating the conversion. The following sections provide the details of those activities.

Simplification List

As we've seen the simplification list in the earlier chapters, those simplifications still hold good for the SAP BW bridge conversion. There is a consolidation note on the simplification, which you can refer during the project (SAP Note 3154420 "Simplification List for SAP Datasphere, SAP BW Bridge").

Support Releases for the Sender System

We'll check the details regarding the supported releases now. Table 7.1 shows the prechecks for the sender system.

Scenario: Prechecks Release	Minimum Support Package	Recommended Support Package
SAP BW 7.3	08	08 or higher
SAP BW 7.31	05	05 or higher
SAP BW 7.4	09	09 or higher
SAP BW 7.5	05	05 or higher
SAP BW 7.51	08	08 or higher
SAP BW/4HANA 2021	00	00 or higher

Table 7.1 SAP BW Bridge: Prechecks Minimum Release

Table 7.2 shows the details of the sender system for shell conversion.

Scenario: Shell Conversion (Sender) Release	Minimum Support Package	Recommended Support Package
SAP BW 7.3	10	08 or higher
SAP BW 7.31	10	05 or higher
SAP BW 7.4	12	09 or higher
SAP BW 7.5	05	05 or higher
SAP BW 7.51	08	08 or higher
SAP BW/4HANA 2021	00	00 or higher

Table 7.2 SAP BW Bridge Shell Conversion: Sender System

Supported Releases for the Target System

You need to have the target system for SAP BW bridge in SAP Datasphere in a supported version; ensure that the release is 22 or higher for the target system.

SAP Source Systems

When establishing a connection between the source system and SAP Datasphere or SAP BW bridge, it's crucial to meet specific minimum release requirements for each source system type. Additionally, it's essential to ensure that ODP-based frameworks are supported for data extraction across all source systems. This ensures compatibility and seamless data integration between the source system and SAP Datasphere or SAP BW bridge. Let's walk through the supported ODP contexts and the supported releases:

- ODP-SAP

  ODP facilitates connections between SAP systems such as SAP ERP or SAP S/4HANA as the DataSource and SAP Datasphere, including SAP BW bridge. Communication occurs via Remote Function Call (RFC), with DataSources accessible through the ODP context for DataSources (extractors) in SAP. This setup enables seamless data integration and access within the SAP ecosystem, enhancing overall data management capabilities. It's essential to ensure that DataSources are released for data provisioning using ODP. For guidance on releasing SAP DataSources and customer-defined DataSources for ODP, as shown in Table 7.3.

Scenario: SAP Source System – SAP_ODP Release	Minimum Support Package	Recommended Support Package
SAP NetWeaver 7.0	24	Latest
SAP NetWeaver 7.0 EhP 1	09	Latest
SAP NetWeaver 7.0 EhP 2	08	Latest
SAP NetWeaver 7.3	08	Latest
SAP NetWeaver 7.3 EhP 1	05	Latest
SAP NetWeaver 7.4	02	Latest
SAP NetWeaver 7.5	00	Latest
SAP BW/4HANA 1.0	00	Latest
SAP BW/4HANA 2.0	00	Latest
SAP BW/4HANA 2021	00	Latest

Table 7.3 ODP-SAP

- **ODP-BW**

  The ODP framework enables leveraging the ODP source system technology to establish a data mart scenario connecting SAP BW or SAP BW/4HANA systems with SAP Datasphere and SAP BW bridge. See Table 7.4 for more details.

Scenario: SAP Source System – ODP-BW Release	Minimum Support Package	Recommended Support Package
SAP NetWeaver 7.0	36	Latest
SAP NetWeaver 7.0 EhP 1	19	Latest
SAP NetWeaver 7.0 EhP 2	19	Latest
SAP NetWeaver 7.3	17	Latest
SAP NetWeaver 7.3 EhP 1	20	Latest
SAP NetWeaver 7.4	05	Latest
SAP NetWeaver 7.5	00	Latest
SAP BW/4HANA 1.0	00	Latest
SAP BW/4HANA 2.0	00	Latest
SAP BW/4HANA 2021	00	Latest

Table 7.4 ODP-BW

- **ODP-CDS**

  The ODP context designated for ABAP CDS views (ABAP_CDS) serves the purpose of integrating ABAP CDS views along with their analytics annotations into SAP Datasphere and SAP BW bridge. This integration enables seamless access to ABAP CDS views originating from SAP S/4HANA and SAP S/4HANA Cloud. When the ABAP CDS view includes the necessary annotations, it becomes usable for both complete extraction and data access, as well as for delta extraction. See Table 7.5 for more details.

Scenario: SAP Source system – ODP-CDS Full Extraction Release	Minimum Support Package	Recommended Support Package
SAP NetWeaver 7.4	08	Latest
SAP NetWeaver 7.5	02	Latest
SAP BW/4HANA 1.0	00	Latest

Table 7.5 ODP-CDS

7 SAP BW Bridge for SAP Datasphere

Scenario: SAP Source system – ODP-CDS Full Extraction Release	Minimum Support Package	Recommended Support Package
SAP BW/4HANA 2.0	00	Latest
SAP BW/4HANA 2021	00	Latest
Delta Extraction		
SAP NetWeaver 7.5	05	Latest
SAP BW/4HANA 1.0	01	Latest
SAP BW/4HANA 2.0	00	Latest
SAP BW/4HANA 2021	00	Latest

Table 7.5 ODP-CDS (Cont.)

- ODP-SLT

    Through the use of ODP infrastructure- and trigger-based replication facilitated by the SAP Landscape Transformation Replication Server, data can be seamlessly transferred from SAP systems to the SAP BW bridge system in real time. Acting as a pivotal provider within this infrastructure, SAP Landscape Transformation Replication Server enables the exposure of tables from SAP sources as delta queues, effectively capturing changes as they occur. Subsequently, the SAP BW bridge system can subscribe to these delta queues, gaining access to the replicated data via the ODP infrastructure for further refinement and processing. This integration streamlines the flow of information, ensuring that critical data is readily available for analysis and decision-making purposes within the SAP BW bridge environment.

    Ensure compliance with the licensing conditions when using SAP Landscape Transformation Replication Server, which require specific minimum release versions, including the Data Migration Integration Server (DMIS) Add-On 2011 SP 05 and SAP NetWeaver versions 7.3 SPS 10, 7.31 SPS 09, or 7.4 SPS 04 with the ODP infrastructure. Additionally, the SAP source system must have DMIS Add-On 2011 SP 03/SP 04 or higher, or 2010 SP 08/SP 09 installed for compatibility and optimal functionality. This will change frequently, so refer to the SAP Landscape Transformation documentation for any changes in the SAP Landscape Transformation–based version updates.

- Other source systems

    Apart from the source system that we discussed earlier, the other source system isn't supported in SAP BW bridge. For example, source systems categorized as file or SAP HANA are unavailable within the SAP Datasphere and SAP BW bridge options.

SAP BW Bridge Sizing

The important step in the prepare phase to determine the sizing for SAP BW bridge is the same as we've covered for the SAP BW/4HANA sizing in Chapter 3, Section 3.7. You can get the Sizing report from SAP 2296290 "New Sizing Report for SAP BW/4HANA." Always use the latest version of this SAP Note for optimal sizing. You can use Transaction SE38 to call sizing report /SDF/HANA_BW_SIZING. In the screen, you need to choose the **Run SAP Datasphere BW Bridge Sizing** checkbox in the **SAP Datasphere BW Bridge Sizing** tab shown in Figure 7.4. You can choose the options for providing InfoProviders based on the input field; if you choose *, then all providers will be considered. Once this option is selected, the tool will perform the sizing for the SAP BW bridge scenario, which you can choose to execute in the background. The job will be created with name **/SDF/HANA_BW_SIZING**.

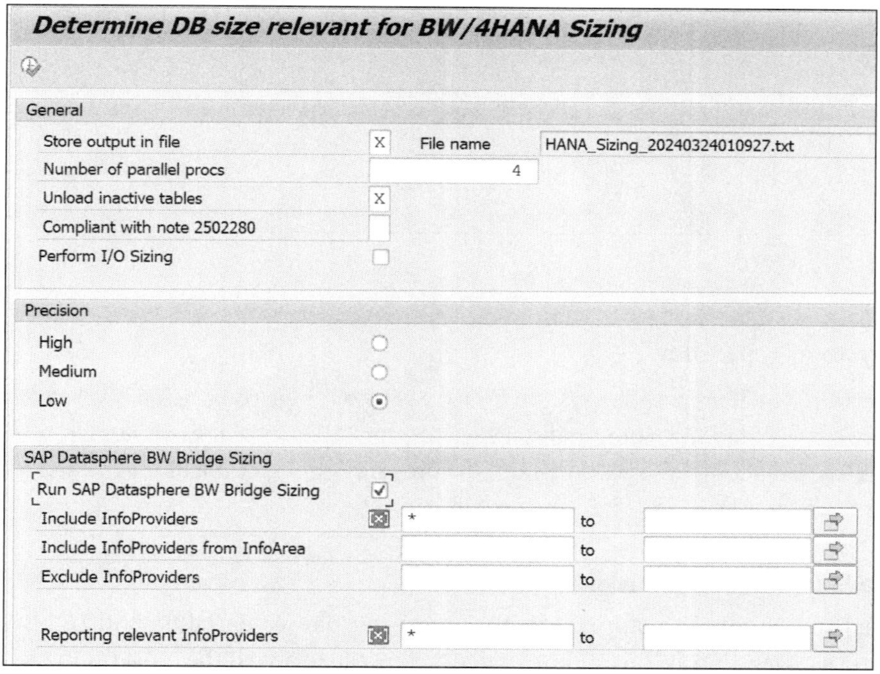

Figure 7.4 SAP BW Bridge Sizing

The result is shown in Figure 7.5.

> **Sizing in the Transfer Cockpit**
>
> If you've installed the transfer cockpit, then you can call the sizing report from Transaction RSB4HCONV. We'll discuss the steps to install the transfer cockpit in the next section.

7 SAP BW Bridge for SAP Datasphere

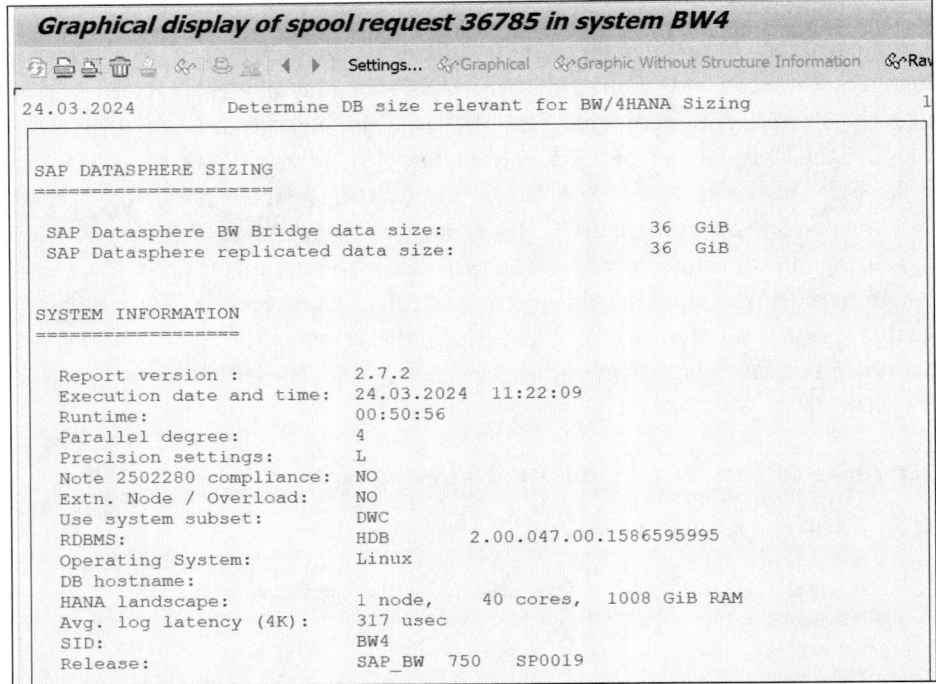

Figure 7.5 Sizing Result

7.3.2 System Preparation

In this section, you'll prepare the system for the SAP BW bridge conversion. We'll cover the steps to download the Note Analyzer files for the SAP BW bridge conversion, and, then you'll learn about the system landscape preparation which includes the sender system and source system preparation activities.

Downloading Note Analyzer Files

The initial step involves acquiring the Note Analyzer files essential for the SAP BW bridge process. Specifically, for SAP BW bridge, it's crucial to access the most up-to-date XML file provided within SAP Note 3141688. This note, tailored for the conversion process from SAP BW or SAP BW/4HANA to SAP Datasphere, offers valuable insights. There's an attachment within this note's contents called *SAP_Bridge_Transfer_Note_Analyzer_<Latest_Date>.XML*. This XML file serves as a pivotal component for the SAP BW bridge procedure and should be promptly downloaded to the user's workstation for further use and reference. Once the ZIP file is extracted, you'll see the files shown in Figure 7.6.

You need to download the Z_SAP_BW_NOTE_ANALYZER report shown in Figure 7.6 and create an ABAP report in all systems, including the source system. Once the report

7.3 Shell Conversion with SAP BW Bridge

is created, load the required XML file from the local workstation via the 📂 icon and then execute (Figure 7.7). The report will list all the SAP Notes with different statuses.

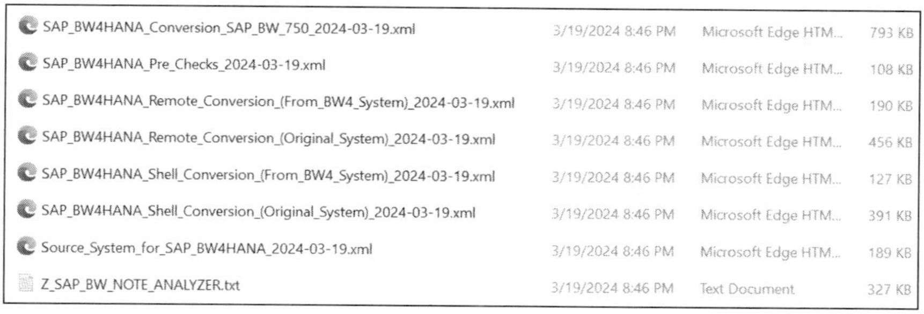

Figure 7.6 Note Analyzer for SAP BW Bridge in SAP Datasphere

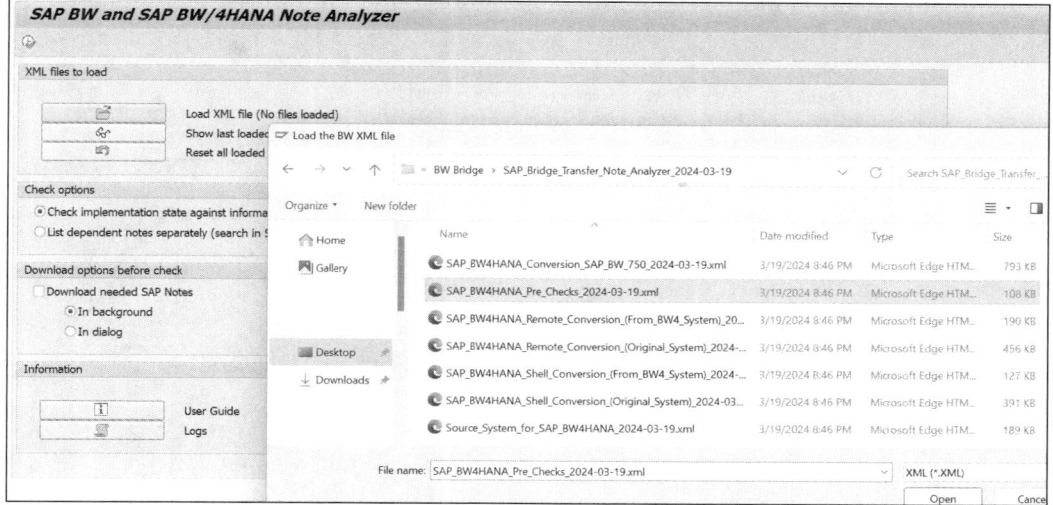

Figure 7.7 XML Files Check for SAP BW Bridge

The sample note result will be as shown in Figure 7.8. Install the red and yellow status notes in serial order, and don't miss any manual steps.

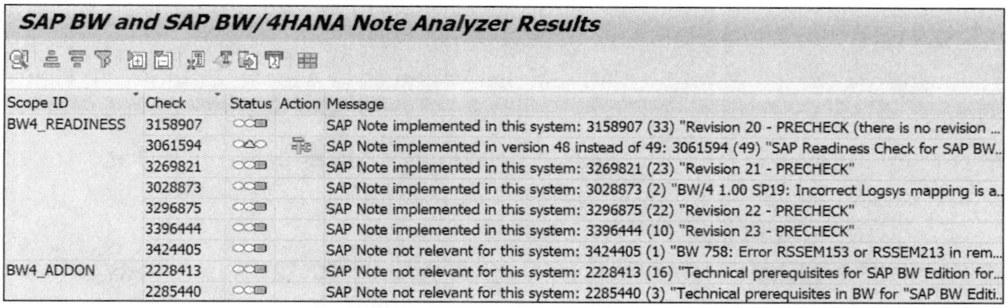

Figure 7.8 Note Analyzer Results

357

System Landscape Preparation

The Note Analyzer provides comprehensive support for landscape preparation and installation processes. By using the Note Analyzer, users are guided through a structured approach to updating specific components within SAP BW or SAP BW/4HANA, eliminating the need for implementing support packages. This analyzer acts as a valuable tool, facilitating the installation of all essential updates and tools crucial for a smooth transition from SAP BW or SAP BW/4HANA to SAP Datasphere. It encompasses various components, including the precheck, transfer cockpit, and necessary ODP updates, ensuring a seamless conversion process while maintaining system integrity and functionality. Figure 7.9 shows the system landscape for shell conversion in SAP BW bridge.

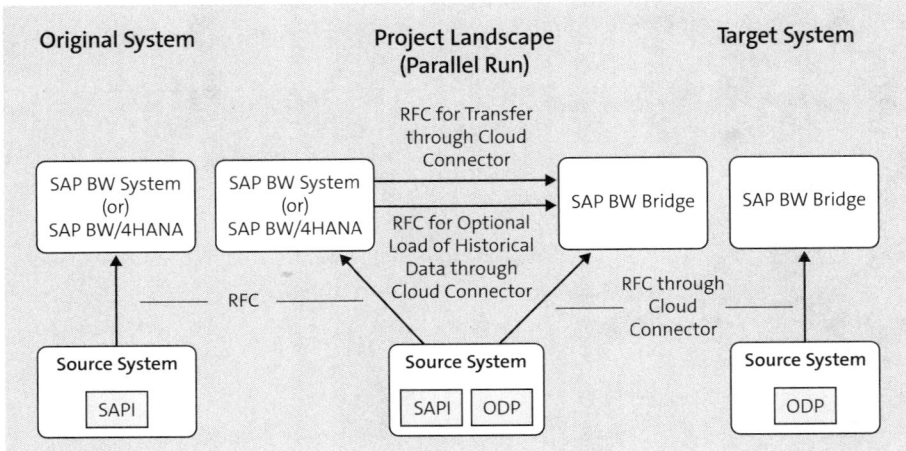

Figure 7.9 System Landscape

As you can see in Figure 7.9, the original system is SAP BW or SAP BW/4HANA, which is connected to a source system that uses the Service API (SAPI) framework for the extraction. During the conversion, we'll have a project landscape that will have sender and target SAP BW bridge systems. The source system will have the ODP framework with the DataSources exposed to ODP. Finally, post-conversion, you'll have the SAP BW bridge connected to the source system, and extraction happens using the ODP framework.

SAP BW Bridge Sender System Preparation

Following are the details of the XML file that you need to choose, as shown in Figure 7.10:

❶ For SAP BW systems running on SAP NetWeaver releases between 7.30 and 7.51, or SAP BW/4HANA 2021, refer to the XML file *SAP_BW4HANA_Readiness_Check_[last_update].xml* provided in SAP Note 2575059.

❷ For all the available source systems that are connected to SAP BW or SAP BW/4HANA, use *Source_System_for_SAP_BW4HANA_[last_update].xml*.

7.3 Shell Conversion with SAP BW Bridge

❸ Install the transfer cockpit and perform a shell conversion in the realization phase:
- For SAP BW systems running on SAP NetWeaver releases from 7.30 to 7.51, acting as sending systems for a shell conversion, you can choose to employ *SAP_BW4HANA_Shell_Conversion_(Original_System)_[last_update].xml*.
- For SAP BW/4HANA 2021, functioning as sending systems for a shell conversion, use *SAP_BW4HANA_Shell_Conversion_(From_BW4_System)_[last_update].xml*.

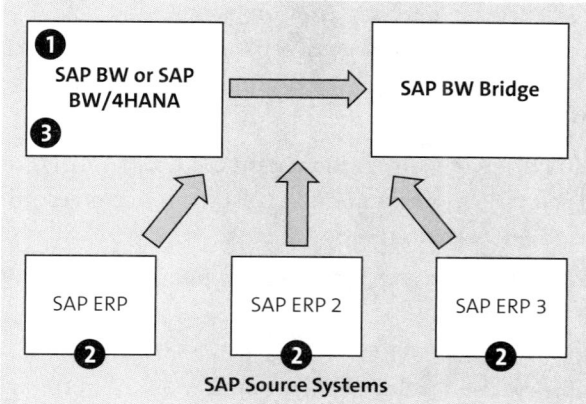

Figure 7.10 System Landscape XML

After the installation of these XML files, the sender system will have the transfer cockpit and all required SAP Notes for the SAP BW bridge conversion. You can use Transaction RSB4HCONV to open the transfer cockpit and perform prechecks and complete code scan, as discussed in detail in Chapter 3.

SAP BW Bridge Source System Preparation

For all the available SAP-based source systems that are connected to SAP BW or SAP BW/4HANA, use *Source_System_for_SAP_BW4HANA_[last_update].xml*. To ensure smooth integration with SAP Datasphere or SAP BW bridge, it's crucial that SAP source systems, including SAP BW systems intended for use as DataSources, are compatible with ODP, as detailed in SAP Note 2473145. Moreover, all DataSources must be released for ODP to facilitate seamless data transfer. Following is a structured approach to ensure compatibility and readiness.

First, verify that all essential source systems meet the general compatibility requirements for ODP, which were outlined in Section 7.3.1 earlier. Next, proceed to update the ODP functionality across all source systems using the Note Analyzer, following the instructions provided in SAP Note 3141688 and additional details available in the same, such as XML prerequisites. Subsequently, ensure that all necessary SAP DataSources are appropriately released for ODP, as specified in SAP Note 2232584. Additionally, don't forget to release generic DataSources for ODP by referring to the guidelines provided in SAP Note 2350464. By diligently following these steps, you can ensure that your SAP

359

source systems are fully compatible and ready for seamless integration with SAP Datasphere or SAP BW bridge, facilitating efficient data provisioning and processing.

7.3.3 Realization Phase

We'll start with SAP Datasphere. In the following, we assume that an SAP Datasphere tenant is already provisioned, and only SAP BW bridge needs to be added. To begin integrating SAP BW bridge into your SAP Datasphere subscription, ensure you have the appropriate capacity units allocated for compute, storage, SAP BW bridge, and data lake components. Assuming your SAP Datasphere tenant isn't yet provisioned, follow these steps to configure and add SAP BW bridge:

1. Configure the size of your SAP BW bridge tenant in the tenant configuration. Navigate to **System Configuration • Tenant Configuration**, and adjust the settings accordingly. Save your configuration, and review before submitting.
2. Create your SAP BW bridge tenant by accessing **System Configuration • SAP BW Bridge**. Click on **Create** to initiate the process.
3. Provide a name and description for the SAP BW bridge space. If configuring a productive tenant, deselect the **Enable system for development** option. For nonproductive tenants, select this option based on the system role, enable for development tenants, and disable for test tenants in a three-tier landscape.
4. Monitor the provisioning process by refreshing the status. A **Create Succeeded** status indicates successful provisioning of the SAP BW bridge tenant, including the creation of the space and the necessary connection with SAP Datasphere. Access this space in **Space Management**.
5. Additionally, ensure you have a Cloud Connector installed, meeting the minimal version requirements. This connector facilitates connections with the sender system for shell conversion and other source systems. Create necessary software components, development packages, and transports to complete the setup.

By following these steps meticulously, you can seamlessly integrate SAP BW bridge into your existing SAP Datasphere subscription, enhancing your data management capabilities and facilitating efficient data processing. If there are any errors during this process, refer to SAP Note 3134262 "How to Get Started with SAP BW Bridge." You can see this space in Figure 7.11.

To establish connectivity between the sender system and your SAP BW bridge tenant, it's imperative to set up a Cloud Connector within your on-premise network with a version of 2.13.1 or higher. Acting as an intermediary, the Cloud Connector facilitates communication between the on-premise source system and your SAP BW bridge tenant, which is technically anchored on an ABAP platform within SAP Business Technology Platform (SAP BTP). The RFC protocol is used for seamless data exchange between the on-premise source systems and SAP BW bridge. Configuring the on-premise source system entails designating it as a communication system, with specific properties tailored to its functionality. The source system connectivity is orchestrated through a series of steps as described next.

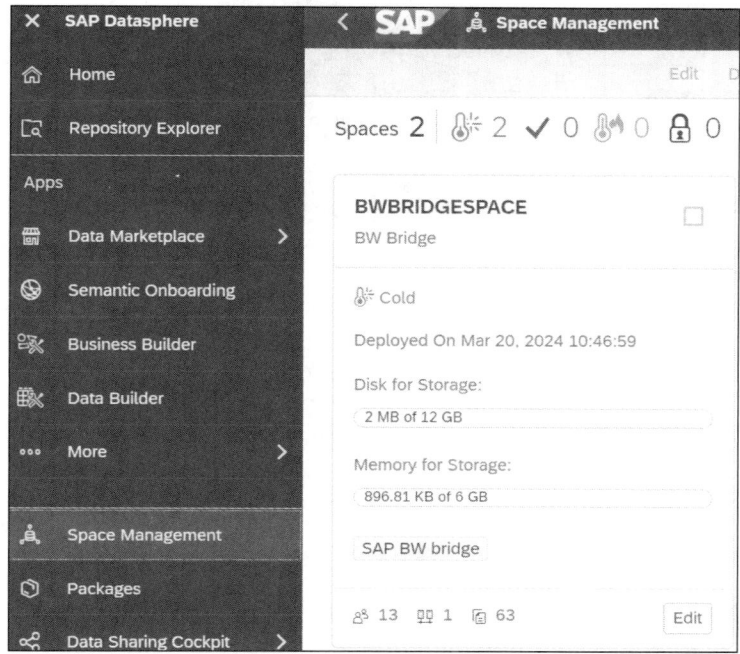

Figure 7.11 SAP BW Bridge Space

First, the SAP Datasphere subaccount must be added to the Cloud Connector to establish a bridge between the cloud environment and on-premise systems. Subsequently, the on-premise source system is created within the Cloud Connector, followed by the addition of relevant resources to this system. Next, a service channel linking to the SAP BW bridge tenant is established within the Cloud Connector to enable seamless data transmission. Within the SAP BW bridge tenant, a communication system is then created to finalize the setup. Additionally, within the SAP BW modeling tools, users have the option to create the source system, further enhancing configurational flexibility. As an optional step, users may select and activate preconfigured SAP BW bridge content objects to streamline the integration process and leverage predefined functionalities. By following these procedures meticulously, organizations can ensure robust connectivity between on-premise source systems and their SAP BW bridge tenant, facilitating efficient data exchange and management. See Figure 7.12 for more details regarding Cloud Connector, which shows how the Cloud Connector is used for SAP BW or SAP BW/4HANA integration to SAP BW bridge using RFC and connectivity service.

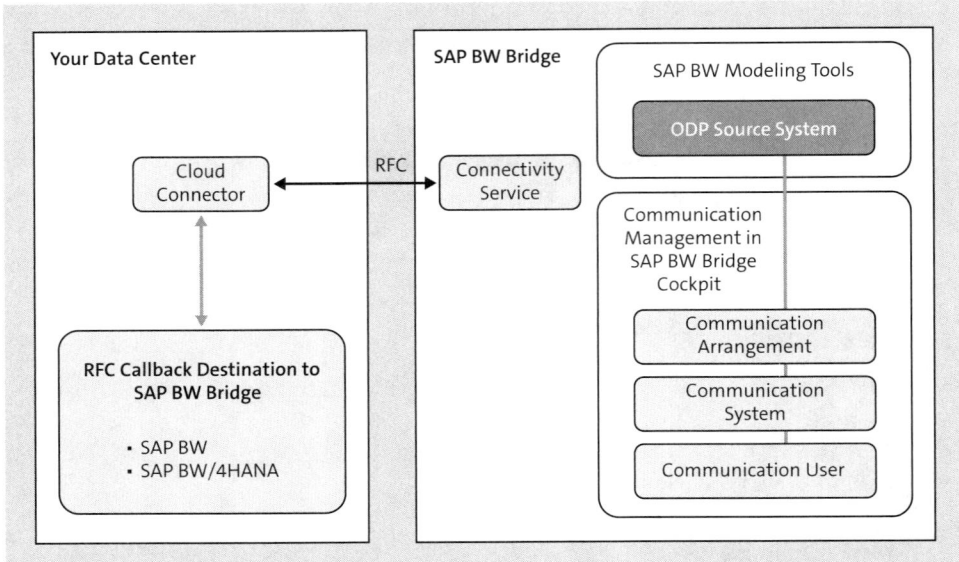

Figure 7.12 Cloud Connector

7.3.4 Connect the Sender System to SAP BW Bridge

You need to work with the Basis team and security team to connect the sender system to SAP BW bridge. This has lots of activities that will be beyond the scope of this book, so we assume that the SAP BW sender system connects to the SAP BW bridge, and you've performed the following activities:

1. Add the SAP Datasphere subaccount in the Cloud Connector.
2. Add a service channel to SAP BW bridge tenant in Cloud Connector.

   To enable communication between the sender system and the SAP BW bridge tenant via RFC, it's essential to establish a service channel within the Cloud Connector. This service channel serves as a conduit, allowing the sender system to initiate calls to the SAP BW bridge tenant seamlessly.

3. Create a communication system in the SAP BW bridge tenant.

   In general, the on-premise system must be configured as a communication system in the SAP BW bridge tenant. A communication system is a defined configuration that represents a specific system serving as a communication partner. It encompasses technical details necessary for communication, including user credentials for inbound/outbound communication. The sender system must be created as the communication system in the SAP BW bridge tenant; for example, in our case, we'll create the SAP BW sender system as the communication system for our demo. You need to have access to the SAP BW bridge cockpit to set up a new communication system. In SAP BW bridge, you can see the **Data Integration Monitor** screen shown in Figure 7.13.

7.3 Shell Conversion with SAP BW Bridge

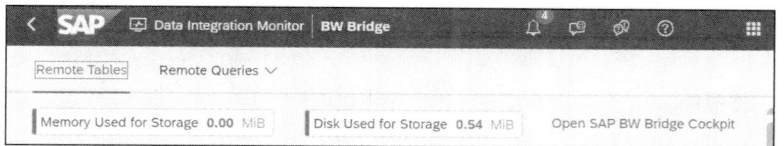

Figure 7.13 Data Integration Monitor

4. In Figure 7.13, choose **Open SAP BW Bridge Cockpit**, log in, and navigate to **Communication Management** (see Figure 7.14).

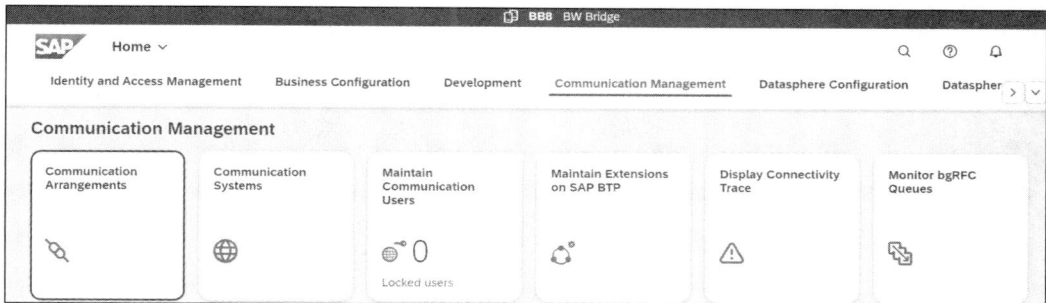

Figure 7.14 Communication Management

5. Choose **Communication Systems** in the SAP Fiori app, and you can see the option to create a new system, as shown in Figure 7.15.

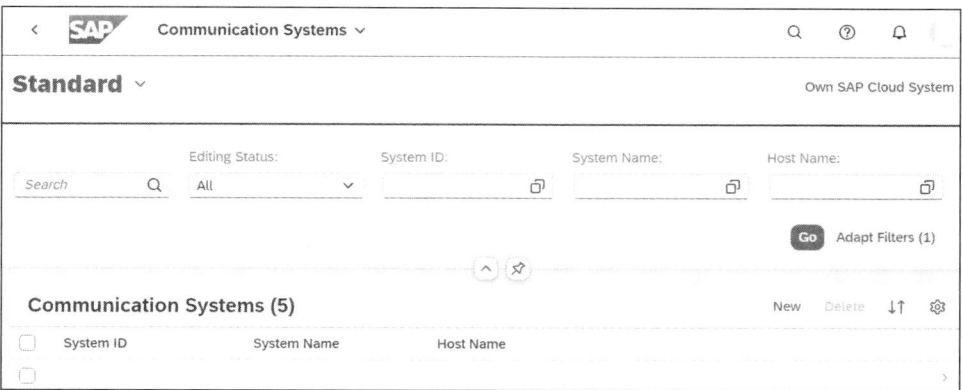

Figure 7.15 Communication Systems

6. Click **New**, and enter the **System ID** and **System Name**. It's recommended to create the name as "MIGRATE<SID>". In our case, the sender system ID is BW4, so we'll create the communication system as "MIGRATE_BW4", as shown in Figure 7.16.

363

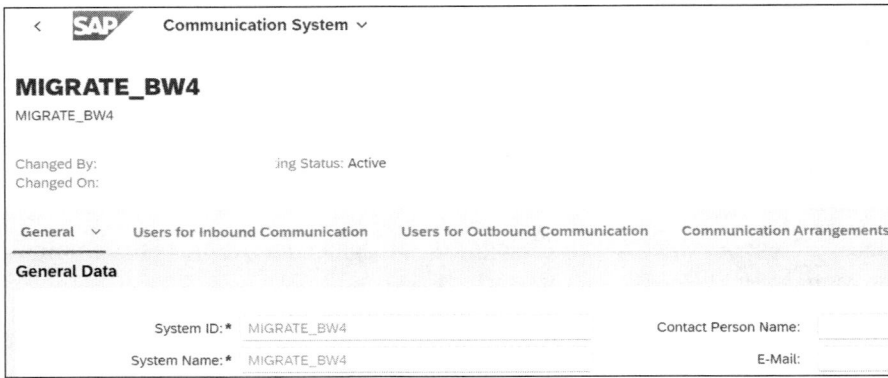

Figure 7.16 System ID

7. In the **General** section, fill in the technical data with an arbitrary **Host Name** and **Port** number as "443". Add the user for inbound communication, granting necessary authorizations. Save the configuration (see Figure 7.17).

Figure 7.17 Host Name

8. In the **User for Inbound Communication** tab, you can create the user for inbound communication from the cockpit. The user for outbound communication is the user in the on-premise system (in our case, BW4 system-based user). Make sure it has all the profiles such as **S_BI_WHM_RFC** if its SAP BW and **S_BI_WX_RFC** if its SAP BW/4HANA or another SAP system (see Figure 7.18).

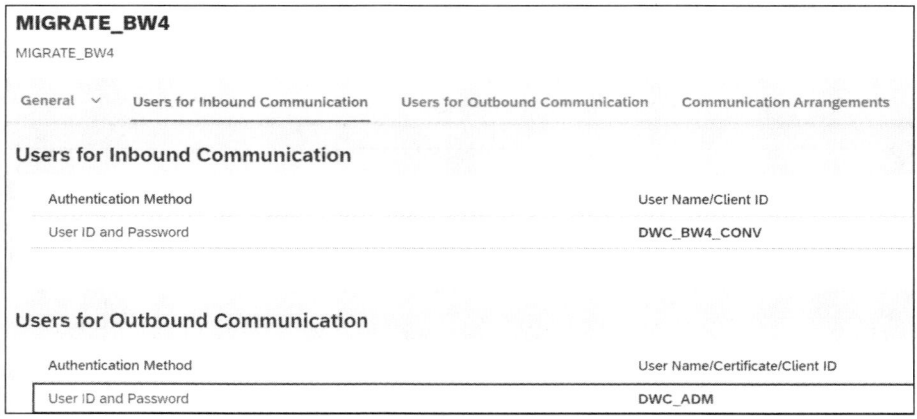

Figure 7.18 RFC User

9. The outbound user is **DWC_ADM**, so you need to ensure that the backend for the outbound user in the SAP BW sender system will have the profiles shown in Figure 7.19. Note **S_BI-WHM_RFC in the Profile column.**

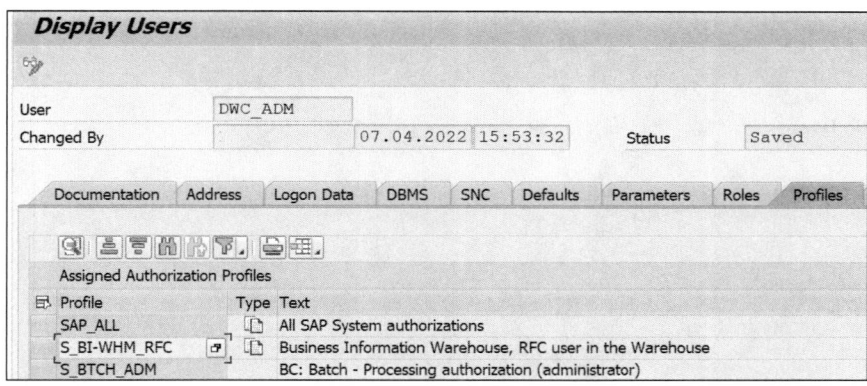

Figure 7.19 Outbound User

10. A communication arrangement outlines how specific communication partners interact in each scenario. It defines who communicates with whom and how they communicate. In the context of the migration scenario, a communication arrangement needs to be established within the SAP BW bridge tenant. To create a communication arrangement for migration within the SAP BW bridge cockpit. Open SAP Datasphere, navigate to **More • Data Integration Monitor • Open SAP BW Bridge Cockpit**, and log in to access the **Communication Arrangements** section. Here, you'll initiate a new arrangement and select the predefined scenario, such as SAP BW Bridge – ODF RFC Source System Integration. Input the technical name of the source system as the arrangement name, and specify the chosen communication system. Once configured, save the arrangement to finalize the setup process.

11. You can see the **Own SAP Cloud System** in Figure 7.20. Create an **RFC Destination** with the same name in the SAP BW sender system. When setting the scope for shell transfer, be sure to select an RFC destination that points to the destination SAP BW bridge tenant.

12. To establish an RFC connection to the SAP BW bridge tenant from the sender system, start by logging into the sender system and accessing Transaction SM59. From there, choose **Create** to initiate the setup of a new RFC connection. Provide a name and a descriptive label for the **RFC Destination**. Then, input the Cloud Connector's hostname (without https:// and port number) as the **Target Host**, along with the local instance number designated for the service channel. On the **Logon & Security** tab, shown in Figure 7.21, specify the language, client, and user credentials. Save the RFC destination, and conduct a connection test to ensure its functionality. If any errors occur during testing, examine error messages in the **Problems** view, and consult with your infrastructure team to verify firewall settings if necessary.

7 SAP BW Bridge for SAP Datasphere

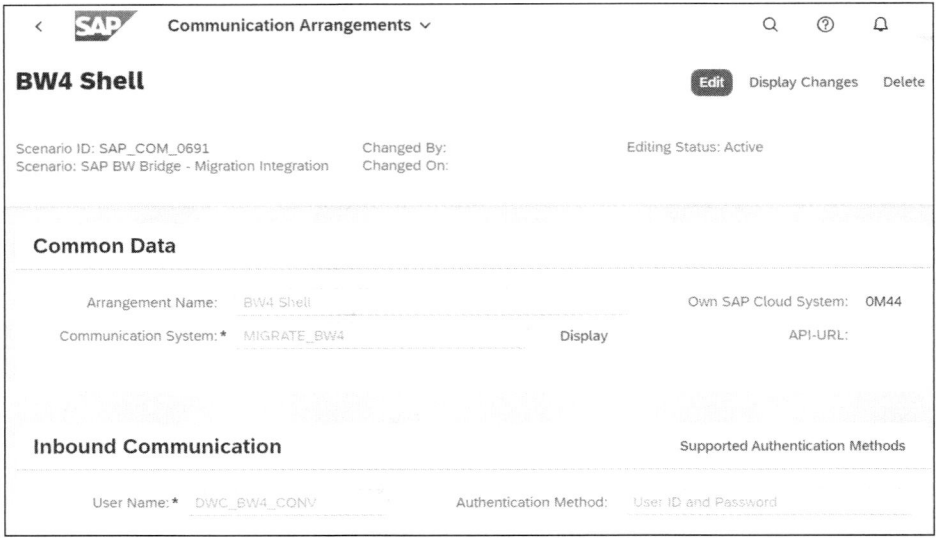

Figure 7.20 Communication Arrangement

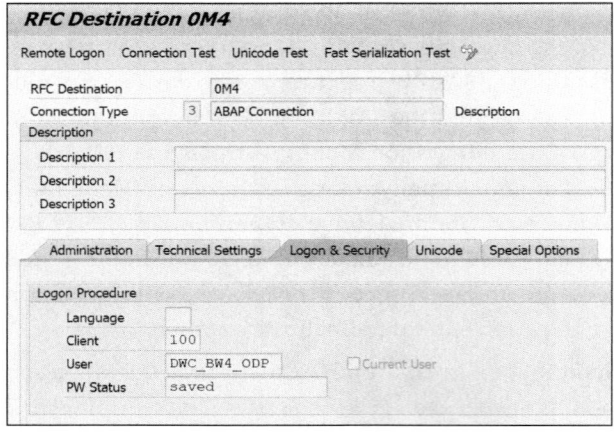

Figure 7.21 RFC Destination to SAP BW Bridge

Once all of these are completed, we can create a SAP BW project in the next step.

7.3.5 Create an SAP BW Bridge Project

Follow these steps to create an SAP BW bridge project:

1. As a prerequisite to SAP BW bridge, you need to have SAP BW modeling tools for the creation of the project. Before creating the project, you need to have a few details in hand. Open the SAP BW bridge space in SAP Datasphere, and in **Apps**, choose **Connections**, select **BW Bridge**, and click on **Edit**. In the window shown in Figure 7.22, use the copy button to copy the **SAP BW Service Key** into a document.

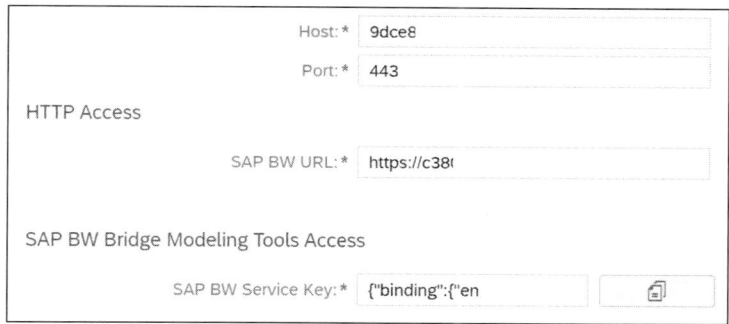

Figure 7.22 SAP BW Service Key

2. Ensure that you have a valid user in SAP Datasphere with all required business roles, and verify that with your security team. Example roles are as follows:
   - Administrator - Data Warehouse Operation (DWC): BR_ADMINISTRATOR_DWC
   - Developer - Data Warehouse (DWC): BR_DEVELOPER_DWC
   - Administrator: SAP_BR_ADMINISTRATOR
   - Developer: SAP_BR_DEVELOPER

3. Open the SAP BW modeling tools, and create the **BW bridge Project** shown in Figure 7.23.

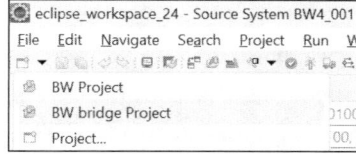

Figure 7.23 SAP BW Bridge Project

4. Now you'll see this option, as shown in Figure 7.24, which will prompt you to **Use a Service Key**.

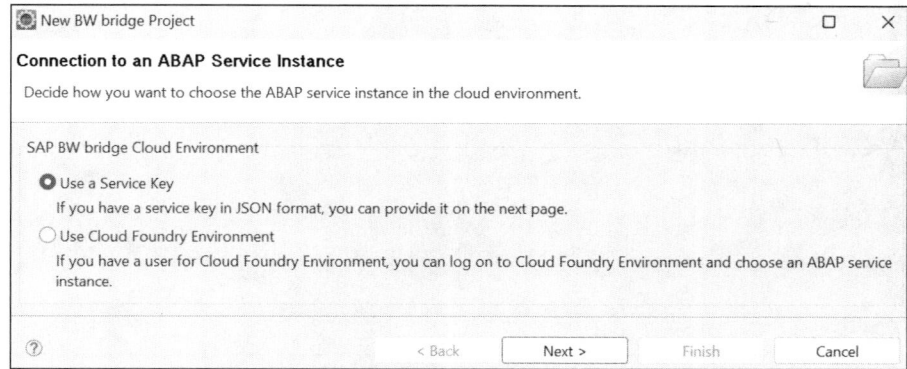

Figure 7.24 Service Key Prompt

5. Choose **Next**, and provide the copied service key, as shown in Figure 7.25.

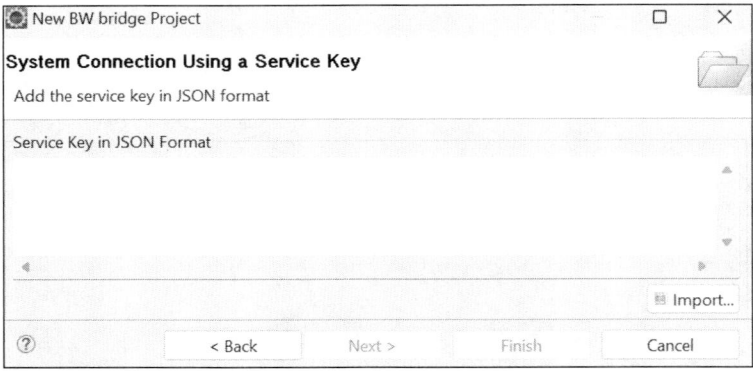

Figure 7.25 Service Key Update

6. If the service key is correct, it will try to open the logon for SAP Datasphere. If that's verified, then you'll get the message **Successfully Logged In**, and you can close the window. Check Eclipse again your SAP BW project is created, as shown in Figure 7.26.

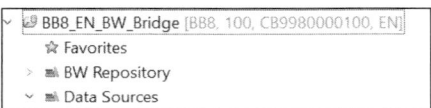

Figure 7.26 SAP BW Bridge Project

7. Now add the SAP BW sender to the SAP BW bridge DataSource. Under **Data Sources**, choose **New • Source system**. Provide the name and description, and choose **ODP**. If the destination is red, maintain the communication arrangement and ensure you have the RFC destination. You'll see the source system for the SAP BW sender, as shown in Figure 7.27. You need to make sure that the ODP context is set correct as SAP BW because we're connecting to an SAP BW system.

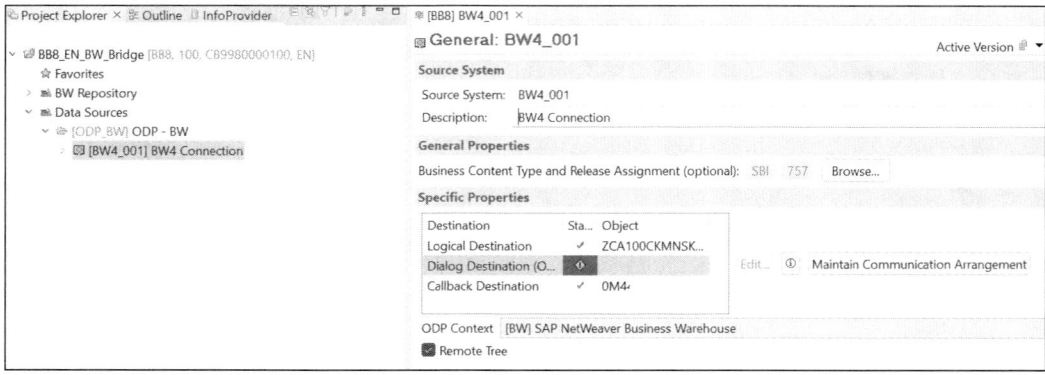

Figure 7.27 SAP BW Sender DataSource in SAP BW Bridge

7.3 Shell Conversion with SAP BW Bridge

8. You can similarly connect the other source system to SAP ERP and SAP S/4HANA in the SAP BW bridge, as shown in Figure 7.28.

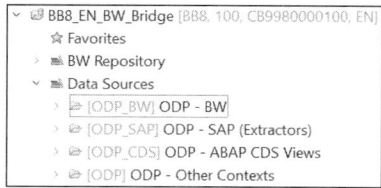

Figure 7.28 SAP BW Bridge ODP Connectivity

This will complete the steps to create an SAP BW bridge project and connect it to the SAP BW sender.

7.3.6 ABAP Cloud Project in ABAP Development Tools

You need to have an ABAP cloud project in the SAP BW bridge as a prerequisite. This can be done in Eclipse using ABAP development tools. With this ABAP cloud project, you can create the ABAP packages further. These packages are very important during the transport phase. Follow these steps:

1. In Eclipse, open the ABAP perspective, and create the ABAP cloud project. In the same process, you need to provide the same service key from the SAP BW bridge space, as shown in Figure 7.29.

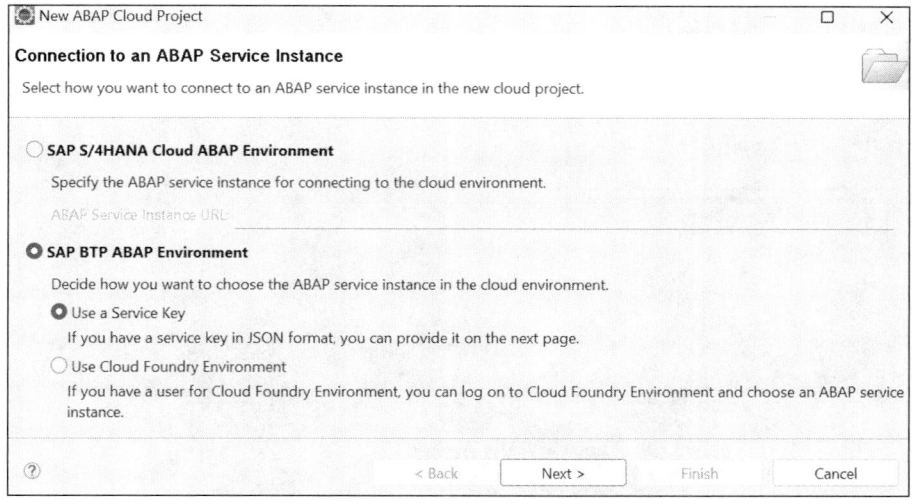

Figure 7.29 Create the ABAP Cloud Project

2. Once you're done with all the steps, you can see the completed project, as shown in Figure 7.30.

369

7 SAP BW Bridge for SAP Datasphere

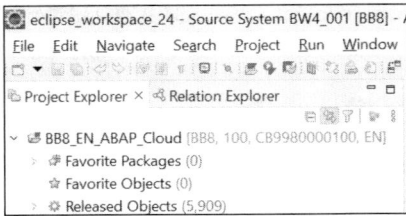

Figure 7.30 Completed ABAP Cloud Project

SAP BW bridge operates on an ABAP platform. To manage and transport its objects, such as DSOs and transformations, you must establish a software component and an ABAP development package. Let's start with the steps to create software components:

1. Open the SAP BW bridge cockpit, and choose **Software Component Life Cycle Management**, as shown in Figure 7.31.

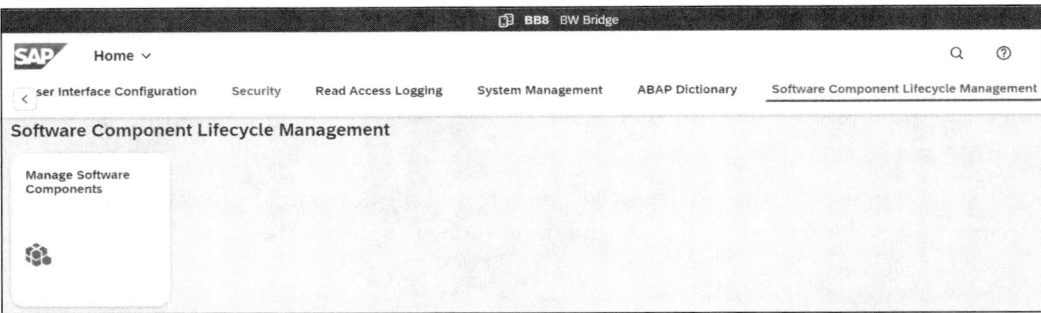

Figure 7.31 Creation of Software Components

2. Enter into **Manage Software Components**, and choose **Create** at the bottom of the screen shown in Figure 7.32.

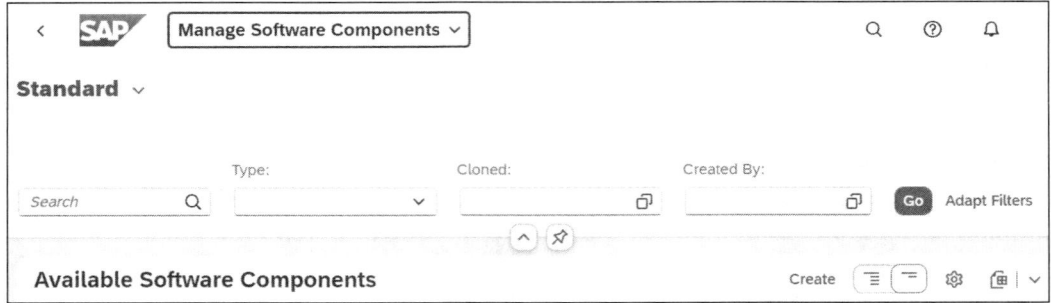

Figure 7.32 Software Components

3. Choosing **Create** creates the component, as shown in Figure 7.33. In our example, we've created component **ZZPRK**.

370

7.3 Shell Conversion with SAP BW Bridge

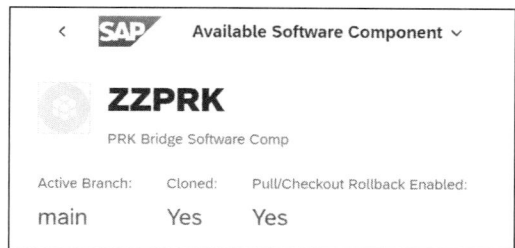

Figure 7.33 Software Component

Next, we'll create an ABAP package using these steps:

1. From the ABAP cloud project, choose **New** • **ABAP Package**, and you can see the screen to provide the **Name** of the package and the **Description**. You need to provide the software component name as your super package; in our case, we've provided **ZZPRK** as the super package (Figure 7.34).

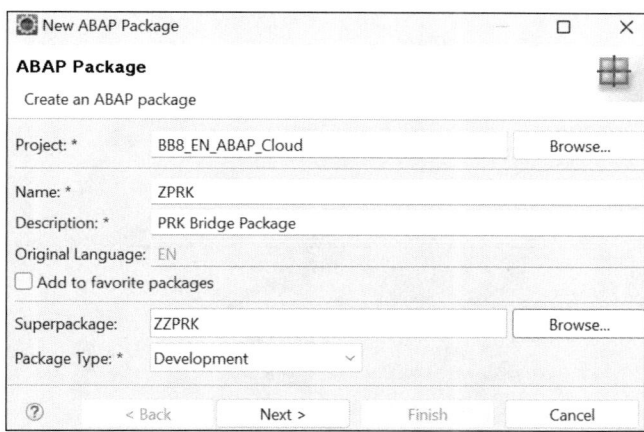

Figure 7.34 ABAP Package

2. Once you choose **Next**, you'll see that the software component and transport layer are selected by default based on the previous input, as shown in Figure 7.35.

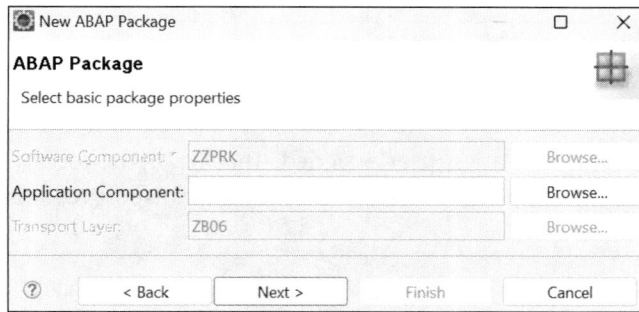

Figure 7.35 Software Component in the ABAP Package

371

3. Choosing **Next** in Figure 7.35 will take you to the screen shown in Figure 7.36 to create the transport request.

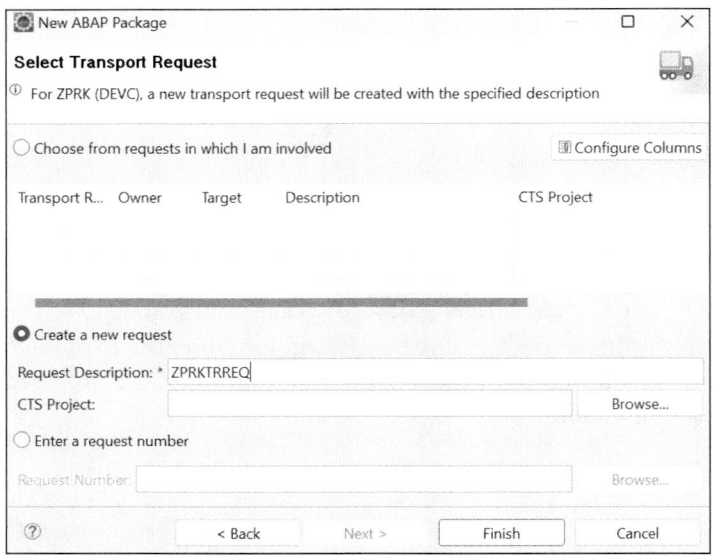

Figure 7.36 Transport Request Creation

The screen shown in Figure 7.37 appears.

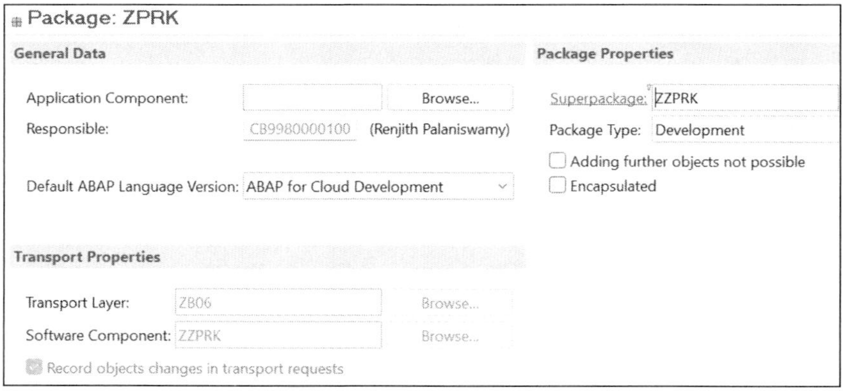

![Package: ZPRK screen]

Figure 7.37 ABAP Package

4. We'll use this package (**ZPRK**) in our conversion. You also need to create a transport request that will have the communication user assigned. That transport request should be of type local in the SAP BW bridge so that you'll use that transport request for the objects, as shown in Figure 7.38.

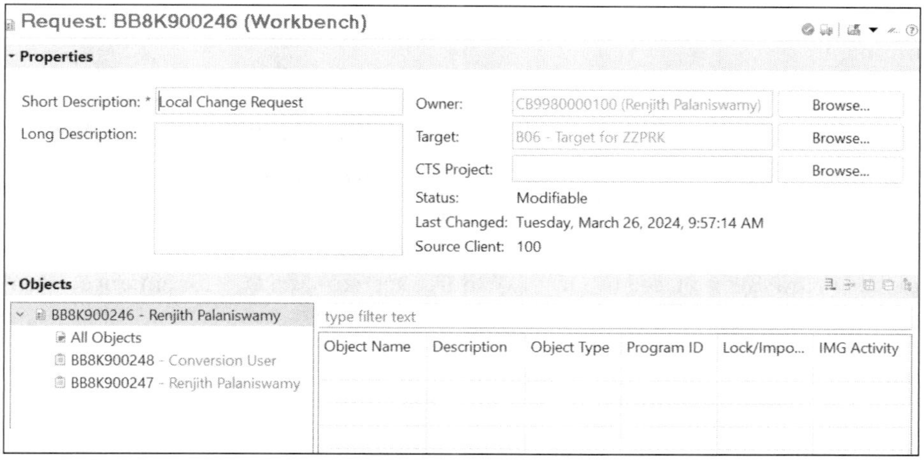

Figure 7.38 Transport Request for the Tool

7.3.7 Object-Specific Simplification

During conversion tasks involving data acquisition from SAP source systems and SAP BW source systems, SAPI DataSources are converted using the transfer cockpit. For data warehousing conversion tasks, which include InfoCubes, classic DSOs, InfoSets, and MultiProviders, the transfer cockpit is used.

In SAP BW bridge, certain object types that aren't available are substituted with other object types or functionalities. The following outlines the object types no longer supported and their replacements, along with corresponding simplifications.

During the conversion process of the SAP BW system, several object types require either replacement or adjustment to ensure compatibility with the updated framework. The classic DSO is replaced by the ADSO, enhancing functionality and efficiency. Similarly, the DSO with old request management is substituted with the advanced version featuring new request management capabilities for improved performance. InfoCubes are replaced by the ADSOs to adapt to the evolving architecture and requirements, while InfoSets are substituted with the CompositeProvider to streamline data organization and access. Semantic Partitioned Objects (SPOs), such as InfoCubes or classic DSOs, are replaced by the semantic group of the ADSO, enhancing semantic modeling capabilities. Additionally, the old CompositeProvider (COPR) is replaced by the updated CompositeProvider (HCPR) to align with the latest standards and functionalities, ensuring compatibility with the new system.

Furthermore, the conversion process entails replacing SAP BW and SAP source systems with the ODP source system paradigm, enhancing data integration and compatibility across systems. InfoPackages and persistent staging areas (PSA) are replaced with the combined use of the DTP and ADSO, optimizing data handling and processing. These replacements and adjustments facilitate a smooth transition to the updated framework, maintaining system integrity and functionality while accommodating new requirements and standards, thereby ensuring seamless operations within the SAP BW environment.

There are requirements for adjustments during the conversion, and here are a few object types that require adjustments or replacements during the conversion process:

- Various object types within the SAP BW workspace necessitate adjustments or replacements to ensure seamless integration with SAP BW/4HANA. Specifically, DataSources must be aligned with changes in source systems, with export DataSources in SAP BW replaced by ODP-BW source systems and DTP for enhanced compatibility. Similarly, adjustments are required in DTP, where error DTPs are substituted with DTIS, managed automatically by the post-conversion task list.

- Objects within the data flow and process chain/process variants require substitutions for continued functionality. For instance, InfoPackages are replaced with DTPs, and the activation process for classic DSOs is substituted with ADSOs. Additionally, transformations necessitate adjustments in both source and target configurations, particularly in the lookup of DSOs, ensuring data integrity throughout the conversion process. These replacements and adjustments facilitate a smooth transition to SAP BW/4HANA, maintaining system integrity and functionality while accommodating updated standards and requirements.

7.3.8 Prework

You need to perform the following tasks before executing shell conversion in SAP BW bridge:

- **Simulating shell conversion**
  You can use task list **SAP_BW4_TRANSFER_CHECK_CLOUD_SHL** to check the cloud shell transfer (see Figure 7.39).

- **Authorizations**
  Authorization is required for the following authorization objects: S_RS_B4H, S_TC, S_RFC, and S_RFC_ADM. Additionally, authorization is needed to read any selected object intended for transfer.

- **Prechecks and code scan**
  Complete the precheck and code scan based on the findings from the transfer cockpit.

Figure 7.39 Transfer Cloud Shell Check

7.3.9 Execution

In this section, we'll focus on the shell conversion execution for the SAP BW bridge conversion. We'll take a sample SAP BW data model and convert it to SAP BW bridge. You should have competed the prerequisites at this point. Let's use the data model in the sender system for the SAP BW bridge shell conversion shown in Figure 7.40.

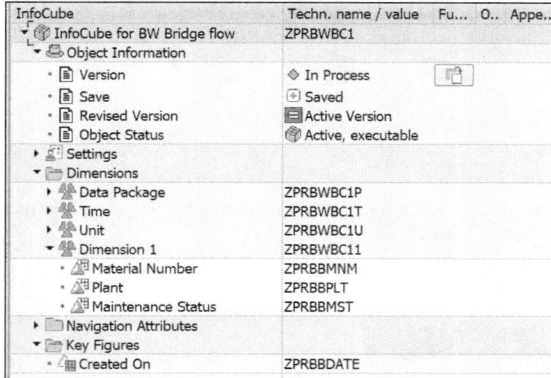

Figure 7.40 InfoCube for Conversion

The data flow for InfoCube **ZPRBWBC1** is shown in Figure 7.41.

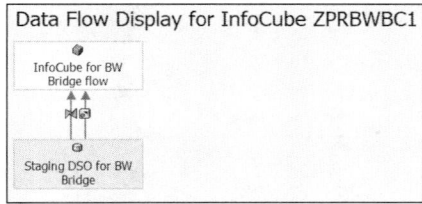

Figure 7.41 Data Flow for the InfoCube

You have the following options to start the SAP BW bridge shell conversion:

- **Option 1: Using the transfer cockpit**
  - Use Transaction RSB4HCONV, and select **Execute Scope Transfer**. Choose the **SAP Datasphere SAP BW Bridge** option, as shown in Figure 7.42.

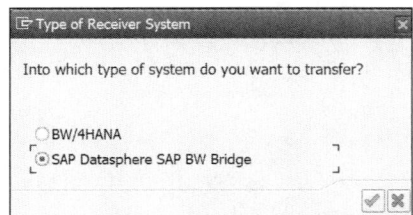

Figure 7.42 Receiver System of Type SAP Datasphere, SAP BW Bridge

7 SAP BW Bridge for SAP Datasphere

– Choose the **Transfer Scenario** next; for our demo, we'll start with **Shell (Metadata only)**, as shown in Figure 7.43.

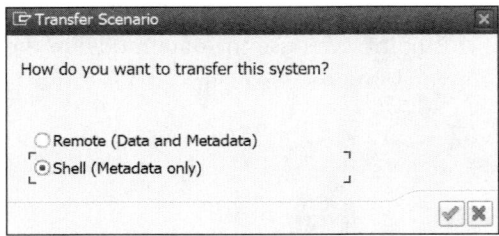

Figure 7.43 Transfer Scenario: Shell

– After confirming, the next popup allows you to create a new task list run or display the existing task list run. Because we're creating for the first time, choose **New** button as shown in image (see Figure 7.44).

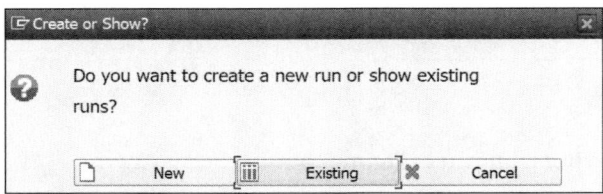

Figure 7.44 Create or Display the Task List run

– The task list run will be created, as shown in Figure 7.45.

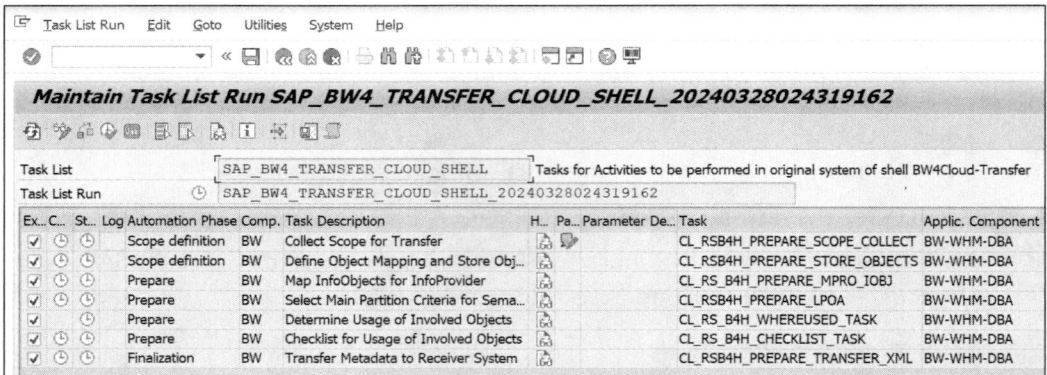

Figure 7.45 Shell Conversion: Task List Run Initial Screen

- **Option 2: Using the task list name**
  You can use Transaction STC01 to execute SAP_BW4_TRANSFER_CLOUD_SHELL. It will take you to the screen with a new task list run, as shown in Figure 7.45.

376

Because we've already created a task list run in option 1, we'll use that for our SAP BW bridge shell conversion process.

We'll now move to the **Collect Scope for Transfer** activity. Execute the activity in the task list run to get to the screen shown in Figure 7.46. Before that, you'll need to click the edit icon and enter the SAP BW object name, as shown in Figure 7.45.

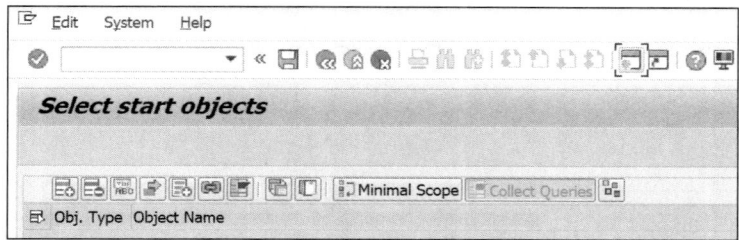

Figure 7.46 Selection of Start Objects for the SAP BW Bridge Shell Conversion

The first step is to choose the icon, and provide the RFC **Destination** of the SAP BW bridge system that we need to transfer, as shown in Figure 7.47.

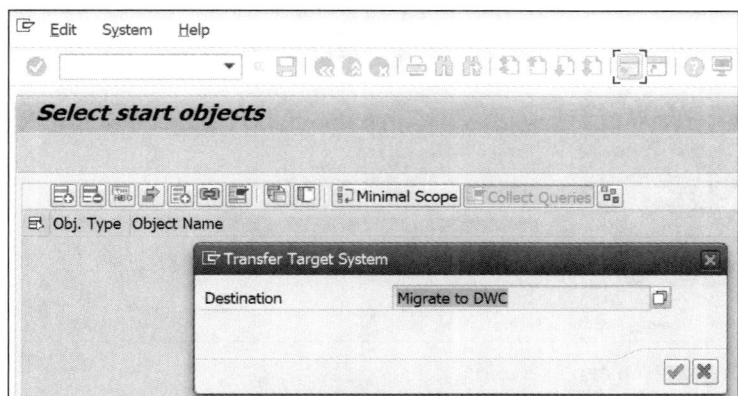

Figure 7.47 RFC Destination for SAP BW Bridge

Choose the icon to select the start objects for the transfer. In this example, choose **CUBE** as the InfoCube object type, and provide the technical name of the InfoCube as "ZPRBWBC1", as shown in Figure 7.48.

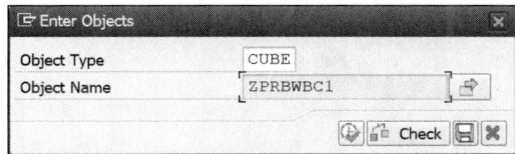

Figure 7.48 Start Object Technical Name

Upon execution of the start object list, the **Select start objects** screen shows the details of the objects included in the scope. You can add multiple start objects with multiple **CUBE**s using the same process discussed previously. Each object will be presented in separate rows, showcasing their respective details, as shown in Figure 7.49.

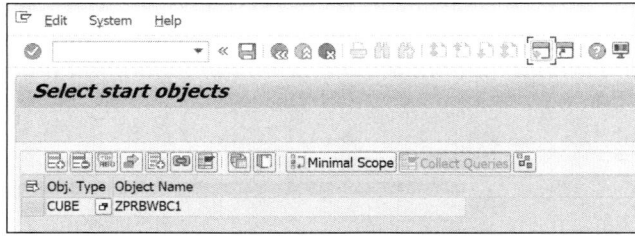

Figure 7.49 Start Object List Display

After you've finalized the start object list, you can save the list and choose the icon, which will take you to the task list run screen. Use the icon to start the activity, and once the job is started, you can see the status as next to the activity name. You can choose to check the final status, as shown in Figure 7.50.

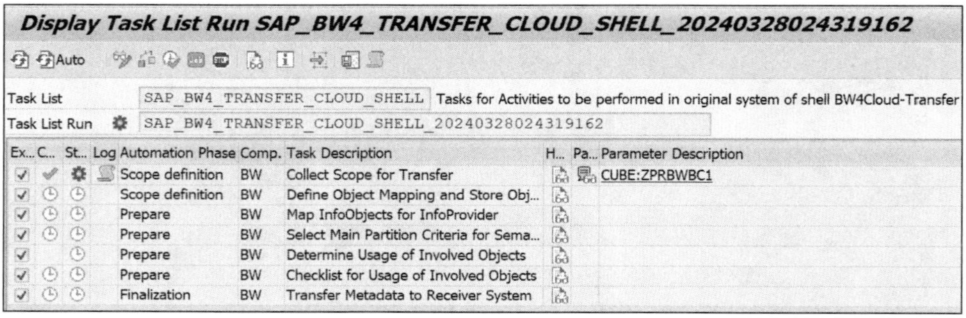

Figure 7.50 Collect Scope for Transfer: Execution

Once the activity is completed successfully, the tool will stop in the next activity, which is **Define Object Mapping and Store Obj**, as shown in Figure 7.51.

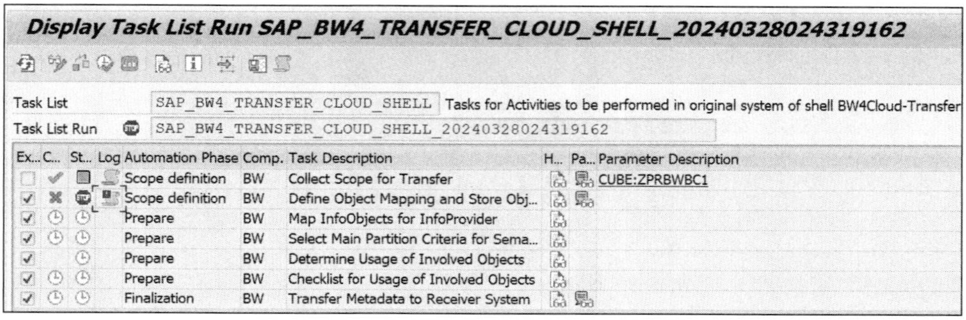

Figure 7.51 Define Object Mapping

7.3 Shell Conversion with SAP BW Bridge

Choose the edit icon to get into the edit mode of the task list and you'll see the parameter icon for object mapping. Click on that icon to open the screen to maintain the object list. Figure 7.52 shows the details of this screen:

❶ **Original TLOGO**
❷ **Original Object Name**
❸ **New TLOGO** (post-conversion InfoCube to ADSO)
❹ **Transfer Object** (selected by default)
❺ **Special Options**

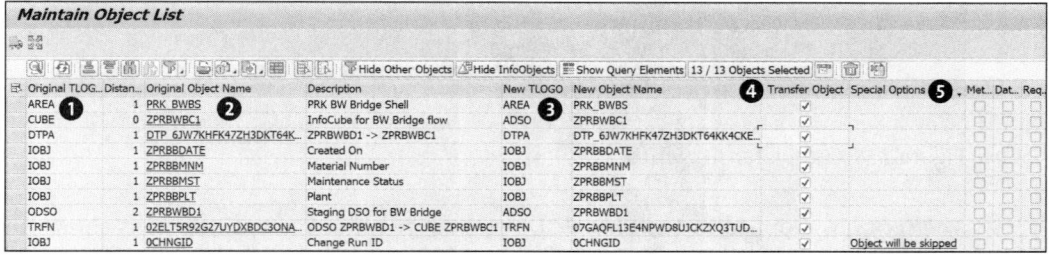

Figure 7.52 Object Mapping: Details

The next step is to create a transport request. Choose the icon to open the popup shown in Figure 7.53. Provide the **Package** name that was created earlier and the **Request/Task** created earlier.

Figure 7.53 Transport Request Details

After assigning the package and request, save and go back using . If you choose to execute, the screen will appear as shown in Figure 7.54.

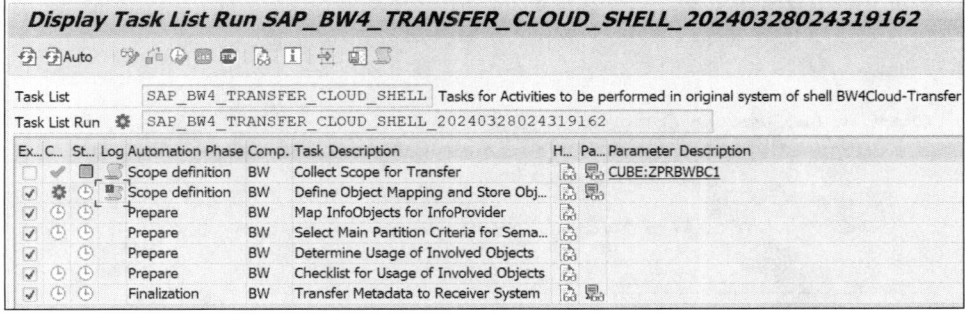

Figure 7.54 Object Mapping: Execution

Upon execution of the start object list, the **Display Task List Run** screen will display the details of the objects included in the scope. Each object will be presented in its own row, showcasing its respective details, as shown in Figure 7.55.

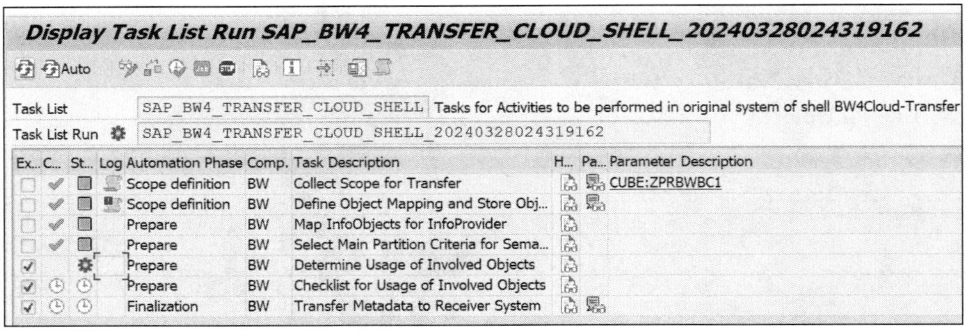

Figure 7.55 Completed Activities

The tool has stopped in the **Checklist** step as it requires manual input, as shown in Figure 7.56.

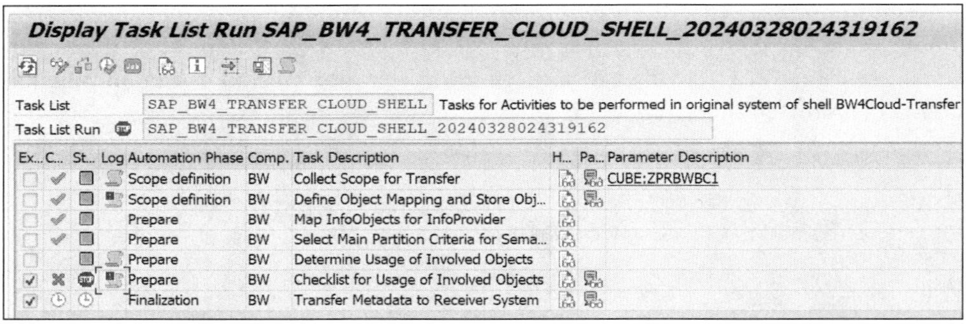

Figure 7.56 Checklist for Usage of Involved Objects

Use the icon for the task list, and click the icon to open the **CheckList for Customer Coding** screen. Ensure that the **BADI** and **AUTH** code are already resolved; if so, you can choose the checkbox for both and save the checklist screen, as shown in Figure 7.57.

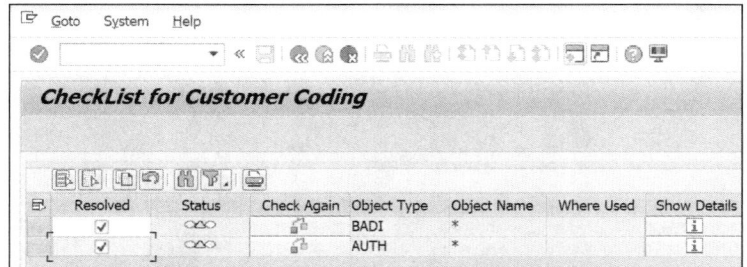

Figure 7.57 Checklist Status for BADI and AUTH

7.3 Shell Conversion with SAP BW Bridge

The preceding step concluded successfully with a green status, initiating the subsequent task of transferring metadata to the receiver system. This process uses the package defined alongside the request number, as shown in Figure 7.58.

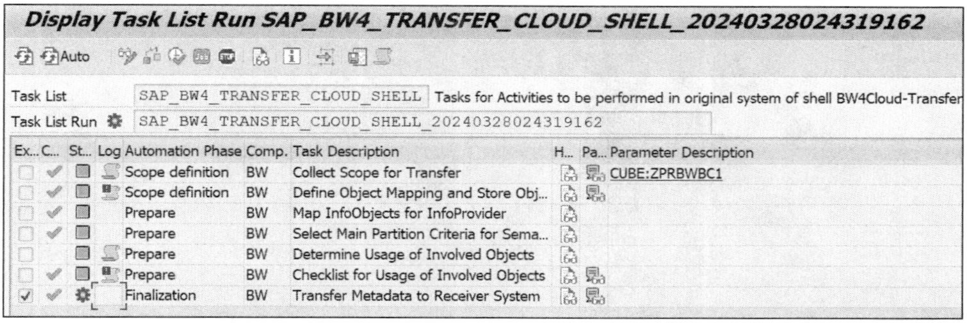

Figure 7.58 Transfer Metadata Status

Upon the successful transfer of metadata in SAP Datasphere and SAP BW bridge, this step will display a green status, as shown in Figure 7.59.

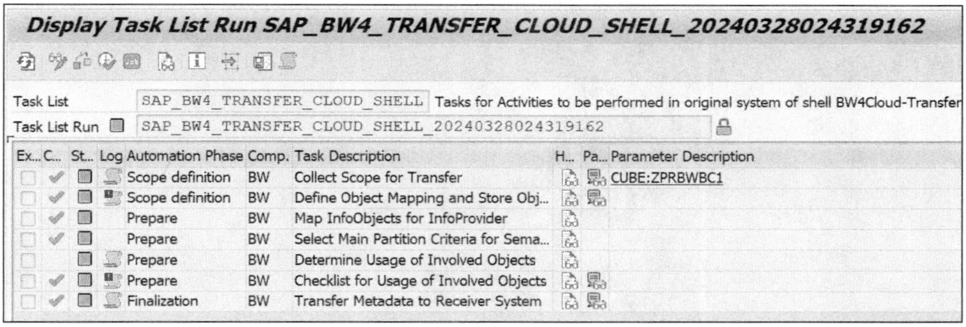

Figure 7.59 Shell Conversion SAP BW Bridge Status

To validate the transfer of metadata you need to log on to SAP Datasphere and SAP BW Bridge. Check for the ADSO (**ZPRBWRC1**) in SAP BW bridge, as shown in Figure 7.60.

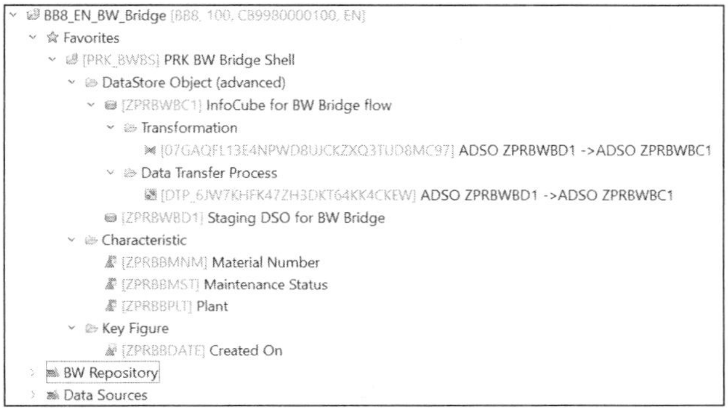

Figure 7.60 SAP BW Bridge Shell Conversion: Metadata Transfer Check

7 SAP BW Bridge for SAP Datasphere

You can see the details of the before ❶ and after ❷ SAP BW bridge shell conversion in Figure 7.61.

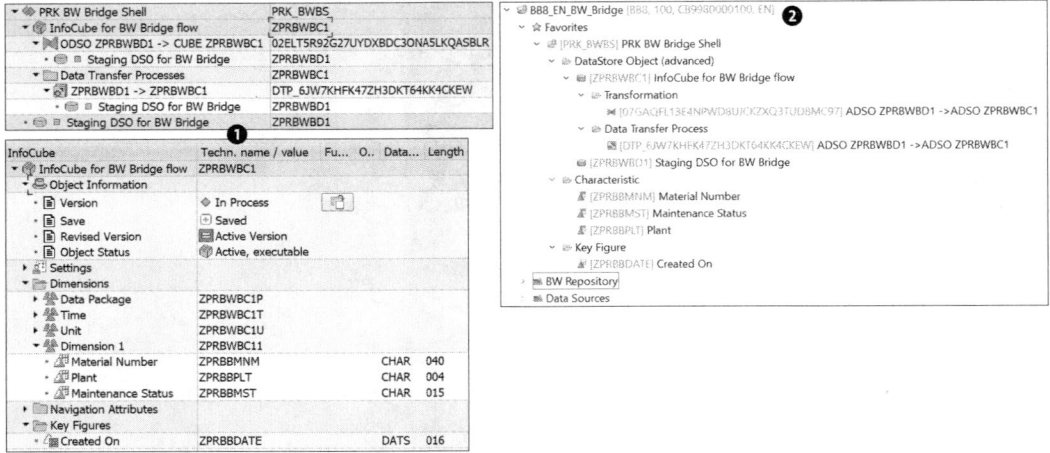

Figure 7.61 Before and After Shell Conversion

Figure 7.62 shows the data flow for ADSO (**ZPRBWBC1**) in SAP BW bridge using the SAP BW modeling tools.

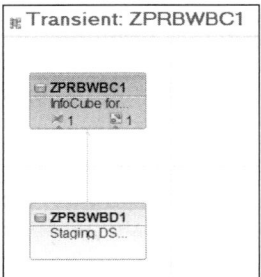

Figure 7.62 Data Flow in SAP BW Bridge

7.4 Remote Conversion with SAP BW Bridge

In the following sections, we'll walk through the steps to perform a remote conversion in SAP BW bridge.

7.4.1 Preparation

Refer to Chapter 5 for the provisioning in the sender system (in our case, SAP BW), in the SAP ERP system, and for the remote conversion; to understand the basics of the Conversion Cockpit; to learn about package creation; and to see how to execute the remote conversion for SAP BW/4HANA.

7.4 Remote Conversion with SAP BW Bridge

We're assuming you have all the settings that we made for the shell conversion already. In the sender system, you need to have the landscape shown in Figure 7.63.

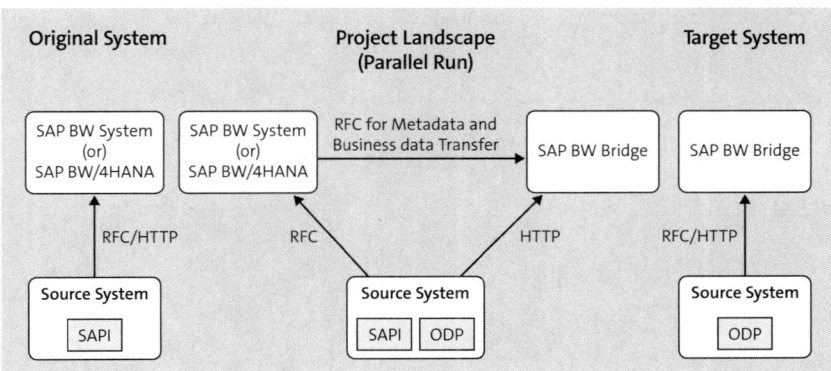

Figure 7.63 Remote Conversion Scope

The XML files you need to install are shown in Figure 7.64.

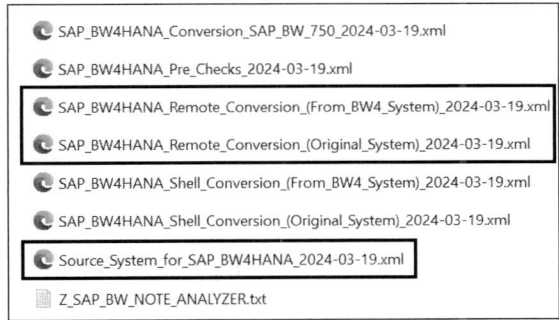

Figure 7.64 Note Analyzer: XML Files

You can import the XML files in your SAP BW/4HANA or SAP BW system. If you need any one of the XML files listed either for the SAP BW/4HANA system or the original system, you need to import the XML notes for all connected SAP ERP and SAP S/4HANA systems using **Source_System_for_SAP_BW/4HANA_<Latest_Date>** from Figure 7.64. There is no SAP-based backend GUI access to SAP BW bridge, and you need to use the SAP BW modeling tools and ADTs to check the objects and tables. You can see the sequence in Figure 7.65.

To prepare for the remote conversion, you must first complete the prechecks using the precheck tool. Do the adjustments based on the precheck result. Use the simulation to check for any errors in the scope objects, as shown in Figure 7.66.

383

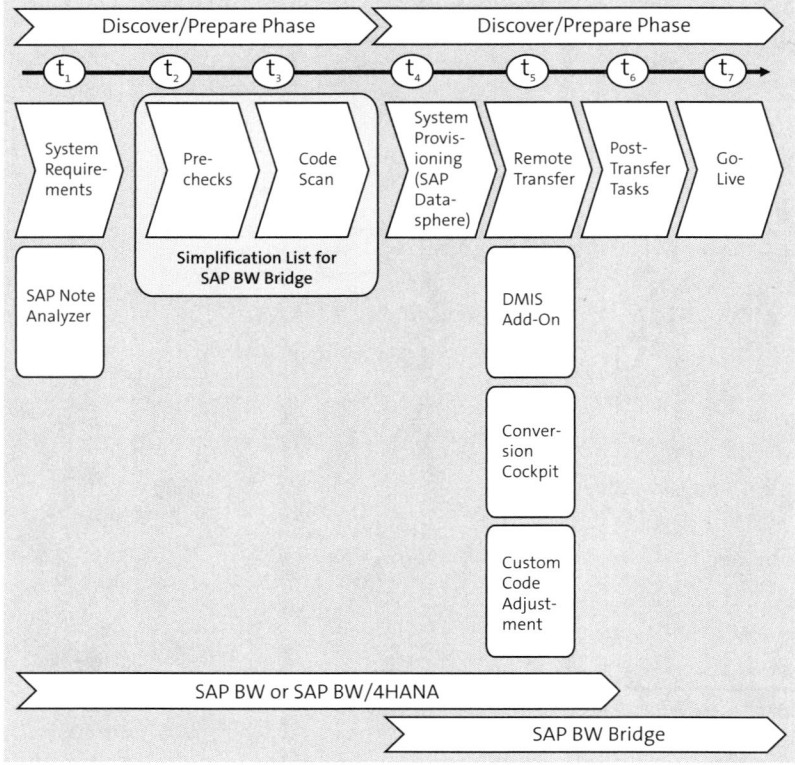

Figure 7.65 Remote Conversion Flow

Figure 7.66 Task List for Transfer Check Remote Conversion

The data model for remote conversion includes the SAP BW InfoCube (**ZPRBWBRC**) in the sender SAP BW system that will be used for the scope selection. You can see the data model in Figure 7.67.

Selecting the **Manage** option in Transaction RSA1 for the InfoCube will display the request details. You can see that there are **15 Transferred Records** and **15 Added Records** in Figure 7.68. In a SAP BW bridge remote conversion, the records will be transferred to the converted ADSO in SAP BW bridge.

7.4 Remote Conversion with SAP BW Bridge

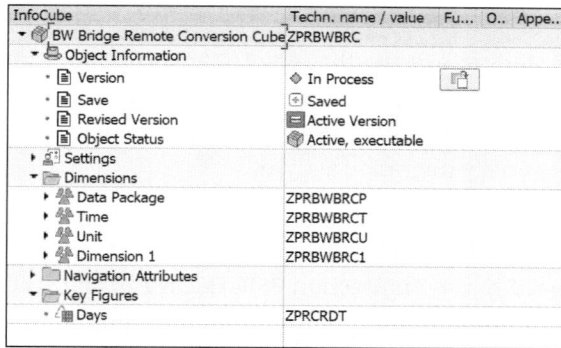

Figure 7.67 SAP BW InfoCube Model

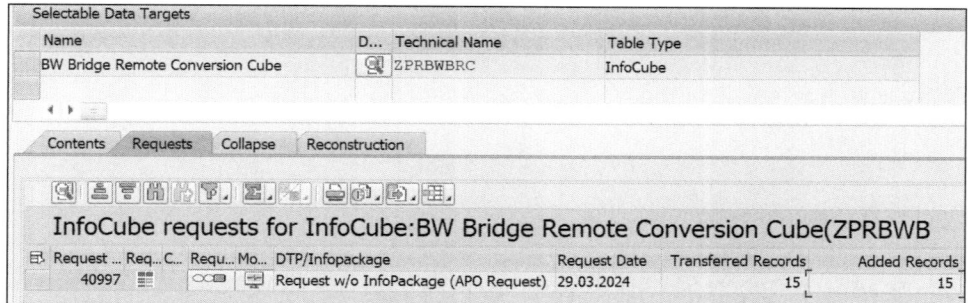

Figure 7.68 SAP BW InfoCube Records

If you check the contents of the SAP BW InfoCube using the **Contents** tab, you can see the data in the InfoCube, as shown in Figure 7.69.

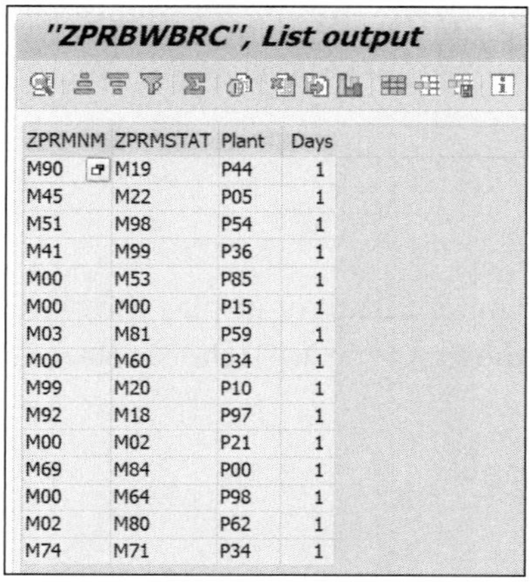

ZPRMNM	ZPRMSTAT	Plant	Days
M90	M19	P44	1
M45	M22	P05	1
M51	M98	P54	1
M41	M99	P36	1
M00	M53	P85	1
M00	M00	P15	1
M03	M81	P59	1
M00	M60	P34	1
M99	M20	P10	1
M92	M18	P97	1
M00	M02	P21	1
M69	M84	P00	1
M00	M64	P98	1
M02	M80	P62	1
M74	M71	P34	1

"ZPRBWBRC", List output

Figure 7.69 Data in the SAP BW InfoCube

In next section, we'll collect this InfoCube and send it to SAP BW bridge using remote conversion.

385

7.4.2 Execution

In this section, you'll learn about the steps to execute the remote conversion for SAP BW bridge. You'll use the Conversion Cockpit to do this activity, which requires the creation of an active package followed by the sequential execution of scope selection, object mapping setup, generation, and finally migration activities.

Package Creation

In the SAP BW sender system, you need to use Transaction RSB4HCONV to start the remote conversion. Choose the **Execute Scope Transfer** button, and the popup shown in Figure 7.70 appears.

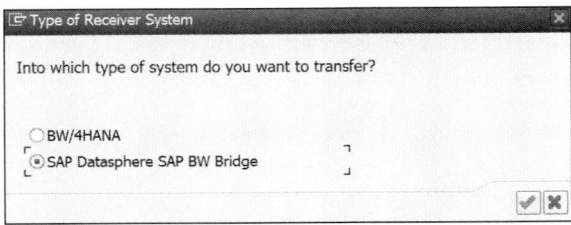

Figure 7.70 Receiver System Selection

Choose **SAP Datasphere SAP BW Bridge** and click on ✓ to see the next popup, as shown in Figure 7.71, prompting you to choose the transfer scenario. In our case, we'll choose the **Remote (Data and Metadata)** radio button.

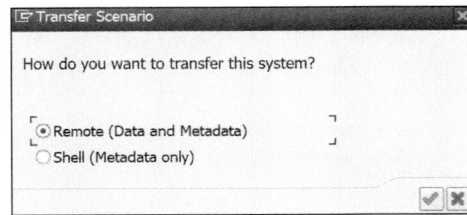

Figure 7.71 Transfer Scenario

In the next popup, you'll have the option to create a new package or display the existing package. Because we're creating the package for the first time, we'll choose **New**, as shown in Figure 7.72.

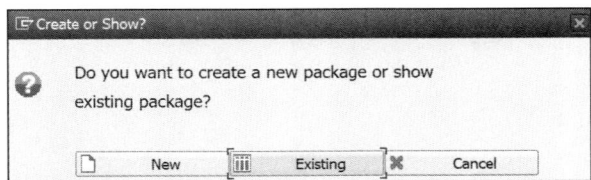

Figure 7.72 Package Type Selection

7.4 Remote Conversion with SAP BW Bridge

After confirming, the package will be created and shown in the process tree shown in Figure 7.73. You can see the **Package Number** as **9001P** and other details such as the **Package Description**. This package number will be same for the entire task list, and you can save this and open it anytime for the task list execution. You'll use **Existing** option if you're opening the same package next time.

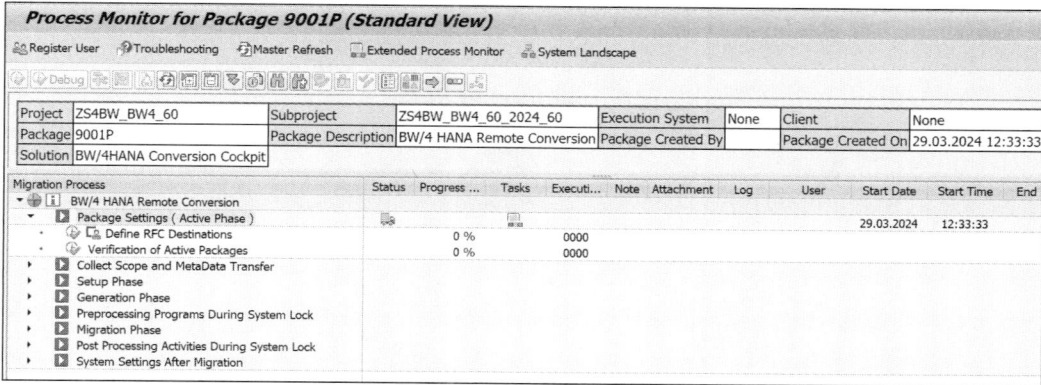

Figure 7.73 Package Settings

Package Settings

During this phase, you'll be able to designate the RFC destination and activate this package number to prevent any other user from initiating a remote conversion. The first step is to start the activity one by one in serial order in the process tree. To see the details, refer to Figure 7.73.

If you choose 🔍 in the **Define RFC Destinations** activity, the screen shown in Figure 7.74 will appear. Provide the RFC destination that was created to connect to the SAP BW bridge in the **Receiver system** input field; in our case, the destination name is **Migrate to DWC Remote**. Make sure to click **Test Connection** to ensure there is no error.

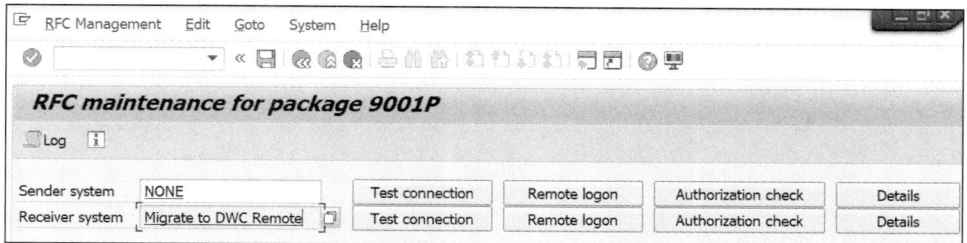

Figure 7.74 RFC for SAP BW Bridge

Once you've provided the RFC destination and the test appears green, you can choose to save the details. You'll see the popup shown in Figure 7.75.

387

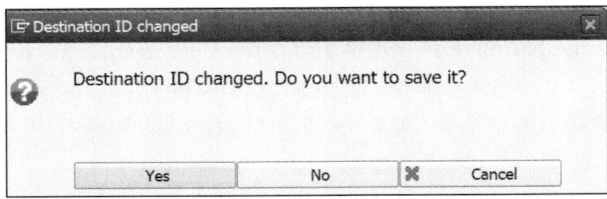

Figure 7.75 RFC Change Confirmation

Choose **Yes** to go to the execution with green status, as shown in Figure 7.76. If there is a red status, then you need to fix your RFC destination and re-execute this step again until it's green.

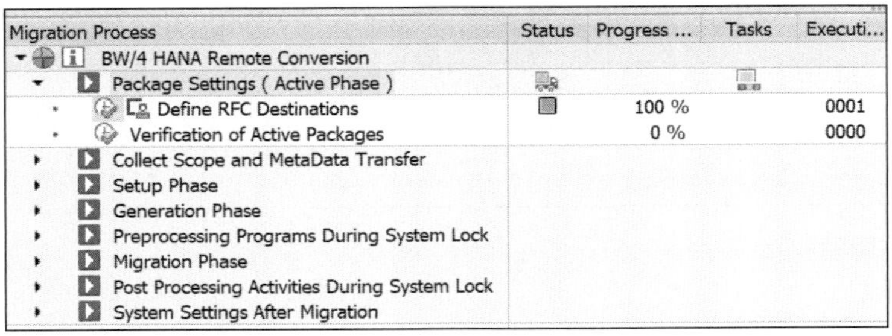

Figure 7.76 Package Settings: RFC Destination Status

The next step is to execute the **Verification of Active Packages** activity. This activity doesn't need any parameters. The activity will check if there is any active package in the system, and, if not, the current package **9001P** will be made active, and this step will be green. In our case, the step is green, as shown in Figure 7.77.

> **Note**
> Once the package is made active, other users can't start remote conversions until this package is deactivated.

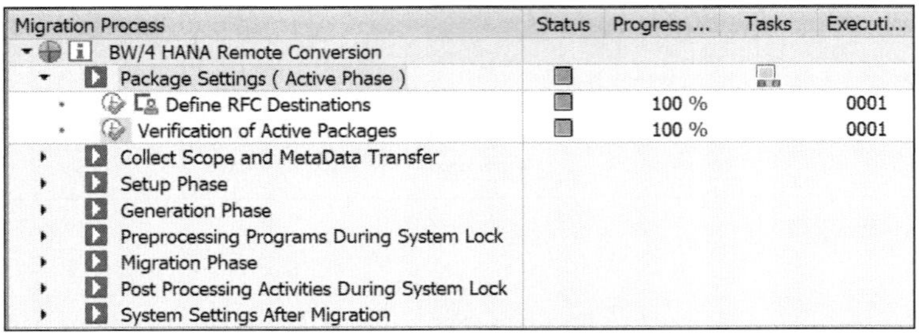

Figure 7.77 Verification of Active Packages: Status

7.4 Remote Conversion with SAP BW Bridge

Collect Scope and Metadata Transfer

The next step is very important as this will collect the scope and then transfer the objects to SAP BW bridge. We'll walk through all the activities in this step in the following sections.

Collect Scope for Transfer (Data Flow)

The first activity, **Collect Scope for Transfer (Data flow)**, can be executed using 🔘 in Figure 7.78.

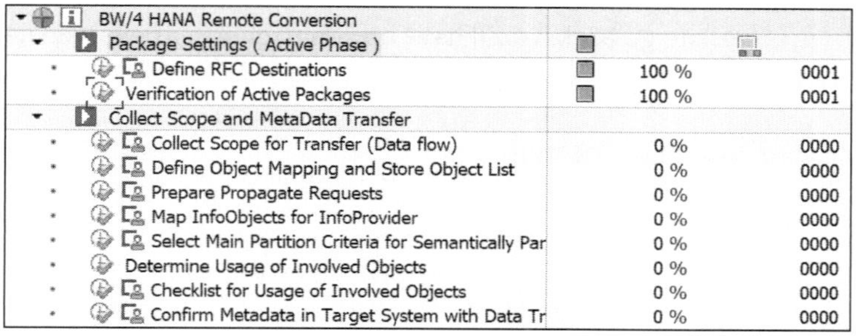

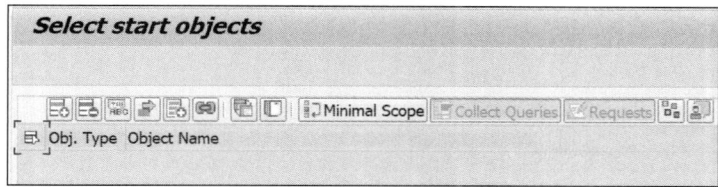

Figure 7.78 Scope and Metadata Transfer

Once you execute **Collect Scope for Transfer (Data flow)**, you'll see the options to add the objects for the scope. Choose the add start objects icon 📄, as shown in Figure 7.79.

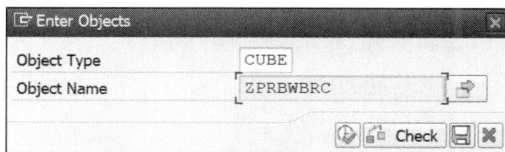

Figure 7.79 Start Object Selection Screen

Selecting the 📄 icon will bring up the popup shown in Figure 7.80. For our demo, we'll add the **Object Type** as **CUBE** and the **Object Name** as **ZPRBWBRC**. You can use the 📄 icon to add multiple objects if required.

Figure 7.80 Adding Scope Objects

Once you choose 🔘, you'll see the status of the added scope object, as shown in Figure 7.81.

389

7 SAP BW Bridge for SAP Datasphere

Figure 7.81 Final Scope Objects

Save the objects and then choose to go back. You'll have option to choose the job selection, as shown in Figure 7.82.

Figure 7.82 Job Type Selection

It's best practice to choose **Background**, which starts the job. You'll see the activity with running status (shown with the icon), as shown in Figure 7.83.

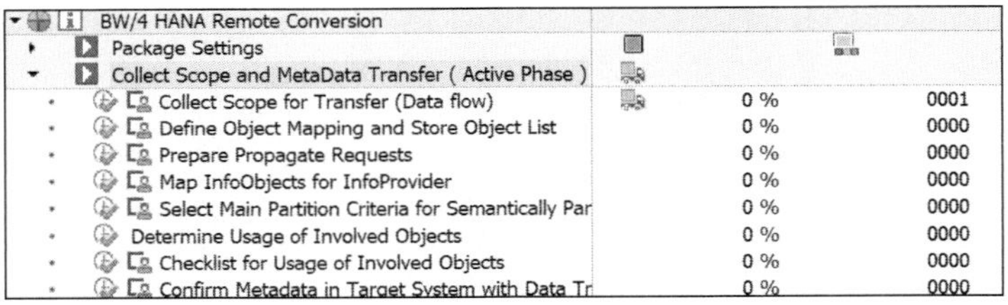

BW/4 HANA Remote Conversion		
▶ Package Settings		
▼ Collect Scope and MetaData Transfer (Active Phase)		
Collect Scope for Transfer (Data flow)	0 %	0001
Define Object Mapping and Store Object List	0 %	0000
Prepare Propagate Requests	0 %	0000
Map InfoObjects for InfoProvider	0 %	0000
Select Main Partition Criteria for Semantically Par	0 %	0000
Determine Usage of Involved Objects	0 %	0000
Checklist for Usage of Involved Objects	0 %	0000
Confirm Metadata in Target System with Data Tr	0 %	0000

Figure 7.83 Scope Transfer: Status

If the checks are successful, you'll see the green status and **100%** for the activity, which means this step is successful, and you can start the next step. The details are shown in Figure 7.84.

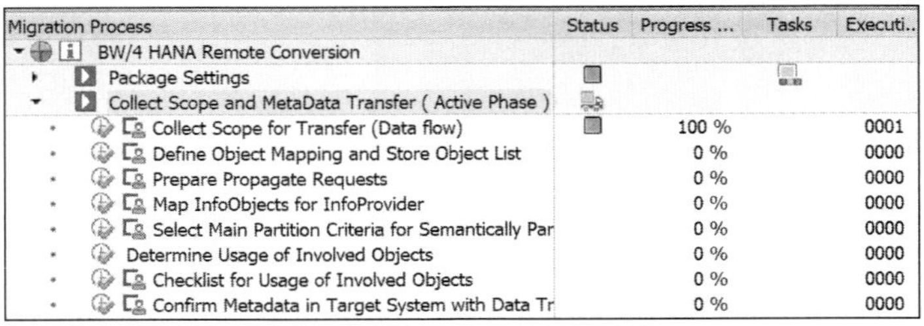

Migration Process	Status	Progress ...	Tasks	Executi..
▼ BW/4 HANA Remote Conversion				
▶ Package Settings				
▼ Collect Scope and MetaData Transfer (Active Phase)				
Collect Scope for Transfer (Data flow)		100 %		0001
Define Object Mapping and Store Object List		0 %		0000
Prepare Propagate Requests		0 %		0000
Map InfoObjects for InfoProvider		0 %		0000
Select Main Partition Criteria for Semantically Par		0 %		0000
Determine Usage of Involved Objects		0 %		0000
Checklist for Usage of Involved Objects		0 %		0000
Confirm Metadata in Target System with Data Tr		0 %		0000

Figure 7.84 Scope Transfer: Post-Execution

7.4 Remote Conversion with SAP BW Bridge

When the **Collect Scope for Transfer (Data flow)** step is initiated, regardless of its status (red or green), a task list run is generated in the background. You can locate the task list in Transaction STC02 based on your username. Our generated task list run is displayed in Figure 7.85.

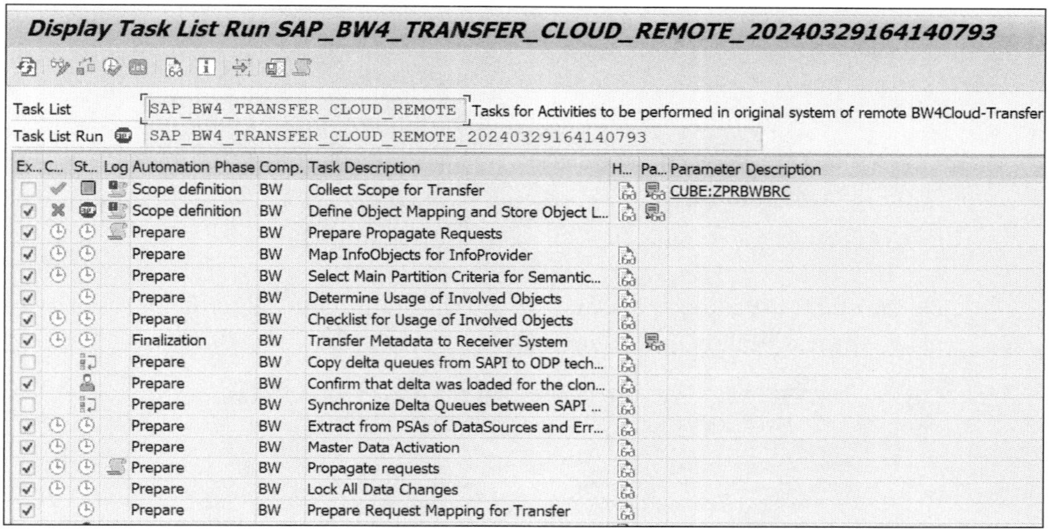

Figure 7.85 Task List for Active Package in Transaction STC02

Define Object Mapping and Store Object List

From the task list, we'll now come back to the process tree to start the next sequential activity, which is **Define Object Mapping and Store Object List**. The execution will show the popup seen in Figure 7.86. You can see the details of the **Original Object TLOGO** and **New TLOGO Object**, and the **Transfer** checkbox is selected, which means those objects will be transferred to the target SAP BW bridge.

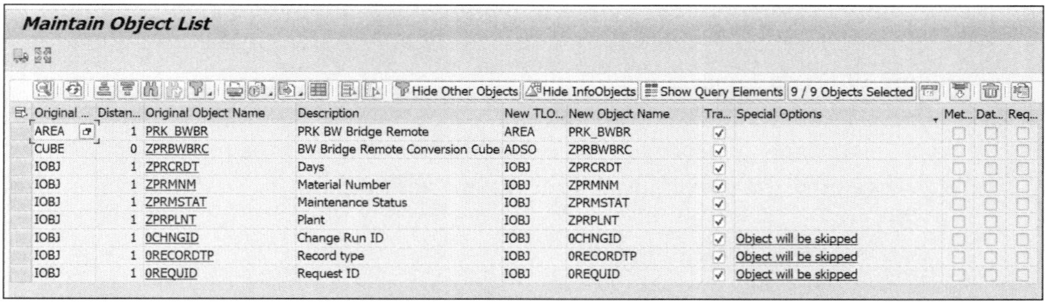

Figure 7.86 Object Mapping for the Selected Scope

Choose ![icon] to provide the **Package** and the associated **Request/Task** for the import of the object in the SAP BW bridge, and then press `Enter` (see Figure 7.87).

7 SAP BW Bridge for SAP Datasphere

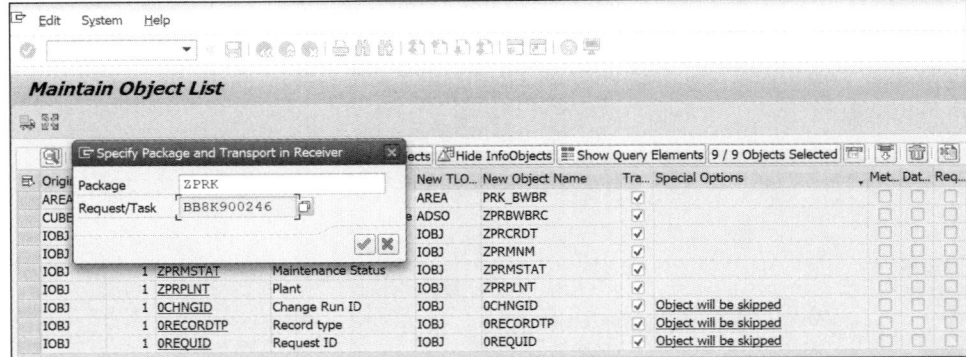

Figure 7.87 Package and Transport Request for SAP BW Bridge

Save the object list and go back, and the job will start in the background. The activity will be shown as running (see Figure 7.88).

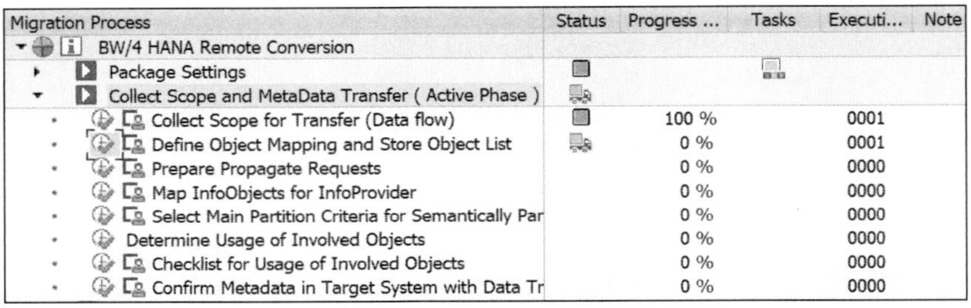

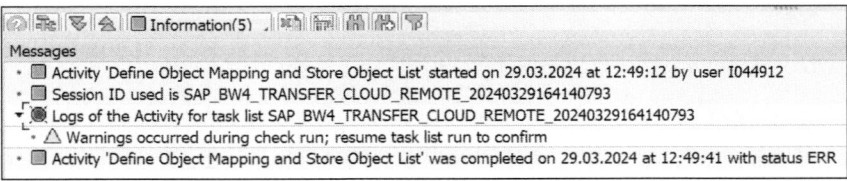

Figure 7.88 Object Mapping: Running Status

The preceding activity ended with a red status, as shown in Figure 7.89.

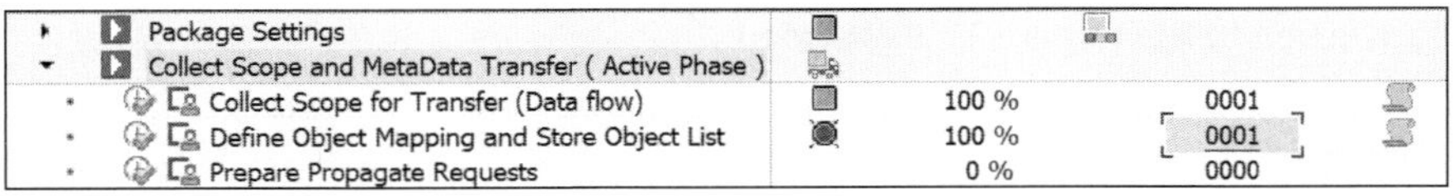

Figure 7.89 Object Mapping: Error

If you choose the icon, you can see the red warning and the reason for the error: **Warnings occurred during check run; resume task list run to confirm**, as shown in Figure 7.90.

Figure 7.90 Object Mapping: Error Text

7.4 Remote Conversion with SAP BW Bridge

Come back to the process tree, and execute the **Define Object Mapping and Store Object List** activity again, and it will be now green, as shown in Figure 7.91.

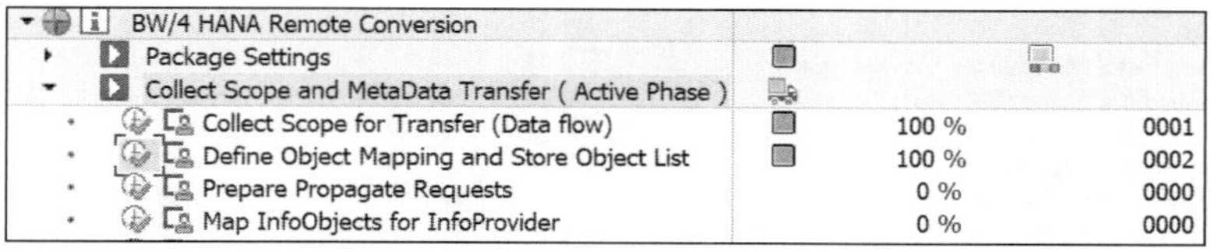

Figure 7.91 Object Mapping: Post Job Resume

Prepare Propagate Request

This step will collect the required DTP for the data mart update. Finish up other activities such as **Map InfoObjects for InfoProvider** and **Select Main Partition Criteria for Semantically Partitioned**, and the **Determine Usage of Involved Objects** activity will finish, as shown in Figure 7.92. Note that based on your scope objects, if there are any manual input requirements, then you need to finish those activities in each step and start the next step manually. If there is no manual input requirement, then the tool will stop before the **Checklist for Usage of Involved Objects** activity.

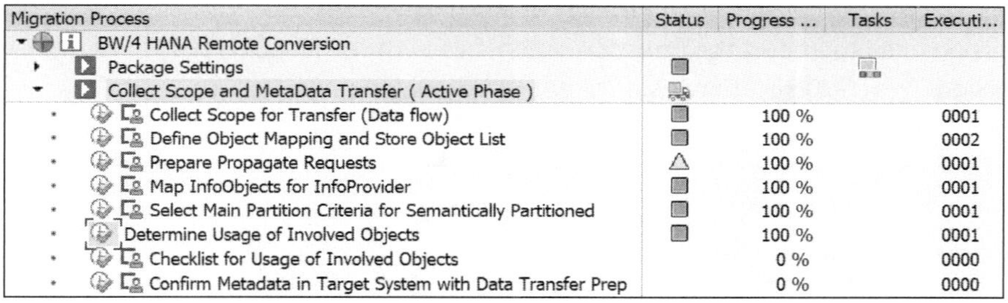

Figure 7.92 Status of Other Activities

Checklist for Usage of Involved Objects

If you execute the **Checklist for Usage of Involved Objects** step, you'll see the screen to choose the status for **BADI** and **AUTH**. Choose the checkbox for both because we've fixed those issues in the prework, as shown in Figure 7.93.

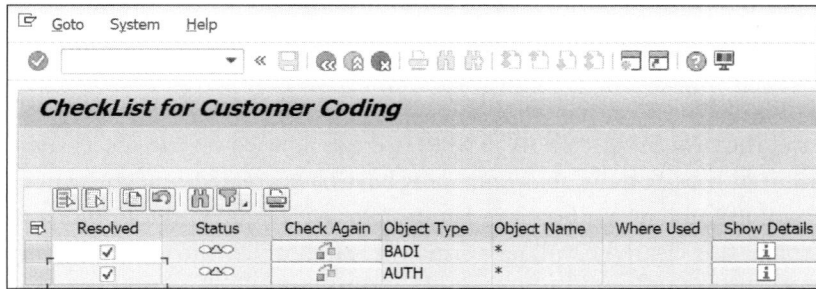

Figure 7.93 Customer Coding Checks

7 SAP BW Bridge for SAP Datasphere

Once completed, you can see the status of the execution, as shown in Figure 7.94.

Migration Process	Status	Progress ...	Tasks	Executi...
▼ 🌐 ⓘ BW/4 HANA Remote Conversion				
▶ ▶ Package Settings	▪		▪	
▼ ▶ Collect Scope and MetaData Transfer (Active Phase)	🔄			
• 🔄 🗂 Collect Scope for Transfer (Data flow)	▪	100 %		0001
• 🔄 🗂 Define Object Mapping and Store Object List	▪	100 %		0002
• 🔄 🗂 Prepare Propagate Requests	⚠	100 %		0001
• 🔄 🗂 Map InfoObjects for InfoProvider	▪	100 %		0001
• 🔄 🗂 Select Main Partition Criteria for Semantically Partitioned	▪	100 %		0001
• 🔄 Determine Usage of Involved Objects	▪	100 %		0001
• 🔄 🗂 Checklist for Usage of Involved Objects	▪	100 %		0001
• 🔄 🗂 Confirm Metadata in Target System with Data Transfer Prep		0 %		0000
▶ ▶ Setup Phase				
▶ ▶ Generation Phase				

Figure 7.94 Usage of Involved Objects Status

The next step is to check the SAP BW bridge and ensure the objects are available. If that's the case, you can choose to execute the **Confirm Metadata in Target System** and choose **Confirm** in the popup. This will set the value to green or yellow, as shown in Figure 7.95.

> **SAP BW Bridge versus SAP BW/4HANA: Confirm Metadata**
>
> There is no need to import the transport request as with the SAP BW/4HANA remote conversion. This step execution is sufficient to manually confirm.

Migration Process	Status	Progress ...	Tasks	Executi...
▼ 🌐 ⓘ BW/4 HANA Remote Conversion				
▶ ▶ Package Settings	▪		▪	
▼ ▶ Collect Scope and MetaData Transfer (Active Phase)	⚠			
• 🔄 🗂 Collect Scope for Transfer (Data flow)	▪	100 %		0001
• 🔄 🗂 Define Object Mapping and Store Object List	▪	100 %		0002
• 🔄 🗂 Prepare Propagate Requests	⚠	100 %		0001
• 🔄 🗂 Map InfoObjects for InfoProvider	▪	100 %		0001
• 🔄 🗂 Select Main Partition Criteria for Semantically Partitioned	▪	100 %		0001
• 🔄 Determine Usage of Involved Objects	▪	100 %		0001
• 🔄 🗂 Checklist for Usage of Involved Objects	▪	100 %		0001
• 🔄 🗂 Confirm Metadata in Target System with Data Transfer Prep	⚠	100 %		0001
▶ ▶ Setup Phase				
▶ ▶ Generation Phase				

Figure 7.95 Confirm Metadata in the Target System: SAP BW Bridge

Check Metadata in SAP BW bridge

Log on to the SAP BW bridge in SAP Datasphere in the SAP BW modeling tools, and you can see the details of the converted objects; for example, the InfoCube is now ADSO (**ZPRBWBRC**), and the new InfoArea (**PRK_BWBR**) is created based on the object mapping. The InfoObjects based on the object mapping have been created, and you can see the details in SAP BW bridge, as shown in Figure 7.96. We've added the InfoArea to favorites to explain this clearly; it will be normally created under the **BW Repository** only.

7.4 Remote Conversion with SAP BW Bridge

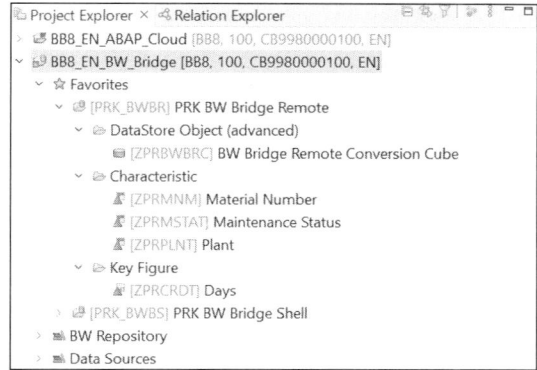

Figure 7.96 SAP BW Bridge: Object Status

Navigate to the SAP BW bridge cockpit, and access the **Datasphere Manage** section to obtain information about the converted InfoCube, now represented as an ADSO (see Figure 7.97).

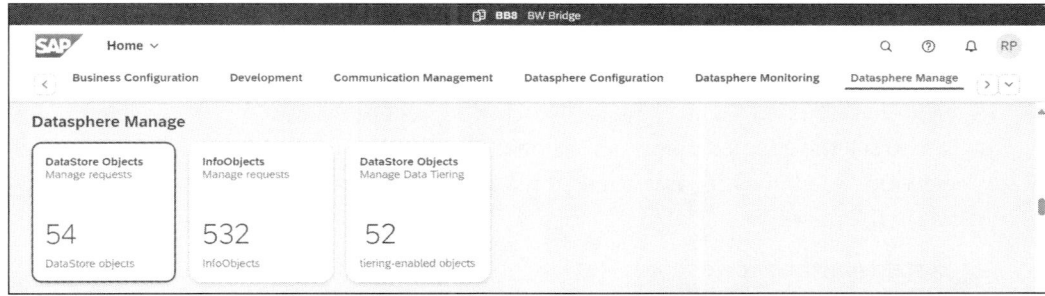

Figure 7.97 SAP BW Bridge Cockpit: Manage

Open the **DataStore Object** tile and search for the ADSO (**ZPRBWBRC**) as shown in Figure 7.98.

Figure 7.98 Manage DSOs in SAP BW Bridge Cockpit

395

You can see that the active table of the ADSO doesn't have any data because the data migration hasn't yet started, as shown in Figure 7.99.

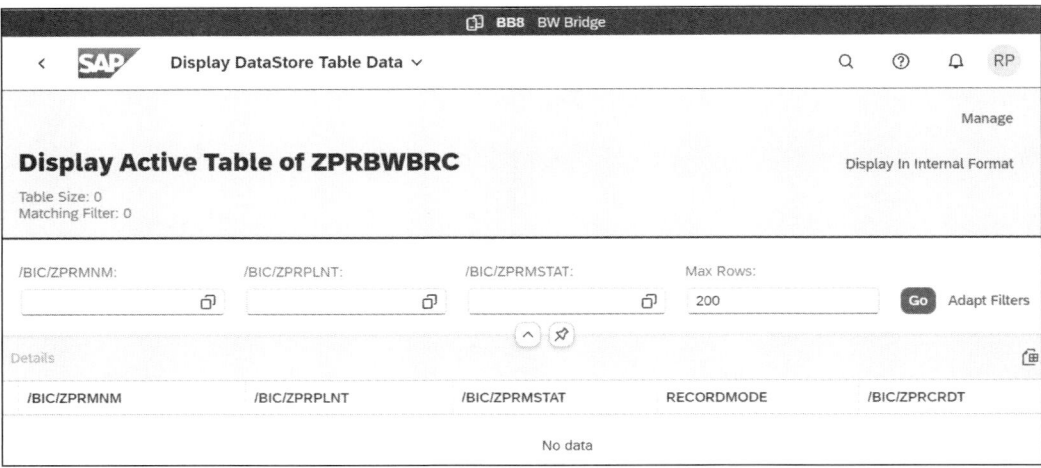

Figure 7.99 Active Table of ADSO in SAP BW Bridge

Setup Phase

We'll now focus on the next phase, which will prepare the system for the data migration, to see the detailed objects in the phase. Select the **Choose View** icon, and then select **Extended View**, as shown in Figure 7.100. This view will show the additional activities in the phase.

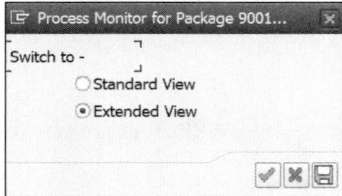

Figure 7.100 View Selection

Figure 7.101 shows the list of activities in the **Setup Phase** when the **Extended View** is selected. It has both group activity and manual activity.

Next, we'll execute the **Make Settings for Selected BW Object Tables** group activity. This group activity will, by default, start other activities under it. You can see that the activity is now running, as shown in Figure 7.102. This will take care of the activities in the original system and the target system.

Once all the group activities are completed, the tool will stop without execution if there is any activity that requires manual input, as you can see in Figure 7.103. The **Conversion Setup (Logical System Rename)** step has a few activities that need manual execution.

7.4 Remote Conversion with SAP BW Bridge

Project	ZS4BW_BW4_60	Subproject	ZS4BW_BW4_60_2024_60	Execution System	
Package	9001P		Package Description	BW/4 HANA Remote Conversion	Package Created B
Solution	BW/4HANA Conversion Cockpit				

Migration Process	Status	Progress ...	Tasks	Executi...	Note
▼ ▶ Setup Phase					
▼ Make Settings for Selected BW Object Tables		0 %		0000	
▼ ▶ Original System				0000	
· Read Object Mappings		0 %		0000	
· Identify BW Table Pools for Data Migration		0 %		0000	
· Collect and Transfer Package Settings Information		0 %		0000	
▼ ▶ Target System				0000	
· Identify Semantic Differences between InfoCubes and ADSOs		0 %		0000	
· Identify InfoCubes for Migration		0 %		0000	
· Generate Dimension and Fact Tables for InfoCubes		0 %		0000	
· Build Table Mappings between Mapped Objects		0 %		0000	
· Display DDIC Differences Between Mapped Objects		0 %		0000	
· Configure MIGSID Conversion Rule for Master Data ID Fields		0 %		0000	
· Make Settings to Facilitate Program Generation		0 %		0000	
▼ ▶ Conversion Setup (Logical System Rename)				0000	
· Select Conversion-Relevant InfoObjects		0 %		0000	
· Define Mapping Values		0 %		0000	
▶ Make Settings for Conversion-Relevant InfoObjects		0 %		0000	
· View BW Objects in the Tablepool Browser (Optional)				0000	

Figure 7.101 Activities in the Setup Phase

Migration Process	Status	Progress ...	Tasks	Executi...	Note
▼ ▶ Setup Phase (Active Phase)	🚚				
▼ Make Settings for Selected BW Object Tables	🚚	0 %		0001	
▼ ▶ Original System				0000	
· Read Object Mappings	🚚	0 %		0001	
· Identify BW Table Pools for Data Migration		0 %		0000	
· Collect and Transfer Package Settings Information		0 %		0000	
▼ ▶ Target System				0000	
· Identify Semantic Differences between InfoCubes and ADSOs		0 %		0000	
· Identify InfoCubes for Migration		0 %		0000	
· Generate Dimension and Fact Tables for InfoCubes		0 %		0000	
· Build Table Mappings between Mapped Objects		0 %		0000	
· Display DDIC Differences Between Mapped Objects		0 %		0000	
· Configure MIGSID Conversion Rule for Master Data ID Fields		0 %		0000	
· Make Settings to Facilitate Program Generation		0 %		0000	
▼ ▶ Conversion Setup (Logical System Rename)				0000	
· Select Conversion-Relevant InfoObjects		0 %		0000	
· Define Mapping Values		0 %		0000	
▶ Make Settings for Conversion-Relevant InfoObjects		0 %		0000	
· View BW Objects in the Tablepool Browser (Optional)				0000	

Figure 7.102 Setup Phase: Execution

Migration Process	Status	Progress ...	Tasks	Executi...
▼ ▶ Target System				0000
· Identify Semantic Differences between InfoCubes and ADSOs	▢	100 %		0001
· Identify InfoCubes for Migration	▢	100 %		0001
· Generate Dimension and Fact Tables for InfoCubes	▢	100 %		0001
· Build Table Mappings between Mapped Objects	△	100 %		0001
· Display DDIC Differences Between Mapped Objects	▢	100 %		0001
· Configure MIGSID Conversion Rule for Master Data ID Fields	▢	100 %		0001
· Make Settings to Facilitate Program Generation	△	100 %		0001
▼ ▶ Conversion Setup (Logical System Rename)				0000
· Select Conversion-Relevant InfoObjects		0 %		0000
· Define Mapping Values		0 %		0000
▶ Make Settings for Conversion-Relevant InfoObjects		0 %		0000
· View BW Objects in the Tablepool Browser (Optional)				0000

Figure 7.103 Setup Phase: Manual Activities

Execute the manual activities in serial order, as shown in Figure 7.104.

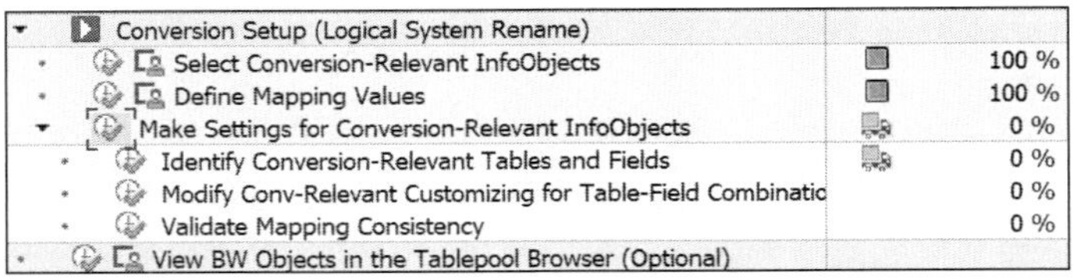

Conversion Setup (Logical System Rename)		
Select Conversion-Relevant InfoObjects	☐	100 %
Define Mapping Values	☐	100 %
Make Settings for Conversion-Relevant InfoObjects		0 %
Identify Conversion-Relevant Tables and Fields		0 %
Modify Conv-Relevant Customizing for Table-Field Combinatio		0 %
Validate Mapping Consistency		0 %
View BW Objects in the Tablepool Browser (Optional)		

Figure 7.104 Setup Phase: Settings for Conversion-Relevant InfoObjects

Once all the activities are completed in the setup phase, you'll see the tool with green and yellow statuses, as shown in Figure 7.105.

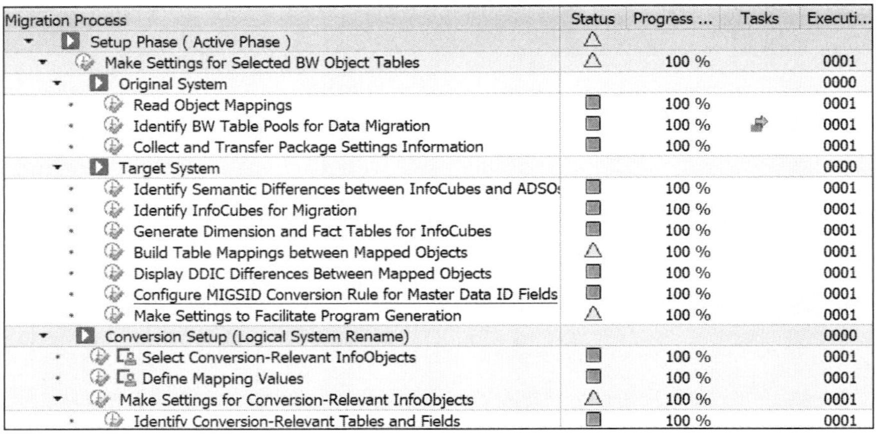

Migration Process	Status	Progress ...	Tasks	Executi...
Setup Phase (Active Phase)	△			
Make Settings for Selected BW Object Tables	△	100 %		0001
Original System				0000
Read Object Mappings	☐	100 %		0001
Identify BW Table Pools for Data Migration	☐	100 %	➡	0001
Collect and Transfer Package Settings Information	☐	100 %		0001
Target System				0000
Identify Semantic Differences between InfoCubes and ADSO	☐	100 %		0001
Identify InfoCubes for Migration	☐	100 %		0001
Generate Dimension and Fact Tables for InfoCubes	☐	100 %		0001
Build Table Mappings between Mapped Objects	△	100 %		0001
Display DDIC Differences Between Mapped Objects	☐	100 %		0001
Configure MIGSID Conversion Rule for Master Data ID Fields	☐	100 %		0001
Make Settings to Facilitate Program Generation	△	100 %		0001
Conversion Setup (Logical System Rename)				0000
Select Conversion-Relevant InfoObjects	☐	100 %		0001
Define Mapping Values	☐	100 %		0001
Make Settings for Conversion-Relevant InfoObjects	△	100 %		0001
Identify Conversion-Relevant Tables and Fields	☐	100 %		0001

Figure 7.105 Setup Phase: Final Status

Generation Phase

The next step is the generation phase, where the report for the migration is generated. The initial activity list is shown in Figure 7.106.

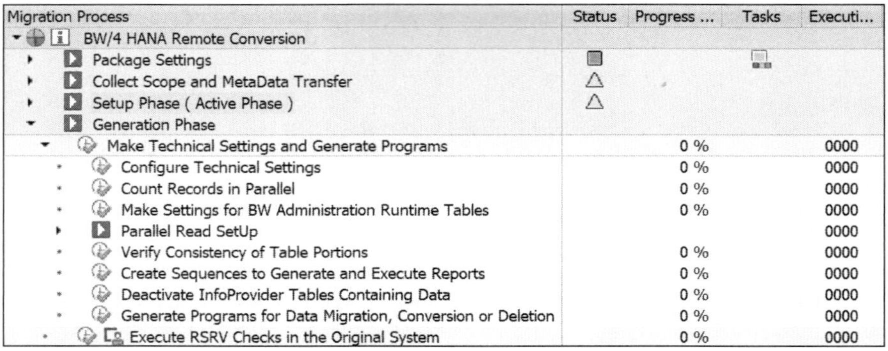

Migration Process	Status	Progress ...	Tasks	Executi...
BW/4 HANA Remote Conversion				
Package Settings	☐			
Collect Scope and MetaData Transfer	△			
Setup Phase (Active Phase)	△			
Generation Phase				
Make Technical Settings and Generate Programs		0 %		0000
Configure Technical Settings		0 %		0000
Count Records in Parallel		0 %		0000
Make Settings for BW Administration Runtime Tables		0 %		0000
Parallel Read SetUp				0000
Verify Consistency of Table Portions		0 %		0000
Create Sequences to Generate and Execute Reports		0 %		0000
Deactivate InfoProvider Tables Containing Data		0 %		0000
Generate Programs for Data Migration, Conversion or Deletion		0 %		0000
Execute RSRV Checks in the Original System		0 %		0000

Figure 7.106 Generation Phase: Before Execution

7.4 Remote Conversion with SAP BW Bridge

You have the option to run the group activity **Make Technical Settings and Generate Programs**. This process handles technical configurations, ensures the consistency of table portions, establishes sequences for generating and executing reports, populates conversion tables to maintain these sequences, deactivates the InfoProvider if data already exists in the target system, and ultimately runs the **Generate Programs for Data Migration, Conversion, or Deletion** activity, as shown in Figure 7.107.

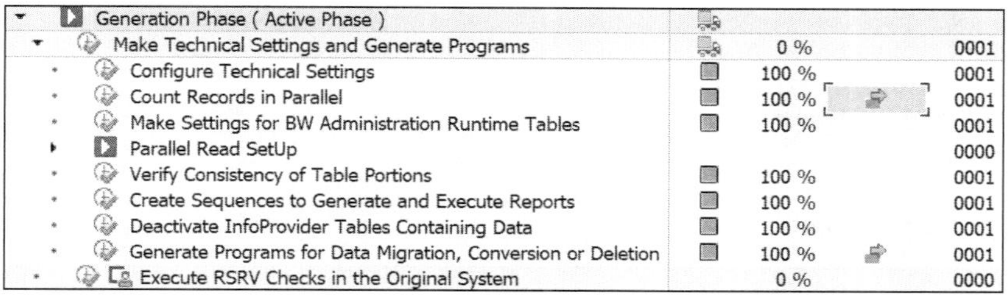

Generation Phase (Active Phase)			
Make Technical Settings and Generate Programs		0 %	0001
Configure Technical Settings	☐	100 %	0001
Count Records in Parallel	☐	100 %	0001
Make Settings for BW Administration Runtime Tables	☐	100 %	0001
Parallel Read SetUp			0000
Verify Consistency of Table Portions	☐	100 %	0001
Create Sequences to Generate and Execute Reports	☐	100 %	0001
Deactivate InfoProvider Tables Containing Data	☐	100 %	0001
Generate Programs for Data Migration, Conversion or Deletion	☐	100 %	0001
Execute RSRV Checks in the Original System		0 %	0000

Figure 7.107 Generation Phase: Post Execution

Once all the group activities are completed, you can execute the manual activity to execute Transaction RSRV. In the popup, you save the **RSRV Package ID** for later execution and execute the activity using ⊕ to confirm it, as shown in Figure 7.108.

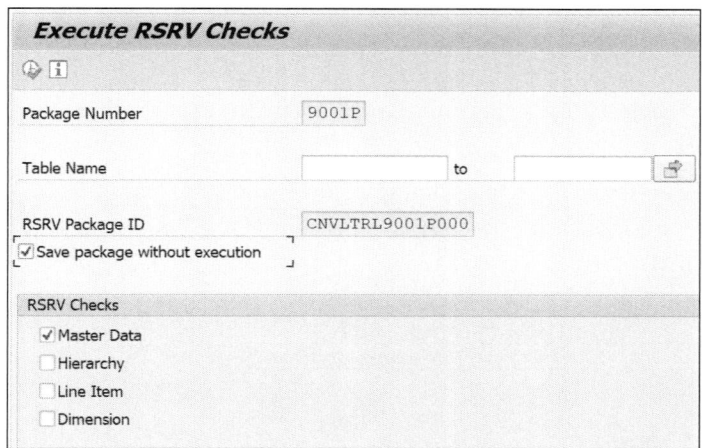

Figure 7.108 Generation Phase: Transaction RSRV Package Selection

Once **Save package without execution** is checked, note down the **RSRV Package ID** for checks, and use ⊕ to make the status green, as shown in Figure 7.109.

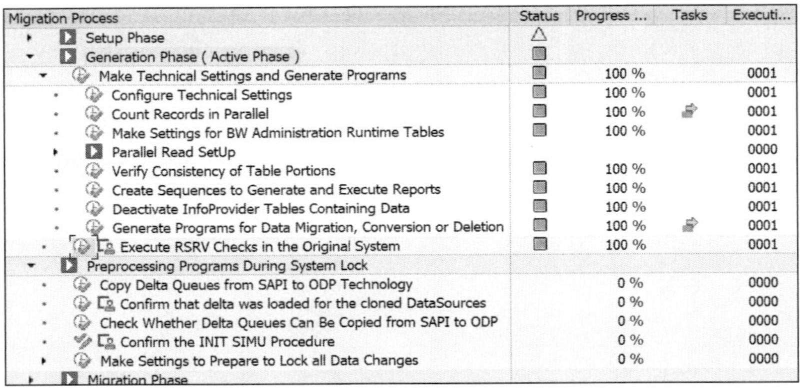

7 SAP BW Bridge for SAP Datasphere

Figure 7.109 Generation Phase: After Completion

Preprocessing Programs during System Lock

During this step, clones for the DataSource in both SAP ERP and SAP S/4HANA will be created initially. Following this, it's necessary to manually confirm the success of the delta for the cloned DataSource in the sender SAP BW system. If any DataSource lacks Init Simulation support, make a note of it for subsequent execution.

In the next group activity, **Make Settings to Prepare to Lock all Data Changes**, the tool will synchronize the delta in Transaction RSA7 and Transaction ODQMON in the sender system. Subsequently, it will propagate the request in the sender SAP BW system, followed by locking all scope objects. This will result in downtime for data loading in the sender SAP BW system.

Technically, your SAP BW system will be in a complete downtime state from this **Lock** status, and it will only be unlocked after the migration is completed. You can find the list of activities in Figure 7.110.

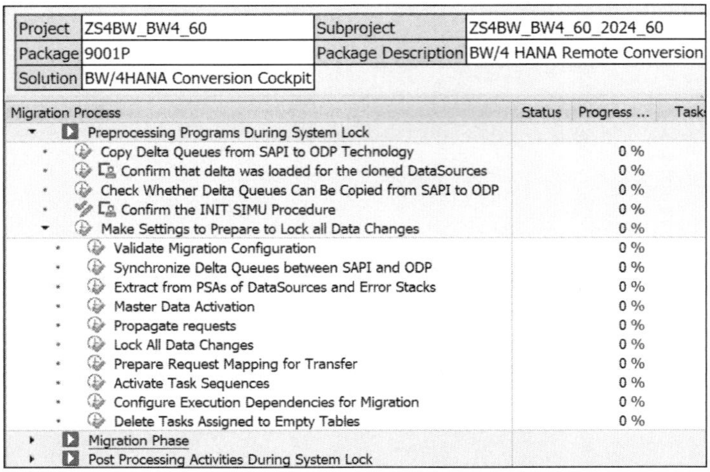

Figure 7.110 Preprocessing Programs during System Lock: Before Execution

7.4 Remote Conversion with SAP BW Bridge

After all the activity is completed, you can see the status as shown in Figure 7.111.

Preprocessing Programs During System Lock (Active Phase)		
Copy Delta Queues from SAPI to ODP Technology	100 %	0001
Confirm that delta was loaded for the cloned DataSources	100 %	0001
Check Whether Delta Queues Can Be Copied from SAPI to ODP	100 %	0001
Confirm the INIT SIMU Procedure	100 %	0001
Make Settings to Prepare to Lock all Data Changes	0 %	0001
Validate Migration Configuration	100 %	0001
Synchronize Delta Queues between SAPI and ODP	100 %	0002
Extract from PSAs of DataSources and Error Stacks	100 %	0001
Master Data Activation	100 %	0001
Propagate requests	100 %	0001
Lock All Data Changes	100 %	0001
Prepare Request Mapping for Transfer	100 %	0001
Activate Task Sequences	100 %	0001
Configure Execution Dependencies for Migration	100 %	0001
Delete Tasks Assigned to Empty Tables	100 %	0001
Migration Phase		

Figure 7.111 Preprocessing Programs during System Lock: Post Completion

Migration Phase

In this phase, you'll do the actual data migration from the sender SAP BW system to SAP BW bridge. The list of activities is shown in Figure 7.112.

Migration Process	Status	Progress ...	Tasks	Executi...
BW/4 HANA Remote Conversion				
Package Settings				
Collect Scope and MetaData Transfer				
Setup Phase				
Generation Phase				
Preprocessing Programs During System Lock				
Migration Phase (Active Phase)				
Make Runtime Settings for Migration		0 %		0000
Reset Cluster Data Deletion Date				0000
Start Migration		0 %		0000
Monitor: Transformation Status (Optional)				0000
Confirm Data Selection and Unlock Data Loading		0 %		0000
Verify Task Completion		0 %		0000
Post Processing Activities During System Lock				
System Settings After Migration				

Figure 7.112 Migration Phase: Before Execution

Execute the **Make Runtime Settings for Migration** step to set the jobs to be used in the sender and receiver system, as shown in Figure 7.113.

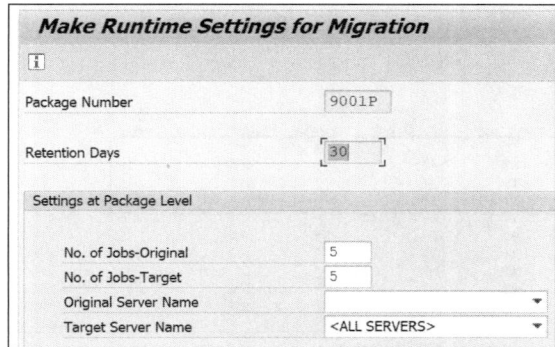

Figure 7.113 Runtime Settings in the Migration Phase

You can maintain the **Deletion Date for Cluster Data** in the next activity; just maintain the date until which you need the cluster tables, as shown in Figure 7.114.

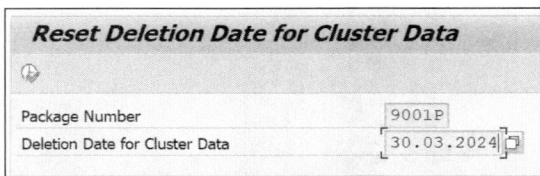

Figure 7.114 Selection of Cluster Data Deletion

Figure 7.115 shows the status after maintaining the **Deletion Date for Cluster Data** activities.

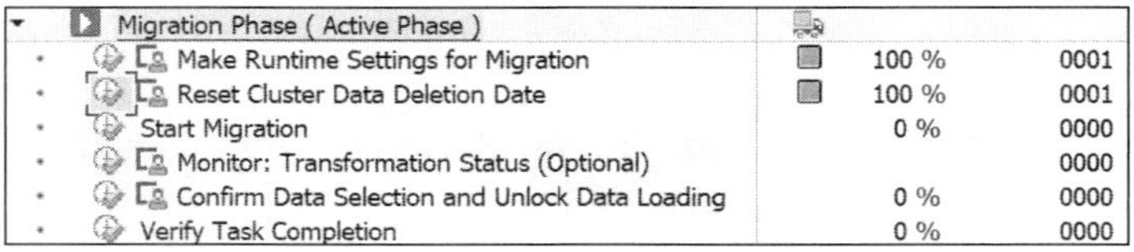

Figure 7.115 Status before Migration

The next step is to start the migration and execute the **Start Migration** activity. And The tool will start the actual data transfer from the sender SAP BW system to SAP BW bridge, and you can see that the step is running in Figure 7.116.

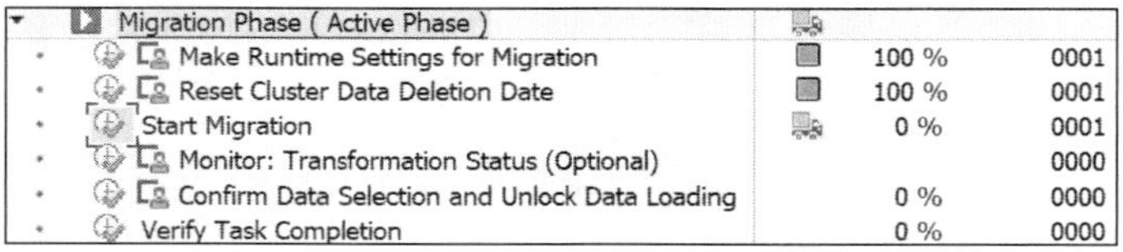

Figure 7.116 Start Migration: Running Status

You can see the status of the migration using the **Monitor Transformation Status** screen with the data shown in Figure 7.117.

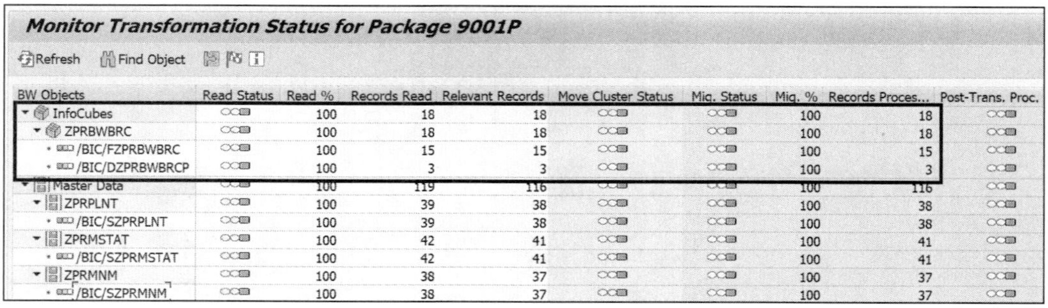

Figure 7.117 Monitor Migration Status

7.4 Remote Conversion with SAP BW Bridge

You can see 15 records are transferred per the monitoring. Now complete all the other steps in the cockpit, and deactivate the package.

Log on to the SAP Datasphere Cockpit, open the ADSO active table, and you can see that **15 records** are now available after the migration, as shown in Figure 7.118.

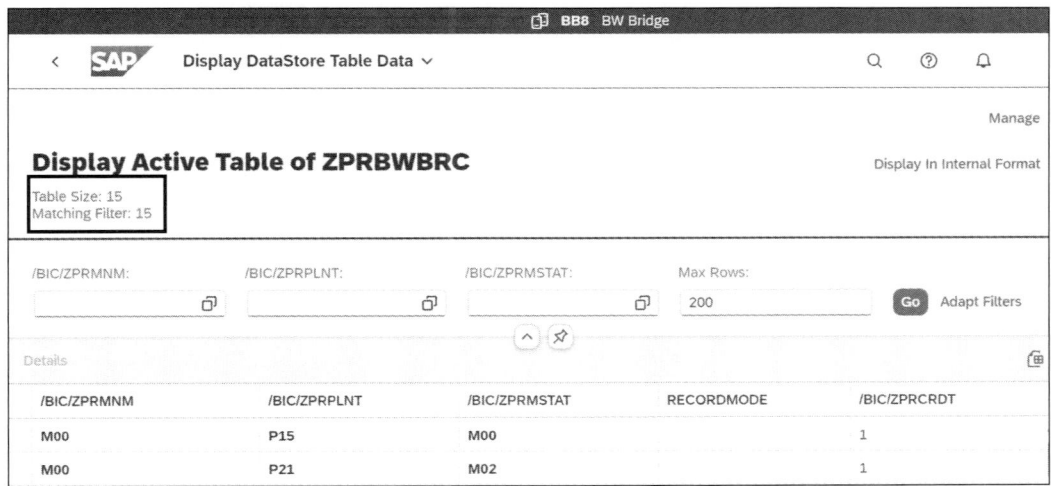

Figure 7.118 SAP BW Bridge ADSO Content: Post Migration

For reconciliation, you can refer to the InfoCube (**ZPRBWBRC**) in the SAP BW sender system and its actual records. You can see that the InfoCube had the same **15** records, as shown in Figure 7.119.

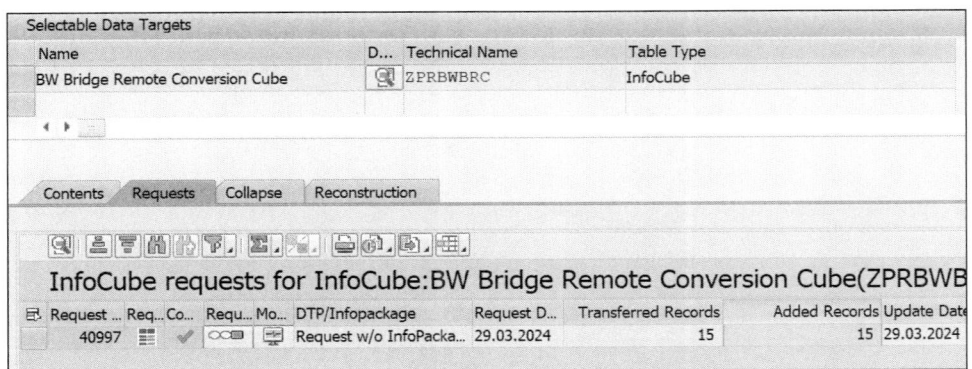

Figure 7.119 InfoCube data in the SAP BW Sender System

Once the package is completed, you can see that the task list is also finalized 🔒, as shown in Figure 7.120.

403

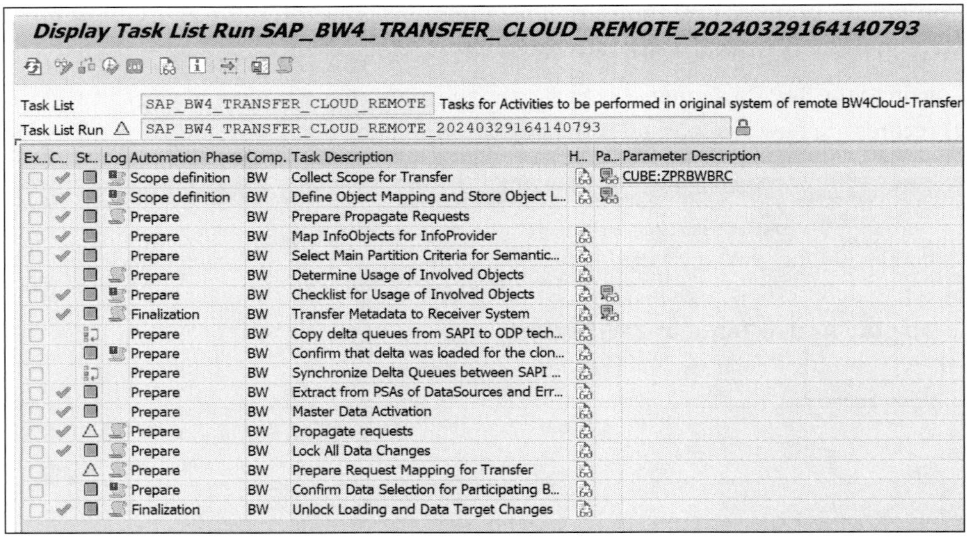

7 SAP BW Bridge for SAP Datasphere

![Display Task List Run SAP_BW4_TRANSFER_CLOUD_REMOTE_20240329164140793](task-list-screenshot)

Figure 7.120 Completed Package and Associated Task List

7.4.3 Post-Migration Options in SAP Datasphere

In SAP Datasphere, you can now import these ADSOs as entities in the Data Builder, as shown in Figure 7.121.

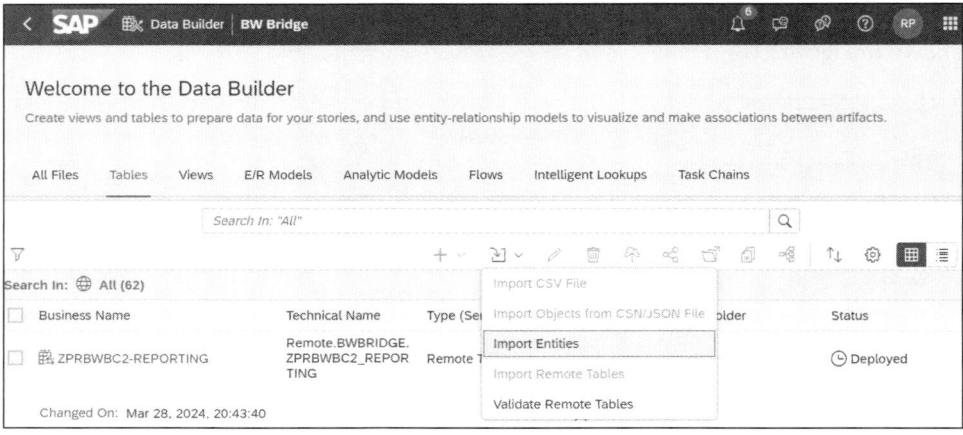

Figure 7.121 Deploying the ADSO from SAP BW Bridge

There will be a series of steps to import and deploy from SAP BW bridge, as shown in Figure 7.122.

You can select the ADSO that you've converted with data from the SAP BW sender. It's selected in this process, as shown in Figure 7.123.

Finally, you can import and deploy the selected entities, as shown in Figure 7.124.

7.4 Remote Conversion with SAP BW Bridge

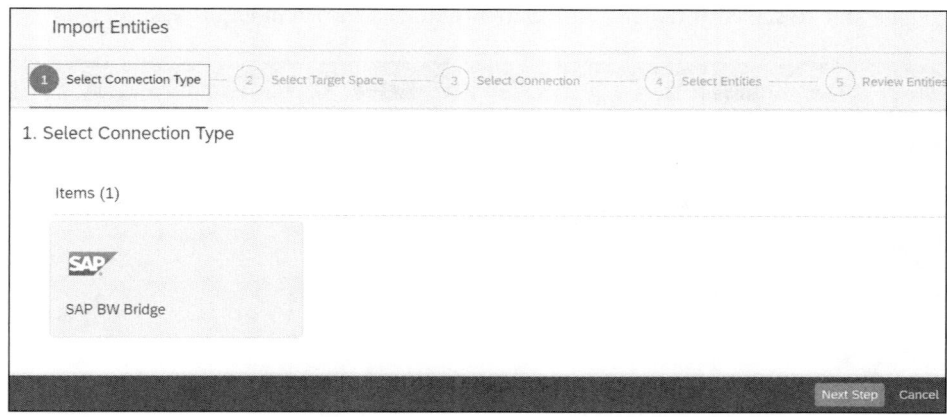

Figure 7.122 Import Process

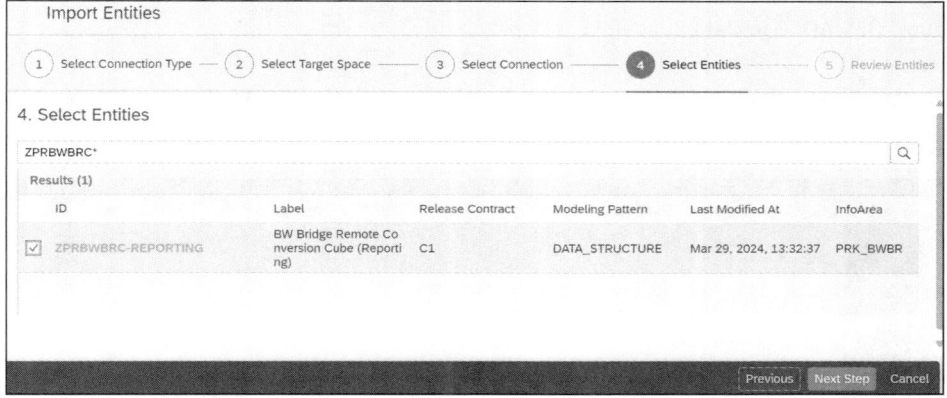

Figure 7.123 ADSO Selection

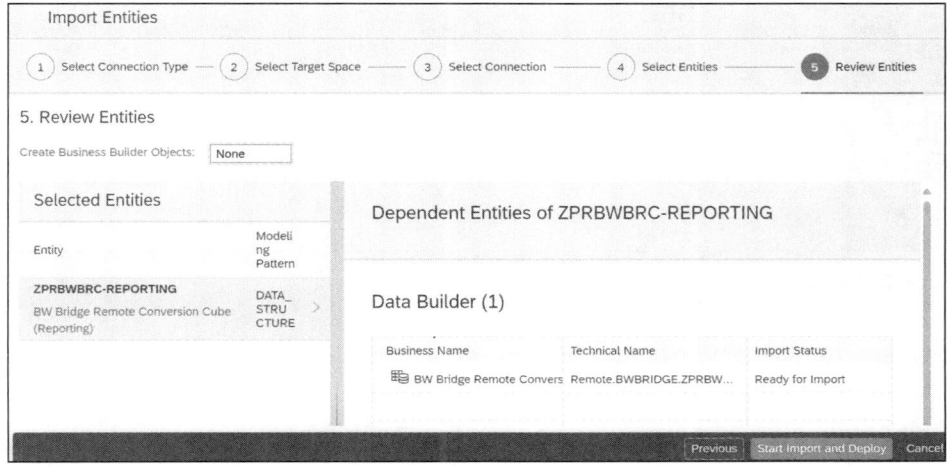

Figure 7.124 Deployment Confirmation

You'll be able to see your tables in the Data Builder, as shown in Figure 7.125.

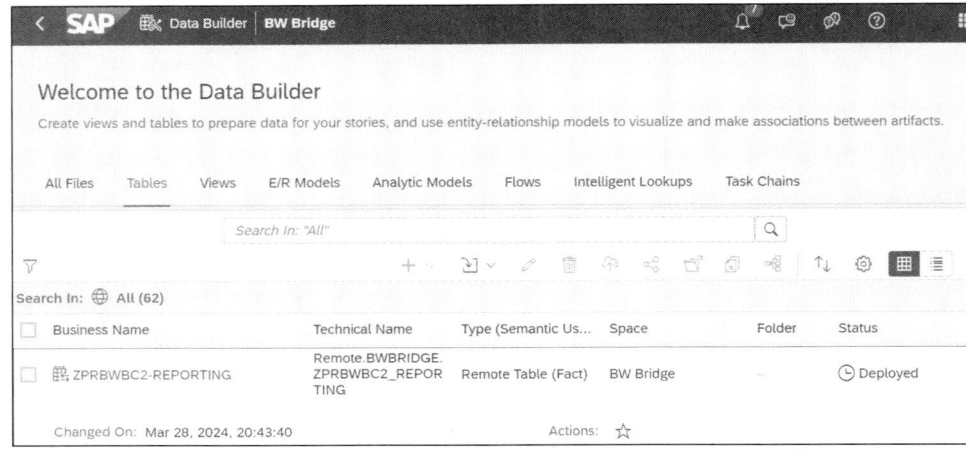

Figure 7.125 ADSO in Data Builder: Post Deployment

You can see the data in Figure 7.126 and use the converted ADSO for any purpose with multiple modeling options in SAP Datasphere.

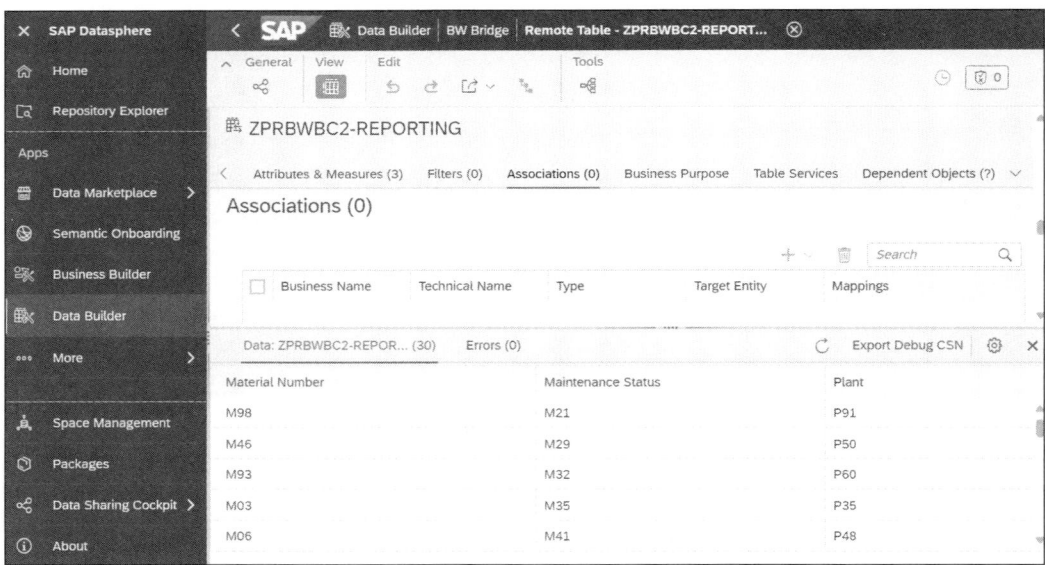

Figure 7.126 Data Builder in SAP Datasphere

There are multiple options to perform the data modeling, which you can find in the Data Modeling part of SAP Datasphere.

7.5 Summary

By completing this chapter, you've learned how to convert from SAP BW or SAP BW/4HANA to SAP BW bridge in SAP Datasphere. You've now reached the end of the book. We hope you found this book useful and that it will be a valuable tool as you continue to learn, explore, and work with SAP conversion to SAP BW/4HANA and SAP BW bridge in SAP Datasphere.

The Author

 **Renjith Kumar Palaniswamy** is an SAP BW/4HANA architect with more than 15 years of experience working with SAP BW, SAP BW/4HANA, SAP HANA, and SAP S/4HANA. Renjith is currently an SAP BW/4HANA services architect at SAP America, Inc., within the SAP Cloud Success Services Center of Expertise (CoE). His expertise encompasses native SAP BW/4HANA data modeling, mixed SAP BW/4HANA modeling, and SAP BW/4HANA greenfield implementations and system conversions.

Index

A

ABAP cloud project 369
ABAP development tools 369
ABAP package 371
Add-on 163
   compatibility 102
   third-party 165
Advanced DataStore object (ADSO) 33, 125, 207, 301, 330
   modeling 41
   remodeling 46
Aggregate 125
Analysis authorization 26
Analytical view 126
Application server 71
Architect data mart 34
Authorization conversion 302
Authorization object 301
Authorization transfer 304

B

Business scenario 49
BWREMOTE user 152

C

Cloud connector 361
Cluster date 236
Collect scope 194
Communication user 171
CompositeProvider 34, 43
   analytical index 126
Conversion phases 63
Conversion scenario 51
Conversion type 53
Core Data Services (CDS) views 45
Corporate memory 34
Custom code compatibility 101

D

Data archiving 130
Data loading 239
Data model 187
Data services source system 133
Data transfer logic 311
Data transfer process 128
Data volume management 83
Database platform 70
Database sizing 102
DataSource 24
   generic/export 129
   replicate 162
DataStore object (DSO) 25, 122
DB Connect source system 132
Delta DTP 233
Delta InfoPackage 229
Delta queue 226
   copy 231
   synchronize 232
Delta run 303
DMIS add-on 163
Dual stack split 71

E

Eclipse
   ABAP perspective 369
   SAP BW/4HANA cockpit 154
Eclipse-based modeling tool 35
EDW propagation 34
End routine 26
Exchange rates 160
Execution user 170
Extraction object 23

G

Generation phase 220
   migration, conversion, deletion 223
   parallel read 221
   RSRV checks 223
   table portions 222
   technical settings 221
Global settings
   transfer 159
Greenfield implementation 54

H

Hierarchy DataSource 130
High performance 32
HybridProvider 125

I

InfoCube	24, 123
InfoObject	24, 122
catalog	122
conversion-relevant	217, 219
main partition	203
modeling	37
InfoPackage	127
InfoProvider	202
InfoSet	26, 124
InfoSource	130
INIT SIMU	231
Initial run	303
In-place conversion	55, 279
business scenarios	279
data model	312
execution	309
execution cleanup	310
features	309
final system conversion	339
landscape considerations	300
mode prerequisites	297
operating modes	295
realization phase	287
scope selection	318
scope transfer	317
sender system provisioning	286
start	315
starter add-on	288
switch modes	296
system landscape	281
transfer standard authorizations	301
unsupported object	298

L

Landscape management database (LMDB)	79
Landscape transformation	62
Layered scalable architecture (LSA)	28
Layered scalable architecture++ (LSA++)	34
Lock data changes	231

M

Maintenance planner	78, 79, 141, 289
Mapping value	218
Metadata	205
Migration	
phase	235
runtime settings	235
start	237

Migration (Cont.)	
time-related statistics	241
Modeling object	24
Modern interface	31
MultiProvider	124
Myself source system	132

N

Near-line storage (NLS)	131
Nonpersistent InfoProvider	25
Note Analyzer	92, 356

O

Object	
compatibility	100
deletion	337
involved	203
mapping	198
Object-specific simplification	373
ODP-BW	353
ODP-CDS	353
ODP-SAP	352
ODP-SLT	354
Open hub destination	26, 130
Open ODS view	33
Openness	31
Operational data provisioning (ODP)	77, 281, 286
delta queues	226
expose DataSource	163
SAP BW source system	158
SAP source system	156
Operational DataStore layer	34

P

Package settings	192
Persistent staging area (PSA)	127
extract	233
Post-conversion	247
Precheck	105
cleanup report	119
code scan	117
execute	107
fix issues	114
in-place conversion	109
install tools	106
remote conversion	113
sizing report	116
Prepare phase	63

Index

Prepare propagate request 201
Preprocessing programs 225
Process chain 26, 47
Project management 67

R

Readiness check 94
   execution 98
   functions 100
   prerequisites 96
   view 102
Realization phase 65
Real-time data acquisition 129
Real-time extraction 47
Remote conversion 58, 135
   authorization objects 172
   collect scope 389
   define object mapping 391
   execution 192, 386
   generation phase 398
   involved objects 393
   migration phase 401
   package creation 190, 386
   package settings 387
   preparation 169, 382
   prepare propagate request 393
   SAP BW bridge 382
   SAW BW bridge metadata 394
   sender system precheck 185
   setup phase 396
   system lock 400
   target system precheck 186
   user roles 170, 171
Remote Function Call (RFC) 162
   destination 152
Report
   BS_ANLY_DS_RELEASE_ODP 286
   CUBE_SAMPLE_CREATE 249
   RC_BW_COLLECT_ANALYSIS_DATA 96
   RODPS_OS_EXPOSE 286
   RS_B4H_CLEANUP_METADATA 248
   RS_B4HANA_INPLACE_CLEANUP 343
   RS_B4HANA_REQUEST_RESET 248
   RS_B4HANA_RESET_DELTA_CLONE 248
   RS_B4HANA_TRANSFER_REM_END 246
   RS_B4HANA_TRANSFER_REM_RESET 248
   RS_B4HANA_WHITELIST_MAINTAIN 298
   RS_DELETE_TLOGO 338
   RSB4H_STAT_MAINTIAN 337
   RSDG_ADSO_ACTIVATE 249
   RSDG_CUBE_ACTIVATE 249

Report (Cont.)
   RSDG_HCPR_ACTIVATE 249
   RSDG_IOBJ_ACTIVATE_ALL 249
   RSDG_MPRO_ACTIVATE 249
   RSDG_ODSO_ACTIVATE 249
   RSDG_TRFN_ACTIVATE 249
   RSDS_DATASOURCE_ACTIVATE_ALL 249
   RSDSO_SAMPLE_ADSO 249
   RSZDELETE 339
   Z_SAP_BW_NOTE_ANALYZER 93
Reporting object 26
RSRV checks 241

S

SAP Analysis for Microsoft Office 27
SAP Analytics Cloud 73
SAP BEx Analyzer 27
SAP BEx SAP BEx Query Designer 26
SAP BEx Web Analyzer 27
SAP Business Planning and
   Consolidation 165
SAP Business Suite 77
   applications 76
SAP Business Warehouse (SAP BW) 76,
   77, 313
   delete objects 338
   delete queries 339
   history 21
   object 23
   precheck tools 283
   sizing 84
   system provisioning 282
   versions 23
SAP BusinessObjects Business
   Intelligence (BI) 72
SAP BusinessObjects Data Services 77
SAP BusinessObjects Web Intelligence 28
SAP BW bridge 87, 345, 347
   cockpit 348, 370
   create project 366
   remote conversion 382
   shell conversion 350
   sizing 355
SAP BW/4HANA
   business content 164
   data modeling object 37
   deployment options 48
   design principles 30
   evolution 29
   install 141
   new features 35

Index

SAP BW/4HANA (Cont.)
   operating system requirements 149
   post-installation 151
   settings 159
   starter add-on 289, 293
   statistics 47
   web cockpit 153
SAP BW/4HANA conversion cockpit 175, 176
   activity type 179
   package type 177
   phase management 178
   process tree 177
   status management 178
SAP Crystal Reports 28
SAP Datasphere 87, 345
   post-migration 404
   roles 367
SAP Fiori 73
SAP HANA
   attribute views 126
   database 136
   lifecycle manager tool 139
   smart data integration 77
   tailored data center integration 139
SAP HANA, platform edition 137
SAP Lumira 27
SAP Lumira, designer edition 27
SAP S/4HANA 77
SAP Solution Manager 79, 83
SAP source system 132
SAP_BW4_SETUP_SIMPLE task list 151
Scope check tool 179
Scope object 187
Scope selection
   delta load 328
   delta queues 326
   execute 318
   extract from PSA 329
   involved objects 324
   lock data changes 330
   main partition 324
   map InfoObjects 324
   master data 330
   metadata 331
   object mapping 320
   post-transfer checks 333
   propagate requests 322, 330
   request mapping 331
   retrieve data 332
   save data 331
   synchronize delta queues 328
   transfer request info 332

Scope selection (Cont.)
   unlock loading 332
Scope transfer 302
Semantic partitioned objects 125
Sender system
   landscape preparation 166
   prechecks 351
   provisioning 166
   SAP BW bridge 362
   SAP BW bridge preparation 358
   shell conversion 351
Setup phase 211
Shell conversion 61, 251
   business case 252
   execution 375
   flow 252
   involved objects 271, 272
   landscape preparation 256, 358
   main partition criteria 271
   map InfoObjects 270
   object mapping 266
   post-conversion 274
   prechecks 257
   prepare phase 351
   prework 374
   realization phase 259, 360
   scope object identification 259
   scope selection execution 263
   system compatibility 254
   system landscape 253
   system preparation 356
   system provisioning 255
Simplicity 30
Simplification list 119, 351
   core data warehouse modeling object 121
   data staging 126
   SAP BW 3.X data flow 120
   source system 131
Simulation mode 303
Sizing report 84
Software Provisioning Manager (SWPM) 149
   components 149
   user interface 149
Source system 76
   create 155
   flat file 158
   minimum support release 74
   Provisioning 168
   SAP 352
   SAP BW bridge preparation 359
Source system readiness 101
ST-A/PI 95

Index

Start routine 26
Store object list 198
System landscape 71
System lock 240
System object 312
System preparation 69, 74
System requirements 69
System settings
   after migration 242

T

Table
   /BIC/AZPRKIDS00 314
   /BIC/FZPRKICUB 314
   ROOSATTR 286
   RSB4H_STAT 337
   RSBKDTP 249
   RSBKREQUEST 249
   RSBOHDEST 249
   RSDCUBE 249
   RSDDSTAT 249
   RSDIOBJ 249
   RSDODSO 249
   RSDS 249
   RSLDPIO 249
   RSOLTPSOURCE 249
   RSOOBJXREF 311
   RSPCCHAIN 249
   RSRREPDIR 249
   RSTRAN 249
Target system
   minimum support release 75
   provisioning 136
   supported releases 352
Technical content 338
Third-party tools 73
Transaction
   AL11 90
   BW4WEB 154
   CNVLTRL_MONI_C 237
   DB02 185
   ODQMON 226, 228, 326, 327
   OMSL 161
   PFCG 155
   RS2HANA_ADMIN 126
   RSA1 .... 96, 316, 317, 384
   RSA7 .... 187, 312

Transaction (Cont.)
   RSB4HCONV 119, 190, 260, 285, 294, 296, 305, 310, 315, 336, 338, 339, 359, 375, 386
   RSKC 161
   RSMIGRATE 115, 120
   RSPC1 316
   RZ04 186
   SAINT 293, 341
   SE01 .... 205, 274
   SE16 189
   SE38 .... 85, 95, 169, 256, 286, 337
   SM37 108
   SM59 162
   STC01 .... 151, 153, 180, 183, 197, 262, 297, 315, 376
   STC02 .... 184, 194, 197, 246, 318
   STCO2 204
   STMS_IMPORT 206, 274
Transfer standard authorization 336
Transformation 129
Transformation rule 26
Transformation status 237
Transport path 162

U

UD Connect source system 132
Unicode 70
User roles 155

V

Validate migration configuration 232
Verify task completion 239
Virtual cube 25
Virtual data mart 34
Virtual time hierarchy 152
VirtualProvider 125

W

Web service source system 133
With an SAP BW landscape 50
Without an SAP BW landscape 49

X

XML 340
XML file 167, 284, 357, 383

Interested in reading more?

Please visit our website for all new book
and e-book releases from SAP PRESS.

www.sap-press.com